MULTIPLICATIVE NUMBER THEORY I: CLASSICAL THEORY

Prime numbers are the multiplicative building blocks of natural numbers. Understanding their overall influence and especially their distribution gives rise to central questions in mathematics and physics. In particular their finer distribution is closely connected with the Riemann hypothesis, the most important unsolved problem in the mathematical world. Assuming only subjects covered in a standard degree in mathematics, the authors comprehensively cover all the topics met in first courses on multiplicative number theory and the distribution of prime numbers. They bring their extensive and distinguished research expertise to bear in preparing the student for intelligent reading of the more advanced research literature. The text, which is based on courses taught successfully over many years at Michigan, Imperial College and Pennsylvania State, is enriched by comprehensive historical notes and references as well as over 500 exercises.

Hugh Montgomery is a Professor of Mathematics at the University of Michigan.

Robert Vaughan is a Professor of Mathematics at Pennsylvannia State University.

CAMBRIDGE STUDIES IN ADVANCED MATHEMATICS

All the titles listed below can be obtained from good booksellers of from Cambridge University Press. For a complete series listing visit:
http://www.cambridge.org/series/sSeries.asp?code=CSAM

Already published

Multiplicative Number Theory
I. Classical Theory

HUGH L. MONTGOMERY
University of Michigan, Ann Arbor

ROBERT C. VAUGHAN
Pennsylvania State University, University Park

CAMBRIDGE
UNIVERSITY PRESS

CAMBRIDGE UNIVERSITY PRESS
Cambridge, New York, Melbourne, Madrid, Cape Town,
Singapore, São Paulo, Delhi, Mexico City

Cambridge University Press
The Edinburgh Building, Cambridge CB2 8RU, UK

Published in the United States of America by Cambridge University Press, New York

www.cambridge.org
Information on this title: www.cambridge.org/9781107405820

First published 2007
First paperback edition 2011

A catalogue record for this publication is available from the British Library

ISBN 978-0-521-84903-6 Hardback
ISBN 978-1-107-40582-0 Paperback

Dedicated to our teachers:

> P. T. Bateman
> J. H. H. Chalk
> H. Davenport
> T. Estermann
> H. Halberstam
> A. E. Ingham

Talet är tänkandets början och slut.
Med tanken föddes talet.
Utöfver talet når tanken icke.

Numbers are the beginning and end of thinking.
With thoughts were numbers born.
Beyond numbers thought does not reach.
MAGNUS GUSTAF MITTAG-LEFFLER, 1903

Contents

Preface

Our object is to introduce the interested student to the techniques, results, and terminology of multiplicative number theory. It is not intended that our discussion will always reach the research frontier. Rather, it is hoped that the material here will prepare the student for intelligent reading of the more advanced research literature.

Analytic number theorists are not very uniformly distributed around the world and it possible that a student may be working without the guidance of an experienced mentor in the area. With this in mind, we have tried to make this volume as self-contained as possible.

We assume that the reader has some acquaintance with the fundamentals of elementary number theory, abstract algebra, measure theory, complex analysis, and classical harmonic analysis. More specialized or advanced background material in analysis is provided in the appendices.

The relationship of exercises to the material developed in a given section varies widely. Some exercises are designed to illustrate the theory directly whilst others are intended to give some idea of the ways in which the theory can be extended, or developed, or paralleled in other areas. The reader is cautioned that papers cited in exercises do not necessarily contain a solution.

This volume is the first instalment of a larger project. We are preparing a second volume, which will cover such topics as uniform distribution, bounds for exponential sums, a wider zero-free region for the Riemann zeta function, mean and large values of Dirichlet polynomials, approximate functional equations, moments of the zeta function and L functions on the line $\sigma = 1/2$, the large sieve, Vinogradov's method of prime number sums, zero density estimates, primes in arithmetic progressions on average, sums of primes, sieve methods, the distribution of additive functions and mean values of multiplicative functions, and the least prime in an arithmetic progression. The present volume was twenty-five years in preparation—we hope to be a little quicker with the second volume.

Many people have assisted us in this work—including P. T. Bateman, E. Bombieri, T. Chan, J. B. Conrey, H. G. Diamond, T. Estermann, J. B. Friedlander, S. W. Graham, S. M. Gonek, A. Granville, D. R. Heath-Brown, H. Iwaniec, H. Maier, G. G. Martin, D. W. Masser, A. M. Odlyzko, G. Peng, C. Pomerance, H.–E. Richert, K. Soundararajan, and U. M. A. Vorhauer. In particular, our doctoral students, and their students also, have been most helpful in detecting errors of all types. We are grateful to them all. We would be most happy to hear from any reader who detects a misprint, or might suggest improvements.

Finally we thank our loved ones and friends for their long term support and the long–suffering David Tranah at Cambridge University Press for his forbearance.

Notation

Symbol	Meaning	Found on page
$\mathbb{C}$	The set of complex numbers.	109
$\mathbb{F}_p$	A field of p elements.	9
$\mathbb{N}$	The set of natural numbers, $1, 2, \ldots$	114
$\mathbb{Q}$	The set of rational numbers.	120
$\mathbb{R}$	The set of real numbers.	43
$\mathbb{T}$	$\mathbb{R}/\mathbb{Z}$, known as the *circle group* or the *one-dimensional torus*, which is to say the real numbers modulo 1.	110
$\mathbb{Z}$	The set of rational integers.	20
B	constant in the Hadamard product for $\xi(s)$	347, 349
B_k	Bernoulli numbers.	496ff
$B_k(x)$	Bernoulli polynomials.	45, 495ff
$B(\chi)$	constant in the Hadamard product for $\xi(s, \chi)$	351, 352
C_0	Euler's constant	26
$c_q(n)$	The sum of $e(an/q)$ with a running over a reduced residue system modulo q; known as *Ramanujan's sum*.	110
$c_\chi(n)$	$= \sum_{a=1}^{q} \chi(a)e(an/q)$.	286, 290
$d(n)$	The number of positive divisors of n, called the *divisor function*.	2
$d_k(n)$	The number of ordered k-tuples of positive integers whose product is n.	43
$E_0(\chi)$	$= 1$ if $\chi = \chi_0$, 0 otherwise.	358

Symbol	Meaning	Found on page
E_k	The *Euler numbers*, also known as the *secant coefficients*.	506
$e(\theta)$	$= e^{2\pi i\theta}$; the complex exponential with period 1.	64, 108ff
$L(s, \chi)$	A Dirichlet *L*-function.	120
$\mathrm{Li}(x)$	$= \int_0^x \frac{du}{\log u}$ with the Cauchy principal value taken at 1; the *logarithmic integral*.	189
$\mathrm{li}(x)$	$= \int_2^x \frac{du}{\log u}$; the *logarithmic integral*.	5
$M(x)$	$= \sum_{n\leq x} \mu(n)$	182
$M(x; q, a)$	The sum of $\mu(n)$ over those $n \leq x$ for which $n \equiv a \pmod q$.	383
$M(x, \chi)$	The sum of $\chi(n)\mu(n)$ over those $n \leq x$.	383
$N(T)$	The number of zeros $\rho = \beta + i\gamma$ of $\zeta(s)$ with $0 < \gamma \leq T$.	348, 452ff
$N(T, \chi)$	The number of zeros $\rho = \beta + i\gamma$ of $L(s, \chi)$ with $\beta > 0$ and $0 \leq \beta \leq T$.	454
$P(n)$	The largest prime factor of n.	202
$Q(x)$	the number of square-free numbers not exceeding x	36
$S(t)$	$= \frac{1}{\pi} \arg\zeta(\frac{1}{2} + it)$.	452
$S(t, \chi)$	$= \frac{1}{\pi} \arg L(\frac{1}{2} + it, \chi)$.	454
$\mathrm{si}(x)$	$= -\int_x^\infty \frac{\sin u}{u} du$; the *sine integral*.	139
T_k	The *tangent coefficients*.	505
$w(u)$	The *Buchstab function*, defined by the equation $(uw(u))' = w(u-1)$ for $u > 2$ together with the initial condition $w(u) = 1/u$ for $1 < u \leq 2$.	216
$Z(t)$	Hardy's function. The function $Z(t)$ is real-valued, and $|Z(t)| = |\zeta(\frac{1}{2} + it)|$.	456ff
β	The real part of a zero of the zeta function or of an *L*-function.	173
$\Gamma(s)$	$= \int_0^\infty e^{-x} x^{s-1} dx$ for $\sigma > 0$; called the *Gamma function*.	30, 520ff

Symbol	Meaning	Found on page
$\Gamma(s,a)$	$= \int_a^\infty e^{-w} w^{s-1}\, dw$; the *incomplete Gamma function*.	327
γ	The imaginary part of a zero of the zeta function or of an L-function.	172
$\Delta_N(\theta)$	$= 1 + 2\sum_{n=1}^{N-1}(1 - n/N)\cos 2\pi n\theta$; known as the *Fejér kernel*.	174
$\varepsilon(\chi)$	$= \tau(\chi)/\left(i^\kappa q^{1/2}\right)$.	332
$\zeta(s)$	$= \sum_{n=1}^\infty n^{-s}$ for $\sigma > 1$, known as *the Riemann zeta function*.	2
$\zeta(s,\alpha)$	$= \sum_{n=0}^\infty (n+\alpha)^{-s}$ for $\sigma > 1$; known as the *Hurwitz zeta function*.	30
$\zeta_K(s)$	$\sum_{\mathfrak{a}} N(\mathfrak{a})^{-s}$; known as the *Dedekind zeta function* of the algebraic number field K.	343
Θ	$= \sup \Re \rho$	430, 463
$\vartheta(x)$	$= \sum_{p \le x} \log p$.	46
$\vartheta(z)$	$= \sum_{n=-\infty}^\infty e^{-\pi n^2 z}$ for $\Re z > 0$.	329
$\vartheta(x;q,a)$	The sum of $\log p$ over primes $p \le x$ for which $p \equiv a \pmod q$.	128, 377ff
$\vartheta(x,\chi)$	$= \sum_{p \le x} \chi(p)\log p$.	377ff
κ	$= (1 - \chi(-1))/2$.	332
$\Lambda(n)$	$= \log p$ if $n = p^k$, $= 0$ otherwise; known as the *von Mangoldt Lambda function*.	23
$\Lambda_2(n)$	$= \Lambda(n)\log n + \sum_{bc=n} \Lambda(b)\Lambda(c)$.	251
$\Lambda(x;q,a)$	The sum of $\lambda(n)$ over those $n \le x$ such that $n \equiv a \pmod q$.	383
$\Lambda(x,\chi)$	$= \sum_{n \le x} \chi(n)\lambda(n)$.	383
$\lambda(n)$	$= (-1)^{\Omega(n)}$; known as the *Liouville lambda function*.	21
$\mu(n)$	$= (-1)^{\omega(n)}$ for square-free n, $= 0$ otherwise. Known as the *Möbius mu function*.	21
$\mu(\sigma)$	the Lindelöf mu function	330
$\xi(s)$	$= \frac{1}{2}s(s-1)\zeta(s)\Gamma(s/2)\pi^{-s/2}$.	328
$\xi(s,\chi)$	$= L(s,\chi)\Gamma((s+\kappa)/2)(q/\pi)^{(s+\kappa)/2}$ where χ is a primitive character modulo q, $q > 1$.	333

Symbol	Meaning	Found on page
$\Pi(x)$	$= \sum_{n \le x} \Lambda(n)/\log n$.	416
$\pi(x)$	The number of primes not exceeding x.	3
$\pi(x; q, a)$	The number of $p \le x$ such that $p \equiv a$ (mod q),.	90, 358
$\pi(x, \chi)$	$= \sum_{p \le x} \chi(p)$.	377ff
ρ	$= \beta + i\gamma$; a zero of the zeta function or of an L-function.	173
$\rho(u)$	The *Dickman function*, defined by the equation $u\rho'(u) = -\rho(u-1)$ for $u > 1$ together with the initial condition $\rho(u) = 1$ for $0 \le u \le 1$.	200
$\sigma(n)$	The sum of the positive divisors of n.	27
$\sigma_a(n)$	$= \sum_{d \mid n} d^a$.	28
τ	$= |t| + 4$.	14
$\tau(\chi)$	$= \sum_{a=1}^{q} \chi(a)e(a/q)$; known as the *Gauss sum* of χ.	286ff
$\Phi_q(z)$	The qth cyclotomic polynomial, which is to say a monic polynomial with integral coefficients, of degree $\varphi(q)$, whose roots are the numbers $e(a/q)$ for $(a, q) = 1$.	64
$\Phi(x, y)$	The number of $n \le x$ such that all prime factors of n are $\ge y$.	215
$\Phi(y)$	$= \frac{1}{\sqrt{2\pi}} \int_{-\infty}^{y} e^{-t^2/2}\, dt$; the cumulative distribution function of a normal random variable with mean 0 and variance 1.	235
$\varphi(n)$	The number of a, $1 \le a \le n$, for which $(a, n) = 1$; known as *Euler's totient function*.	27
$\chi(n)$	A Dirichlet character.	115
$\psi(x)$	$= \sum_{n \le x} \Lambda(n)$.	46
$\psi(x, y)$	The number of $n \le x$ composed entirely of primes $p \le y$.	199
$\psi(x; q, a)$	The sum of $\Lambda(n)$ over $n \le x$ for which $n \equiv a \pmod{q}$.	128, 377ff
$\psi(x, \chi)$	$= \sum_{n \le x} \chi(n)\Lambda(n)$.	377ff
$\Omega(n)$	The number of prime factors of n, counting multiplicity.	21
$\omega(n)$	The number of distinct primes dividing n.	21

Symbol	Meaning	Found on page
$[x]$	The unique integer such that $[x] \le x < [x]+1$; called the *integer part* of x.	15, 24
$\{x\}$	$= x - [x]$; called the *fractional part* of x.	24
$\|x\|$	The distance from x to the nearest integer.	477
$f(x) = O(g(x))$	$|f(x)| \le Cg(x)$ where C is an absolute constant.	3
$f(x) = o(g(x))$	$\lim f(x)/g(x) = 0$.	3
$f(x) \ll g(x)$	$f(x) = O(g(x))$.	3
$f(x) \gg g(x)$	$g(x) = O(f(x))$, g non-negative.	4
$f(x) \asymp g(x)$	$cf(x) \le g(x) \le Cf(x)$ for some positive absolute constants c, C.	4
$f(x) \sim g(x)$	$\lim f(x)/g(x) = 1$.	3

1

Dirichlet series: I

1.1 Generating functions and asymptotics

The general rationale of analytic number theory is to derive statistical information about a sequence $\{a_n\}$ from the analytic behaviour of an appropriate generating function, such as a power series $\sum a_n z^n$ or a Dirichlet series $\sum a_n n^{-s}$.
The type of generating function employed depends on the problem being in-
vestigated. There are no rigid rules governing the kind of generating function
that is appropriate – the success of a method justifies its use – but we usually
deal with additive questions by means of power series or trigonometric sums,
and with multiplicative questions by Dirichlet series. For example, if

$$f(z) = \sum_{n=1}^{\infty} z^{n^k}$$

for $|z| < 1$, then the nth power series coefficient of $f(z)^s$ is the number $r_{k,s}(n)$
of representations of n as a sum of s positive k^{th} powers,

$$n = m_1^k + m_2^k + \cdots + m_s^k.$$

We can recover $r_{k,s}(n)$ from $f(z)^s$ by means of Cauchy's coefficient formula:

$$r_{k,s}(n) = \frac{1}{2\pi i} \oint \frac{f(z)^s}{z^{n+1}} \, dz.$$

By choosing an appropriate contour, and estimating the integrand, we can determine the asymptotic size of $r_{k,s}(n)$ as $n \to \infty$, provided that s is sufficiently
large, say $s > s_0(k)$. This is the germ of the Hardy–Littlewood circle method,
but considerable effort is required to construct the required estimates.

To appreciate why power series are useful in dealing with additive problems, note that if $A(z) = \sum a_k z^k$ and $B(z) = \sum b_m z^m$ then the power series

1

coefficients of $C(z) = A(z)B(z)$ are given by the formula

$$c_n = \sum_{k+m=n} a_k b_m. \tag{1.1}$$

The terms are grouped according to the sum of the indices, because $z^k z^m = z^{k+m}$.

A *Dirichlet series* is a series of the form $\alpha(s) = \sum_{n=1}^{\infty} a_n n^{-s}$ where s is a complex variable. If $\beta(s) = \sum_{m=1}^{\infty} b_m m^{-s}$ is a second Dirichlet series and $\gamma(s) = \alpha(s)\beta(s)$, then (ignoring questions relating to the rearrangement of terms of infinite series)

$$\gamma(s) = \sum_{k=1}^{\infty} a_k k^{-s} \sum_{m=1}^{\infty} b_m m^{-s} = \sum_{k=1}^{\infty} \sum_{m=1}^{\infty} a_k b_m (km)^{-s} = \sum_{n=1}^{\infty} \left(\sum_{km=n} a_k b_m \right) n^{-s}. \tag{1.2}$$

That is, we expect that $\gamma(s)$ is a Dirichlet series, $\gamma(s) = \sum_{n=1}^{\infty} c_n n^{-s}$, whose coefficients are

$$c_n = \sum_{km=n} a_k b_m. \tag{1.3}$$

This corresponds to (1.1), but the terms are now grouped according to the product of the indices, since $k^{-s} m^{-s} = (km)^{-s}$.

Since we shall employ the complex variable s extensively, it is useful to have names for its real and complex parts. In this regard we follow the rather peculiar notation that has become traditional: $s = \sigma + it$.

Among the Dirichlet series we shall consider is the *Riemann zeta function*, which for $\sigma > 1$ is defined by the absolutely convergent series

$$\zeta(s) = \sum_{n=1}^{\infty} n^{-s}. \tag{1.4}$$

As a first application of (1.3), we note that if $\alpha(s) = \beta(s) = \zeta(s)$ then the manipulations in (1.3) are justified by absolute convergence, and hence we see that

$$\sum_{n=1}^{\infty} d(n) n^{-s} = \zeta(s)^2 \tag{1.5}$$

for $\sigma > 1$. Here $d(n)$ is the *divisor function*, $d(n) = \sum_{d|n} 1$.

From the rate of growth or analytic behaviour of generating functions we glean information concerning the sequence of coefficients. In expressing our findings we employ a special system of notation. For example, we say, '$f(x)$ is asymptotic to $g(x)$' as x tends to some limiting value (say $x \to \infty$), and write

$f(x) \sim g(x) \ (x \to \infty)$, if

$$\lim_{x \to \infty} \frac{f(x)}{g(x)} = 1.$$

An instance of this arises in the formulation of the Prime Number Theorem (PNT), which concerns the asymptotic size of the number $\pi(x)$ of prime numbers not exceeding x; $\pi(x) = \sum_{p \le x} 1$. Conjectured by Legendre in 1798, and finally proved in 1896 independently by Hadamard and de la Vallée Poussin, the Prime Number Theorem asserts that

$$\pi(x) \sim \frac{x}{\log x}.$$

Alternatively, we could say that

$$\pi(x) = (1 + o(1)) \frac{x}{\log x},$$

which is to say that $\pi(x)$ is $x/\log x$ plus an error term that is in the limit negligible compared with $x/\log x$. More generally, we say, '$f(x)$ is small oh of $g(x)$', and write $f(x) = o(g(x))$, if $f(x)/g(x) \to 0$ as x tends to its limit.

The Prime Number Theorem can be put in a quantitative form,

$$\pi(x) = \frac{x}{\log x} + O\left(\frac{x}{(\log x)^2}\right). \tag{1.6}$$

Here the last term denotes an implicitly defined function (the difference between the other members of the equation); the assertion is that this function has absolute value not exceeding $Cx(\log x)^{-2}$. That is, the above is equivalent to asserting that there is a constant $C > 0$ such that the inequality

$$\left| \pi(x) - \frac{x}{\log x} \right| \le \frac{Cx}{(\log x)^2}$$

holds for all $x \ge 2$. In general, we say that $f(x)$ is 'big oh of $g(x)$', and write $f(x) = O(g(x))$ if there is a constant $C > 0$ such that $|f(x)| \le Cg(x)$ for all x in the appropriate domain. The function f may be complex-valued, but g is necessarily non-negative. The constant C is called the *implicit constant*; it is an absolute constant unless the contrary is indicated. For example, if C is liable to depend on a parameter α, we might say, 'For any fixed value of α, $f(x) = O(g(x))$'. Alternatively, we might say, '$f(x) = O(g(x))$ where the implicit constant may depend on α', or more briefly, $f(x) = O_\alpha(g(x))$.

When there is no main term, instead of writing $f(x) = O(g(x))$ we save a pair of parentheses by writing instead $f(x) \ll g(x)$. This is read, '$f(x)$ is less-than-less-than $g(x)$', and we write $f(x) \ll_\alpha g(x)$ if the implicit constant may depend on α. To provide an example of this notation, we recall that Chebyshev

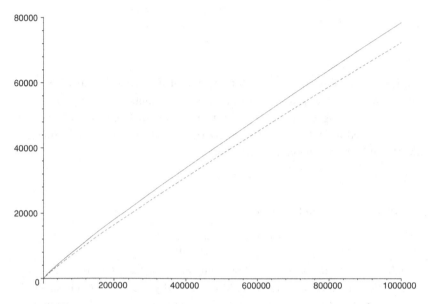

Figure 1.1 Graph of $\pi(x)$ (solid) and $x/\log x$ (dotted) for $2 \leq x \leq 10^6$.

proved that $\pi(x) \ll x/\log x$. This is of course weaker than the Prime Number Theorem, but it was derived much earlier, in 1852. Chebyshev also showed that $\pi(x) \gg x/\log x$. In general, we say that $f(x) \gg g(x)$ if there is a positive constant c such that $f(x) \geq cg(x)$ and g is non-negative. In this situation both f and g take only positive values. If both $f \ll g$ and $f \gg g$ then we say that f and g have the same order of magnitude, and write $f \asymp g$. Thus Chebyshev's estimates can be expressed as a single relation,

$$\pi(x) \asymp \frac{x}{\log x}.$$

The estimate (1.6) is best possible to the extent that the error term is not $o(x(\log x)^{-2})$. We have also a special notation to express this:

$$\pi(x) - \frac{x}{\log x} = \Omega\left(\frac{x}{(\log x)^2}\right).$$

In general, if $\limsup_{x \to \infty} |f(x)|/g(x) > 0$ then we say that $f(x)$ is 'Omega of $g(x)$', and write $f(x) = \Omega(g(x))$. This is precisely the negation of the statement '$f(x) = o(g(x))$'. When studying numerical values, as in Figure 1.1, we find that the fit of $x/\log x$ to $\pi(x)$ is not very compelling. This is because the error term in the approximation is only one logarithm smaller than the main term. This error term is not oscillatory – rather there is a second main term of this size:

$$\pi(x) = \frac{x}{\log x} + \frac{x}{(\log x)^2} + O\left(\frac{x}{(\log x)^3}\right).$$

This is also best possible, but the main term can be made still more elaborate to give a smaller error term. Gauss was the first to propose a better approximation to $\pi(x)$. Numerical studies led him to observe that the density of prime numbers in the neighbourhood of x is approximately $1/\log x$. This suggests that the number of primes not exceeding x might be approximately equal to the *logarithmic integral*,

$$\mathrm{li}(x) = \int_2^x \frac{1}{\log u}\, du.$$

(Orally, 'li' rhymes with 'pi'.) By repeated integration by parts we can show that

$$\mathrm{li}(x) = x \sum_{k=1}^{K-1} \frac{(k-1)!}{(\log x)^k} + O_K\left(\frac{x}{(\log x)^K}\right)$$

for any positive integer K; thus the secondary main terms of the approximation to $\pi(x)$ are contained in $\mathrm{li}(x)$.

In Chapter 6 we shall prove the Prime Number Theorem in the sharper quantitative form

$$\pi(x) = \mathrm{li}(x) + O\left(\frac{x}{\exp(c\sqrt{\log x})}\right)$$

for some suitable positive constant c. Note that $\exp(c\sqrt{\log x})$ tends to infinity faster than any power of $\log x$. The error term above seems to fall far from what seems to be the truth. Numerical evidence, such as that in Table 1.1, suggests that the error term in the Prime Number Theorem is closer to $\sqrt{x}$ in size. Gauss noted the good fit, and also that $\pi(x) < \mathrm{li}(x)$ for all x in the range of his extensive computations. He proposed that this might continue indefinitely, but the numerical evidence is misleading, for in 1914 Littlewood showed that

$$\pi(x) - \mathrm{li}(x) = \Omega_\pm\left(\frac{x^{1/2}\log\log\log x}{\log x}\right).$$

Here the subscript $\pm$ indicates that the error term achieves the stated order of magnitude infinitely often, and in both signs. In particular, the difference $\pi - \mathrm{li}$ has infinitely many sign changes. More generally, we write $f(x) = \Omega_+(g(x))$ if $\limsup_{x\to\infty} f(x)/g(x) > 0$, we write $f(x) = \Omega_-(g(x))$ if $\liminf_{x\to\infty} f(x)/g(x) < 0$, and we write $f(x) = \Omega_\pm(g(x))$ if both these relations hold.

Table 1.1 Values of $\pi(x)$, li(x), $x/\log x$ for $x = 10^k$, $1 \le k \le 22$.

x	$\pi(x)$	li(x)	$x/\log x$
10	4	5.12	4.34
10^2	25	29.08	21.71
10^3	168	176.56	144.76
10^4	1229	1245.09	1085.74
10^5	9592	9628.76	8685.89
10^6	78498	78626.50	72382.41
10^7	664579	664917.36	620420.69
10^8	5761455	5762208.33	5428681.02
10^9	50847534	50849233.90	48254942.43
10^{10}	455052511	455055613.54	434294481.90
10^{11}	4118054813	4118066399.58	3948131653.67
10^{12}	37607912018	37607950279.76	36191206825.27
10^{13}	346065536839	346065458090.05	334072678387.12
10^{14}	3204941750802	3204942065690.91	3102103442166.08
10^{15}	29844570422669	29844571475286.54	28952965460216.79
10^{16}	279238341033925	279238344248555.75	271434051189532.39
10^{17}	2623557157654233	2623557165610820.07	2554673422960304.87
10^{18}	24739954287740860	24739954309690413.98	24127471216847323.76
10^{19}	234057667276344607	234057667376222382.22	228576043106974646.13
10^{20}	2220819602560918840	2220819602783663483.55	2171472409516259138.26
10^{21}	21127269486018731928	21127269486616126182.33	20680689614440563221.48
10^{22}	201467286689315906290	201467286691248261498.15	197406582683296285295.97

In the exercises below we give several examples of the use of generating functions, mostly power series, to establish relations between various counting functions.

1.1.1 Exercises

1. Let $r(n)$ be the number of ways that n cents of postage can be made, using only 1 cent, 2 cent, and 3 cent stamps. That is, $r(n)$ is the number of ordered triples (x_1, x_2, x_3) of non-negative integers such that $x_1 + 2x_2 + 3x_3 = n$.
   (a) Show that

   $$\sum_{n=0}^{\infty} r(n)z^n = \frac{1}{(1-z)(1-z^2)(1-z^3)}$$

   for $|z| < 1$.
   (b) Determine the partial fraction expansion of the rational function above.

That is, find constants $a, b, \ldots, f$ so that the above is

$$\frac{a}{(z-1)^3} + \frac{b}{(z-1)^2} + \frac{c}{z-1} + \frac{d}{z+1} + \frac{e}{z-\omega} + \frac{f}{z-\overline{\omega}}$$

where $\omega = e^{2\pi i/3}$ and $\overline{\omega} = e^{-2\pi i/3}$ are the primitive cube roots of unity.

(c) Show that $r(n)$ is the integer nearest $(n+3)^2/12$.

(d) Show that $r(n)$ is the number of ways of writing $n = y_1 + y_2 + y_3$ with $y_1 \geq y_2 \geq y_3 \geq 0$.

2. Explain why

$$\prod_{k=0}^{\infty} \left(1 + z^{2^k}\right) = 1 + z + z^2 + \cdots$$

for $|z| < 1$.

3. (L. Mirsky & D. J. Newman) Suppose that $0 \leq a_k < m_k$ for $1 \leq k \leq K$, and that $m_1 < m_2 < \cdots < m_K$. This is called a *family of covering congruences* if every integer x satisfies at least one of the congruences $x \equiv a_k \pmod{m_k}$. A system of covering congruences is called *exact* if for every value of x there is exactly one value of k such that $x \equiv a_k \pmod{m_k}$. Show that if the system is exact then

$$\sum_{k=1}^{K} \frac{z^{a_k}}{1 - z^{m_k}} = \frac{1}{1-z}$$

for $|z| < 1$. Show that the left-hand side above is

$$\sim \frac{e^{2\pi i a_K/m_K}}{m_K(1-r)}$$

when $z = re^{2\pi i/m_K}$ and $r \to 1^-$. On the other hand, the right-hand side is bounded for z in a neighbourhood of $e^{2\pi i/m_K}$ if $m_K > 1$. Deduce that a family of covering congruences is not exact if $m_k > 1$.

4. Let $p(n;k)$ denote the number of partitions of n into at most k parts, that is, the number of ordered k-tuples $(x_1, x_2, \ldots, x_k)$ of non-negative integers such that $n = x_1 + x_2 + \cdots + x_k$ and $x_1 \geq x_2 \geq \cdots \geq x_k$. Let $p(n) = p(n;n)$ denote the total number of partitions of n. Also let $p_o(n)$ be the number of partitions of n into an odd number of parts, $p_o(n) = \sum_{2\nmid k} p(n;k)$. Finally, let $p_d(n)$ denote the number of partitions of n into distinct parts, so that $x_1 > x_2 > \cdots > x_k$. By convention, put $p(0) = p_o(0) = p_d(0) = 1$.

(a) Show that there are precisely $p(n;k)$ partitions of n into parts not exceeding k.

(b) Show that

$$\sum_{n=0}^{\infty} p(n;k)z^n = \prod_{j=1}^{k}(1 - z^j)^{-1}$$

for $|z| < 1$.

(c) Show that

$$\sum_{n=0}^{\infty} p(n)z^n = \prod_{k=1}^{\infty}(1 - z^k)^{-1}$$

for $|z| < 1$.

(d) Show that

$$\sum_{n=0}^{\infty} p_d(n)z^n = \prod_{k=1}^{\infty}(1 + z^k)$$

for $|z| < 1$.

(e) Show that

$$\sum_{n=0}^{\infty} p_o(n)z^n = \prod_{k=1}^{\infty}(1 - z^{2k-1})^{-1}$$

for $|z| < 1$.

(f) By using the result of Exercise 2, or otherwise, show that the last two generating functions above are identically equal. Deduce that $p_o(n) = p_d(n)$ for all n.

5. Let $A(n)$ denote the number of ways of associating a product of n terms; thus $A(1) = A(2) = 1$ and $A(3) = 2$. By convention, $A(0) = 0$.

(a) By considering the possible positionings of the outermost parentheses, show that

$$A(n) = \sum_{k=1}^{n-1} A(k)A(n - k)$$

for all $n \geq 2$.

(b) Let $P(z) = \sum_{n=0}^{\infty} A(n)z^n$. Show that

$$P(z)^2 = P(z) - z.$$

Deduce that

$$P(z) = \frac{1 - \sqrt{1 - 4z}}{2} = \sum_{n=1}^{\infty} \binom{1/2}{n} 2^{2n-1}(-1)^{n-1}z^n.$$

(c) Conclude that $A(n) = \binom{2n-2}{n-1}/n$ for all $n \geq 1$. These are called the *Catalan numbers*.

(d) What needs to be said concerning the convergence of the series used above?

6. (a) Let n_k denote the total number of monic polynomials of degree k in $\mathbb{F}_p[x]$. Show that $n_k = p^k$.

(b) Let $P_1, P_2, \ldots$ be the irreducible monic polynomials in $\mathbb{F}_p[x]$, listed in some (arbitrary) order. Show that

$$\prod_{r=1}^{\infty}(1 + z^{\deg P_r} + z^{2 \deg P_r} + z^{3 \deg P_r} + \cdots) = 1 + pz + p^2 z^2$$

$$+ p^3 z^3 + \cdots$$

for $|z| < 1/p$.

(c) Let g_k denote the number of irreducible monic polynomials of degree k in $\mathbb{F}_p[x]$. Show that

$$\prod_{k=1}^{\infty}(1 - z^k)^{-g_k} = (1 - pz)^{-1} \qquad (|z| < 1/p).$$

(d) Take logarithmic derivatives to show that

$$\sum_{k=1}^{\infty} k g_k \frac{z^{k-1}}{1 - z^k} = \frac{p}{1 - pz} \qquad (|z| < 1/p).$$

(e) Show that

$$\sum_{k=1}^{\infty} k g_k \sum_{m=1}^{\infty} z^{mk} = \sum_{n=1}^{\infty} p^n z^n \qquad (|z| < 1/p).$$

(f) Deduce that

$$\sum_{k \mid n} k g_k = p^n$$

for all positive integers n.

(g) (Gauss) Use the Möbius inversion formula to show that

$$g_n = \frac{1}{n} \sum_{k \mid n} \mu(k) p^{n/k}$$

for all positive integers n.

(h) Use (f) (not (g)) to show that

$$\frac{p^n}{n} - \frac{2p^{n/2}}{n} \leq g_n \leq \frac{p^n}{n}.$$

(i) If a monic polynomial of degree n is chosen at random from $\mathbb{F}_p[x]$, about how likely is it that it is irreducible? (Assume that p and/or n is large.)

(j) Show that $g_n > 0$ for all p and all $n \geq 1$. (If $P \in \mathbb{F}_p[x]$ is irreducible and has degree n, then the quotient ring $\mathbb{F}_p[x]/(P)$ is a field of p^n elements. Thus we have proved that there is such a field, for each prime p and integer $n \geq 1$. It may be further shown that the order of a finite field is necessarily a prime power, and that any two finite fields of the same order are isomorphic. Hence the field of order p^n, whose existence we have proved, is essentially unique.)

7. (E. Berlekamp) Let p be a prime number. We recall that polynomials in a single variable (mod p) factor uniquely into irreducible polynomials. Thus a monic polynomial $f(x)$ can be expressed uniquely (mod p) in the form $g(x)h(x)^2$ where $g(x)$ is square-free (mod p) and both g and h are monic. Let s_n denote the number of monic square-free polynomials (mod p) of degree n. Show that

$$\left(\sum_{k=0}^{\infty} s_k z^k \right) \left(\sum_{m=0}^{\infty} p^m z^{2m} \right) = \sum_{n=0}^{\infty} p^n z^n$$

for $|z| < 1/p$. Deduce that

$$\sum_{k=0}^{\infty} s_k z^k = \frac{1 - pz^2}{1 - pz},$$

and hence that $s_0 = 1$, $s_1 = p$, and that $s_k = p^k(1 - 1/p)$ for all $k \geq 2$.

8. (cf Wagon 1987) (a) Let $\mathcal{I} = [a, b]$ be an interval. Show that $\int_{\mathcal{I}} e^{2\pi i x}\, dx = 0$ if and only if the length $b - a$ of $\mathcal{I}$ is an integer.

   (b) Let $\mathcal{R} = [a, b] \times [c, d]$ be a rectangle. Show that $\iint_{\mathcal{R}} e^{2\pi i(x+y)}\, dx\, dy = 0$ if and only if at least one of the edge lengths of $\mathcal{R}$ is an integer.

   (c) Let $\mathcal{R}$ be a rectangle that is a union of finitely many rectangles $\mathcal{R}_i$; the $\mathcal{R}_i$ are disjoint apart from their boundaries. Show that if all the $\mathcal{R}_i$ have the property that at least one of their side lengths is an integer, then $\mathcal{R}$ also has this property.

9. (L. Moser) If $\mathcal{A}$ is a set of non-negative integers, let $r_A(n)$ denote the number of representations of n as a sum of two distinct members of $\mathcal{A}$. That is, $r_A(n)$ is the number of ordered pairs (a_1, a_2) for which $a_1 \in \mathcal{A}, a_2 \in \mathcal{A}, a_1 + a_2 = n$, and $a_1 \neq a_2$. Let $A(z) = \sum_{a \in \mathcal{A}} z^a$.

   (a) Show that $\sum_n r_A(n) z^n = A(z)^2 - A(z^2)$ for $|z| < 1$.

   (b) Suppose that the non-negative integers are partitioned into two sets $\mathcal{A}$ and $\mathcal{B}$ in such a way that $r_A(n) = r_B(n)$ for all non-negative integers n. Without loss of generality, $0 \in \mathcal{A}$. Show that $1 \in \mathcal{B}$, that $2 \in \mathcal{B}$, and that $3 \in \mathcal{A}$.

   (c) With $\mathcal{A}$ and $\mathcal{B}$ as above, show that $A(z) + B(z) = 1/(1 - z)$ for $|z| < 1$.

   (d) Show that $A(z) - B(z) = (1 - z)\big(A(z^2) - B(z^2)\big)$, and hence by induction that

$$A(z) - B(z) = \prod_{k=0}^{\infty} \left(1 - z^{2^k}\right)$$

for $|z| < 1$.

(e) Let the *binary weight* of n, denoted $w(n)$, be the number of 1's in the binary expansion of n. That is, if $n = 2^{k_1} + \cdots + 2^{k_r}$ with $k_1 > \cdots > k_r$, then $w(n) = r$. Show that $\mathcal{A}$ consists of those non-negative integers n for which $w(n)$ is even, and that $\mathcal{B}$ is the set of those integers for which $w(n)$ is odd.

1.2 Analytic properties of Dirichlet series

Having provided some motivation for the use of Dirichlet series, we now turn to the task of establishing some of their basic analytic properties, corresponding to well-known facts concerning power series.

Theorem 1.1 *Suppose that the Dirichlet series $\alpha(s) = \sum_{n=1}^{\infty} a_n n^{-s}$ converges at the point $s = s_0$, and that $H > 0$ is an arbitrary constant. Then the series $\alpha(s)$ is uniformly convergent in the sector $S = \{s : \sigma \geq \sigma_0, |t - t_0| \leq H(\sigma - \sigma_0)\}$.*

By taking H large, we see that the series $\alpha(s)$ converges for all s in the half-plane $\sigma > \sigma_0$, and hence that the domain of convergence is a half-plane. More precisely, we have

Corollary 1.2 *Any Dirichlet series $\alpha(s) = \sum_{n=1}^{\infty} a_n n^{-s}$ has an abscissa of convergence σ_c with the property that $\alpha(s)$ converges for all s with $\sigma > \sigma_c$, and for no s with $\sigma < \sigma_c$. Moreover, if s_0 is a point with $\sigma_0 > \sigma_c$, then there is a neighbourhood of s_0 in which $\alpha(s)$ converges uniformly.*

In extreme cases a Dirichlet series may converge throughout the plane ($\sigma_c = -\infty$), or nowhere ($\sigma_c = +\infty$). When the abscissa of convergence is finite, the series may converge everywhere on the line $\sigma_c + it$, it may converge at some but not all points on this line, or nowhere on the line.

Proof of Theorem 1.1   Let $R(u) = \sum_{n>u} a_n n^{-s_0}$ be the remainder term of the series $\alpha(s_0)$. First we show that for any s,

$$\sum_{n=M+1}^{N} a_n n^{-s} = R(M)M^{s_0-s} - R(N)N^{s_0-s} + (s_0 - s)\int_{M}^{N} R(u)u^{s_0-s-1}\,du.$$

$$(1.7)$$

To see this we note that $a_n = (R(n-1) - R(n))n^{s_0}$, so that by partial summation

$$\sum_{n=M+1}^{N} a_n n^{-s} = \sum_{n=M+1}^{N} (R(n-1) - R(n))n^{s_0-s}$$

$$= R(M)M^{s_0-s} - R(N)N^{s_0-s} - \sum_{n=M+1}^{N} R(n-1)((n-1)^{s_0-s} - n^{s_0-s}).$$

The second factor in this last sum can be expressed as an integral,

$$(n-1)^{s_0-s} - n^{s_0-s} = -(s_0 - s) \int_{n-1}^{n} u^{s_0-s-1} \, du,$$

and hence the sum is

$$(s - s_0) \sum_{n=M+1}^{N} R(n-1) \int_{n-1}^{n} u^{s_0-s-1} \, du = (s - s_0) \sum_{n=M+1}^{N} \int_{n-1}^{n} R(u) u^{s_0-s-1} \, du$$

since $R(u)$ is constant in the interval $[n-1, n)$. The integrals combine to give (1.7).

If $|R(u)| \leq \varepsilon$ for all $u \geq M$ and if $\sigma > \sigma_0$, then from (1.7) we see that

$$\left| \sum_{n=M+1}^{N} a_n n^{-s} \right| \leq 2\varepsilon + \varepsilon|s - s_0| \int_{M}^{\infty} u^{\sigma_0-\sigma-1} \, du \leq \left(2 + \frac{|s - s_0|}{\sigma - \sigma_0}\right)\varepsilon.$$

For s in the prescribed region we see that

$$|s - s_0| \leq \sigma - \sigma_0 + |t - t_0| \leq (H + 1)(\sigma - \sigma_0),$$

so that the sum $\sum_{M+1}^{N} a_n n^{-s}$ is uniformly small, and the result follows by the uniform version of Cauchy's principle. □

In deriving (1.7) we used partial summation, although it would have been more efficient to use the properties of the Riemann–Stieltjes integral (see Appendix A):

$$\sum_{n=M+1}^{N} a_n n^{-s} = -\int_{M}^{N} u^{s_0-s} \, dR(u) = -u^{s_0-s} R(u) \Big|_{M}^{N} + \int_{M}^{N} R(u) \, du^{s_0-s}$$

by Theorems A.1 and A.2. By Theorem A.3 this is

$$= M^{s_0-s} R(M) - N^{s_0-s} R(N) + (s_0 - s) \int_{M}^{N} R(u) u^{s_0-s-1} \, du.$$

In more complicated situations it is an advantage to use the Riemann–Stieltjes integral, and subsequently we shall do so without apology.

The series $\alpha(s) = \sum a_n n^{-s}$ is locally uniformly convergent for $\sigma > \sigma_c$, and each term is an analytic function, so it follows from a general principle of

Weierstrass that $\alpha(s)$ is analytic for $\sigma > \sigma_c$, and that the differentiated series is locally uniformly convergent to $\alpha'(s)$:

$$\alpha'(s) = -\sum_{n=1}^{\infty} a_n (\log n) n^{-s} \tag{1.8}$$

for s in the half-plane $\sigma > \sigma_c$.

Suppose that s_0 is a point on the line of convergence (i.e., $\sigma_0 = \sigma_c$), and that the series $\alpha(s_0)$ converges. It can be shown by example that

$$\lim_{\substack{s \to s_0 \\ \sigma > \sigma_c}} \alpha(s)$$

need not exist. However, $\alpha(s)$ is continuous in the sector $\mathcal{S}$ of Theorem 1.1, in view of the uniform convergence there. That is,

$$\lim_{\substack{s \to s_0 \\ s \in \mathcal{S}}} \alpha(s) = \alpha(s_0), \tag{1.9}$$

which is analogous to Abel's theorem for power series.

We now express a convergent Dirichlet series as an absolutely convergent integral.

Theorem 1.3 *Let $A(x) = \sum_{n \le x} a_n$. If $\sigma_c < 0$, then $A(x)$ is a bounded function, and*

$$\sum_{n=1}^{\infty} a_n n^{-s} = s \int_1^{\infty} A(x) x^{-s-1} \, dx \tag{1.10}$$

for $\sigma > 0$. If $\sigma_c \ge 0$, then

$$\limsup_{x \to \infty} \frac{\log |A(x)|}{\log x} = \sigma_c, \tag{1.11}$$

and (1.10) holds for $\sigma > \sigma_c$.

Proof We note that

$$\sum_{n=1}^{N} a_n n^{-s} = \int_{1^-}^{N} x^{-s} \, dA(x) = A(x) x^{-s} \Big|_{1^-}^{N} - \int_{1^-}^{N} A(x) \, dx^{-s}$$

$$= A(N) N^{-s} + s \int_1^{N} A(x) x^{-s-1} \, dx.$$

Let ϕ denote the left-hand side of (1.11). If $\theta > \phi$ then $A(x) \ll x^{\theta}$ where the implicit constant may depend on the a_n and on θ. Thus if $\sigma > \theta$, then the integral in (1.10) is absolutely convergent. Thus we obtain (1.10) by letting $N \to \infty$, since the first term above tends to 0 as $N \to \infty$.

Suppose that $\sigma_c < 0$. By Corollary 1.2 we know that $A(x)$ tends to a finite limit as $x \to \infty$, and hence $\phi \le 0$, so that (1.10) holds for all $\sigma > 0$.

Now suppose that $\sigma_c \geq 0$. By Corollary 1.2 we know that the series in (1.10) diverges when $\sigma < \sigma_c$. Hence $\phi \geq \sigma_c$. To complete the proof it suffices to show that $\phi \leq \sigma_c$. Choose $\sigma_0 > \sigma_c$. By (1.7) with $s = 0$ and $M = 0$ we see that

$$A(N) = -R(N)N^{\sigma_0} + \sigma_0 \int_0^N R(u)u^{\sigma_0-1}du.$$

Since $R(u)$ is a bounded function, it follows that $A(N) \ll N^{\sigma_0}$ where the implicit constant may depend on the a_n and on σ_0. Hence $\phi \leq \sigma_0$. Since this holds for any $\sigma_0 > \sigma_c$, we conclude that $\phi \leq \sigma_c$. $\square$

The terms of a power series are majorized by a geometric progression at points strictly inside the circle of convergence. Consequently power series converge very rapidly. In contrast, Dirichlet series are not so well behaved. For example, the series

$$\sum_{n=1}^{\infty}(-1)^{n-1}n^{-s} \tag{1.12}$$

converges for $\sigma > 0$, but it is absolutely convergent only for $\sigma > 1$. In general we let σ_a denote the infimum of those σ for which $\sum_{n=1}^{\infty} |a_n|n^{-\sigma} < \infty$. Then σ_a, the *abscissa of absolute convergence*, is the abscissa of convergence of the series $\sum_{n=1}^{\infty} |a_n|n^{-s}$, and we see that $\sum a_n n^{-s}$ is absolutely convergent if $\sigma > \sigma_a$, but not if $\sigma < \sigma_a$. We now show that the strip $\sigma_c \leq \sigma \leq \sigma_a$ of conditional convergence is never wider than in the example (1.12).

Theorem 1.4   *In the above notation, $\sigma_c \leq \sigma_a \leq \sigma_c + 1$.*

Proof   The first inequality is obvious. To prove the second, suppose that $\varepsilon > 0$. Since the series $\sum a_n n^{-\sigma_c-\varepsilon}$ is convergent, the summands tend to 0, and hence $a_n \ll n^{\sigma_c+\varepsilon}$ where the implicit constant may depend on the a_n and on ε. Hence the series $\sum a_n n^{-\sigma_c-1-2\varepsilon}$ is absolutely convergent by comparison with the series $\sum n^{-1-\varepsilon}$. $\square$

Clearly a Dirichlet series $\alpha(s)$ is uniformly bounded in the half-plane $\sigma > \sigma_a + \varepsilon$, but this is not generally the case in the strip of conditional convergence. Nevertheless, we can limit the rate of growth of $\alpha(s)$ in this strip.

To aid in formulating our next result we introduce a notational convention that arises because many estimates relating to Dirichlet series are expressed in terms of the size of $|t|$. Our interest is in large values of this quantity, but in order that the statements be valid for small $|t|$ we sometimes write $|t| + 4$. Since this is cumbersome in complicated expressions, we introduce a shorthand: $\tau = |t| + 4$.

Theorem 1.5 *Suppose that $\alpha(s) = \sum a_n n^{-s}$ has abscissa of convergence σ_c. If δ and ε are fixed, $0 < \varepsilon < \delta < 1$, then*

$$\alpha(s) \ll \tau^{1-\delta+\varepsilon}$$

uniformly for $\sigma \geq \sigma_c + \delta$. The implicit constant may depend on the coefficients a_n, on δ, and on ε.

By the example found in Exercise 8 at the end of this section, we see that the bound above is reasonably sharp.

Proof Let s be a complex number with $\sigma \geq \sigma_c + \delta$. By (1.7) with $s_0 = \sigma_c + \varepsilon$ and $N \to \infty$, we see that

$$\alpha(s) = \sum_{n=1}^{M} a_n n^{-s} + R(M)M^{\sigma_c+\varepsilon-s} + (\sigma_c + \varepsilon - s) \int_{M}^{\infty} R(u)u^{\sigma_c+\varepsilon-s-1} \, du.$$

Since the series $\alpha(\sigma_c + \varepsilon)$ converges, we know that $a_n \ll n^{\sigma_c+\varepsilon}$, and also that $R(u) \ll 1$. Thus the above is

$$\ll \sum_{n=1}^{M} n^{-\delta+\varepsilon} + M^{-\delta+\varepsilon} + \frac{|\sigma_c + \varepsilon - s|}{\sigma - \sigma_c - \varepsilon} M^{\sigma_c+\varepsilon-\sigma}.$$

By the integral test the sum here is

$$< \int_{0}^{M} u^{-\delta+\varepsilon} \, du = \frac{M^{1-\delta+\varepsilon}}{1-\delta+\varepsilon} \ll M^{1-\delta+\varepsilon}.$$

Hence on taking $M = [\tau]$ we obtain the stated estimate. $\qquad\square$

We know that the power series expansion of a function is unique; we now show that the same is true for Dirichlet series expansions.

Theorem 1.6 *If $\sum a_n n^{-s} = \sum b_n n^{-s}$ for all s with $\sigma > \sigma_0$ then $a_n = b_n$ for all positive integers n.*

Proof We put $c_n = a_n - b_n$, and consider $\sum c_n n^{-s}$. Suppose that $c_n = 0$ for all $n < N$. Since $\sum c_n n^{-\sigma} = 0$ for $\sigma > \sigma_0$ we may write

$$c_N = -\sum_{n>N} c_n (N/n)^{\sigma}.$$

By Theorem 1.4 this sum is absolutely convergent for $\sigma > \sigma_0 + 1$. Since each term tends to 0 as $\sigma \to \infty$, we see that the right-hand side tends to 0, by the principle of dominated convergence. Hence $c_N = 0$, and by induction we deduce that this holds for all N. $\qquad\square$

Suppose that f is analytic in a domain $\mathcal{D}$, and that $0 \in \mathcal{D}$. Then f can be expressed as a power series $\sum_{n=0}^{\infty} a_n z^n$ in the disc $|z| < r$ where r is the distance from 0 to the boundary $\partial\mathcal{D}$ of $\mathcal{D}$. Although Dirichlet series are analytic functions, the situation regarding Dirichlet series expansions is very different: The collection of functions that may be expressed as a Dirichlet series in some half-plane is a very special class. Moreover, the line $\sigma_c + it$ of convergence need not contain a singular point of $\alpha(s)$. For example, the Dirichlet series (1.12) has abscissa of convergence $\sigma_c = 0$, but it represents the entire function $(1 - 2^{1-s})\zeta(s)$. (The connection of (1.12) to the zeta function is easy to establish, since

$$\sum_{n=1}^{\infty}(-1)^{n-1}n^{-s} = \sum_{n=1}^{\infty}n^{-s} - 2\sum_{\substack{n=1 \\ n \text{ even}}}^{\infty} n^{-s} = \zeta(s) - 2^{1-s}\zeta(s)$$

for $\sigma > 1$. That this is an entire function follows from Theorem 10.2.) Since a Dirichlet series does not in general have a singularity on its line of convergence, it is noteworthy that a Dirichlet series with non-negative coefficients not only has a singularity on the line $\sigma_c + it$, but actually at the point σ_c.

Theorem 1.7   (Landau) *Let $\alpha(s) = \sum a_n n^{-s}$ be a Dirichlet series whose abscissa of convergence σ_c is finite. If $a_n \geq 0$ for all n then the point σ_c is a singularity of the function $\alpha(s)$.*

It is enough to assume that $a_n \geq 0$ for all sufficiently large n, since any finite sum $\sum_{n=1}^{N} a_n n^{-s}$ is an entire function.

Proof   By replacing a_n by $a_n n^{-\sigma_c}$, we may assume that $\sigma_c = 0$. Suppose that $\alpha(s)$ is analytic at $s = 0$, so that $\alpha(s)$ is analytic in the domain $\mathcal{D} = \{s : \sigma > 0\} \cup \{|s| < \delta\}$ if $\delta > 0$ is sufficiently small. We expand $\alpha(s)$ as a power series at $s = 1$:

$$\alpha(s) = \sum_{k=0}^{\infty} c_k(s - 1)^k. \tag{1.13}$$

The coefficients c_k can be calculated by means of (1.8),

$$c_k = \frac{\alpha^{(k)}(1)}{k!} = \frac{1}{k!}\sum_{n=1}^{\infty} a_n(-\log n)^k n^{-1}.$$

The radius of convergence of the power series (1.13) is the distance from 1 to the nearest singularity of $\alpha(s)$. Since $\alpha(s)$ is analytic in $\mathcal{D}$, and since the nearest points not in $\mathcal{D}$ are $\pm i\delta$, we deduce that the radius of convergence is at least $\sqrt{1 + \delta^2} = 1 + \delta'$, say. That is,

$$\alpha(s) = \sum_{k=0}^{\infty}\frac{(1 - s)^k}{k!}\sum_{n=1}^{\infty} a_n(\log n)^k n^{-1}$$

for $|s - 1| < 1 + \delta'$. If $s < 1$ then all terms above are non-negative. Since series of non-negative numbers may be arbitrarily rearranged, for $-\delta' < s < 1$ we may interchange the summations over k and n to see that

$$\alpha(s) = \sum_{n=1}^{\infty} a_n n^{-1} \sum_{k=0}^{\infty} \frac{(1 - s)^k (\log n)^k}{k!}$$

$$= \sum_{n=1}^{\infty} a_n n^{-1} \exp\left((1 - s) \log n\right) = \sum_{n=1}^{\infty} a_n n^{-s}.$$

Hence this last series converges at $s = -\delta'/2$, contrary to the assumption that $\sigma_c = 0$. Thus $\alpha(s)$ is not analytic at $s = 0$. □

1.2.1 Exercises

1. Suppose that $\alpha(s)$ is a Dirichlet series, and that the series $\alpha(s_0)$ is boundedly oscillating. Show that $\sigma_c = \sigma_0$.
2. Suppose that $\alpha(s) = \sum_{n=1}^{\infty} a_n n^{-s}$ is a Dirichlet series with abscissa of convergence σ_c. Suppose that $\alpha(0)$ converges, and put $R(x) = \sum_{n>x} a_n$. Show that σ_c is the infimum of those numbers θ such that $R(x) \ll x^{\theta}$.
3. Let $A_k(x) = \sum_{n \leq x} a_n (\log n)^k$.
   (a) Show that

   $$A_0(x) - \frac{A_1(x)}{\log x} = a_1 + \int_2^x \frac{A_1(u)}{u(\log u)^2} \, du.$$

   (b) Suppose that $A_1(x) \ll x^{\theta}$ where $\theta > 0$ and the implicit constant may depend on the sequence $\{a_n\}$. Show that

   $$A_0(x) = \frac{A_1(x)}{\log x} + O(x^{\theta}(\log x)^{-2}).$$

   (c) Let σ_c denote the abscissa of convergence of $\sum a_n n^{-s}$, and σ_c' the abscissa of convergence of $\sum a_n (\log n) n^{-s}$. Show that $\sigma_c' = \sigma_c$. (The remarks following the proof of Theorem 1.1 imply only that $\sigma_c' \leq \sigma_c$.)
4. (Landau 1909b) Let $\alpha(s) = \sum a_n n^{-s}$ be a Dirichlet series with abscissa of convergence σ_c and abscissa of absolute convergence $\sigma_a > \sigma_c$. Let $C(x) = \sum_{n \leq x} a_n n^{-\sigma_c}$ and $A(x) = \sum_{n \leq x} |a_n| n^{-\sigma_c}$.
   (a) By a suitable application of Theorem 1.3, or otherwise, show that $C(x) \ll x^{\varepsilon}$ and that $A(x) \ll x^{\sigma_a - \sigma_c + \varepsilon}$ for any $\varepsilon > 0$, where the implicit constants may depend on ε and on the sequence $\{a_n\}$.
   (b) Show that if $\sigma > \sigma_c$ then

   $$\sum_{n>N} a_n n^{-s} = -C(N) N^{\sigma_c - s} + (s - \sigma_c) \int_N^{\infty} C(u) u^{\sigma_c - s - 1} \, du.$$

Deduce that the above is $\ll \tau N^{\sigma_c - \sigma + \varepsilon}$ uniformly for s in the half-plane $\sigma \geq \sigma_c + \varepsilon$ where the implicit constant may depend on ε and on the sequence $\{a_n\}$.

(c) Show that

$$\sum_{n=1}^{N} |a_n| n^{-\sigma} = A(N) N^{-\sigma + \sigma_c} + (\sigma - \sigma_c) \int_1^N A(u) u^{-\sigma + \sigma_c - 1} \, du$$

for any σ. Deduce that the above is $\ll N^{\sigma_a - \sigma + \varepsilon}$ uniformly for σ in the interval $\sigma_c \leq \sigma \leq \sigma_a$, for any given $\varepsilon > 0$. Here the implicit constant may depend on ε and on the sequence $\{a_n\}$.

(d) Let $\theta(\sigma) = (\sigma_a - \sigma)/(\sigma_a - \sigma_c)$. By making a suitable choice of N, show that

$$\alpha(s) \ll \tau^{\theta(\sigma) + \varepsilon}$$

uniformly for s in the strip $\sigma_c + \varepsilon \leq \sigma \leq \sigma_a$.

5. (a) Show that if $\alpha(s) = \sum a_n n^{-s}$ has abscissa of convergence $\sigma_c < \infty$, then

$$\lim_{\sigma \to \infty} \alpha(\sigma) = a_1.$$

(b) Show that $\zeta'(s) = -\sum_{n=1}^{\infty} (\log n) n^{-s}$ for $\sigma > 1$.

(c) Show that $\lim_{\sigma \to \infty} \zeta'(\sigma) = 0$.

(d) Show that there is no half-plane in which $1/\zeta'(s)$ can be written as a convergent Dirichlet series.

6. Let $\alpha(s) = \sum a_n n^{-s}$ be a Dirichlet series with $a_n \geq 0$ for all n. Show that $\sigma_c = \sigma_a$, and that

$$\sup_t |\alpha(s)| = \alpha(\sigma)$$

for any given $\sigma > \sigma_c$.

7. (Vivanti 1893; Pringsheim 1894) Suppose that $f(z) = \sum_{n=0}^{\infty} a_n z^n$ has radius of convergence 1 and that $a_n \geq 0$ for all n. Show that $z = 1$ is a singular point of f.

8. (Bohr 1910, p. 32) Let $t_1 = 4$, $t_{r+1} = 2^{t_r}$ for $r \geq 1$. Put $\alpha(s) = \sum a_n n^{-s}$ where $a_n = 0$ unless $n \in [t_r, 2t_r]$ for some r, in which case put

$$a_n = \begin{cases} t_r^{it_r} & (n = t_r), \\ n^{it_r} - (n-1)^{it_r} & (t_r < n < 2t_r), \\ -(2t_r - 1)^{it_r} & (n = 2t_r). \end{cases}$$

(a) Show that $\sum_{t_r}^{2t_r} a_n = 0$.

(b) Show that if $t_r \leq x < 2t_r$ for some r, then $A(x) = [x]^{it_r}$ where $A(x) = \sum_{n \leq x} a_n$.

(c) Show that $A(x) \ll 1$ uniformly for $x \geq 1$.

(d) Deduce that $\alpha(s)$ converges for $\sigma > 0$.

(e) Show that $\alpha(it)$ does not converge; conclude that $\sigma_c = 0$.

(f) Show that if $\sigma > 0$, then

$$\alpha(s) = \sum_{r=1}^{R} \sum_{n=t_r}^{2t_r} a_n n^{-s} + s \int_{t_{R+1}}^{\infty} A(x)x^{-s-1} \, dx \, .$$

(g) Suppose that $\sigma > 0$. Show that the above is

$$\sum_{n=t_R}^{2t_R} a_n n^{-s} + O(t_{R-1}) + O\left(\frac{|s|}{\sigma t_{R+1}^{\sigma}}\right).$$

(h) Show that if $\sigma > 0$, then

$$\sum_{n=t_R}^{2t_R} a_n n^{-s} = s \int_{t_R}^{2t_R} [x]^{it_R} x^{-s-1} \, dx \, .$$

(i) Show that if $n \leq x < n+1$, then $\Re(n^{it_R} x^{-it_R}) \geq 1/2$. Deduce that

$$\left| \int_{t_R}^{2t_R} [x]^{it_R} x^{-\sigma - it_R - 1} \, dx \right| \gg t_R^{-\sigma} \, .$$

(j) Suppose that $\delta > 0$ is fixed. Conclude that if $R \geq R_0(\delta)$, then $|\alpha(\sigma + it_R)| \gg t_R^{1-\sigma}$ uniformly for $\delta \leq \sigma \leq 1 - \delta$.

(k) Show that $\sum |a_n| n^{-\sigma} < \infty$ when $\sigma > 1$. Deduce that $\sigma_a = 1$.

1.3 Euler products and the zeta function

The situation regarding products of Dirichlet series is somewhat complicated, but it is useful to note that the formal calculation in (2) is justified if the series are absolutely convergent.

Theorem 1.8 *Let $\alpha(s) = \sum a_n n^{-s}$ and $\beta(s) = \sum b_n n^{-s}$ be two Dirichlet series, and put $\gamma(s) = \sum c_n n^{-s}$ where the c_n are given by (1.3). If s is a point at which the two series $\alpha(s)$ and $\beta(s)$ are both absolutely convergent, then $\gamma(s)$ is absolutely convergent and $\gamma(s) = \alpha(s)\beta(s)$.*

The mere convergence of $\alpha(s)$ and $\beta(s)$ is not sufficient to justify (1.2). Indeed, the square of the series (1.12) can be shown to have abscissa of convergence $\geq 1/4$.

A function is called an *arithmetic function* if its domain is the set $\mathbb{Z}$ of integers, or some subset of the integers such as the natural numbers. An arithmetic function $f(n)$ is said to be *multiplicative* if $f(1) = 1$ and if $f(mn) = f(m)f(n)$ whenever $(m, n) = 1$. Also, an arithmetic function $f(n)$ is called *totally multiplicative* if $f(1) = 1$ and if $f(mn) = f(m)f(n)$ for all m and n. If f is multiplicative then the Dirichlet series $\sum f(n)n^{-s}$ factors into a product over primes. To see why this is so, we first argue formally (i.e., we ignore questions of convergence). When the product

$$\prod_p (1 + f(p)p^{-s} + f(p^2)p^{-2s} + f(p^3)p^{-3s} + \cdots)$$

is expanded, the generic term is

$$\frac{f\left(p_1^{k_1}\right)f\left(p_2^{k_2}\right)\cdots f\left(p_r^{k_r}\right)}{\left(p_1^{k_1} p_2^{k_2} \cdots p_r^{k_r}\right)^s}.$$

Set $n = p_1^{k_1} p_2^{k_2} \cdots p_r^{k_r}$. Since f is multiplicative, the above is $f(n)n^{-s}$. Moreover, this correspondence between products of prime powers and positive integers n is one-to-one, in view of the fundamental theorem of arithmetic. Hence after rearranging the terms, we obtain the sum $\sum f(n)n^{-s}$. That is, we expect that

$$\sum_{n=1}^{\infty} f(n)n^{-s} = \prod_p (1 + f(p)p^{-s} + f(p^2)p^{-2s} + \cdots). \qquad (1.14)$$

The product on the right-hand side is called the *Euler product* of the Dirichlet series. The mere convergence of the series on the left does not imply that the product converges; as in the case of the identity (1.2), we justify (1.14) only under the stronger assumption of absolute convergence.

Theorem 1.9 *If f is multiplicative and $\sum |f(n)|n^{-\sigma} < \infty$, then (1.14) holds.*

If f is totally multiplicative, then the terms on the right-hand side in (1.14) form a geometric progression, in which case the identity may be written more concisely,

$$\sum_{n=1}^{\infty} f(n)n^{-s} = \prod_p (1 - f(p)p^{-s})^{-1}. \qquad (1.15)$$

Proof For any prime p,

$$\sum_{k=0}^{\infty} |f(p^k)|p^{-k\sigma} \le \sum_{n=1}^{\infty} |f(n)|n^{-\sigma} < \infty,$$

so each sum on the right-hand side of (1.14) is absolutely convergent. Let y be a positive real number, and let $\mathcal{N}$ be the set of those positive integers composed entirely of primes not exceeding y, $\mathcal{N} = \{n : p|n \Rightarrow p \leq y\}$. (Note that $1 \in \mathcal{N}$.) Since a product of finitely many absolutely convergent series may be arbitrarily rearranged, we see that

$$\Pi_y = \prod_{p \leq y} \left(1 + f(p)p^{-s} + f(p^2)p^{-2s} + \cdots\right) = \sum_{n \in \mathcal{N}} f(n)n^{-s}.$$

Hence

$$\left| \Pi_y - \sum_{n=1}^{\infty} f(n)n^{-s} \right| \leq \sum_{n \notin \mathcal{N}} |f(n)|n^{-\sigma}.$$

If $n \leq y$ then all prime factors of n are $\leq y$, and hence $n \in \mathcal{N}$. Consequently the sum on the right above is

$$\leq \sum_{n > y} |f(n)|n^{-\sigma},$$

which is small if y is large. Thus the partial products Π_y tend to $\sum f(n)n^{-s}$ as $y \to \infty$. □

Let $\omega(n)$ denote the number of distinct primes dividing n, and let $\Omega(n)$ be the number of distinct prime powers dividing n. That is,

$$\omega(n) = \sum_{p|n} 1, \qquad \Omega(n) = \sum_{p^k|n} 1 = \sum_{p^k\|n} k. \qquad (1.16)$$

It is easy to distinguish these functions, since $\omega(n) \leq \Omega(n)$ for all n, with equality if and only if n is square-free. These functions are examples of *additive functions* because they satisfy the functional relation $f(mn) = f(m) + f(n)$ whenever $(m, n) = 1$. Moreover, $\Omega(n)$ is *totally additive* because this functional relation holds for all pairs m, n. An exponential of an additive function is a multiplicative function. In particular, the *Liouville lambda function* is the totally multiplicative function $\lambda(n) = (-1)^{\Omega(n)}$. Closely related is the *Möbius mu function*, which is defined to be $\mu(n) = (-1)^{\omega(n)}$ if n is square-free, $\mu(n) = 0$ otherwise. By the fundamental theorem of arithmetic we know that a multiplicative (or additive) function is uniquely determined by its values at prime powers, and similarly that a totally multiplicative (or totally additive) function is uniquely determined by its values at the primes. Thus $\mu(n)$ is the unique multiplicative function that takes the value -1 at every prime, and the value 0 at every higher power of a prime, while $\lambda(n)$ is the unique totally multiplicative function that takes the value -1 at every prime. By using Theorem 1.9 we can determine the Dirichlet series generating functions of $\lambda(n)$ and of $\mu(n)$ in terms of the Riemann zeta function.

Corollary 1.10 *For $\sigma > 1$,*

$$\sum_{n=1}^{\infty} n^{-s} = \zeta(s) = \prod_p (1 - p^{-s})^{-1}, \tag{1.17}$$

$$\sum_{n=1}^{\infty} \mu(n)n^{-s} = \frac{1}{\zeta(s)} = \prod_p (1 - p^{-s}), \tag{1.18}$$

and

$$\sum_{n=1}^{\infty} \lambda(n)n^{-s} = \frac{\zeta(2s)}{\zeta(s)} = \prod_p (1 + p^{-s})^{-1}. \tag{1.19}$$

Proof   All three series are absolutely convergent, since $\sum n^{-\sigma} < \infty$ for $\sigma > 1$, by the integral test. Since the coefficients are multiplicative, the Euler product formulae follow by Theorem 1.9. In the first and third cases use the variant (1.15). On comparing the Euler products in (1.17) and (1.18), it is immediate that the second of these Dirichlet series is $1/\zeta(s)$. As for (1.19), from the identity $1 + z = (1 - z^2)/(1 - z)$ we deduce that

$$\prod_p (1 + p^{-s}) = \frac{\prod_p (1 - p^{-2s})}{\prod_p (1 - p^{-s})} = \frac{\zeta(s)}{\zeta(2s)}.$$

$\square$

The manipulation of Euler products, as exemplified above, provides a powerful tool for relating one Dirichlet series to another.

In (1.17) we have expressed $\zeta(s)$ as an absolutely convergent product; hence in particular $\zeta(s) \neq 0$ for $\sigma > 1$. We have not yet defined the zeta function outside this half-plane, but we shall do so shortly, and later we shall find that the zeta function does have zeros in the half-plane $\sigma \leq 1$. These zeros play an important role in determining the distribution of prime numbers.

Many important relations involving arithmetic functions can be expressed succinctly in terms of Dirichlet series. For example, the fundamental elementary identity

$$\sum_{d|n} \mu(d) = \begin{cases} 1 & \text{if } n = 1, \\ 0 & \text{if } n > 1. \end{cases} \tag{1.20}$$

is equivalent to the identity

$$\zeta(s) \cdot \frac{1}{\zeta(s)} = 1,$$

in view of (1.3), (1.17), (1.18), and Theorem 1.6. More generally, if

$$F(n) = \sum_{d|n} f(d) \qquad (1.21)$$

for all n, then, apart from questions of convergence,

$$\sum F(n)n^{-s} = \zeta(s) \sum f(n)n^{-s}.$$

By Möbius inversion, the identity (1.21) is equivalent to the relation

$$f(n) = \sum_{d|n} \mu(d)F(n/d),$$

which is to say that

$$\sum f(n)n^{-s} = \frac{1}{\zeta(s)} \sum F(n)n^{-s}.$$

Such formal manipulations can be used to suggest (or establish) many useful elementary identities.

For $\sigma > 1$ the product (1.17) is absolutely convergent. Since $\log(1 - z)^{-1} = \sum_{k=1}^{\infty} z^k/k$ for $|z| < 1$, it follows that

$$\log \zeta(s) = \sum_p \log(1 - p^{-s})^{-1} = \sum_p \sum_{k=1}^{\infty} k^{-1} p^{-ks}.$$

On differentiating, we find also that

$$\frac{\zeta'(s)}{\zeta(s)} = -\sum_p \sum_{k=1}^{\infty} (\log p) p^{-ks}$$

for $\sigma > 1$. This is a Dirichlet series, whose n^{th} coefficient is the von Mangoldt lambda function: $\Lambda(n) = \log p$ if n is a power of p, $\Lambda(n) = 0$ otherwise.

Corollary 1.11 *For $\sigma > 1$,*

$$\log \zeta(s) = \sum_{n=1}^{\infty} \frac{\Lambda(n)}{\log n} n^{-s}$$

and

$$-\frac{\zeta'(s)}{\zeta(s)} = \sum_{n=1}^{\infty} \Lambda(n)n^{-s}.$$

The quotient $f'(s)/f(s)$, obtained by differentiating the logarithm of $f(s)$, is known as the *logarithmic derivative* of f. Subsequently we shall often write it more concisely as $\frac{f'}{f}(s)$.

The important elementary identity

$$\sum_{d\mid n} \Lambda(d) = \log n \tag{1.22}$$

is reflected in the relation

$$\zeta(s)\left(-\frac{\zeta'}{\zeta}(s)\right) = -\zeta'(s),$$

since

$$-\zeta'(s) = \sum_{n=1}^{\infty}(\log n)n^{-s}$$

for $\sigma > 1$.

We now continue the zeta function beyond the half-plane in which it was initially defined.

Theorem 1.12 *Suppose that $\sigma > 0$, $x > 0$, and that $s \neq 1$. Then*

$$\zeta(s) = \sum_{n \leq x} n^{-s} + \frac{x^{1-s}}{s-1} + \frac{\{x\}}{x^s} - s\int_x^{\infty} \{u\}u^{-s-1}\, du. \tag{1.23}$$

Here $\{u\}$ denotes the fractional part of u, so that $\{u\} = u - [u]$ where $[u]$ denotes the integral part of u.

Proof of Theorem 1.12   For $\sigma > 1$ we have

$$\zeta(s) = \sum_{n=1}^{\infty} n^{-s} = \sum_{n \leq x} n^{-s} + \sum_{n > x} n^{-s}.$$

This second sum we write as

$$\int_x^{\infty} u^{-s}\, d[u] = \int_x^{\infty} u^{-s}\, du - \int_x^{\infty} u^{-s}\, d\{u\}.$$

We evaluate the first integral on the right-hand side, and integrate the second one by parts. Thus the above is

$$= \frac{x^{1-s}}{s-1} + \{x\}x^{-s} + \int_x^{\infty} \{u\}\, du^{-s}.$$

Since $(u^{-s})' = -su^{-s-1}$, the desired formula now follows by Theorem A.3. The integral in (1.23) is convergent in the half-plane $\sigma > 0$, and uniformly so for $\sigma \geq \delta > 0$. Since the integrand is an analytic function of s, it follows that the integral is itself an analytic function for $\sigma > 0$. By the uniqueness of analytic continuation the formula (1.23) holds in this larger half-plane. $\square$

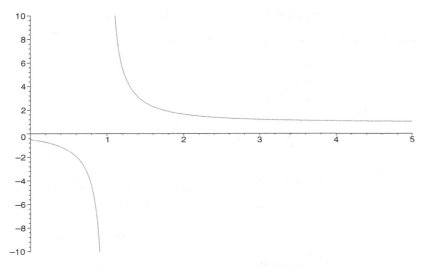

Figure 1.2 The Riemann zeta function $\zeta(s)$ for $0 < s \leq 5$.

By taking $x = 1$ in (1.23) we obtain in particular the identity

$$\zeta(s) = \frac{s}{s-1} - s \int_1^\infty \{u\} u^{-s-1}\, du \tag{1.24}$$

for $\sigma > 0$. Hence we have

Corollary 1.13 *The Riemann zeta function has a simple pole at $s = 1$ with residue 1, but is otherwise analytic in the half-plane $\sigma > 0$.*

A graph of $\zeta(s)$ that exhibits the pole at $s = 1$ is provided in Figure 1.2. By repeatedly integrating by parts we can continue $\zeta(s)$ into successively larger half-planes; this is systematized by using the Euler–Maclaurin summation formula (see Theorem B.5). In Chapter 10 we shall continue the zeta function by a different method. For the present we note that (1.24) yields useful inequalities for the zeta function on the real line.

Corollary 1.14 *The inequalities*

$$\frac{1}{\sigma - 1} < \zeta(\sigma) < \frac{\sigma}{\sigma - 1}$$

hold for all $\sigma > 0$. In particular, $\zeta(\sigma) < 0$ for $0 < \sigma < 1$.

Proof From the inequalities $0 \leq \{u\} < 1$ it follows that

$$0 \leq \int_1^\infty \{u\} u^{-\sigma-1}\, du < \int_1^\infty u^{-\sigma-1}\, du = \frac{1}{\sigma}.$$

This suffices. □

We now put the parameter x in (1.23) to good use.

Corollary 1.15 *Let δ be fixed, $\delta > 0$. Then for $\sigma \geq \delta$, $s \neq 1$,*

$$\sum_{n \leq x} n^{-s} = \frac{x^{1-s}}{1-s} + \zeta(s) + O(\tau x^{-\sigma}). \tag{1.25}$$

In addition,

$$\sum_{n \leq x} \frac{1}{n} = \log x + C_0 + O(1/x) \tag{1.26}$$

where C_0 is Euler's constant,

$$C_0 = 1 - \int_1^{\infty} \{u\} u^{-2} \, du = 0.5772156649\ldots. \tag{1.27}$$

Proof The first estimate follows by crudely estimating the integral in (1.23):

$$\int_x^{\infty} \{u\} u^{-s-1} \, du \ll \int_x^{\infty} u^{-\sigma-1} \, du = \frac{x^{-\sigma}}{\sigma}.$$

As for the second estimate, we note that the sum is

$$\int_{1-}^{x} u^{-1} \, d[u] = \int_{1-}^{x} u^{-1} \, du - \int_{1-}^{x} u^{-1} \, d\{u\}$$

$$= \log x + 1 - \{x\}/x - \int_1^x \{u\} u^{-2} \, du.$$

The result now follows by writing $\int_1^x = \int_1^\infty - \int_x^\infty$, and noting that

$$\int_x^{\infty} \{u\} u^{-2} \, du \ll \int_x^{\infty} u^{-2} \, du = 1/x.$$

$\square$

By letting $s \to 1$ in (1.25) and comparing the result with (1.26), or by letting $s \to 1$ in (1.24) and comparing the result with (1.27), we obtain

Corollary 1.16 *Let*

$$\zeta(s) = \frac{1}{s-1} + \sum_{k=0}^{\infty} a_k(s-1)^k \tag{1.28}$$

be the Laurent expansion of $\zeta(s)$ at $s = 1$. Then a_0 is Euler's constant, $a_0 = C_0$.

Euler's constant also arises in the theory of the gamma function. (See Appendix C and Chapter 10.)

Corollary 1.17 *Let $\delta > 0$ be fixed. Then*

$$\zeta(s) = \frac{1}{s-1} + O(1)$$

uniformly for s in the rectangle $\delta \leq \sigma \leq 2$, $|t| \leq 1$, *and*

$$\zeta(s) \ll (1 + \tau^{1-\sigma}) \min \left(\frac{1}{|\sigma - 1|}, \log \tau \right)$$

uniformly for $\delta \leq \sigma \leq 2$, $|t| \geq 1$.

Proof   The first assertion is clear from (1.24). When $|t|$ is larger, we obtain a bound for $|\zeta(s)|$ by estimating the sum in (1.25). Assume that $x \geq 2$. We observe that

$$\sum_{n \leq x} n^{-s} \ll \sum_{n \leq x} n^{-\sigma} \ll 1 + \int_1^x u^{-\sigma} \, du$$

uniformly for $\sigma \geq 0$. If $0 \leq \sigma \leq 1 - 1/\log x$, then this integral is $(x^{1-\sigma} - 1)/(1 - \sigma) < x^{1-\sigma}/(1 - \sigma)$. If $|\sigma - 1| \leq 1/\log x$, then $u^{-\sigma} \asymp u^{-1}$ uniformly for $1 \leq u \leq x$, and hence the integral is $\asymp \int_1^x u^{-1} \, du = \log x$. If $\sigma \geq 1 + 1/\log x$, then the integral is $< \int_1^\infty u^{-\sigma} \, du = 1/(\sigma - 1)$. Thus

$$\sum_{n \leq x} n^{-s} \ll (1 + x^{1-\sigma}) \min \left(\frac{1}{|\sigma - 1|}, \log x \right) \tag{1.29}$$

uniformly for $0 \leq \sigma \leq 2$. The second assertion now follows by taking $x = \tau$ in (1.25).   □

1.3.1   Exercises

1. Suppose that $f(mn) = f(m)f(n)$ whenever $(m, n) = 1$, and that f is not identically 0. Deduce that $f(1) = 1$, and hence that f is multiplicative.
2. (Stieltjes 1887) Suppose that $\sum a_n$ converges, that $\sum |b_n| < \infty$, and that c_n is given by (1.3). Show that $\sum c_n$ converges to $(\sum a_n)(\sum b_n)$. (Hint: Write $\sum_{n \leq x} c_n = \sum_{n \leq x} b_n A(x/n)$ where $A(y) = \sum_{n \leq y} a_n$.)
3. Determine $\sum \varphi(n) n^{-s}$, $\sum \sigma(n) n^{-s}$, and $\sum |\mu(n)| n^{-s}$ in terms of the zeta function. Here $\varphi(n)$ is Euler's 'totient function', which is the number of a, $1 \leq a \leq n$, such that $(a, n) = 1$.
4. Let q be a positive integer. Show that if $\sigma > 1$, then

$$\sum_{\substack{n=1 \\ (n,q)=1}}^{\infty} n^{-s} = \zeta(s) \prod_{p|q} (1 - p^{-s}).$$

5. Show that if $\sigma > 1$, then

$$\sum_{n=1}^{\infty} d(n)^2 n^{-s} = \zeta(s)^4 / \zeta(2s).$$

6. Let $\sigma_a(n) = \sum_{d|n} d^a$. Show that

$$\sum_{n=1}^{\infty} \sigma_a(n)\sigma_b(n)n^{-s} = \zeta(s)\zeta(s-a)\zeta(s-b)\zeta(s-a-b)/\zeta(2s-a-b)$$

when $\sigma > \max(1, 1 + \Re a, 1 + \Re b, 1 + \Re(a+b))$.

7. Let $F(s) = \sum_p (\log p)p^{-s}$, $G(s) = \sum_p p^{-s}$ for $\sigma > 1$. Show that in this half-plane,

$$-\frac{\zeta'}{\zeta}(s) = \sum_{k=1}^{\infty} F(ks),$$

$$F(s) = -\sum_{d=1}^{\infty} \mu(d)\frac{\zeta'}{\zeta}(ds),$$

$$\log \zeta(s) = \sum_{k=1}^{\infty} G(ks)/k,$$

$$G(s) = \sum_{d=1}^{\infty} \frac{\mu(d)}{d} \log \zeta(ds).$$

8. Let $F(s)$ and $G(s)$ be defined as in the preceding problem. Show that if $\sigma > 1$, then

$$\sum_{n=1}^{\infty} \omega(n)n^{-s} = \zeta(s)G(s) = \zeta(s)\sum_{d=1}^{\infty} \frac{\mu(d)}{d} \log \zeta(ds),$$

$$\sum_{n=1}^{\infty} \Omega(n)n^{-s} = \zeta(s)\sum_{k=1}^{\infty} G(ks) = \zeta(s)\sum_{k=1}^{\infty} \frac{\varphi(k)}{k} \log \zeta(ks).$$

9. Let t be a fixed real number, $t \neq 0$. Describe the limit points of the sequence of partial sums $\sum_{n \leq x} n^{-1-it}$.

10. Show that $\sum_{n=1}^{N} n^{-1} > \log N + C_0$ for all positive integers N, and that $\sum_{n \leq x} n^{-1} > \log x$ for all positive real numbers x.

11. (a) Show that if a_n is totally multiplicative, and if $\alpha(s) = \sum a_n n^{-s}$ has abscissa of convergence σ_c, then

$$\sum_{n=1}^{\infty} (-1)^{n-1} a_n n^{-s} = (1 - 2a_2 2^{-s})\alpha(s)$$

for $\sigma > \sigma_c$.

(b) Show that

$$\sum_{n=1}^{\infty} (-1)^{n-1} n^{-s} = (1 - 2^{1-s})\zeta(s)$$

for $\sigma > 0$.

(c) (Shafer 1984) Show that

$$\sum_{n=1}^{\infty}(-1)^n(\log n)n^{-1} = C_0 \log 2 - \frac{1}{2}(\log 2)^2.$$

12. (Stieltjes 1885) Show that if k is a positive integer, then

$$\sum_{n\leq x}\frac{(\log n)^k}{n} = \frac{(\log x)^{k+1}}{k+1} + C_k + O_k\Big(\frac{(\log x)^k}{x}\Big)$$

for $x \geq 1$ where

$$C_k = \int_1^{\infty}\{u\}(\log u)^{k-1}(k - \log u)u^{-2}\,du.$$

Show that the numbers a_k in (1.28) are given by $a_k = (-1)^k C_k/k!$.

13. Let $\mathcal{D}$ be the disc of radius 1 and centre 2. Suppose that the numbers ε_k tend monotonically to 0, that the numbers t_k tend monotonically to 0, and that the numbers N_k tend monotonically to infinity. We consider the Dirichlet series $\alpha(s) = \sum_n a_n n^{-s}$ with coefficients $a_n = \varepsilon_k n^{it_k}$ for $N_{k-1} < n \leq N_k$. For suitable choices of the ε_k, t_k, and N_k we show that the series converges at $s = 1$ but that it is not uniformly convergent in $\mathcal{D}$.

(a) Suppose that $\sigma_k = 2 - \sqrt{1 - t_k^2}$, so that $s_k = \sigma_k + it_k \in \mathcal{D}$. Show that if

$$N_k^{t_k^2} \ll 1, \tag{1.30}$$

then

$$\Big|\sum_{N_{k-1} < n \leq N_k} a_n n^{-s_k}\Big| \gg \varepsilon_k \log \frac{N_k}{N_{k-1}}.$$

Thus if

$$\varepsilon_k \log \frac{N_k}{N_{k-1}} \gg 1 \tag{1.31}$$

then the series is not uniformly convergent in $\mathcal{D}$.

(b) By using Corollary 1.15, or otherwise, show that if $(a, b] \subseteq (N_{k-1}, N_k]$, then

$$\sum_{a < n \leq b} a_n n^{-1} \ll \frac{\varepsilon_k}{t_k}.$$

Hence if

$$\sum_{k=1}^{\infty}\frac{\varepsilon_k}{t_k} < \infty, \tag{1.32}$$

then the series $\alpha(1)$ converges.

(c) Show that the parameters can be chosen so that (1.30)–(1.32) hold, say by taking $N_k = \exp(1/\varepsilon_k)$ and $t_k = \varepsilon_k^{1/2}$ with ε_k tending rapidly to 0.

14. Let $t(n) = (-1)^{\Omega(n)-\omega(n)} \prod_{p|n}(p-1)^{-1}$, and put $T(s) = \sum_n t(n)n^{-s}$.

(a) Show that for $\sigma > 0$, $T(s)$ has the absolutely convergent Euler product

$$T(s) = \prod_p \left(1 + \frac{1}{(p-1)(p^s + 1)}\right).$$

(b) Determine all zeros of the function $1 + 1/((p-1)(p^s + 1))$.

(c) Show that the line $\sigma = 0$ is a natural boundary of the function $T(s)$.

15. Suppose throughout that $0 < \alpha \le 1$. For $\sigma > 1$ we define the *Hurwitz zeta function* by the formula

$$\zeta(s, \alpha) = \sum_{n=0}^{\infty} (n + \alpha)^{-s}.$$

Thus $\zeta(s, 1) = \zeta(s)$.

(a) Show that $\zeta(s, 1/2) = (2^s - 1)\zeta(s)$.

(b) Show that if $x \ge 0$ then

$$\zeta(s, \alpha) = \sum_{0 \le n \le x} (n + \alpha)^{-s} + \frac{(x + \alpha)^{1-s}}{s - 1} + \frac{\{x\}}{(x + \alpha)^s}$$
$$- s \int_x^{\infty} \{u\}(u + \alpha)^{-s-1}\, du.$$

(c) Deduce that $\zeta(s, \alpha)$ is an analytic function of s for $\sigma > 0$ apart from a simple pole at $s = 1$ with residue 1.

(d) Show that

$$\lim_{s \to 1}\left(\zeta(s, \alpha) - \frac{1}{s - 1}\right) = 1/\alpha - \log\alpha - \int_0^{\infty} \frac{\{u\}}{(u + \alpha)^2}\, du.$$

(e) Show that

$$\lim_{s \to 1}\left(\zeta(s, \alpha) - \frac{1}{s - 1}\right) = \sum_{0 \le n \le x} \frac{1}{n + \alpha} - \log(x + \alpha) + \frac{\{x\}}{x + \alpha}$$
$$- \int_x^{\infty} \frac{\{u\}}{(u + \alpha)^2}\, du.$$

(f) Let $x \to \infty$ in the above, and use (C.2), (C.10) to show that

$$\lim_{s \to 1}\left(\zeta(s, \alpha) - \frac{1}{s - 1}\right) = -\frac{\Gamma'}{\Gamma}(\alpha).$$

(This is consistent with Corollary 1.16, in view of (C.11).)

1.4 Notes

Section 1.1. For a brief introduction to the Hardy–Littlewood circle method, including its application to Waring's problem, see Davenport (2005). For a comprehensive account of the method, see Vaughan (1997). Other examples of the fruitful use of generating functions are found in many sources, such as Andrews (1976) and Wilf (1994).

Algorithms for the efficient computation of $\pi(x)$ have been developed by Meissel (Lehmer, 1959), Mapes (1963), Lagarias, Miller & Odlyzko (1985), Deléglise & Rivat (1996), and by X. Gourdon. For discussion of these methods, see Chapter 1 of Riesel (1994) and the web page of Gourdon & Sebah at http://numbers.computation.free.fr/Constants/Primes/countingPrimes.html.

The 'big oh' notation was introduced by Paul Bachmann (1894, p. 401). The 'little oh' was introduced by Edmund Landau (1909a, p. 61). The $\asymp$ notation was introduced by Hardy (1910, p. 2). Our notation $f \sim g$ also follows Hardy (1910). The Omega notation was introduced by G. H. Hardy and J. E. Littlewood (1914, p. 225). Ingham (1932) replaced the Ω_R and Ω_L of Hardy and Littlewood by Ω_+ and Ω_-. The $\ll$ notation is due to I. M. Vinogradov.

Section 1.2. The series $\sum a_n n^{-s}$ is called an *ordinary* Dirichlet series, to distinguish it from a *generalized* Dirichlet series, which is a sum of the form $\sum a_n e^{-\lambda_n s}$ where $0 < \lambda_1 < \lambda_2 < \cdots, \lambda_n \to \infty$. We see that generalized Dirichlet series include both ordinary Dirichlet series ($\lambda_n = \log n$) and power series ($\lambda_n = n$). Theorems 1.1, 1.3, 1.6, and 1.7 extend naturally to generalized Dirichlet series, and even to the more general class of functions $\int_0^\infty e^{-us} \, dA(u)$ where $A(u)$ is assumed to have finite variation on each finite interval $[0, U]$. The proof of the general form of Theorem 1.6 must be modified to depend on uniform, rather than absolute, convergence, since a generalized Dirichlet series may be never more than conditionally convergent (e.g., $\sum(-1)^n (\log n)^{-s}$). If we put $a = \limsup(\log n)/\lambda_n$, then the general form of Theorem 1.4 reads $\sigma_c \le \sigma_a \le \sigma_c + a$. Hardy & Riesz (1915) have given a detailed account of this subject, with historical attributions. See also Bohr & Cramér (1923).

Jensen (1884) showed that the domain of convergence of a generalized Dirichlet series is always a half-plane. The more precise information provided by Theorem 1.1 is due to Cahen (1894) who proved it not only for ordinary Dirichlet series but also for generalized Dirichlet series.

The construction in Exercise 1.2.8 would succeed with the simpler choice $a_n = n^{it_r}$ for $t_r \le n \le 2t_r$, $a_n = 0$ otherwise, but then to complete the argument one would need a further tool, such as the Kusmin–Landau inequality (cf. Mordell 1958). The square of the Dirichlet series in Exercise 1.2.8 has abscissa of convergence $1/2$; this bears on the result of Exercise 2.1.9. Information concerning the convergence of the product of two Dirichlet series is found in Exercises 1.3.2, 2.1.9, 5.2.16, and in Hardy & Riesz (1915).

Theorem 1.7 originates in Landau (1905). The analogue for power series had been proved earlier by Vivanti (1893) and Pringsheim (1894). Landau's proof extends to generalized Dirichlet series (including power series).

Section 1.3. The hypothesis $\sum |f(n)|n^{-\sigma} < \infty$ of Theorem 1.9 is equivalent to the assertion that

$$\prod_p (1 + |f(p)|p^{-\sigma} + |f(p^2)|p^{-2\sigma} + \cdots) < \infty,$$

which is slightly stronger than merely asserting that the Euler product converges absolutely. We recall that a product $\prod_n (1 + a_n)$ is said to be absolutely convergent if $\prod_n (1 + |a_n|) < \infty$. To see that the hypothesis $\prod_p (1 + |f(p)p^{-s} + \cdots|) < \infty$ is not sufficient, consider the following example due to Ingham: For every prime p we take $f(p) = 1$, $f(p^2) = -1$, and $f(p^k) = 0$ for $k > 2$. Then the product is absolutely convergent at $s = 0$, but the terms $f(n)$ do not tend to 0, and hence the series $\sum f(n)$ diverges. Indeed, it can be shown that $\sum_{n \le x} f(n) \sim cx$ as $x \to \infty$ where $c = \prod_p (1 - 2p^{-2} + p^{-3}) > 0$.

Euler (1735) defined the constant C_0, which he denoted C. Mascheroni (1790) called the constant γ, which is in common use, but we wish to reserve this symbol for the imaginary part of a zero of the zeta function or an L-function. It is conjectured that Euler's constant C_0 is irrational. The early history of the determination of the initial digits of C_0 has been recounted by Nielsen (1906, pp. 8–9). More recently, Wrench (1952) computed 328 digits, Knuth (1963) computed 1,271 digits, Sweeney (1963) computed 3,566 digits, Beyer & Waterman (1974) computed 4,879 digits, Brent (1977) computed 20,700 digits, Brent & McMillan (1980) computed 30,100 digits. At this time, it seems that more than 10^8 digits have been computed – see the web page of X. Gourdon & P. Sebah at http://numbers.computation.free.fr/Constants/Gamma/gamma.html. To 50 places, Euler's constant is

$$C_0 = 0.57721\ 56649\ 01532\ 86060\ 65120\ 90082\ 40243\ 10421\ 59335\ 93992.$$

Statistical analysis of the continued fraction coefficients of C_0 suggest that it satisfies the Gauss–Kusmin law, which is to say that C_0 seems to be a typical irrational number.

Landau & Walfisz (1920) showed that the functions $F(s)$ and $G(s)$ of Exercise 1.3.7 have the imaginary axis $\sigma = 0$ as a natural boundary. For further work on Dirichlet series with natural boundaries see Estermann (1928a,b) and Kurokawa (1987).

1.5 References

Andrews, G. E. (1976). *The Theory of Partitions*, Reprint. Cambridge: Cambridge University Press (1998).

Bachmann, P. (1894). *Zahlentheorie, II, Die analytische Zahlentheorie*, Leipzig: Teubner.

Beyer, W. A. & Waterman, M. S. (1974). Error analysis of a computation of Euler's constant and ln 2, *Math. Comp.* **28**, 599–604.

Bohr, H. (1910). *Bidrag til de Dirichlet'ske Rækkers theori*, København: G. E. C. Gad; *Collected Mathematical Works*, Vol. I, København: Danske Mat. Forening, 1952. A3.

Bohr, H. & Cramér, H. (1923). *Die neuere Entwicklung der analytischen Zahlentheorie*, Enzyklopädie der Mathematischen Wissenschaften, 2, C8, 722–849; H. Bohr, *Collected Mathematical Works*, Vol. III, København: Dansk Mat. Forening, 1952, H; H. Cramér, *Collected Works*, Vol. 1, Berlin: Springer-Verlag, 1952, pp. 289–416.

Brent, R. P. (1977). Computation of the regular continued fraction of Euler's constant, *Math. Comp.* **31**, 771–777.

Brent, R. P. & McMillan, E. M. (1980). Some new algorithms for high-speed computation of Euler's constant, *Math. Comp.* **34**, 305–312.

Cahen, E. (1894). Sur la fonction $\zeta(s)$ de Riemann et sur des fonctions analogues, *Ann. de l'École Normale* (3) **11**, 75–164.

Davenport, H. (2005). *Analytic Methods for Diophantine Equations and Diophantine Inequalities*. Second edition, Cambridge: Cambridge University Press.

Deléglise, M. & Rivat, J. (1996). Computing $\pi(x)$: the Meissel, Lehmer, Lagarias, Miller, Odlyzko method, *Math. Comp.* **65**, 235–245.

Estermann, T. (1928a). On certain functions represented by Dirichlet series, *Proc. London Math. Soc.* (2) **27**, 435–448.

  (1928b). On a problem of analytic continuation, *Proc. London Math. Soc.* (2) **27**, 471–482.

Euler, L. (1735). De Progressionibus harmonicus observationes, *Comm. Acad. Sci. Imper. Petropol.* **7**, 157; Opera Omnia, ser. 1, vol. 14, Teubner, 1914, pp. 93–95.

Hardy, G. H. (1910). *Orders of Infinity*. Cambridge Tract 12, Cambridge: Cambridge University Press.

Hardy, G. H. & Littlewood, J. E. (1914). Some problems of Diophantine approximation (II), *Acta Math.* **37**, 193–238; *Collected Papers*, Vol I. Oxford: Oxford University Press. 1966, pp. 67–112.

Hardy, G. H. & Riesz, M. (1915). *The General Theory of Dirichlet's Series*, Cambridge Tract No. 18. Cambridge: Cambridge University Press. Reprint: Stechert–Hafner (1964).

Ingham, A. E. (1932). *The Distribution of Prime Numbers*, Cambridge Tract 30. Cambridge: Cambridge University Press.

Jensen, J. L. W. V. (1884). Om Rækkers Konvergens, *Tidsskrift for Math.* (5) **2**, 63–72.

(1887). Sur la fonction $\zeta(s)$ de Riemann, *Comptes Rendus Acad. Sci.* Paris **104**, 1156–1159.

Knuth, D. E. (1962). Euler's constant to 1271 places, *Math. Comp.* **16**, 275–281.

Kurokawa, N. (1987). On certain Euler products, *Acta Arith.* **48**, 49–52.

Lagarias, J. C., Miller, V. S., & Odlyzko, A. M. (1985). Computing $\pi(x)$: The Meissel–Lehmer method, *Math. Comp.* **44**, 537–560.

Lagarias, J. C. & Odlyzko, A. M. (1987). Computing $\pi(x)$: An analytic method, *J. Algorithms* **8**, 173–191.

Landau, E. (1905). Über einen Satz von Tschebyschef, *Math. Ann.* **61**, 527–550; *Collected Works*, Vol. 2, Essen: Thales, 1986, pp. 206–229.

(1909a). *Handbuch der Lehre von der Verteilung der Primzahlen*, Leipzig: Teubner. Reprint: Chelsea (1953).

(1909b). Über das Konvergenzproblem der Dirichlet'schen Reihen, *Rend. Circ. Mat. Palermo* **28**, 113–151; *Collected Works*, Vol. 4, Essen: Thales, 1986, pp. 181–220.

Landau, E. & Walfisz, A. (1920). Über die Nichtfortsetzbarkeit einiger durch Dirichletsche Reihen definierte Funktionen, *Rend. Circ. Mat. Palermo* **44**, 82–86; *Collected Works*, Vol. 7, Essen: Thales, 1986, pp. 252–256.

Lehmer, D. H. (1959). On the exact number of primes less than a given limit, *Illinois J. Math.* **3**, 381–388.

Mapes, D. C. (1963). Fast method for computing the number of primes less than a given limit, *Math. Comp.* **17**, 179–185.

Mascheroni, L. (1790). *Abnotationes ad calculum integrale Euleri*, Vol. 1. Ticino: Galeatii. Reprinted in the Opera Omnia of L. Euler, Ser. 1, Vol 12, Teubner, 1914, pp. 415–542.

Mordell, L. J. (1958). On the Kusmin–Landau inequality for exponential sums, *Acta Arith.* **4**, 3–9.

Nielsen, N. (1906). *Handbuch der Theorie der Gammafunktion.* Leipzig: Teubner.

Pringsheim, A. (1894). Über Functionen, welche in gewissen Punkten endliche Differentialquotienten jeder endlichen Ordnung, aber kein Taylorsche Reihenentwickelung besitzen, *Math. Ann.* **44**, 41–56.

Riesel, H. (1994). *Prime Numbers and Computer Methods for Factorization*, Second ed., Progress in Math. 126. Boston: Birkhäuser.

Shafer, R. E. (1984). Advanced problem 6456, *Amer. Math. Monthly* **91**, 205.

Stieltjes, T. J. (1885). Letter 75 in *Correspondance d'Hermite et de Stieltjes*, B. Baillaud & H. Bourget, eds., Paris: Gauthier-Villars, 1905.

(1887). Note sur la multiplication de deux séries, *Nouvelles Annales* (3) **6**, 210–215.

Sweeney, D. W. (1963). On the computation of Euler's constant, *Math. Comp.* **17**, 170–178.

Vaughan, R. C. (1997). *The Hardy–Littlewood Method*, Second edition, Cambridge Tract 125. Cambridge: Cambridge University Press.

Vivanti, G. (1893). Sulle serie di potenze, *Rivista di Mat.* **3**, 111–114.

Wagon, S. (1987). Fourteen proofs of a result about tiling a rectangle, *Amer. Math. Monthly* **94**, 601–617.

Widder, D. V. (1971). *An Introduction to Transform Theory.* New York: Academic Press.

Wilf, H. (1994). *Generatingfunctionology*, Second edition. Boston: Academic Press.

Wrench, W. R. Jr (1952). A new calculation of Euler's constant, *MTAC* **6**, 255.

2

The elementary theory of arithmetic functions

2.1   Mean values

We say that an arithmetic function $F(n)$ has a mean value c if

$$\lim_{N \to \infty} \frac{1}{N} \sum_{n=1}^{N} F(n) = c.$$

In this section we develop a simple method by which mean values can be shown to exist in many interesting cases.

If two arithmetic functions f and F are related by the identity

$$F(n) = \sum_{d|n} f(d), \qquad (2.1)$$

then we can write f in terms of F:

$$f(n) = \sum_{d|n} \mu(d) F(n/d). \qquad (2.2)$$

This is the *Möbius inversion formula*. Conversely, if (2.2) holds for all n then so also does (2.1). If f is generally small then F has an asymptotic mean value. To see this, observe that

$$\sum_{n \leq x} F(n) = \sum_{n \leq x} \sum_{d|n} f(d).$$

By iterating the sums in the reverse order, we see that the above is

$$= \sum_{d \leq x} f(d) \sum_{\substack{n \leq x \\ d|n}} 1 = \sum_{d \leq x} f(d) [x/d].$$

35

Since $[y] = y + O(1)$, this is

$$= x \sum_{d \leq x} \frac{f(d)}{d} + O \left(\sum_{d \leq x} |f(d)| \right). \tag{2.3}$$

Thus F has the mean value $\sum_{d=1}^{\infty} f(d)/d$ if this series converges and if $\sum_{d \leq x} |f(d)| = o(x)$. This approach, though somewhat crude, often yields useful results.

Theorem 2.1   *Let $\varphi(n)$ be Euler's totient function. Then for $x \geq 2$,*

$$\sum_{n \leq x} \frac{\varphi(n)}{n} = \frac{6}{\pi^2} x + O(\log x).$$

Proof   We recall that $\varphi(n) = n \prod_{p|n} (1 - 1/p)$. On multiplying out the product, we see that

$$\frac{\varphi(n)}{n} = \sum_{d|n} \frac{\mu(d)}{d}.$$

On taking $f(d) = \mu(d)/d$ in (2.3), it follows that

$$\sum_{n \leq x} \frac{\varphi(n)}{n} = x \sum_{d \leq x} \frac{\mu(d)}{d^2} + O(\log x).$$

Since $\sum_{d > x} d^{-2} \ll x^{-1}$, we see that

$$\sum_{d \leq x} \frac{\mu(d)}{d^2} = \sum_{d=1}^{\infty} \frac{\mu(d)}{d^2} + O\left(\frac{1}{x}\right) = \frac{1}{\zeta(2)} + O\left(\frac{1}{x}\right)$$

by Corollary 1.10. From Corollary B.3 we know that $\zeta(2) = \pi^2/6$; hence the proof is complete. □

Let $Q(x)$ denote the number of square-free integers not exceeding x, $Q(x) = \sum_{n \leq x} \mu(n)^2$. We now calculate the asymptotic density of these numbers.

Theorem 2.2   *For all $x \geq 1$,*

$$Q(x) = \frac{6}{\pi^2} x + O\left(x^{1/2}\right).$$

Proof   Every positive integer n is uniquely of the form $n = ab^2$ where a is square-free. Thus n is square-free if and only if $b = 1$, so that by (1.20)

$$\sum_{d^2|n} \mu(d) = \sum_{d|b} \mu(d) = \mu(n)^2. \tag{2.4}$$

This is a relation of the shape (2.1) where $f(d) = \mu(\sqrt{d})$ if d is a perfect square, and $f(d) = 0$ otherwise. Hence by (2.3),

$$Q(x) = x \sum_{d^2 \le x} \frac{\mu(d)}{d^2} + O\left(\sum_{d^2 \le x} 1\right).$$

The error term is $\ll x^{1/2}$, and the sum in the main term is treated as in the preceding proof. □

We note that the argument above is routine once the appropriate identity (2.4) is established. This relation can be discovered by considering (2.2), or by using Dirichlet series: Let Q denote the class of square-free numbers. Then for $\sigma > 1$,

$$\sum_{n \in Q} n^{-s} = \prod_p (1 + p^{-s}) = \prod_p \frac{1 - p^{-2s}}{1 - p^{-s}} = \frac{\zeta(s)}{\zeta(2s)}.$$

Now $1/\zeta(2s)$ can be written as a Dirichlet series in s, with coefficients $f(n) = \mu(d)$ if $n = d^2$, $f(n) = 0$ otherwise. Hence the convolution equation (2.4) gives the coefficients of the product Dirichlet series $\zeta(s) \cdot 1/\zeta(2s)$.

Suppose that a_k, b_m, c_n are joined by the convolution relation

$$c_n = \sum_{km=n} a_k b_m, \tag{2.5}$$

and that $A(x)$, $B(x)$, $C(x)$ are their respective summatory functions. Then

$$C(x) = \sum_{km \le x} a_k b_m, \tag{2.6}$$

and it is useful to note that this double sum can be iterated in various ways. On one hand we see that

$$C(x) = \sum_{k \le x} a_k B(x/k); \tag{2.7}$$

this is the line of reasoning that led to (2.3) (take $a_k = f(k)$, $b_m = 1$). At the opposite extreme,

$$C(x) = \sum_{m \le x} b_m A(x/m), \tag{2.8}$$

and between these we have the more general identity

$$C(x) = \sum_{k \le y} a_k B(x/k) + \sum_{m \le x/y} b_m A(x/m) - A(y)B(x/y) \tag{2.9}$$

for $0 < y \le x$. This is obvious once it is observed that the first term on the right sums those terms $a_k b_m$ for which $km \le x$, $k \le y$, the second sum includes the pairs (k, m) for which $km \le x, m \le x/y$, and the third term subtracts those $a_k b_m$ for which $k \le y, m \le x/y$, since these (k, m) were included in both the previous terms. The advantage of (2.9) over (2.7) is that the number of terms is reduced ($\ll y + x/y$ instead of $\ll x$), and at the same time A and B are evaluated only at large values of the argument, so that asymptotic formulæ for these quantities may be expected to be more accurate. For example, if we wish to estimate the average size of $d(n)$ we take $a_k = b_m = 1$, and then from (2.3) we see that

$$\sum_{n \le x} d(n) = x \log x + O(x).$$

To obtain a more accurate estimate we observe that the first term on the right-hand side of (2.9) is

$$\sum_{k \le y} [x/k] = x \sum_{k \le y} 1/k + O(y).$$

By Corollary 1.15 this is

$$x \log y + C_0 x + O(x/y + y).$$

Here the error term is minimized by taking $y = x^{1/2}$. The second term on the right in (2.9) is then identical to the first, and the third term is $[x^{1/2}]^2 = x + O(x^{1/2})$, and we have

Theorem 2.3   *For $x \ge 2$.*

$$\sum_{n \le x} d(n) = x \log x + (2C_0 - 1)x + O\left(x^{1/2}\right).$$

We often construct estimates with one or more parameters, and then choose values of the parameters to optimize the result. The instance above is typical – we minimized $x/y + y$ by taking $y = x^{1/2}$. Suppose, more generally, that we wish to minimize $T_1(y) + T_2(y)$ where T_1 is a decreasing function, and T_2 is an increasing function. We could differentiate and solve for a root of $T_1'(y) + T_2'(y) = 0$, but there is a quicker method: Find y_0 so that $T_1(y_0) = T_2(y_0)$. This does not necessarily yield the exact minimum value of $T_1(y) + T_2(y)$, but it is easy to see that

$$T_1(y_0) \le \min_y (T_1(y) + T_2(y)) \le 2T_1(y_0),$$

so the bound obtained in this way is at most twice the optimal bound.

Despite the great power of analytic techniques, the 'method of the hyperbola' used above is a valuable tool. The sequence c_n given by (2.5) is called the *Dirichlet convolution* of a_k and b_m; in symbols, $c = a * b$. Arithmetic functions form a ring when equipped with pointwise addition, $(a + b)_n = a_n + b_n$, and

Dirichlet convolution for multiplication. This ring is called the *ring of formal Dirichlet series*. Manipulations of arithmetic functions in this way correspond to manipulations of Dirichlet series without regard to convergence. This is analogous to the ring of formal power series, in which multiplication is provided by Cauchy convolution, $c_n = \sum_{k+m=n} a_k b_m$.

In the ring of formal Dirichlet series we let O denote the arithmetic function that is identically 0; this is the additive identity. The multiplicative identity is i where $i_1 = 1, i_n = 0$ for $n > 1$. The arithmetic function that is identically 1 we denote by $\mathbf{1}$, and we similarly abbreviate $\mu(n)$, $\Lambda(n)$, and $\log n$ by μ, Λ, and L. In this notation, the characteristic property of $\mu(n)$ is that $\mu * \mathbf{1} = i$, which is to say that μ and $\mathbf{1}$ are convolution inverses of each other, and the Möbius inversion formula takes the compact form

$$a * \mathbf{1} = b \quad \Longleftrightarrow \quad a = b * \mu.$$

In the elementary study of prime numbers the relations $\Lambda * \mathbf{1} = L, L * \mu = \Lambda$ are fundamental.

2.1.1 Exercises

1. (de la Vallée Poussin 1898; cf. Landau 1911) Show that

$$\sum_{n \le x} \{x/n\} = (1 - C_0)x + O\left(x^{1/2}\right)$$

where C_0 is Euler's constant, and $\{u\} = u - [u]$ is the fractional part of u.

2. (Duncan 1965; cf. Rogers 1964, Orr 1969) Let $Q(x)$ be defined as in Theorem 2.2.

   (a) Show that $Q(N) \ge N - \sum_p [N/p^2]$ for every positive integer N.

   (b) Justify the relations

   $$\sum_p \frac{1}{p^2} < \frac{1}{4} + \sum_{k=1}^{\infty} \frac{1}{(2k+1)^2} < \frac{1}{4} + \frac{1}{2} \sum_{k=1}^{\infty} \left(\frac{1}{2k} - \frac{1}{2k+2}\right) = 1/2.$$

   (c) Show that $Q(N) > N/2$ for all positive integers N.

   (d) Show that every positive integer $n > 1$ can be written as a sum of two square-free numbers.

3. (Linfoot & Evelyn 1929) Let $\mathcal{Q}_k$ denote the set of positive k^{th} power free integers (i.e., $q \in \mathcal{Q}_k$ if and only if $m^k | q \Rightarrow m = 1$).

   (a) Show that

   $$\sum_{n \in \mathcal{Q}_k} n^{-s} = \frac{\zeta(s)}{\zeta(ks)}$$

   for $\sigma > 1$.

(b) Show that for any fixed integer $k > 1$

$$\sum_{\substack{n \le x \\ n \in Q_k}} 1 = \frac{x}{\zeta(k)} + O\left(x^{1/k}\right)$$

for $x \ge 1$.

4. (cf. Evelyn & Linfoot 1930) Let N be a positive integer, and suppose that P is square-free.

(a) Show that the number of residue classes $n \pmod{P^2}$ for which (n, P^2) is square-free and $(N - n, P^2)$ is square-free is

$$P^2 \prod_{\substack{p|P \\ p^2|N}} \left(1 - \frac{1}{p^2}\right) \prod_{\substack{p|P \\ p^2 \nmid N}} \left(1 - \frac{2}{p^2}\right).$$

(b) Show that the number of integers n, $0 < n < N$, for which (n, P^2) is square-free and $(N - n, P^2)$ is square-free is

$$N \prod_{\substack{p|P \\ p^2|N}} \left(1 - \frac{1}{p^2}\right) \prod_{\substack{p|P \\ p^2 \nmid N}} \left(1 - \frac{2}{p^2}\right) + O(P^2).$$

(c) Show that the number of n, $0 < n < N$, such that n is divisible by the square of a prime $> y$ is $\ll N/y$.

(d) Take P to be the product of all primes not exceeding y. By letting y tend to infinity slowly, show that the number of ways of writing N as a sum of two square-free integers is $\sim c(N)N$ where

$$c(N) = a \prod_{p^2|N} \left(1 + \frac{1}{p^2 - 2}\right), \qquad a = \prod_{p} \left(1 - \frac{2}{p^2}\right).$$

5. (cf. Hille 1937) Suppose that $f(x)$ and $F(x)$ are complex-valued functions defined on $[1, \infty)$. Show that

$$F(x) = \sum_{n \le x} f(x/n)$$

for all x if and only if

$$f(x) = \sum_{n \le x} \mu(n) F(x/n)$$

for all x.

6. (cf. Hartman & Wintner 1947) Suppose that $\sum |f(n)|d(n) < \infty$, and that $\sum |F(n)|d(n) < \infty$. Show that

$$F(n) = \sum_{\substack{m \\ n|m}} f(m)$$

for all n if and only if

$$f(n) = \sum_{\substack{m \\ n|m}} \mu(m/n)F(m).$$

7. (Jarník 1926; cf. Bombieri & Pila 1989) Let C be a simple closed curve in the plane, of arc length L. Show that the number of 'lattice points' (m, n), $m, n \in \mathbb{Z}$, lying on C is at most $L + 1$. Show that if C is strictly convex then the number of lattice points on C is $\ll 1 + L^{2/3}$, and that this estimate is best possible.

8. Let C be a simple closed curve in the plane, of arc length L that encloses a region of area A. Let N be the number of lattice points inside C. Show that $|N - A| \le 3(L + 1)$.

9. Let $r(n)$ be the number of pairs (j, k) of integers such that $j^2 + k^2 = n$. Show that

$$\sum_{n \le x} r(n) = \pi x + O\left(x^{1/2}\right).$$

10. (Stieltjes 1887) Suppose that $\sum a_n$, $\sum b_n$ are convergent series, and that $c_n = \sum_{km=n} a_k b_m$. Show that $\sum c_n n^{-1/2}$ converges. (Hence if two Dirichlet series have abscissa of convergence $\le \sigma$ then the product series $\gamma(s) = \alpha(s)\beta(s)$ has abscissa of convergence $\sigma_c \le \sigma + 1/2$.)

11. (a) Show that $\sum_{n \le x} \varphi(n) = (3/\pi^2)x^2 + O(x \log x)$ for $x \ge 2$.
    (b) Show that

$$\sum_{\substack{m \le x \\ n \le x \\ (m,n)=1}} 1 = -1 + 2\sum_{n \le x} \varphi(n)$$

for $x \ge 1$. Deduce that the expression above is $(6/\pi^2)x^2 + O(x \log x)$.

12. Let $\sigma(n) = \sum_{d|n} d$. Show that

$$\sum_{n \le x} \sigma(n) = \frac{\pi^2}{12}x^2 + O(x \log x)$$

for $x \ge 2$.

13. (Landau 1900, 1936; cf. Sitaramachandrarao 1982, 1985, Nowak 1989)
    (a) Show that $n/\varphi(n) = \sum_{d|n} \mu(d)^2/\varphi(d)$.
    (b) Show that

$$\sum_{n \le x} \frac{n}{\varphi(n)} = \frac{\zeta(2)\zeta(3)}{\zeta(6)}x + O(\log x)$$

for $x \ge 2$.

(c) Show that

$$\sum_{d=1}^{\infty} \frac{\mu(d)^2 \log d}{d\varphi(d)} = \left(\sum_p \frac{\log p}{p^2 - p + 1}\right) \prod_p \left(1 + \frac{1}{p(p-1)}\right).$$

(d) Show that for $x \geq 2$,

$$\sum_{n \leq x} \frac{1}{\varphi(n)} = \frac{\zeta(2)\zeta(3)}{\zeta(6)} \left(\log x + C_0 - \sum_p \frac{\log p}{p^2 - p + 1}\right) + O((\log x)/x).$$

14. Let κ be a fixed real number. Show that

$$\sum_{n \leq x} \left(\frac{\varphi(n)}{n}\right)^{\kappa} = c(\kappa)x + O(x^{\varepsilon})$$

where

$$c(\kappa) = \prod_p \left(1 - \frac{1}{p}(1 - (1 - 1/p)^{\kappa})\right).$$

15. (cf. Grosswald 1956, Bateman 1957)
    (a) By using Euler products, or otherwise, show that

$$2^{\omega(n)} = \sum_{d^2 m = n} \mu(d)d(m).$$

    (b) Deduce that

$$\sum_{n \leq x} 2^{\omega(n)} = \frac{6}{\pi^2}x \log x + cx + O\left(x^{1/2} \log x\right)$$

    for $x \geq 2$ where $c = 2C_0 - 1 - 2\zeta'(2)/\zeta(2)^2$.
    (c) Show also that

$$\sum_{n \leq x} 2^{\Omega(n)} = Cx(\log x)^2 + O(x \log x)$$

    where

$$C = \frac{1}{8 \log 2} \prod_{p > 2} \left(1 + \frac{1}{p(p-2)}\right).$$

16. (a) Show that for any positive integer q,

$$\sum_{d|q} \frac{\mu(d) \log d}{d} = -\frac{\varphi(q)}{q} \sum_{p|q} \frac{\log p}{p - 1}.$$

    (b) Show that for any real number $x \geq 1$ and any positive integer q,

$$\sum_{\substack{m \leq x \\ (m,q)=1}} \frac{1}{m} = \left(\log x + C_0 + \sum_{p|q} \frac{\log p}{p - 1}\right) \frac{\varphi(q)}{q} + O\left(2^{\omega(q)}/x\right).$$

(c) Show that for any real number $x \geq 2$ and any positive integer q,

$$\sum_{\substack{n \leq x \\ (n,q)=1}} \frac{1}{\varphi(n)} = \frac{\zeta(2)\zeta(3)}{\zeta(6)} \prod_{p|q} \left(1 - \frac{p}{p^2 - p + 1}\right) \left(\log x + C_0 + \sum_{p|q} \frac{\log p}{p - 1}\right)$$

$$- \sum_{p \nmid q} \frac{\log p}{p^2 - p + 1}\right) + O\left(2^{\omega(q)} \frac{\log x}{x}\right).$$

17. (cf. Ward 1927) Show that for $x \geq 2$,

$$\sum_{n \leq x} \frac{\mu(n)^2}{\varphi(n)} = \log x + C_0 + \sum_{p} \frac{\log p}{p(p - 1)} + O\left(x^{-1/2} \log x\right).$$

18. Let $d_k(n)$ be the number of ordered k-tuples $(d_1, \ldots, d_k)$ of positive integers such that $d_1 d_2 \cdots d_k = n$.
   (a) Show that $d_k(n) = \sum_{d|n} d_{k-1}(d)$.
   (b) Show that $\sum_{n=1}^{\infty} d_k(n) n^{-s} = \zeta(s)^k$ for $\sigma > 1$.
   (c) Show that for every fixed positive integer k,

$$\sum_{n \leq x} d_k(n) = x P_k(\log x) + O\left(x^{1-1/k} (\log x)^{k-2}\right)$$

   for $x \geq 2$, where $P \in \mathbb{R}[z]$ has degree $k - 1$ and leading coefficient $1/(k - 1)!$.

19. (cf. Erdős & Szekeres 1934, Schmidt 1967/68) Let A_n denote the number of non-isomorphic Abelian groups of order n.
   (a) Show that $\sum_{n=1}^{\infty} A_n n^{-s} = \prod_{k=1}^{\infty} \zeta(ks)$ for $\sigma > 1$.
   (b) Show that

$$\sum_{n \leq x} A_n = cx + O\left(x^{1/2}\right)$$

   where $c = \prod_{k=2}^{\infty} \zeta(k)$.

20. (Wintner 1944, p. 46) Suppose that $\sum_d |g(d)|/d < \infty$. Show that $\sum_{d \leq x} |g(d)| = o(x)$. Suppose also that $\sum_{n \leq x} f(n) = cx + o(x)$, and put $h(n) = \sum_{d|n} f(d)g(n/d)$. Show that

$$\sum_{n \leq x} h(n) = cgx + o(x)$$

   where $g = \sum_d g(d)/d$.

21. (a) Show that if a^2 is the largest perfect square $\leq x$ then $x - a^2 \leq 2\sqrt{x}$.
   (b) Let a^2 be as above, and let b^2 be the least perfect square such that $a^2 + b^2 > x$. Show that $a^2 + b^2 < x + 6x^{1/4}$. Thus for any $x \geq 1$, there is a sum of two squares in the interval $(x, x + 6x^{1/4})$. (It is somewhat

embarrassing that this is the best-known upper bound for gaps between sums of two squares.)

22. (Feller & Tornier 1932) Let $f(n)$ denote the multiplicative function such that $f(p) = 1$ for all p, and $f(p^k) = -1$ whenever $k > 1$.

    (a) Show that

$$\sum_{n=1}^{\infty} \frac{f(n)}{n^s} = \zeta(s) \prod_p \left(1 - \frac{2}{p^{2s}}\right)$$

    for $\sigma > 1$.

    (b) Deduce that

$$f(n) = \sum_{d^2 \mid n} \mu(d) 2^{\omega(d)}.$$

    (c) Explain why $2^{\omega(n)} \leq d(n)$ for all n.

    (d) Show that

$$\sum_{n \leq x} f(n) = ax + O\left(x^{1/2} \log x\right)$$

    where a is the constant of Exercise 3.

    (e) Let $g(n)$ denote the number of primes p such that $p^2 \mid n$. Show that the set of n for which $g(n)$ is even has asymptotic density $(1 + a)/2$.

    (f) Put

$$e_k = \frac{1}{k} \sum_{d \mid k} \mu(d) 2^{k/d}.$$

    Show that if $|z| < 1$, then

$$\log(1 - 2z) = \sum_{k=1}^{\infty} e_k \log\left(1 - z^k\right).$$

    (g) Deduce that

$$a = \prod_{k=1}^{\infty} \zeta(2k)^{e_k}.$$

Note that the kth factor here differs from 1 by an amount that is $\ll 1/(k2^k)$. Hence the product converges very rapidly. Since $\zeta(2k)$ can be calculated very accurately by the Euler–Maclaurin formula (see Appendix B), the formula above permits the rapid calculation of the constant a.

23. Let $B_1(x) = x - 1/2$, as in Appendix B.
    (a) Show that

$$\sum_{n \le x} \frac{1}{n} = \log x + C_0 - B_1(\{x\})/x + O(1/x^2).$$

(b) Write $\sum_{n \le x} d(n) = x \log x + (2C_0 - 1)x + \Delta(x)$. Show that

$$\Delta(x) = -2 \sum_{n \le \sqrt{x}} B_1(\{x/n\}) + O(1).$$

(c) Show that $\int_0^X \Delta(x)\, dx \ll X$.
(d) Deduce that

$$\sum_{n \le X} d(n)(X - n) = \int_0^X \left(\sum_{n \le x} d(n) \right) dx$$

$$= \frac{1}{2}X^2 \log X + \left(C_0 - \frac{3}{4} \right) X^2 + O(X).$$

24. Let $r(n)$ be the number of ordered pairs (a, b) of integers for which $a^2 + b^2 = n$.
    (a) Show that

$$\sum_{n \le x} r(n) = 1 + 4[\sqrt{x}] + 8 \sum_{1 \le n \le \sqrt{x/2}} \left[\sqrt{x - n^2} \right] - 4 \left[\sqrt{x/2} \right]^2.$$

(b) Show that

$$\sum_{1 \le n \le \sqrt{x/2}} \sqrt{x - n^2} = \left(\frac{\pi}{8} + \frac{1}{2} \right) x - B_1 \left(\{ \sqrt{x/2} \} \right) - \frac{1}{2}\sqrt{x} + O(1).$$

(c) Write $\sum_{0 \le n \le x} r(n) = \pi x + R(x)$. Show that

$$R(x) = -8 \sum_{1 \le n \le \sqrt{x/2}} B_1 \left(\{ \sqrt{x - n^2} \} \right) + O(1).$$

25. (a) Show that if $(a, q) = 1$, and β is real, then

$$\sum_{n=1}^{q} B_1 \left(\left\{ \frac{a}{q} n + \beta \right\} \right) = B_1(\{ q\beta \}).$$

(b) Show that if $A \ge 1, |f'(x) - a/q| \le A/q^2$ for $1 \le x \le q$, and $(a, q) = 1$, then

$$\sum_{n=1}^{q} B_1(\{ f(n) \}) \ll A.$$

(c) Suppose that $Q \geq 1$ is an integer, $B \geq 1$, and that $1/Q^3 \leq \pm f''(x) \leq B/Q^3$ for $0 \leq x \leq N$ where the choice of sign is independent of x. Show that numbers a_r, q_r, N_r can be determined, $0 \leq r \leq R$ for some R, so that (i) $(a_r, q_r) = 1$, (ii) $q_r \leq Q$, (iii) $|f'(N_r) - a_r/q_r| \leq 1/(q_r Q)$, and (iv) $N_0 = 0, N_r = N_{r-1} + q_{r-1}$ for $1 \leq r \leq R, N - Q \leq N_R \leq N$.

(d) Show that under the above hypotheses

$$\sum_{n=0}^{N} B_1(\{f(n)\}) \ll B(R+1) + Q.$$

(e) Show that the number of s for which $a_s/q_s = a_r/q_r$ is $\ll Q^2/q^2$. Let $1 \leq q \leq Q$. Show that the number of r for which $q_r = q$ is $\ll (Q/q)^2(BNq/Q^3 + 1)$.

(f) Conclude that under the hypotheses of (c),

$$\sum_{n=0}^{N} B_1(\{f(n)\}) \ll B^2 N Q^{-1} \log 2Q + BQ^2.$$

26. Show that if $U \leq \sqrt{x}$, then

$$\sum_{U < n \leq 2U} B_1(\{x/n\}) \ll x^{1/3} \log x.$$

Let $\Delta(x)$ be as in Exercise 23(b). Show that $\Delta(x) \ll x^{1/3}(\log x)^2$.

27. Let $R(x)$ be as in Exercise 24(c). Show that $R(x) \ll x^{1/3} \log x$.

2.2   The prime number estimates of Chebyshev and of Mertens

Because of the irregular spacing of the prime numbers, it seems hopeless to give a useful exact formula for the nth prime. As a compromise we estimate the nth prime, or equivalently, estimate the number $\pi(x)$ of primes not exceeding x. Similarly we put $\vartheta(x) = \sum_{p \leq x} \log p$, and $\psi(x) = \sum_{n \leq x} \Lambda(n)$. As we shall see, these three summatory functions are closely related. We estimate $\psi(x)$ first.

Theorem 2.4   (Chebyshev) *For $x \geq 2$, $\psi(x) \asymp x$.*

The proof we give below establishes only that there is an x_0 such that $\psi(x) \asymp x$ uniformly for $x \geq x_0$. However, both $\psi(x)$ and x are bounded away from 0 and from ∞ in the interval $[2, x_0]$, and hence the implicit constants can be adjusted so that $\psi(x) \asymp x$ uniformly for $x \geq 2$. In subsequent situations of this sort, we shall assume without comment that the reader understands that it suffices to prove the result for all sufficiently large x.

Proof By applying the Möbius inversion formula to (1.22) we find that

$$\Lambda(n) = \sum_{d|n} \mu(d) \log n/d \, .$$

Thus by (2.7) it follows that

$$\psi(x) = \sum_{d \leq x} \mu(d) T(x/d) \tag{2.10}$$

where $T(x) = \sum_{n \leq x} \log n$. By the integral test we see that

$$\int_1^N \log u \, du \leq T(N) \leq \int_1^{N+1} \log u \, du$$

for any positive integer N. Since $\int \log x \, dx = x \log x - x$, it follows easily that

$$T(x) = x \log x - x + O(\log 2x) \tag{2.11}$$

for $x \geq 1$. Despite the precision of this estimate, we encounter difficulties when we substitute this in (2.10), since we have no useful information concerning the sums

$$\sum_{d \leq x} \frac{\mu(d)}{d}, \qquad \sum_{d \leq x} \frac{\mu(d) \log d}{d},$$

which arise in the main terms. To avoid this problem we introduce an idea that is fundamental to much of prime number theory, namely we replace $\mu(d)$ by an arithmetic function a_d that in some way forms a truncated approximation to $\mu(d)$. Suppose that $\mathcal{D}$ is a finite set of numbers, and that $a_d = 0$ when $d \notin \mathcal{D}$. Then by (2.11) we see that

$$\sum_{d \in \mathcal{D}} a_d T(x/d) = (x \log x - x) \sum_{d \in \mathcal{D}} a_d/d - x \sum_{d \in \mathcal{D}} \frac{a_d \log d}{d} + O(\log 2x).$$

$$\tag{2.12}$$

Here the implicit constant depends on the choice of a_d, which we shall consider to be fixed. Since we want the above to approximate the relation (2.10), and since we are hoping that $\psi(x) \asymp x$, we restrict our attention to a_d that satisfy the condition

$$\sum_{d \in \mathcal{D}} \frac{a_d}{d} = 0, \tag{2.13}$$

and hope that

$$-\sum_{d \in \mathcal{D}} \frac{a_d \log d}{d} \quad \text{is near 1.} \tag{2.14}$$

By the definition of $T(x)$ we see that the left-hand side of (2.12) is

$$\sum_{dn \leq x} a_d \log n = \sum_{dn \leq x} a_d \sum_{k|n} \Lambda(k) = \sum_{dkm \leq x} a_d \Lambda(k)$$

$$= \sum_{k \leq x} \Lambda(k) E(x/k) \tag{2.15}$$

where $E(y) = \sum_{dm \leq y} a_d = \sum_d a_d[y/d]$. The expression above will be near $\psi(x)$ if $E(y)$ is near 1. If $y \geq 1$ then

$$\sum_d \mu(d)[y/d] = \sum_d \mu(d) \sum_{k \leq y/d} 1 = \sum_{dk \leq y} \mu(d) = \sum_{n \leq y} \sum_{d|n} \mu(d) = 1,$$

in view of (1.20). Thus $E(y)$ will be near 1 for y not too large if a_d is near $\mu(d)$ for small d. Moreover, by (2.13) we see that $E(y) = -\sum_{d \in \mathcal{D}} a_d\{y/d\}$, so that $E(y)$ is periodic with period dividing $\operatorname{lcm}_{d \in \mathcal{D}} d$. Hence for a given choice of the a_d, the behaviour of $E(y)$ can be determined by a finite calculation.

The simplest realization of this approach involves taking $a_1 = 1$, $a_2 = -2$, $a_d = 0$ for $d > 2$. Then (2.13) holds, the expression (2.14) is $\log 2$, $E(y)$ has period 2 and $E(y) = 0$ for $0 \leq y < 1$, $E(y) = 1$ for $1 \leq y < 2$. Hence for this choice of the a_d the sum in (2.15) satisfies the inequalities

$$\psi(x) - \psi(x/2) = \sum_{x/2 < k \leq x} \Lambda(k) \leq \sum_{k \leq x} \Lambda(k) E(x/k) \leq \sum_{k \leq x} \Lambda(k) = \psi(x).$$

Thus $\psi(x) \geq (\log 2)x + O(\log x)$, which is a lower bound of the desired shape. In addition,

$$\psi(x) - \psi(x/2) \leq (\log 2)x + O(\log x).$$

On replacing x by $x/2^r$ and summing over r we deduce that

$$\psi(x) \leq 2(\log 2)x + O((\log x)^2),$$

so the proof is complete. $\qquad \qquad \square$

Chebyshev obtained better constants than above, by taking $a_1 = a_{30} = 1$, $a_2 = a_3 = a_5 = -1$, $a_d = 0$ otherwise. Then (2.13) holds, the expression (2.14) is $0.92129\ldots$, $E(y) = 1$ for $1 \leq y < 6$, and $0 \leq E(y) \leq 1$ for all y, with the result that

$$\psi(x) \geq (0.9212)x + O(\log x)$$

and

$$\psi(x) \le (1.1056)x + O((\log x)^2).$$

By computing the implicit constants one can use this method to determine a constant x_0 such that $\psi(2x) - \psi(x) > x/2$ for all $x > x_0$. Since the contribution of the proper prime powers is small, it follows that there is at least one prime in the interval $(x, 2x]$, when $x > x_0$. After separate consideration of $x \le x_0$, one obtains Bertrand's postulate: For each real number $x > 1$, there is a prime number in the interval $(x, 2x)$.

> *Chebyshev said it, but I'll say it again:*
> *There's always a prime between n and 2n.*
> N. J. Fine

Corollary 2.5 *For $x \ge 2$,*

$$\vartheta(x) = \psi(x) + O\left(x^{1/2}\right)$$

and

$$\pi(x) = \frac{\psi(x)}{\log x} + O\left(\frac{x}{(\log x)^2}\right).$$

Proof Clearly

$$\psi(x) = \sum_{p^k \le x} \log p = \sum_{k=1}^{\infty} \vartheta\left(x^{1/k}\right).$$

But $\vartheta(y) \le \psi(y) \ll y$, so that

$$\psi(x) - \vartheta(x) = \sum_{k \ge 2} \vartheta(x^{1/k}) \ll x^{1/2} + x^{1/3} \log x \ll x^{1/2}.$$

As for $\pi(x)$, we note that

$$\pi(x) = \int_{2^-}^x (\log u)^{-1} \, d\vartheta(u) = \frac{\vartheta(x)}{\log x} + \int_2^x \frac{\vartheta(u)}{u(\log u)^2} \, du.$$

This last integral is

$$\ll \int_2^x (\log u)^{-2} \, du \ll x(\log x)^{-2},$$

so we have the stated result. □

Corollary 2.6 *For $x \ge 2$, $\vartheta(x) \asymp x$ and $\pi(x) \asymp x/\log x$.*

In Chapters 6 and 8 we shall give several proofs of the Prime Number Theorem (PNT), which asserts that $\pi(x) \sim x/\log x$. By Corollary 2.5 this is equivalent to the estimates $\vartheta(x) \sim x$, $\psi(x) \sim x$. By partial summation it is easily seen that the PNT implies that

$$\sum_{p \leq x} \frac{\log p}{p} \sim \log x,$$

and that

$$\sum_{p \leq x} \frac{1}{p} \sim \log \log x.$$

However, these assertions are weaker than PNT, as we can derive them from Theorem 2.4.

Theorem 2.7 *For $x \geq 2$,*

(a) $\displaystyle\sum_{n \leq x} \frac{\Lambda(n)}{n} = \log x + O(1)$,

(b) $\displaystyle\sum_{p \leq x} \frac{\log p}{p} = \log x + O(1)$,

(c) $\displaystyle\int_1^x \psi(u)u^{-2}\, du = \log x + O(1)$,

(d) $\displaystyle\sum_{p \leq x} \frac{1}{p} = \log \log x + b + O(1/\log x)$,

(e) $\displaystyle\prod_{p \leq x} \left(1 - \frac{1}{p}\right)^{-1} = e^{C_0} \log x + O(1)$

where C_0 is Euler's constant and

$$b = C_0 - \sum_p \sum_{k=2}^{\infty} \frac{1}{kp^k}.$$

Proof Taking $f(d) = \Lambda(d)$ in (2.1), we see from (2.3) that

$$T(x) = \sum_{n \leq x} \log n = x \sum_{d \leq x} \frac{\Lambda(d)}{d} + O\left(\psi(x)\right).$$

By Theorem 2.4 the error term is $\ll x$. Thus (2.11) gives (a). The sum in (b) differs from that in (a) by the amount

$$\sum_{\substack{p^k \leq x \\ k \geq 2}} \frac{\log p}{p^k} \leq \sum_p \frac{\log p}{p(p-1)} \ll 1.$$

To derive (c) we note that the sum in (a) is

$$\int_{2^-}^x u^{-1}\, d\psi(u) = \frac{\psi(u)}{u}\Big|_{2^-}^x + \int_2^x \psi(u)u^{-2}\, du = \int_2^x \psi(u)u^{-2}\, du + O(1)$$

by Theorem 2.4. We now prove (d) without determining the value of the constant b. We express (b) in the form $L(x) = \log x + R(x)$ where $R(x) \ll 1$. Then

$$
\sum_{p \leq x} \frac{1}{p} = \int_{2-}^{x} (\log u)^{-1} \, dL(u) = \int_{2-}^{x} \frac{1}{\log u} \, d\log u + \int_{2-}^{x} \frac{dR(u)}{\log u}
$$

$$
= \int_{2-}^{x} \frac{du}{u \log u} + \left[\frac{R(u)}{\log u} \right]_{2-}^{x} - \int_{2-}^{x} R(u) \, d(\log u)^{-1}
$$

$$
= \log \log x - \log \log 2 + 1 + \frac{R(x)}{\log x} + \int_{2}^{x} \frac{R(u)}{u(\log u)^2} \, du.
$$

The penultimate term is $\ll 1/\log x$, and the integral is $\int_{2}^{\infty} - \int_{x}^{\infty} = \int_{2}^{\infty} + O(1/\log x)$, so we have (d) with

$$
b = 1 - \log \log 2 + \int_{2}^{\infty} \frac{R(u)}{u(\log u)^2} \, du.
$$

As for (e), we note that

$$
\sum_{p \leq x} \log \left(1 - \frac{1}{p} \right)^{-1} = \sum_{p \leq x} \frac{1}{p} + \sum_{p \leq x} \left(\log \left(1 - \frac{1}{p} \right)^{-1} - \frac{1}{p} \right).
$$

The second sum on the right is

$$
\sum_{p} \sum_{k=2}^{\infty} \frac{1}{kp^k} + O\left(\sum_{p > x} p^{-2} \right)
$$

and the error term here is $\ll \sum_{n > x} n^{-2} \ll x^{-1}$, so from (d) we have

$$
\sum_{p \leq x} \log \left(1 - \frac{1}{p} \right)^{-1} = \log \log x + c + O(1/\log x) \tag{2.16}
$$

where $c = b + \sum_{p} \sum_{k \geq 2} (kp^k)^{-1}$. Since $e^z = 1 + O(|z|)$ for $|z| \leq 1$, on exponentiating we deduce that

$$
\prod_{p \leq x} \left(1 - \frac{1}{p} \right)^{-1} = e^c \log x + O(1).
$$

To complete the proof it suffices to show that $c = C_0$. To this end we first note that if $p \leq x$ and $p^k > x$, then $k \geq (\log x)/\log p$. Hence

$$
\sum_{\substack{p \leq x \\ p^k > x}} \frac{1}{kp^k} \ll \sum_{\substack{p \leq x \\ p^k > x}} \frac{\log p}{(\log x)p^k} \ll \sum_{p} \frac{\log p}{\log x} \sum_{k \geq 2} p^{-k} \ll \frac{1}{\log x} \sum_{p} \frac{\log p}{p^2} \ll \frac{1}{\log x},
$$

so that from (2.16) we have

$$\sum_{1 < n \le x} \frac{\Lambda(n)}{n \log n} = \log \log x + c + O(1/\log x).$$

By Corollary 1.15 this can be written

$$\sum_{1 < n \le x} \frac{\Lambda(n)}{n \log n} = \sum_{n \le \log x} \frac{1}{n} + (c - C_0) + O(1/\log 2x).$$

Since this is trivial when $1 \le x < 2$, the above holds for all $x \ge 1$. We express this briefly as $T_1 = T_2 + T_3 + T_4$, and estimate the quantities $I_i = \delta \int_1^\infty x^{-1-\delta} T_i(x) \, dx$. On comparing the results as $\delta \to 0^+$ we shall deduce that $c = C_0$. By Theorem 1.3, Corollary 1.11, and Corollary 1.13 we see that

$$I_1 = \log \zeta(1 + \delta) = \log \frac{1}{\delta} + O(\delta)$$

as $\delta \to 0^+$. Secondly,

$$I_2 = \delta \sum_{n=1}^\infty \frac{1}{n} \int_{e^n}^\infty x^{-1-\delta} \, dx = \sum_{n=1}^\infty \frac{1}{n} e^{-\delta n} = \log(1 - e^{-\delta})^{-1}$$

$$= \log(\delta + O(\delta^2))^{-1} = \log 1/\delta + O(\delta).$$

Thirdly,

$$I_3 = c - C_0,$$

and finally

$$I_4 \ll \delta \int_1^\infty x^{-1-\delta} \frac{dx}{\log 2x} \ll \delta + \delta \int_2^{e^{1/\delta}} \frac{dx}{x \log x} + \delta^2 \int_{e^{1/\delta}}^\infty x^{-1-\delta} \, dx \ll \delta \log 1/\delta.$$

Since the main terms cancel, on letting $\delta \to 0^+$ we see that $c = C_0$.   □

Corollary 2.8   *We have*

$$\limsup_{x \to \infty} \frac{\pi(x)}{x/\log x} \ge 1$$

and

$$\liminf_{x \to \infty} \frac{\pi(x)}{x/\log x} \le 1.$$

Proof   By Corollary 2.5 it suffices to show that $\limsup \psi(u)/u \ge 1$, and that $\liminf \psi(u)/u \le 1$. Suppose that $\limsup \psi(u)/u = a$, and suppose that $\varepsilon > 0$.

2.2 Estimates of Chebyshev and of Mertens

Then there is an x_0 such that $\psi(x) \le (a+\varepsilon)x$ for all $x \ge x_0$, and hence

$$\int_1^x \psi(u)u^{-2}\,du \le \int_1^{x_0} \psi(u)u^{-2}\,du + (a+\varepsilon)\int_{x_0}^x u^{-1}\,du \le (a+\varepsilon)\log x + O_\varepsilon(1).$$

Since this holds for arbitrary $\varepsilon > 0$, it follows that $\int_1^x \psi(u)u^{-2}\,du \le (a + o(1))\log x$. Thus by Theorem 2.7(c) we have $a \ge 1$. Similarly $\liminf \psi(u)/u \le 1$. $\qquad\square$

2.2.1 Exercises

1. (a) Let $d_n = [1, 2, \ldots, n]$. Show that $d_n = e^{\psi(n)}$.
   (b) Let $P \in \mathbb{Z}[x]$, $\deg P \le n$. Put $I = I(P) = \int_0^1 P(x)\,dx$. Show that $Id_{n+1} \in \mathbb{Z}$, and hence that $d_{n+1} \ge 1/|I|$ if $I \ne 0$.
   (c) Show that there is a polynomial P as above so that $Id_{n+1} = 1$.
   (d) Verify that $\max_{0 \le x \le 1} |x^2(1-x)^2(2x-1)| = 5^{-5/2}$.
   (e) For $P(x) = \left(x^2(1-x)^2(2x-1)\right)^{2n}$, verify that $0 < I < 5^{-5n}$.
   (f) Show that $\psi(10n+1) \ge (\frac{1}{2}\log 5)\cdot 10n$.

2. Let $\mathcal{A}$ be the set of integers composed entirely of primes $p \le A_1$, and let $\mathcal{B}$ be the set of integers composed entirely of primes $p > A_1$. Then n is uniquely of the form $n = ab$, $a \in \mathcal{A}$, $b \in \mathcal{B}$. Let $\delta(A_1, A_2)$ denote the density of those n such that $a \le A_2$.
   (a) Give a formula for $\delta(A_1, A_2)$.
   (b) Show that $\delta(A_1, A_2) \gg (\log A_2)/\log A_1$ for $2 \le A_2 \le A_1$.

3. Let $a_n = 1 + \cos\log n$, and note that $a_n \ge 0$ for all n.
   (a) Show that
   $$\sum_{n=1}^\infty a_n n^{-s} = \zeta(s) + \frac{1}{2}\zeta(s+i) + \frac{1}{2}\zeta(s-i)$$
   for $\sigma > 1$.
   (b) By Corollary 1.15, or otherwise, show that
   $$\sum_{n \le x} \frac{a_n}{n} = \log x + O(1).$$
   (c) By integrating by parts as in the proof of Theorem 1.12, show that
   $$\sum_{n \le x} a_n = \left(1 + \frac{x^i}{2(1+i)} + \frac{x^{-i}}{2(1-i)}\right)x + O(\log x).$$
   (d) Deduce that
   $$\liminf_{x \to \infty} \frac{1}{x}\sum_{n \le x} a_n = 1 - \frac{1}{\sqrt{2}}, \qquad \limsup_{x \to \infty} \frac{1}{x}\sum_{n \le x} a_n = 1 + \frac{1}{\sqrt{2}}.$$

Thus for the coefficients a_n we have an analogue of Mertens' estimate of Theorem 2.7(b), but not an analogue of the Prime Number Theorem.

4. (Golomb 1992) Let d_x denote the least common multiple of the positive integers not exceeding x. Show that

$$\binom{2n}{n} = \prod_{k=1}^{\infty} d_{2n/k}^{(-1)^{k-1}}.$$

5. (Chebyshev 1850) From Corollaries 2.5 and 2.8 we see that if there is a number a such that $\psi(x) = (a + o(1))x$ as $x \to \infty$, then we must have $a = 1$. We now take this a step further.

(a) Suppose that there is a number a such that

$$\psi(x) = x + (a + o(1))x/\log x \qquad (2.17)$$

as $x \to \infty$. Deduce that

$$\int_2^x \frac{\psi(u)}{u^2}\, du = \log x + (a + o(1))\log\log x$$

as $x \to \infty$.

(b) By comparing the above with Theorem 2.7(c), deduce that if (2.17) holds, then necessarily $a = 0$.

(c) Suppose that there is a constant A such that

$$\pi(x) = \frac{x}{\log x - A} + o\left(\frac{x}{(\log x)^2}\right) \qquad (2.18)$$

as $x \to \infty$. By writing $\vartheta(x) = \int_{2-}^x \log u\, d\pi(u)$, integrating by parts, and estimating the expressions that arise, show that if (2.18) holds, then

$$\psi(x) = x + (A - 1 + o(1))x/\log x$$

as $x \to \infty$.

(d) Deduce that if (2.18) holds, then $A = 1$.

2.3 Applications to arithmetic functions

The results above are useful in determining the extreme values of familiar arithmetic functions. We consider three instances.

Theorem 2.9 *For all $n \geq 3$,*

$$\varphi(n) \geq \frac{n}{\log \log n} \left(e^{-C_0} + O(1/\log \log n) \right),$$

and there are infinitely many n for which the above relation holds with equality.

Proof Let $\mathcal{R}$ be the set of those n for which $\varphi(n)/n < \varphi(m)/m$ for all $m < n$. We first prove the inequality for these 'record-breaking' $n \in \mathcal{R}$. Suppose that $\omega(n) = k$, and let n^* be the product of the first k primes. If $n \neq n^*$ then $n^* < n$ and $\varphi(n^*)/n^* < \varphi(n)/n$. Hence $\mathcal{R}$ is the set of n of the form

$$n = \prod_{p \leq y} p. \tag{2.19}$$

Taking logarithms, we see that $\log n = \vartheta(y) \asymp y$ by Corollary 2.6. On taking logarithms a second time, it follows that $\log \log n = \log y + O(1)$. Thus by Mertens' formula (Theorem 2.7(e)) we see that

$$\frac{\varphi(n)}{n} = \prod_{p \leq y} \left(1 - \frac{1}{p} \right) = \frac{e^{-C_0}}{\log y} (1 + O(1/\log y)),$$

which gives the desired result for $n \in \mathcal{R}$. If $n \notin \mathcal{R}$ then there is an $m < n$ such that $m \in \mathcal{R}$, $\varphi(m)/m < \varphi(n)/n$. Hence

$$\frac{\varphi(n)}{n} > \frac{\varphi(m)}{m} = \frac{1}{\log \log m} \left(e^{-C_0} + O\left(\frac{1}{\log \log m} \right) \right)$$

$$\geq \frac{1}{\log \log n} \left(e^{-C_0} + O\left(\frac{1}{\log \log n} \right) \right).$$

We note that equality holds for n of the type (2.19), so the proof is complete. $\square$

Theorem 2.10 *For all $n \geq 3$,*

$$1 \leq \omega(n) \leq \frac{\log n}{\log \log n} (1 + O(1/\log \log n)).$$

Proof As in the preceding proof we see that record-breaking values of $\omega(n)$ occur when n is of the form (2.19), and that it suffices to prove the bound for these n. As in the preceding proof, for n given by (2.19) we have $\vartheta(y) = \log n$ and $\log y = \log \log n + O(1)$. This gives the result, and we note that the bound is sharp for these n. $\square$

We now consider the maximum order of $d(n)$. From the pairing $d \leftrightarrow n/d$ of divisors, and the fact that at least one of these is $\leq \sqrt{n}$, it is immediate that $d(n) \leq 2\sqrt{n}$. On the other hand, if n is square-free then $d(n) = 2^{\omega(n)}$, which can be large, but not nearly as large as $\sqrt{n}$. Indeed, for each $\varepsilon > 0$ there is a constant $C(\varepsilon)$ such that

$$d(n) \leq C(\varepsilon)n^\varepsilon \qquad (2.20)$$

for all $n \geq 1$. To see this we express n in terms of its canonical factorization, $n = \prod_p p^a$, so that

$$\frac{d(n)}{n^\varepsilon} = \prod_p \frac{a+1}{p^{a\varepsilon}} = \prod_p f_p(a),$$

say. Let α_p be an integral value of a for which $f_p(a)$ is maximized. From the inequalities $f_p(\alpha_p) \geq f_p(\alpha_p \pm 1)$ we see that

$$(p^\varepsilon - 1)^{-1} - 1 \leq \alpha_p \leq (p^\varepsilon - 1)^{-1},$$

so that we may take $\alpha_p = [(p^\varepsilon - 1)^{-1}]$. Hence (2.20) holds with

$$C(\varepsilon) = \prod_p f_p(\alpha_p).$$

This constant is best possible, since equality holds when $n = \prod_p p^{\alpha_p}$. By analysing the rate at which $C(\varepsilon)$ grows as $\varepsilon \to 0^+$, we derive

Theorem 2.11 *For all $n \geq 3$*

$$\log d(n) \leq \frac{\log n}{\log \log n} \left(\log 2 + O(1/\log \log n)\right).$$

We note that this bound is sharp for n of the form in (2.19).

Proof   It suffices to show that there is an absolute constant K such that

$$C(\varepsilon) \leq \exp\left(K\varepsilon^2 2^{1/\varepsilon}\right), \qquad (2.21)$$

since the stated bound then follows by taking $\varepsilon = (\log 2)/\log \log n$. We observe that $\alpha_p = 0$ if $p > 2^{1/\varepsilon}$, that $\alpha_p = 1$ if $(3/2)^{1/\varepsilon} < p \leq 2^{1/\varepsilon}$, and that $\alpha_p \ll 1/\varepsilon$ when $p \leq (3/2)^{1/\varepsilon}$. Hence

$$\log C(\varepsilon) \ll \sum_{p \leq 2^{1/\varepsilon}} \log(2/p^\varepsilon) + \sum_{p \leq (3/2)^{1/\varepsilon}} \log(1/\varepsilon).$$

Here the second sum is $\pi\left((3/2)^{1/\varepsilon}\right) \log 1/\varepsilon \ll \varepsilon^2 2^{1/\varepsilon}$. The first sum is $(\log 2)\pi(2^{1/\varepsilon}) - \varepsilon\vartheta(2^{1/\varepsilon})$, and by Corollary 2.5 this is $\ll \varepsilon^2 2^{1/\varepsilon}$. Thus we have (2.21), and the proof is complete. $\square$

It is very instructive to consider our various results from the perspective of elementary probability theory. Let d be a fixed integer. Then the set of n that are divisible by d has asymptotic density $1/d$, and we might say, loosely, that the 'probability' that $d|n$ when n is 'randomly chosen' is $1/d$. If d_1 and d_2 are two fixed numbers then the 'probability' that $d_1|n$ and $d_2|n$ is $1/[d_1, d_2]$. If $(d_1, d_2) = 1$ then this 'probability' is $1/(d_1 d_2)$, and we see that the 'events' $d_1|n$, $d_2|n$ are 'independent.' To make this rigourous we consider the integers $1 \le n \le N$, and assign probability $1/N$ to each of the N numbers n. Then

$$\mathbf{P}(d|n) = [N/d]/N = \frac{1}{d} - \frac{1}{N}\{N/d\}.$$

This is $1/d$ if $d|N$; otherwise it is close to $1/d$ if d is small compared to N. Similarly the events $d_1|n$, $d_2|n$ are not independent in general, but are nearly independent if $N/(d_1 d_2)$ is large. The probabilistic heuristic, in which independence is assumed, provides a useful means of constructing conjectures. Many of our investigations can be considered to be directed toward determining whether the cumulative effect of the error terms $\{N/d\}/N$ have a discernible effect.

As an example of the probabilistic approach, we note that n is square-free if and only if none of the numbers $2^2, 3^2, 5^2, \ldots, p^2, \ldots$ divide n. The 'probability' that $p^2 \nmid n$ is approximately $1 - 1/p^2$. Since these events are nearly independent, we predict that the probability that a random integer $n \in [1, N]$ is square-free is approximately $\prod_{p \le N}(1 - 1/p^2)$. This was confirmed in Theorem 2.2. On the other hand, the sieve of Eratosthenes asserts that

$$\sum_{\substack{n \le N \\ (n, P)=1}} 1 = \pi(N) - \pi(\sqrt{N}) + 1$$

where $P = \prod_{p \le \sqrt{N}} p$. For a random $n \in [1, N]$ we expect that the probability that $(n, P) = 1$ should be approximately

$$\frac{\varphi(P)}{P} = \prod_{p \le \sqrt{N}}\left(1 - \frac{1}{p}\right) \sim \frac{2e^{-C_0}}{\log N}$$

by Mertens' formula (Theorem 2.7(e)). This would suggest that perhaps

$$\pi(x) \sim 2e^{-C_0}\frac{x}{\log x}.$$

However, since $2e^{-C_0} = 1.1229189\ldots$, this conflicts with the Prime Number Theorem, and also with Corollary 2.8. Thus the probabilistic model is misleading in this case.

Suppose now that $X_p(n)$ is the arithmetic function

$$X_p(n) = \begin{cases} 1 & \text{if } p|n, \\ 0 & \text{otherwise}, \end{cases}$$

so that $\omega(n) = \sum_p X_p(n)$. If we were to treat the X_p as though they were independent random variables then we would have $\mathbf{E}(X_p) = 1/p$, $\mathrm{Var}(X_p) = (1 - 1/p)/p$. Hence we expect that the average of $\omega(n)$ should be approximately

$$\mathbf{E}\left(\sum_{p \leq n} X_p\right) = \sum_{p \leq n} \mathbf{E}(X_p) = \sum_{p \leq n} \frac{1}{p} = \log\log n + O(1),$$

and that its variance is approximately

$$\mathrm{Var}\left(\sum_{p \leq n} X_p\right) = \sum_{p \leq n} \mathrm{Var}(X_p) = \sum_{p \leq n} \left(1 - \frac{1}{p}\right)\frac{1}{p} = \log\log n + O(1).$$

The first of these is easily confirmed, since by (2.3) we have

$$\sum_{n \leq x} \omega(n) = x \sum_{p \leq x} \frac{1}{p} + O\left(\pi(x)\right).$$

By Mertens' formula (Theorem 2.7(d)) and Chebyshev's bound (Corollary 2.6) this is

$$= x \log\log x + bx + O(x/\log x). \qquad (2.22)$$

As for the variance, we have

Theorem 2.12 (Turán) *For $x \geq 3$,*

$$\sum_{n \leq x} (\omega(n) - \log\log x)^2 \ll x \log\log x \qquad (2.23)$$

and

$$\sum_{1 < n \leq x} (\omega(n) - \log\log n)^2 \ll x \log\log x. \qquad (2.24)$$

These estimates also hold with $\omega(n)$ replaced by $\Omega(n)$.

Let $\mathcal{E}$ be the set of 'exceptional' n for which

$$|\omega(n) - \log\log n| > (\log\log n)^{3/4}.$$

By Theorem 2.12 we see that

$$\sum_{\substack{n \in \mathcal{E} \\ x < n \leq 2x}} 1 \leq (\log\log x)^{-3/2} \sum_{n \leq 2x} (\omega(n) - \log\log n)^2 \ll \frac{x}{(\log\log x)^{1/2}} = o(x),$$

so we have

Corollary 2.13 (Hardy–Ramanujan) *For almost all* n, $\omega(n) \sim \Omega(n) \sim \log\log n$.

Note that in analytic number theory we say 'almost all' when the exceptional set has asymptotic density 0; this conflicts with the usage in some parts of algebra, where the term means that there are at most finitely many exceptions.

Proof of Theorem 2.12 To prove (2.23) we first multiply out the square on the left, and write the sum as

$$\Sigma_2 - 2(\log\log x)\Sigma_1 + [x](\log\log x)^2. \tag{2.25}$$

We have already determined the size of Σ_1 in (2.22). The new sum is

$$\Sigma_2 = \sum_{n\leq x} \omega(n)^2 = \sum_{n\leq x}\left(\sum_{p_1|n}1\right)\left(\sum_{p_2|n}1\right) = \sum_{\substack{p_1\leq x\\p_2\leq x}}\sum_{\substack{n\leq x\\p_i|n}}1.$$

The terms for which $p_1 = p_2$ contribute

$$\sum_{p\leq x}[x/p] = x\sum_{p\leq x}\frac{1}{p} + O(\pi(x)) = x\log\log x + O(x).$$

The terms $p_1 \neq p_2$ contribute

$$\sum_{p_1\neq p_2}\left[\frac{x}{p_1 p_2}\right] \leq x\sum_{\substack{p_1 p_2\leq x\\p_1\neq p_2}}\frac{1}{p_1 p_2} \leq x\left(\sum_{p\leq x}\frac{1}{p}\right)^2 = x(\log\log x)^2 + O(x\log\log x) \tag{2.26}$$

by Mertens' formula (Theorem 2.7(d)). Thus

$$\Sigma_2 \leq x(\log\log x)^2 + O(x\log\log x).$$

The estimate (2.23) now follows by inserting this and (2.22) in (2.25).

We derive (2.24) from (2.23) by applying the triangle inequality $\big|\|\mathbf{x}\| - \|\mathbf{y}\|\big| \leq \|\mathbf{x}-\mathbf{y}\|$ for vectors. This gives

$$\left|\left(\sum_{1<n\leq x}(\omega(n)-\log\log n)^2\right)^{1/2} - \left(\sum_{1<n\leq x}(\omega(n)-\log\log x)^2\right)^{1/2}\right|$$
$$\leq \left(\sum_{1<n\leq x}(\log\log x - \log\log n)^2\right)^{1/2}.$$

By the integral test the sum on the right is

$$= \int_e^x (\log\log x - \log\log u)^2 \, du + O((\log\log x)^2).$$

By integrating by parts twice we find that this integral is

$$-e(\log\log x)^2 - 2e\log\log x + 2\int_2^x \frac{1 + \log\log x - \log\log u}{(\log u)^2} \, du \ll \frac{x}{(\log x)^2}.$$

Thus

$$\left(\sum_{1<n\leq x} (\omega(n) - \log\log n)^2\right)^{1/2} = \left(\sum_{n\leq x}(\omega(n) - \log\log x)^2\right)^{1/2} + O\left(x^{1/2}/\log x\right),$$

and (2.24) follows by squaring both sides and applying (2.23). We omit the similar argument for $\Omega(n)$. □

Since $2^{\omega(n)} \leq d(n) \leq 2^{\Omega(n)}$ for all n, Corollary 2.13 carries an interesting piece of information for $d(n)$:

$$d(n) = (\log n)^{(\log 2 + o(1))}$$

for almost all n. Since this is smaller than the average size of $d(n)$, we see that the average is determined not by the usual size of $d(n)$ but by a sparse set of n for which $d(n)$ is disproportionately large. Since the first moment (i.e., average) of $d(n)$ is inflated by the 'tail' in its distribution, it is not surprising that this effect is more pronounced for the higher moments. As was originally suggested by Ramanujan, it can be shown that for any fixed real number κ there is a positive constant $c(\kappa)$ such that

$$\sum_{n\leq x} d(n)^\kappa \sim c(\kappa)x(\log x)^{2^\kappa - 1} \tag{2.27}$$

as $x \to \infty$.

In order to handle the error terms that arise in our arguments we are frequently led to estimate the mean value of multiplicative functions. In most such cases the method of the hyperbola or the simpler identity (2.3) will suffice, but the labour involved quickly becomes tiresome. It will therefore be convenient to have the following result on record, as it is very readily applied.

Theorem 2.14 *Let f be a non-negative multiplicative function. Suppose that A is a constant such that*

$$\sum_{p\leq x} f(p)\log p \leq Ax \tag{2.28}$$

for all $x \geq 1$, and that

$$\sum_{\substack{p^k \\ k \geq 2}} \frac{f(p^k)k \log p}{p^k} \leq A. \tag{2.29}$$

Then for $x \geq 2$,

$$\sum_{n \leq x} f(n) \ll (A+1)\frac{x}{\log x} \sum_{n \leq x} \frac{f(n)}{n}.$$

We note that this is sharper than the trivial estimate

$$\sum_{n \leq x} f(n) \leq x \sum_{n \leq x} f(n)/n \tag{2.30}$$

that holds whenever $f \geq 0$.

If $f \geq 0$ and f is multiplicative, then

$$\sum_{n \leq x} \frac{f(n)}{n} \leq \prod_{p \leq x} \left(1 + \frac{f(p)}{p} + \frac{f(p^2)}{p^2} + \cdots \right).$$

On combining this with Theorem 2.14 we obtain

Corollary 2.15 *Under the above hypotheses*

$$\sum_{n \leq x} f(n) \ll (A+1)\frac{x}{\log x} \prod_{p \leq x} \left(1 + \frac{f(p)}{p} + \frac{f(p^2)}{p^2} + \cdots \right).$$

Suppose for example that $f(n) = d(n)^\kappa$. We write

$$\prod_{p \leq x} \left(1 + \frac{2^\kappa}{p} + \frac{3^\kappa}{p^2} + \cdots \right) = \left(\prod_{p \leq x} \left(1 - \frac{1}{p} \right)^{-2^\kappa} \right) \left(\prod_{p \leq x} \left(1 - \frac{1}{p} \right)^{2^\kappa} \right.$$
$$\left. \times \left(1 + \frac{2^\kappa}{p} + \frac{3^\kappa}{p^2} + \cdots \right) \right)$$

and observe that the second product tends to a finite limit as $x \to \infty$, so that by Mertens' formula (Theorem 2.7(e)) we have

$$\sum_{n \leq x} d(n)^\kappa \ll x(\log x)^{2^\kappa - 1} \tag{2.31}$$

for any fixed κ. Though weaker than (2.27), this is all that is needed in many cases. We can similarly show that for any fixed real κ,

$$\sum_{n \leq x} \left(\frac{n}{\varphi(n)} \right)^\kappa \ll x. \tag{2.32}$$

Thus we see that $\varphi(n)/n$ is not often very small.

Proof of Theorem 2.14   The desired bound is obtained by adding the two estimates

$$\sum_{n \leq x} f(n) \log \frac{x}{n} \ll x \sum_{n \leq x} \frac{f(n)}{n}, \tag{2.33}$$

$$\sum_{n \leq x} f(n) \log n \ll Ax \sum_{n \leq x} \frac{f(n)}{n}. \tag{2.34}$$

The first of these is immediate, since $f \geq 0$ and $\log x/n \ll x/n$ uniformly for $1 \leq n \leq x$. Since $\log n = \sum_{d|n} \Lambda(d)$, the second sum is

$$\sum_{d \leq x} \Lambda(d) \sum_{m \leq x/d} f(md).$$

Writing $d = p^i$, $m = p^j r$ where $p \nmid r$, we see that this is

$$\sum_{\substack{p,i \geq 1, j \geq 0 \\ p^{i+j} \leq x}} (\log p) f(p^{i+j}) \sum_{\substack{r \leq x/p^{i+j} \\ p \nmid r}} f(r) = \sum_{\substack{p,k \\ p^k \leq x}} k(\log p) f(p^k) \sum_{\substack{r \leq x/p^k \\ p \nmid r}} f(r).$$

Here we have put $i + j = k$. We now drop the condition $p \nmid r$ on the right-hand side, and consider first the contribution of the proper prime powers (i.e., $k \geq 2$). By (2.30) with x replaced by x/p we see that the terms for which $k \geq 2$ contribute

$$\ll x \sum_{p,k \geq 2} (\log p^k) f(p^k) p^{-k} \sum_{r \leq x/p^k} f(r)/r \leq Ax \sum_{n \leq x} f(n)/n$$

by (2.29). It remains to bound

$$\sum_{p \leq x} (\log p) f(p) \sum_{r \leq x/p} f(r) = \sum_{r \leq x} f(r) \sum_{p \leq x/r} f(p) \log p.$$

By (2.28) this is $\leq Ax \sum_{r \leq x} f(r)/r$, so we have (2.34) and the proof is complete.   □

In the above proof we made no use of prime number estimates, but as we have seen the estimates of Chebyshev are useful in verifying the hypotheses and Mertens' formula is helpful in estimating the sum $\sum_{n \leq x} f(n)/n$.

2.3.1 Exercises

1. Let $\sigma(n) = \sum_{d|n} d$.
   (a) Show that $\sigma(n) \varphi(n) \leq n^2$ for all $n \geq 1$.
   (b) Deduce that $n + 1 \leq \sigma(n) \leq e^{C_0} n (\log \log n + O(1))$ for all $n \geq 3$.

2. Show that $d(n) \leq \sqrt{3n}$ with equality if and only if $n = 12$.
3. Let $f(n) = \prod_{p|n}(1 + p^{-1/2})$.
    (a) Show that there is a constant a such that if $n \geq 3$, then

    $$f(n) < \exp\left(a(\log n)^{1/2}(\log \log n)^{-1}\right).$$

    (b) Show that $\sum_{n \leq x} f(n) = cx + O\left(x^{1/2}\right)$ where $c = \prod_p(1 + p^{-3/2})$.
4. Let $d_k(n)$ be as in Exercise 2.1.18. Show that if k and κ are fixed, then

    $$\sum_{n \leq x} d_k(n)^\kappa \ll x(\log x)^{k^\kappa - 1}.$$

    for $x \geq 2$.
5. (Davenport 1932) Let

    $$f(n) = -\sum_{d|n} \frac{\mu(d) \log d}{d}.$$

    (a) By recalling Exercise 2.1.16(a), or otherwise, show that $f(n) \geq 0$ for all n.
    (b) Show that $f(n) \ll \log \log n$ for $n \geq 3$.
    (c) Show that $f(n) \sim \frac{1}{4} \log \log n$ if $n = \prod_{y < p \leq y^2} p$.
    (d) Show that $f(n) \leq \left(\frac{1}{4} + o(1)\right) \log \log n$ as $n \to \infty$.
6. (cf. Bateman & Grosswald 1958) Let $\mathcal{F}$ be the set of 'power-full' numbers where n is power-full if $p|n \Rightarrow p^2|n$.
    (a) Show that

    $$\sum_{n \in \mathcal{F}} n^{-s} = \frac{\zeta(2s)\zeta(3s)}{\zeta(6s)}$$

    for $\sigma > 1/2$.
    (b) Show that

    $$\sum_{\substack{a,b,c \\ a^2 b^3 c^6 = n}} \mu(c) = \begin{cases} 1 & \text{if } n \in \mathcal{F}, \\ 0 & \text{otherwise.} \end{cases}$$

    (c) Show that

    $$\sum_{a^2 b^3 \leq x} 1 = \zeta(3/2)y^{1/2} + \zeta(2/3)y^{1/3} + O\left(y^{1/5}\right).$$

    (d) Show that

    $$\sum_{\substack{n \leq x \\ n \in \mathcal{F}}} 1 = \frac{\zeta(3/2)}{\zeta(3)}x^{1/2} + \frac{\zeta(2/3)}{\zeta(2)}x^{1/3} + O\left(x^{1/5}\right).$$

7. (Bateman 1949) Let $\Phi_q(z)$ denote the qth cyclotomic polynomial,

$$\Phi_q(z) = \prod_{\substack{a=1 \\ (a,q)=1}}^{q} (z - e(a/q))$$

where $e(\theta) = e^{2\pi i \theta}$.

(a) Show that

$$\prod_{d|q} \Phi_d(z) = z^q - 1.$$

(b) Show that

$$\Phi_q(z) = \prod_{d|q} (z^d - 1)^{\mu(q/d)}.$$

(c) If $P(z) = \sum p_n z^n$ and $Q(z) = \sum q_n z^n$ are polynomials with real coefficients, then we say that $P \preccurlyeq Q$ if $|p_n| \leq q_n$ for all non-negative integers n. Show that if $P_1 \preccurlyeq Q_1$ and $P_2 \preccurlyeq Q_2$, then $P_1 + P_2 \preccurlyeq Q_1 + Q_2$ and $P_1 P_2 \preccurlyeq Q_1 Q_2$.

(d) Show that $\Phi_q(z) \preccurlyeq Q_q(z)$ where

$$Q_q(z) = \prod_{d|q} (1 + z^d + z^{2d} + \cdots + z^{q-d}).$$

(e) Show that $Q_q(1) = q^{d(q)/2}$.

(f) Show that for any $\varepsilon > 0$ there is a $q_0(\varepsilon)$ such that if $q > q_0(\varepsilon)$, then all coefficients of Φ_q have absolute value not exceeding

$$\exp\left(q^{(\log 2 + \varepsilon)/\log\log q}\right).$$

8. (Turán 1934) (a) Show that the first sum in (2.26) is

$$= x \sum_{p_1 p_2 \leq x} \frac{1}{p_1 p_2} + O(x).$$

(b) Explain why the sum above is

$$\left(\sum_{p \leq x} \frac{1}{p}\right)^2 - 2 \sum_{p_1 \leq \sqrt{x}} \frac{1}{p_1} \sum_{x/p_1 < p_2 \leq x} \frac{1}{p_2} + \left(\sum_{\sqrt{x} < p \leq x} \frac{1}{p}\right)^2. \qquad (2.35)$$

(c) Show that if $y \leq \sqrt{x}$, then

$$\sum_{x/y < p \leq x} \frac{1}{p} = \log\log x - \log\log(x/y) + O(1/\log x).$$

(d) Show that the right-hand side above is $\asymp (\log y)/\log x$.

(e) Deduce that the second and third terms in (2.35) are $\ll 1$.
(f) Conclude that

$$\Sigma_2 = x(\log\log x)^2 + (2b + 1)\log\log x + O(x)$$

where b is the constant in Theorem 2.7(d).
(g) Show that the left-hand side of (2.23) is $= x\log\log x + O(x)$.
(h) Show that the left-hand side of (2.24) is $= x\log\log x + O(x)$.

9. (cf. Pomerance 1977, Shan 1985) Note that $\varphi(n)|(n-1)$ when n is prime. An old – and still unsolved – problem of D. H. Lehmer asks whether there exists a composite integer n such that $\varphi(n)|(n-1)$. Let S denote the (presumably empty) set of such numbers.

(a) Show that if $n \in S$, then n is square-free.
(b) Suppose that $mp \in S$. Show that $m \equiv 1 \pmod{p-1}$.
(c) Let p be given. Show that the number of m such that $mp \leq x$ and $mp \in S$ is $\ll x/p^2$.
(d) Show that the number of $n \in S$, $n \leq x$, such that n has a prime factor $> y$ is $\ll x/(y\log y)$.
(e) Suppose that $x/y < n \leq x$ and that n is composed entirely of primes $p \leq y$. Show that $\omega(n) \geq (\log x)/(\log y) - 1$.
(f) By Exercise 4, or otherwise, show that the number of $n \leq x$ such that $\omega(n) \geq z$ is $\ll x(\log x)^2/3^z$.
(g) Conclude that the number of $n \leq x$ such that $n \in S$ is $\ll x/\exp(\sqrt{\log x})$.

2.4   The distribution of $\Omega(n) - \omega(n)$

In order to illustrate further the use of elementary techniques we now discuss an elegant result of Rényi, which asserts that the set of numbers n such that $\Omega(n) - \omega(n) = k$ has density d_k, where the d_k are the power series coefficients of the meromorphic function

$$F(z) = \sum_{k=0}^{\infty} d_k z^k = \prod_p \left(1 - \frac{1}{p}\right)\left(1 + \frac{1}{p-z}\right). \tag{2.36}$$

By examining this product we see that F has simple poles at the points $z = p$ ($p \neq 3$), and simple zeros at the points $z = p + 1$ ($p \neq 2$), so that the power series converges for $|z| < 2$. We let $N_k(x)$ denote the number of $n \leq x$ for which $\Omega(n) - \omega(n) = k$; our object is to show that $N_k(x) \sim d_k x$. If this holds for each k then we can deduce that $\sum d_k \leq 1$. By taking $z = 1$ in (2.36) we see that $\sum d_k = 1$, which gives us hope that the asymptotic relation may be fairly uniform in k. This is indeed the case, as we see from the following quantitative form of Rényi's theorem.

Theorem 2.16   *For any non-negative integer k, and any $x \geq 2$,*

$$N_k(x) = d_k x + O\left(\left(\tfrac{3}{4}\right)^k x^{1/2} (\log x)^{4/3} \right).$$

In preparation for the proof of this result we first establish a subsidiary estimate.

Lemma 2.17   *For any $y \geq 0$ and any natural number f,*

$$\sum_{\substack{n \leq y \\ (n,f)=1}} \mu(n)^2 = \frac{6}{\pi^2} \left(\prod_{p|f} \left(1 + \frac{1}{p}\right)^{-1} \right) y + O\left(y^{1/2} \prod_{p|f} (1 - p^{-1/2})^{-1} \right).$$

Proof   Let $\mathcal{D} = \{d : p|d \Rightarrow p|f\}$. By considering the Dirichlet series identity

$$\sum_{\substack{n=1 \\ (n,f)=1}}^{\infty} \mu(n)^2 n^{-s} = \prod_{p \nmid f} (1 + p^{-s}) = \frac{\zeta(s)}{\zeta(2s)} \prod_{p|f} (1 + p^{-s})^{-1} = \frac{\zeta(s)}{\zeta(2s)} \sum_{d \in \mathcal{D}} \lambda(d) d^{-s},$$

or by elementary considerations, we see that the characteristic function of the set of those square-free n such that $(n, f) = 1$ may be written

$$\sum_{\substack{dm=n \\ d \in \mathcal{D}}} \lambda(d) \mu(m)^2.$$

Hence the sum in question is

$$\sum_{d \in \mathcal{D}} \lambda(d) \sum_{m \leq y/d} \mu(m)^2 = \sum_{d \in \mathcal{D}} \lambda(d) \left(\frac{6}{\pi^2} \cdot \frac{y}{d} + O\left(y^{1/2} d^{-1/2}\right) \right)$$

by Theorem 2.2. But $\sum_{d \in \mathcal{D}} \lambda(d)/d = \prod_{p|f} (1 + 1/p)^{-1}$ and $\sum_{d \in \mathcal{D}} d^{-1/2} = \prod_{p|f} (1 - p^{-1/2})^{-1}$, so that the proof is complete. $\square$

Proof of Theorem 2.16   Let $\mathcal{Q}$ denote the set of square-free numbers and $\mathcal{F}$ denote the set of 'power-full' numbers (i.e., those f such that $p|f \Rightarrow p^2|f$). Every number is uniquely expressible in the form $n = qf$, $q \in \mathcal{Q}$, $f \in \mathcal{F}$, $(q, f) = 1$. Hence

$$N_k = \sum_{\substack{f \leq x \\ f \in \mathcal{F} \\ \Omega(f) - \omega(f) = k}} \sum_{\substack{q \leq x/f \\ q \in \mathcal{Q} \\ (q,f)=1}} 1.$$

By Lemma 2.17 this is

$$\frac{6}{\pi^2}x \sum_{\substack{f \le x \\ f \in \mathcal{F} \\ \Omega(f)-\omega(f)=k}} \frac{1}{f}\prod_{p|f}(1+p^{-1})^{-1} + O\left(x^{1/2} \sum_{\substack{f \le x \\ f \in \mathcal{F} \\ \Omega(f)-\omega(f)=k}} f^{-1/2}\prod_{p|f}(1-p^{-1/2})^{-1}\right).$$

In order to appreciate the nature of these sums it is helpful to observe that each member of $\mathcal{F}$ is uniquely of the form $a^2 b^3$ with b square-free, so that there are $\asymp x^{1/2}$ members of $\mathcal{F}$ not exceeding x. Suppose that $z \ge 1$. Then the sum in the error term is

$$\le z^{-k} \sum_{\substack{f \le x \\ f \in \mathcal{F}}} z^{\Omega(f)-\omega(f)} f^{-1/2}\prod_{p|f}(1-p^{-1/2})^{-1}.$$

Since $\Omega(f) - \omega(f)$ is an additive function, it follows that $z^{\Omega(f)-\omega(f)}$ is a multiplicative function. Hence the above is

$$\le z^{-k} \prod_{p \le x}\left(1+\left(1-p^{-1/2}\right)^{-1}\left(\frac{z}{p}+\frac{z^2}{p^{3/2}}+\frac{z^3}{p^2}+\cdots\right)\right).$$

When $p = 2$ the sum converges only for $z < \sqrt{2}$. Hence we take $z = 4/3$, and then the product is

$$\le \prod_{p \le x}\left(1+\frac{4}{3p}+\frac{C}{p^{3/2}}\right) \ll (\log x)^{4/3}$$

by Mertens' formula. Thus

$$\sum_{\substack{f \le x \\ f \in \mathcal{F} \\ \Omega(f)-\omega(f)=k}} f^{-1/2}\prod_{p|f}(1-p^{-1/2})^{-1} \ll \left(\frac{3}{4}\right)^k (\log x)^{4/3}$$

which suffices for the error term.

We now consider the effect of dropping the condition $f \le x$ in the main term. Since

$$\sum_{\substack{U < f \le 2U \\ f \in \mathcal{F} \\ \Omega(f)-\omega(f)=k}} \frac{1}{f}\prod_{p|f}\left(1+\frac{1}{p}\right)^{-1} \le U^{-1/2} \sum_{\substack{U < f \le 2U \\ f \in \mathcal{F} \\ \Omega(f)-\omega(f)=k}} f^{-1/2}\prod_{p|f}(1-p^{-1/2})^{-1}$$

$$\ll U^{-1/2}\left(\frac{3}{4}\right)^k (\log 2U)^{4/3},$$

on taking $U = x2^r$ and summing over $r \ge 0$ we see that

$$\sum_{\substack{f \le x \\ f \in \mathcal{F} \\ \Omega(f)-\omega(f)=k}} \frac{1}{f}\prod_{p|f}\left(1+\frac{1}{p}\right)^{-1} \ll x^{-1/2}\left(\frac{3}{4}\right)^k (\log x)^{4/3}.$$

Hence we have the stated result with

$$d_k = \frac{6}{\pi^2} \sum_{\substack{f \in \mathcal{F} \\ \Omega(f)-\omega(f)=k}} \frac{1}{f} \prod_{p|f} \left(1 + \frac{1}{p}\right)^{-1}.$$

To see that (2.36) holds, it suffices to multiply this by z^k and sum over k.   □

2.4.1   Exercise

1. Let d_k be as in (2.36). Show that

$$d_k = c2^{-k} + O(5^{-k})$$

where

$$c = \frac{1}{4} \prod_{p>2} \left(1 - \frac{1}{(p-1)^2}\right)^{-1}.$$

2.5   Notes

Section 2.1. Mertens (1874 a) showed that $\sum_{n \le x} \varphi(n) = 3x^2/\pi^2 + O(x \log x)$. This refines an earlier estimate of Dirichlet, and is equivalent to Theorem 2.1, by partial summation. Let $R(x)$ denote the error term in Theorem 2.1. Chowla (1932) showed that

$$\int_1^x R(u)^2 \, du \sim \frac{x}{2\pi^2}$$

as $x \to \infty$, and Walfisz (1963, p. 144) showed that

$$R(x) \ll (\log x)^{2/3} (\log \log x)^{4/3}.$$

In the opposite direction, Pillai & Chowla (1930) showed (cf. Exercise 7.3.6) that $R(x) = \Omega(\log \log \log x)$. That the error term changes sign infinitely often was first proved by Erdős & Shapiro (1951), who showed that $R(x) = \Omega_{\pm}(\log \log \log \log x)$. More recently, Montgomery (1987) showed that $R(x) = \Omega_{\pm}(\sqrt{\log \log x})$. It may be speculated that $R(x) \ll \log \log x$ and that $R(x) = \Omega_{\pm}(\log \log x)$.

Theorem 2.2 is due to Gegenbauer (1885).

Theorem 2.3 is due to Dirichlet (1849). The problem of improving the error term in this theorem is known as the *Dirichlet divisor problem*. Let $\Delta(x)$ denote the error term. Voronoï (1903) showed that $\Delta(x) \ll x^{1/3} \log x$ (see Exercises 2.1.23, 2.1.25, 2.1.26). van der Corput (1922) used estimates of exponential sums to show that $\Delta(x) \ll x^{33/100+\varepsilon}$. This exponent has since been reduced by van der Corput (1928), Chih (1950), Richert (1953), Kolesnik (1969, 1973, 1982, 1985), Iwaniec & Mozzochi (1988), and by Huxley (1993), who showed that $\Delta(x) \ll x^{23/73+\varepsilon}$. In the opposite direction, Hardy (1916) showed that $\Delta(x) = \Omega_{\pm}(x^{1/4})$. Soundararajan (2003) showed that

$$\Delta(x) = \Omega\left(x^{1/4}(\log x)^{1/4}(\log\log x)^b(\log\log\log x)^{-5/8}\right)$$

with $b = \frac{3}{4}(2^{4/3} - 1)$, and it is plausible that the first three exponents above are optimal.

The result of Exercise 2.1.12 generalizes to $\mathbb{R}^n$: A lattice point $(a_1, a, \ldots, a_n \in \mathbb{Z}^n)$ is said to be *primitive* if $\gcd(a_1, a_2, \ldots, a_n) = 1$. The asymptotic density of primitive lattice points is easily shown to be $1/\zeta(n)$. In addition, Cai & Bach (2003) have shown that the density of lattice points $a \in \mathbb{Z}^n$ such that $\gcd(a_i, a_j) = 1$ for all pairs with $1 \le i < j \le n$ is

$$\prod_p \left(\left(1 - \frac{1}{p}\right)^n + \frac{n}{p}\left(1 - \frac{1}{p}\right)^{n-1}\right).$$

Section 2.2. Chebyshev (1848) used the asymptotics of $\log \zeta(\sigma)$ as $\sigma \to 1^+$ to obtain Corollary 2.8. In his second paper on prime numbers, Chebyshev (1850) introduced the notations $\vartheta(x)$, $\psi(x)$, $T(x)$, and proved Theorem 2.4, Corollaries 2.5, 2.6, Theorem 2.7(a), and the results of Exercise 2.2.5. Sylvester (1881) devised a more complicated choice of the a_d that gave better constants than those of Chebyshev. Diamond & Erdős (1980) have shown that for any $\varepsilon > 0$ it is possible to choose numbers a_d as in the proof of Theorem 2.4 to show that $(1 - \varepsilon)x < \psi(x) < (1 + \varepsilon)x$ for all sufficiently large x. This does not constitute a proof of the Prime Number Theorem, because the PNT is used in the proof. Chebyshev (1850) also used his main results to prove Bertrand's postulate. Simpler proofs have been devised by various authors. For an easy exposition, see Theorem 8.7 of Niven, Zuckerman & Montgomery (1991). Richert (1949a, b) (cf. Mąkowski 1960) used Bertrand's postulate to show that every integer > 6 can be expressed as a sum of distinct primes. Rosser & Schoenfeld (1962, 1975) and Schoenfeld (1976) have given a large number of very useful explicit estimates for primes and for the Chebyshev functions, of which one example is that $\pi(x) > x/\log x$ for all $x \ge 17$. For the kth prime number, p_k, Dusart (1999) has given the lower bound

$$p_k > k(\log k + \log\log k - 1)$$

for $k \ge 2$. For further explicit estimates, see Schoenfeld (1969), Costa Pereira (1989), and Massias & Robin (1996). In Exercise 2.2.1 we find that $\psi(x) \ge cx + O(1)$ with $c = \frac{1}{2}\log 5 = 0.8047\ldots$. This approach is mentioned by Gel'-fond, in his editorial remarks in the Collected Works of Chebyshev (1946, pp. 285–288). Polynomials can be found that produce better constants, but Gorshkov (1956) showed that the supremum of such constants is < 1, so the Prime Number Theorem cannot be established by this method. For more on this subject, see Montgomery (1994, Chapter 10), Pritsker (1999), and Borwein (2002, Chapter 10).

Theorem 2.7(b)–(e) is due to Mertens (1874a, b). Our determination of the constant in Theorem 2.7(e) incorporates an expository finesse due to Heath-Brown.

Section 2.3. Theorem 2.9 is due to Landau (1903). Runge (1885) proved (2.20), and Wigert (1906/7) showed that $d(n) < n^{(\log 2+\varepsilon)/\log\log n}$ for $n > n_0(\varepsilon)$. Ramanujan (1915a, b) established the upper bound of Theorem 2.11, first with an extra $\log\log\log n$ in the error term, and then without. Ramanujan (1915b) also proved that

$$\frac{\log d(n)}{\log 2} < \mathrm{li}(n) + O\left(n\exp\left(-c\sqrt{\log n}\right)\right)$$

for all $n \geq 2$, and that

$$\frac{\log d(n)}{\log 2} > \mathrm{li}(n) + O\left(n\exp\left(-c\sqrt{\log n}\right)\right)$$

for infinitely many n. For a survey of extreme value estimates of arithmetic functions, see Nicolas (1988).

Theorem 2.12 is due to Turán (1934), although Corollary 2.13 and the estimate (2.22) used in the proof of Theorem 2.12 were established earlier by Hardy & Ramanujan (1917). Kubilius (1956) generalized Turán's inequality to arbitrary additive functions. See Tenenbaum (1995, pp. 302–304) for a proof, and discussion of the sharpest constants.

Theorem 2.14 is due to Hall & Tenenbaum (1988, pp. 2, 11). It represents a weakening of sharper estimates that can be derived with more work. For example, Wirsing (1961) showed that if f is a multiplicative function such that $f(n) \geq 0$ for all n, if there is a constant $C < 2$ such that $f(p^k) \ll C^k$ for all $k \geq 2$, and if

$$\sum_{p\leq x} f(p) \sim \kappa x/\log x$$

as $x \to \infty$ where κ is a positive real number, then

$$\sum_{n\leq x} f(n) \sim \frac{e^{-C_0\kappa}x}{\Gamma(\kappa)\log x} \prod_{p\leq x}\left(1 + \frac{f(p)}{p} + \frac{f(p^2)}{p^2} + \cdots\right).$$

For more information concerning non-negative multiplicative functions, see Wirsing (1967), Hall (1974), Halberstam & Richert (1979), and Hildebrand

(1984, 1986, 1987). For a comprehensive account of the mean values of (not necessarily non-negative) multiplicative functions, see Tenenbaum (1995, pp. 48–50, 308–310, 325–357). The two sides of (2.31) are of the same order of magnitude, and with more work one can derive a more precise asymptotic estimate; see Wilson (1922).

Section 2.4. Rényi (1955) gave a qualitative form of Theorem 2.16. Robinson (1966) gave formulæ for the densities d_k. Kac (1959, pp. 64–71) gave a proof by probabilistic techniques. Generalizations have been given by Cohen (1964) and Kubilius (1964). Sharper estimates for the error term have been derived by Delange (1965, 1967/68, 1973), Kátai (1966), Saffari (1970), and Schwarz (1970).

For a much more detailed historical account of the development of prime number theory, see Narkiewicz (2000).

2.6   References

Bateman, P. T. (1949). Note on the coefficients of the cyclotomic polynomial, *Bull. Amer. Math. Soc.* **55**, 1180–1181.

Bateman, P. T. & Grosswald, E. (1958). On a theorem of Erdős and Szekeres, *Illinois J. Math.* **2**, 88–98.

Bombieri, E. & Pila, J. (1989). The number of integral points on arcs and ovals, *Duke Math. J.* **59**, 337–357.

Borwein, P. (2002). Computational excursions in analysis and number theory. *Canadian Math. Soc.*, New York: Springer.

Cai, J.-Y. & Bach, E. (2003). On testing for zero polynomials by a set of points with bounded precision, *Theoret. Comp. Sci.* **296**, 15–25.

Chebyshev, P. L. (1848). Sur la fonction qui détermine la totalité des nombres premiers inférieurs à une limite donné, *Mem. Acad. Sci.* St. Petersburg **6**, 1–19.

(1850). Mémoire sur nombres premiers, *Mem. Acad. Sci.* St. Petersburg **7**, 17–33.

(1946). Collected works of P. L. Chebyshev, Vol. 1, *Akad. Nauk SSSR*, Moscow–Leningrad.

Chih, T.-T. (1950). A divisor problem, *Acta Sinica Sci. Record* **3**, 177–182.

Chowla, S. (1932). Contributions to the analytic theory of numbers, *Math. Zeit.* **35**, 279–299.

Cohen, E. (1964). Some asymptotic formulas in the theory of numbers, *Trans. Amer. Math. Soc.* **112**, 214–227.

van der Corput, J. G. (1922). Vereschärfung der Abschätzung beim Teilerproblem, *Math. Ann.* **87**, 39–65.

(1928). Zum Teilerproblem, *Math. Ann.* **98**, 697–716.

Costa Pereira, N. (1989). Elementary estimates for the Chebyshev function $\psi(x)$ and for the Möbius function $M(x)$, *Acta Arith.* **52**, 307–337.

Davenport, H. (1932). On a generalization of Euler's function $\phi(n)$, *J. London Math. Soc.* **7**, 290–296; *Collected Works*, Vol. IV. London: Academic Press, pp. 1827–1833.

Delange, H. (1965). Sur un théorème de Rényi, *Acta Arith.* **11**, 241–252.

(1967/68). Sur un théorème de Rényi, II, *Acta Arith.* **13**, 339–362.

(1973). Sur un théorème de Rényi, III, *Acta Arith.* **23**, 157–182.

Diamond, H. G. & Erdős, P. (1980). On sharp elementary prime number estimates, *Enseignement Math.* (2) **26**, 313–321.

Dirichlet, L. (1849). Über die Bestimmung der mittleren Werthe in der Zahlentheorie, *Math. Abhandl. Königl. Akad. Wiss.* Berlin, 69–83; *Werke*, Vol. 2, pp. 49–66.

Duncan, R. L. (1965). The Schnirelmann density of the k-free integers, *Proc. Amer. Math. Soc.* **16**, 1090–1091.

Dusart, P. (1999). The kth prime is greater than $k(\log k + \log \log k - 1)$ for $k \geq 2$, *Math. Comp.* **68**, 411–415.

Erdős, P. & Shapiro, H. N. (1951). On the change of sign of a certain error function, *Canadian J. Math.* **3**, 375–385.

Erdős, P. & Szekeres, G. (1934). Über die Anzahl der Abelschen Gruppen gegebener Ordnung und über ein verwandtes zahlentheoretisches Problem, *Acta Litt. Sci. Szeged* **7**, 95–102.

Evelyn, C. J. A. & Linfoot, E. H. (1930). On a problem in the additive theory of numbers, II, *J. Reine Angew. Math.* **164**, 131–140.

Feller, W. & Tornier, E. (1932). Mengentheoretische Untersuchungen von Eigenschaften der Zahlenreihe, *Math. Ann.* **107**, 188–232.

Gegenbauer, L. (1885). Asymptotische Gesetse der Zahlentheorie, *Denkschriften Österreich. Akad. Wiss. Math.-Natur. Cl.* **49**, 37–80.

Golomb, S. (1992). An inequality for $\binom{2n}{n}$, *Amer. Math. Monthly* **99**, 746–748.

Gorshkov, L. S. (1956). On the deviation of polynomials with rational integer coefficients from zero on the interval [0, 1]. *Proceedings of the 3rd All-union congress of Soviet mathematicians*, Vol. 3, Moscow, pp. 5–7.

Grosswald, E. (1956). The average order of an arithmetic function, *Duke Math. J.* **23**, 41–44.

Halberstam, H. & Richert, H.-E. (1979). On a result of R. R. Hall, *J. Number Theory* **11**, 76–89.

Hall, R. R. (1974). Halving an estimate obtained from the Selberg upper bound method, *Acta Arith.* **25**, 487–500.

Hall, R. R. & Tenenbaum, G. (1988). *Divisors*, Cambridge Tract 90. Cambridge: Cambridge University Press.

Hardy, G. H. (1916). On Dirichlet's divisor problem, *Proc. London Math. Soc.* (2) **15**, 1–25; *Collected Papers*, Vol. 2. Cambridge: Cambridge University Press, pp. 268–292.

Hardy, G. H. & Ramanujan, S. (1917). The normal order of prime factors of a number n, *Quart. J. Math.* **48**, 76–92; *Collected Papers*, Vol. II. Oxford: Oxford University Press, 100–113.

Hartman, P. & Wintner, A. (1947). On Möbius' inversion, *Amer. J. Math.* **69**, 853–858.

Hildebrand, A. (1984). Quantitative mean value theorems for non-negative multiplicative functions I, *J. London Math. Soc.* (2) **30**, 394–406.

(1986). On Wirsing's mean value theorem for multiplicative functions, *Bull. London Math. Soc.* **18**, 147–152.

(1987). Quantitative mean value theorems for non-negative multiplicative functions II, *Acta Arith.* **48**, 209–260.

Hille, E. (1937). The inversion problem of Möbius, *Duke Math. J.* **3**, 549–568.

Huxley, M. N. (1993). Exponential sums and lattice points II. *Proc. London Math. Soc.* (3) **66**, 279–301.

Iwaniec, H. & Mozzochi, C. J. (1988). On the divisor and circle problems, *J. Number Theory* **29**, 60–93.

Jarník, V. (1926). Über die Gitterpunkte auf konvexen Curven, *Math. Z.* **24**, 500–518.

Kac, M. (1959). Statistical Independence in Probability, Analysis and Number Theory, Carus Monograph 12. Washington: Math. Assoc. Amer.

Kátai, I. (1966). A remark on H. Delange's paper "Sur un théorème de Rényi", *Magyar Tud. Akad. Mat. Fiz. Oszt. Közl.* **16**, 269–273.

Kolesnik, G. (1969). The improvement of the error term in the divisor problem, *Mat. Zametki* **6**, 545–554.

(1973). On the estimation of the error term in the divisor problem, *Acta Arith.* **25**, 7–30.

(1982). On the order of $\zeta(\frac{1}{2} + it)$ and $\Delta(R)$, *Pacific J. Math.* **82**, 107–122.

(1985). On the method of exponent pairs, *Acta Arith.* **45**, 115–143.

Kubilius, J. (1956). Probabilistic methods in the theory of numbers (in Russian), *Uspehi Mat. Nauk* **11**, 31–66; *Amer. Math. Soc. Transl.* (2) **19** (1962), 47–85.

(1964). *Probabilistic Methods in the Theory of Numbers*, Translations of Mathematical Monographs, Vol. 11. Providence: American Mathematical Society.

Landau, E. (1900). Ueber die zahlentheoretische Function $\varphi(n)$ und ihre Beziehung zum Goldbachschen Satz, *Nachr. Akad. Wiss. Göttingen*, 177–186; *Collected Works*, Vol. 1. Essen: Thales Verlag, 1985, pp. 106–115.

(1903). Über den Verlauf der zahlentheoretischen Funktion $\varphi(x)$, *Arch. Math. Phys.* (3) **5**, 86–91; *Collected Works*, Vol. 1. Essen: Thales Verlag, 1985, pp. 378–383.

(1911). Sur les valeurs moyennes de certaines fonctions arithmétiques, *Bull. Acad. Royale Belgique*, 443–472; *Collected Works*, Vol. 4. Essen: Thales Verlag, 1986, pp. 377–406.

(1936). On a Titchmarsh–Estermann sum, *J. London Math. Soc.* **11**, 242–245; *Collected Works*, Vol. 9. Essen: Thales Verlag, 1987, pp. 393–396.

Linfoot, E. H. & Evelyn, C. J. A. (1929). On a problem in the additive theory of numbers, I, *J. Reine Angew. Math.* **164**, 131–140.

Mąkowski, A. (1960). Partitions into unequal primes, *Bull. Acad. Pol. Sci.* **8**, 125–126.

Massias, J.-P. & Robin, G. (1996). Bornes effectives pour certaines fonctions concernant les nombres premiers, *J. Théor. Nombres Bordeaux* **8**, 215–242.

Mertens, F. (1874a). Ueber einige asymptotische Gesetze der Zahlentheorie, *J. Reine Angew. Math.* **77**, 289–338.

(1874b). Ein Beitrag zur analytischen Zahlentheorie, *J. Reine Angew. Math.* **78**, 46–62.

Montgomery, H. L. (1987). Fluctuations in the mean of Euler's phi function, *Proc. Indian Acad. Sci. (Math. Sci.)* **97**, 239–245.

(1994). *Ten Lectures on the Interface of Analytic Number Theory and Harmonic Analysis*, CBMS 84. Providence: Amer. Math. Soc.

Narkiewicz, W. (2000). *The Development of Prime Number Theory*. Berlin: Springer-Verlag.

Nicolas, J.-L. (1988). On Highly Composite Numbers. *Ramanujan Revisited* (G. E. Andrews, R. A. Askey, B. C. Berndt, K. G. Ramanathan, R. A. Rankin, eds.). New York: Academic Press, pp. 215–244.

Niven, I. Zuckerman, H. S. & Montgomery, H. L. (1991). *An Introduction to the Theory of Numbers*, Fifth edition. New York: Wiley & Sons.

Nowak, W. G. (1989). On an error term involving the totient function, *Indian J. Pure Appl. Math.* **20**, 537–542.

Orr, R. C. (1969). On the Schnirelmann density of the sequence of k-free integers, *J. London Math. Soc.* **44**, 313–319.

Pillai, S. S. & Chowla, S. D. (1930). On the error term in some formulae in the theory of numbers (I), *J. London Math. Soc.* **5**, 95–101.

Pomerance, C. (1977). On composite n for which $\varphi(n)|(n-1)$, II, *Pacific J. Math.* **69**, 177–186.

Pritsker, I. E. (1999). Chebyshev Polynomials with Integer Coefficients, in *Analytic and Geometric Inequalities and Applications*, Math. Appl. 478. Dordrecht: Kluwer, pp. 335–348.

Ramanujan, S. (1915a). On the number of divisors of a number, *J. Indian Math. Soc.* **7**, 131–133; *Collected Papers*, Cambridge: Cambridge University Press, 1927, pp. 44–46.

(1915b). Highly composite numbers, *Proc. London Math. Soc.* (2) **14**, 347–409; *Collected Papers*, Cambridge: Cambridge University Press, 1927, pp. 78–128.

Rényi, A. (1955). On the density of certain sequences of integers, *Acad. Serbe Sci. Publ. Inst. Math.* **8**, 157–162.

Richert, H.-E. (1949a). Über Zerfällungen in ungleiche Primzahlen, *Math. Z.* **52**, 342–343.

(1949b). Über Zerlegungen in paarweise verschiedene Zahlen, *Norsk Mat. Tidsskr.* **31**, 120–122.

(1953). Verschärfung der Abschätzung beim Dirichletschen Teilerproblem, *Math. Z.* **58**, 204–218.

Robinson, R. L. (1966). An estimate for the enumerative functions of certain sets of integers, *Proc. Amer. Math. Soc.* **17**, 232–237; *Errata*, 1474.

Rogers, K. (1964). The Schnirelmann density of the square-free integers, *Proc. Amer. Math. Soc.* **15**, 515–516.

Rosser, J. B. & Schoenfeld, L. (1962). Approximate formulas for some functions of prime numbers, *Illinois J. Math.* **6**, 64–94.

(1975). Sharper bounds for the Chebyshev functions $\theta(x)$ and $\psi(x)$, *Math. Comp.* **29**, 243–269.

Runge, C. (1885). Über die auflösbaren Gleichungen von der Form $x^5 + ux + v = 0$, *Acta Math.* **7**, 173–186.

Saffari, B. (1970). Sur quelques applications de la "méthode de l'hyperbole" de Dirichlet à la théorie des nombres premiers, *Enseignement Math.* (2) **14**, 205–224.

Schmidt, P. G. (1967/68). Zur Anzahl Abelscher Gruppen gegebener Ordnung, II, *Acta Arith.* **13**, 405–417.

Schoenfeld, L. (1969). An improved estimate for the summatory function of the Möbius function, *Acta Arith.* **15**, 221–233.

(1976). Sharper bounds for the Chebyshev functions $\theta(x)$ and $\psi(x)$, II, *Math. Comp.* **30**, 337–360.

Schwarz, W. (1970). Eine Bemerkung zu einer asymptotischen Formel von Herrn Rényi, *Arch. Math.* (Basel) **21**, 157–166.

Shan, Z. (1985). On composite n for which $\varphi(n)|(n-1)$, *J. China Univ. Sci. Tech.* **15**, 109–112.

Sitaramachandrarao, R. (1982). On an error term of Landau, *Indian J. Pure Appl. Math.* **13**, 882–885.

——— (1985). On an error term of Landau, II, *Rocky Mountain J. Math.* **15**, 579–588.

Soundararajan, K. (2003). Omega results for the divisor and circle problems, *Int. Math. Res. Not.*, 1987–1998.

Stieltjes, T. J. (1887). Note sur la multiplication de deux séries, *Nouvelles Annales* (3) **6**, 210–215.

Sylvester, J. J. (1881). On Tchebycheff's theory of the totality of the prime numbers comprised within given limits, *Amer. J. Math.* **4**, 230–247.

Tenenbaum, G. (1995). *Introduction to Analytic and Probabilistic Number Theory*, Cambridge Studies 46, Cambridge: Cambridge University Press.

Turán, P. (1934). On a theorem of Hardy and Ramanujan, *J. London Math. Soc.* **9**, 274–276.

de la Vallée Poussin, C. J. (1898). Sur les valeurs moyennes de certaines fonctions arithmétiques, *Ann. Soc. Sci. Bruxelles* **22**, 84–90.

Voronoï, G. (1903). Sur un problème du calcul des fonctions asymptotiques, *J. Reine Angew. Math.* **126**, 241–282.

Walfisz, A. (1963). Weylsche Exponentialsummen in der neueren Zahlentheorie, *Mathematische Forschungsberichte* 15, Berlin: VEB Deutscher Verlag Wiss.

Ward, D. R. (1927). Some series involving Euler's function, *J. London Math. Soc.* **2**, 210–214.

Wigert, S. (1906/7). Sur l'ordre de grandeur du nombre des diviseurs d'un entier, *Ark. Mat.* **3**, 1–9.

Wilson, B. M. (1922). Proofs of some formulæ enunciated by Ramanujan, *Proc. London Math. Soc.* **21**, 235–255.

Wintner, A. (1944). The Theory of Measure in Arithmetic Semigroups. Baltimore: Waverly Press.

Wirsing, E. (1961). Das asymptotische Verhalten von Summen über multiplikative Funktionen, *Math. Ann.* **143**, 75–102.

——— (1967). Das asymptotische Verhalten von Summen über multiplikative Funktionen, II, *Acta Math. Acad. Sci. Hungar.* **18**, 411–467.

3

Principles and first examples of sieve methods

3.1 Initiation

The aim of sieve theory is to construct estimates for the number of integers remaining in a set after members of certain arithmetic progressions have been discarded. If P is given, then the asymptotic density of the set of integers relatively prime to P is $\varphi(P)/P$; with the aid of sieves we can estimate how quickly this asymptotic behaviour is approached. Throughout this chapter we let $S(x, y; P)$ denote the numbers of integers n in the interval $x < n \leq x + y$ for which $(n, P) = 1$. A first (weak) result is provided by

Theorem 3.1 (Eratosthenes–Legendre) *For any real x, and any $y \geq 0$,*

$$S(x, y; P) = \frac{\varphi(P)}{P}y + O\left(2^{\omega(P)}\right).$$

Of course if y is an integral multiple of P then the above holds with no error term. Since $2^{\omega(P)} \leq d(P) \ll P^\varepsilon$, the main term above is larger than the error term if $y \geq P^\varepsilon$; thus the reduced residues are roughly uniformly distributed in the interval $(0, P]$.

Proof From the characteristic property (1.20) of the Möbius μ-function, and the fact that $d|(n, P)$ if and only if $d|n$ and $d|P$, we see that

$$S(x, y; P) = \sum_{\substack{x < n \leq x+y}} \sum_{\substack{d|n \\ d|P}} \mu(d)$$

$$= \sum_{d|P} \mu(d) \sum_{\substack{x < n \leq x+y \\ d|n}} 1$$

$$= \sum_{d|P} \mu(d) \left(\left[\frac{x+y}{d}\right] - \left[\frac{x}{d}\right]\right). \qquad (3.1)$$

Removing the square brackets, we see that this is

$$= y \sum_{d|P} \frac{\mu(d)}{d} + O\left(\sum_{d|P} |\mu(d)|\right),$$

which is the desired result. $\qquad\square$

The identity (3.1) can be considered to be an instance of Sylvester's principle of inclusion–exclusion, which in general asserts that if S is a finite set and $S_1, \ldots, S_R$ are subsets of S, then

$$\text{card}\left(S \setminus \bigcup_{r=1}^{R} S_r\right) = \text{card}(S) - \Sigma_1 + \Sigma_2 - \cdots + (-1)^R \Sigma_R \qquad (3.2)$$

where

$$\Sigma_s = \sum_{1 \le r_1 < \cdots < r_s \le R} \text{card}\left(\bigcap_{j=1}^{s} S_{r_j}\right).$$

To obtain (3.1) we take $S = \{n \in \mathbb{Z} : x < n \le x + y\}$, $R = \omega(P)$, we let $p_1, \ldots, p_R$ be the distinct primes dividing P, and we put $S_r = \{n : x < n \le x + y, p_r | n\}$. Here we see that the Möbius μ-function has an important combinatorial significance, namely that it enables us to present the inclusion–exclusion identity in a compact manner, in arithmetic situations such as (3.1) above.

To prove (3.2) it suffices to note that if an element of S is not in any of the S_r, then it is counted once on the right-hand side, while if it is in precisely $t > 0$ of the sets S_r then it is counted $\binom{t}{s}$ times in Σ_s, and hence it contributes altogether

$$\sum_{s=0}^{P} (-1)^s \binom{t}{s} = \sum_{s=0}^{t} (-1)^s \binom{t}{s} = (1-1)^t = 0.$$

If p is a prime, then either $p|P$ or $(p, P) = 1$. Hence

$$\pi(x+y) - \pi(x) \le \omega(P) + S(x, y; P), \qquad (3.3)$$

so that a bound for $S(x, y; P)$ can be used to bound the number of prime numbers in an interval. In view of the main term in Theorem 3.1, it is reasonable to expect that it will be best to take P of the form

$$P = \prod_{p \le z} p. \qquad (3.4)$$

On taking $z = \log y$, we see immediately that

$$\pi(x+y) - \pi(x) \le \left(e^{-C_0} + \varepsilon(y)\right) \frac{y}{\log \log y}$$

where $\varepsilon(y) \to 0$ as $y \to \infty$. This bound is very weak, but has the interesting property of being uniform in x. Since the bound for the error term in Theorem 3.1 is very crude, we might expect that more is true, so that perhaps

$$S(x, y; P) \sim \frac{\varphi(P)}{P} y$$

even when z is fairly large. However, as we have already noted in our remarks following Theorem 2.11, this asymptotic formula fails when $z = y^{1/2}$.

In order to derive a sharper estimate for $S(x, y; P)$, we replace $\mu(d)$ by a more general arithmetic function λ_d that in some sense is a truncated approximation to $\mu(d)$. This is reminiscent of our derivation of the Chebyshev bounds, but in fact the specific properties required of the λ_d are now rather different. Suppose that we seek an upper bound for $S(x, y; P)$. Let λ_n^+ be a function such that

$$\sum_{d|n} \lambda_d^+ \geq \begin{cases} 1 & \text{if } n = 1, \\ 0 & \text{otherwise.} \end{cases} \tag{3.5}$$

Such a λ_d^+ we call an 'upper bound sifting function', and by arguing as in the proof of Theorem 3.1 we see that

$$S(x, y; P) \leq \sum_{x < n \leq x+y} \sum_{\substack{d|n \\ d|P}} \lambda_d^+ = y \sum_{d|P} \lambda_d^+/d + O\left(\sum_{d|P} |\lambda_d^+|\right). \tag{3.6}$$

This will be useful if $\sum_{d|P} \lambda_d^+/d$ is not much larger than $\varphi(P)/P$, and if $\sum_{d|P} |\lambda_d^+|$ is much smaller than $2^{\omega(P)}$. Brun (1915) was the first to succeed with an argument of this kind. He took his λ_n^+ to be of the form

$$\lambda_n^+ = \begin{cases} \mu(n) & \text{if } n \in \mathcal{D}^+, \\ 0 & \text{otherwise,} \end{cases}$$

where $\mathcal{D}^+$ is a judiciously chosen set of integers. A sieve of this kind is called 'combinatorial'. With Brun's choice of $\mathcal{D}^+$ it is easy to verify (3.5), and it is not hard to bound $\sum_{d|P} |\lambda_d^+|$, but the determination of the asymptotic size of the main term $\sum_{d|P} \lambda_d^+/d$ presents some technical difficulties. We do not develop a detailed account of Brun's method, but the spirit of the approach can be appreciated by considering the following simple choice of $\mathcal{D}^+$: Let r be an integer at our disposal, and put

$$\mathcal{D}^+ = \{n : \omega(n) \leq 2r\}.$$

We observe that

$$\sum_{d|P} \lambda_d^+ = \sum_{j=0}^{2r} \sum_{\substack{d|P \\ \omega(d)=j}} \mu(d) = \sum_{j=0}^{2r} (-1)^j \binom{\omega(P)}{j}.$$

Then (3.5) follows on taking $J = 2r$, $h = \omega(P)$ in the binomial coefficient identity

$$\sum_{j=0}^{J}(-1)^j \binom{h}{j} = (-1)^J \binom{h-1}{J}.$$

This identity can in turn be proved by induction, or by equating coefficients in the power series identity

$$\left(\sum_{i=0}^{\infty} x^i\right)\left(\sum_{j=0}^{h}(-1)^j \binom{h}{j} x^j\right) = (1-x)^{h-1} = \sum_{J=0}^{h-1}(-1)^J \binom{h-1}{J} x^J.$$

Lower bounds for $S(x, y; P)$ can be derived in a parallel manner, by introducing a lower bound sifting function λ_n^-. That is, λ_n^- is an arithmetic function such that

$$\sum_{d|n} \lambda_d^- \le \begin{cases} 1 & \text{if } n = 1, \\ 0 & \text{otherwise.} \end{cases} \tag{3.7}$$

Corresponding to the upper bound (3.6) we have

$$S(x, y; P) \ge y \sum_{d|P} \lambda_d^-/d - O\left(\sum_{d|P} |\lambda_d^-|\right). \tag{3.8}$$

Unfortunately, this lower bound may be negative, in which case it is useless, since trivially $S(x, y; P) \ge 0$. Brun determined λ_d^- combinatorially by constructing a set $\mathcal{D}^-$ similar to his $\mathcal{D}^+$. Indeed, an admissible set can be obtained by taking

$$\mathcal{D}^- = \{n : \omega(n) \le 2r - 1\}.$$

By Brun's method it can be shown that

$$\pi(x + y) - \pi(x) \ll \frac{y}{\log y}. \tag{3.9}$$

When $x = 0$ this is merely a weak form of the Chebyshev upper bound. The main utility of the above is that it holds uniformly in x. We shall establish a refined form of (3.9) in the next section (cf. Corollary 3.4).

3.1.1 Exercises

1. (Charles Dodgson) In a very hotly fought battle, at least 70% of the combatants lost an eye, at least 75% an ear, at least 80% an arm, and at least 85% a leg. What can you say about the percentage that lost all four members?

2. (P. T. Bateman) Would you believe a market investigator who reports that of 1000 people, 816 like candy, 723 like ice cream, 645 like cake, while 562 like both candy and ice cream, 463 like both candy and cake, 470 like both ice cream and cake, while 310 like all three?

3. (Erdős 1946) For $x > 0$ write

$$\sum_{\substack{1 \leq n \leq x \\ (n,k)=1}} 1 = \frac{\varphi(k)}{k} x + E_k(x).$$

(a) Show that if $k > 1$, then

$$E_k(x) = -\sum_{d|k} \mu(d) B_1(\{x/d\})$$

where $B_1(z) = z - 1/2$ is the first Bernoulli polynomial. Let $E_k(x)$ be defined by this formula when $x < 0$.

(b) Show that if $k > 1$, then $E_k(x)$ is periodic with period k, that $E_k(x)$ is an odd function (apart from values at discontinuities), and that

$$\int_0^k E_k(x)\, dx = 0.$$

(c) By using the result of Exercise B.10, or otherwise, show that if $d|k$ and $e|k$, then

$$\int_0^k B_1(\{x/d\}) B_1(\{x/e\})\, dx = \frac{(d,e)^2}{12de} k.$$

(d) Show that if $k > 1$, then

$$\int_0^k E_k(x)^2\, dx = \frac{1}{12} 2^{\omega(k)} \varphi(k).$$

(e) Deduce that if $k > 1$, then

$$\max_x |E_k(x)| \gg 2^{\omega(k)/2} \left(\frac{\varphi(k)}{k} \right)^{1/2}.$$

4. (Lehmer 1955; cf. Vijayaraghavan 1951) Let $E_k(x)$ be defined as above.

(a) Show that $|E_k(x)| \leq 2^{\omega(k)-1}$ for all $k > 1$.

(b) Suppose that k is composed of distinct primes $p \equiv 3 \pmod 4$, and that $\omega(k)$ is even. Show that if $d|k$, then $\mu(d) B_1(\{k/(4d)\}) = -1/4$.

(c) Show that there exist infinitely many numbers k for which

$$\max_x |E_k(x)| \geq 2^{\omega(k)-2}.$$

5. (Behrend 1948; cf. Heilbronn 1937, Rohrbach 1937, Chung 1941, van der
   Corput 1958) Let $a_1, \ldots, a_J$ be positive integers, and let $T(a_1, \ldots, a_J)$ denote the asymptotic density of the set of those positive integers that are not
   divisible by any of the a_i.

   (a) Show that $T(a_1, \ldots, a_J) = \sum_{j=0}^{J}(-1)^j \Sigma_j$ where

   $$\Sigma_j = \sum_{1 \leq i_1 < \cdots < i_j \leq J} \frac{1}{[a_{i_1}, \ldots, a_{i_j}]}.$$

   (b) Show that if $a_1, \ldots, a_J$ are pairwise relatively prime, then

   $$T(a_1, \ldots, a_J) = \prod_{j=1}^{J}\left(1 - \frac{1}{a_j}\right).$$

   (c) Show if $(d, v_s) = 1$ for $1 \leq s \leq S$, then

   $$T(du_1, \ldots, du_R, v_1, \ldots, v_S) = \frac{1}{d}T(u_1, \ldots, u_R, v_1, \ldots, v_S)$$
   $$+ \left(1 - \frac{1}{d}\right)T(v_1, \ldots, v_S).$$

   (d) Suppose that $d|a_j$ for $1 \leq j \leq j_0$, that $(d, a_j) = 1$ for $j > j_0$, that $d|b_k$
   for $1 \leq k \leq k_0$, and that $(d, b_k) = 1$ for $k_0 < k \leq K$. Put $a'_j = a_j/d$ for
   $1 \leq j \leq j_0$, and $b'_k = b_k/d$ for $1 \leq k \leq k_0$. Explain why

   $$T(a_1, \ldots, a_J)T(b_1, \ldots, b_K)$$
   $$= \frac{1}{d}T(a'_1, \ldots, a'_{j_0}, a_{j_0+1}, \ldots, a_J)T(b'_1, \ldots, b'_{k_0}, b_{k_0+1}, \ldots, b_K)$$
   $$+ \left(1 - \frac{1}{d}\right)T(a_{j_0+1}, \ldots, a_J)T(b_{k_0+1}, \ldots, b_K)$$
   $$- \frac{1}{d}\left(1 - \frac{1}{d}\right)(T(a_{j_0+1}, \ldots, a_J) - T(a'_1, \ldots, a'_{j_0}, a_{j_0+1}, \ldots, a_J))$$
   $$\cdot \left(T(b_{k_0+1}, \ldots, b_K) - T(b'_1, \ldots, b'_{k_0}, b_{k_0+1}, \ldots, b_K)\right).$$

   (e) Explain why the factors that constitute the last term above are all non-
   negative.

   (f) Show that

   $$T(a_1, \ldots, a_J, b_1, \ldots, b_K) \geq T(a_1, \ldots, a_J)T(b_1, \ldots, b_K).$$

   (g) Show that

   $$T(a_1, \ldots, a_J) \geq \prod_{j=1}^{J}\left(1 - \frac{1}{a_j}\right).$$

3.2   The Selberg lambda-squared method

Let Λ_n be a real-valued arithmetic function such that $\Lambda_1 = 1$. Then

$$\left(\sum_{d|n} \Lambda_d\right)^2 \geq \begin{cases} 1 & \text{if } n = 1, \\ 0 & \text{if } n > 1. \end{cases}$$

This simple observation can be used to obtain an upper bound for $S(x, y; P)$; namely

$$
\begin{aligned}
S(x, y; P) &\leq \sum_{\substack{x < n \leq x+y}} \left(\sum_{\substack{d|n \\ d|P}} \Lambda_d\right)^2 \\
&= \sum_{\substack{d|P \\ e|P}} \Lambda_d \Lambda_e \sum_{\substack{x < n \leq x+y \\ d|n, e|n}} 1 \\
&= \sum_{\substack{d|P \\ e|P}} \Lambda_d \Lambda_e \left(\left[\frac{x+y}{[d,e]}\right] - \left[\frac{x}{[d,e]}\right]\right) \\
&= y \sum_{\substack{d|P \\ e|P}} \frac{\Lambda_d \Lambda_e}{[d,e]} + O\left(\left(\sum_{d|P} |\Lambda_d|\right)^2\right).
\end{aligned}
\tag{3.10}
$$

In the general framework of the preceding section this amounts to taking

$$\lambda_n^+ = \sum_{\substack{d,e \\ [d,e]=n}} \Lambda_d \Lambda_e,$$

since it then follows that

$$\sum_{d|n} \lambda_d^+ = \left(\sum_{d|n} \Lambda_d\right)^2.$$

We now suppose that $\Lambda_n = 0$ for $n > z$ where z is a parameter at our disposal, in the hope that this will restrict the size of the error term. As for the main term, we see that we wish to minimize a quadratic form subject to the constraint $\Lambda_1 = 1$. In fact we can diagonalize this quadratic form and determine the optimal Λ_n exactly; this permits us to prove

Theorem 3.2   *Let x, y, and z be real numbers such that $y > 0$ and $z \geq 1$. For any positive integer P we have*

$$S(x, y; P) \leq \frac{y}{L_P(z)} + O(z^2 L_P(z)^{-2})$$

where

$$L_P(z) = \sum_{\substack{n \le z \\ n|P}} \frac{\mu(n)^2}{\varphi(n)}.$$

Proof Clearly we may assume that P is square-free. Since $[d, e](d, e) = de$ and $\sum_{d|n} \varphi(d) = n$, we see that

$$\frac{1}{[d, e]} = \frac{(d, e)}{de} = \frac{1}{de} \sum_{f|d, f|e} \varphi(f).$$

Hence

$$\sum_{d|P, e|P} \frac{\Lambda_d \Lambda_e}{[d, e]} = \sum_{f|P} \varphi(f) \sum_{\substack{d \\ f|d|P}} \frac{\Lambda_d}{d} \sum_{\substack{e \\ f|e|P}} \frac{\Lambda_e}{e}$$

$$= \sum_{f|P} \varphi(f) y_f^2$$

where

$$y_f = \sum_{\substack{d \\ f|d|P}} \frac{\Lambda_d}{d}. \tag{3.11}$$

This linear change of variables, from Λ_d to y_f, is non-singular. That is, if the y_f are given then there exist unique Λ_d such that the above holds. Indeed, by a form of the Möbius inversion formula (cf. Exercise 2.1.6) the above is equivalent to the relation

$$\Lambda_d = d \sum_{\substack{f \\ d|f|P}} y_f \mu(f/d). \tag{3.12}$$

Moreover, from these formulæ we see that $\Lambda_d = 0$ for all $d > z$ if and only if $y_f = 0$ for all $f > z$. Thus we have diagonalized the quadratic form in (3.10), and by (3.12) we see that the constraint $\Lambda_1 = 1$ is equivalent to the linear condition

$$\sum_{f|P} y_f \mu(f) = 1. \tag{3.13}$$

We determine the value of the constrained minimum by completing squares. If the y_f satisfy (3.13), then

$$\sum_{f|P} \varphi(f) y_f^2 = \sum_{\substack{f|P \\ f \le z}} \varphi(f) \left(y_f - \frac{\mu(f)}{\varphi(f) L_P(z)} \right)^2 + \frac{1}{L_P(z)}. \tag{3.14}$$

Here the right-hand side is minimized by taking

$$y_f = \frac{\mu(f)}{\varphi(f)L_P(z)} \tag{3.15}$$

for $f \le z$, and we note that these y_f satisfy (3.13). Hence the minimum of the quadratic form in (3.10), subject to $\Lambda_1 = 1$, is precisely $1/L_P(z)$; this gives the main term.

We now treat the error term. Since P is square-free, from (3.12) and (3.15) we see that

$$\Lambda_d = \frac{d}{L_P(z)} \sum_{\substack{f \\ d|f|P \\ f \le z}} \frac{\mu(f)\mu(f/d)}{\varphi(f)} = \frac{d\mu(d)}{L_P(z)\varphi(d)} \sum_{\substack{m|P \\ (m,d)=1 \\ m \le z/d}} \frac{\mu(m)^2}{\varphi(m)}; \tag{3.16}$$

here we have put $m = f/d$. Thus

$$\sum_{d \le z} |\Lambda_d| \le \frac{1}{L_P(z)} \sum_{d \le z} \frac{d}{\varphi(d)} \sum_{m \le z/d} \frac{1}{\varphi(m)} = \frac{1}{L_P(z)} \sum_{m \le z} \frac{1}{\varphi(m)} \sum_{d \le z/m} \frac{d}{\varphi(d)}.$$

Since $d/\varphi(d) = \sum_{r|d} \mu^2(r)/\varphi(r)$, it follows by the method of Section 2.1 that

$$\sum_{d \le y} \frac{d}{\varphi(d)} = \sum_{r \le y} \frac{\mu^2(r)}{\varphi(r)}[y/r] \le y \sum_r \frac{\mu^2(r)}{r\varphi(r)} \ll y.$$

On inserting this in our former estimate, we find that

$$\sum_{d \le z} |\Lambda_d| \ll \frac{z}{L_P(z)} \sum_{m \le z} \frac{1}{m\varphi(m)} \ll \frac{z}{L_P(z)}. \tag{3.17}$$

This gives the stated error term, so the proof is complete. $\square$

In order to apply Theorem 3.2, we require a lower bound for the sum $L_P(z)$. To this end we show that

$$\sum_{n \le z} \frac{\mu(n)^2}{\varphi(n)} > \log z \tag{3.18}$$

for all $z \ge 1$. Let $s(n)$ denote the largest square-free number dividing n (sometimes called the 'square-free kernel of n'). Then for square-free n,

$$\frac{1}{\varphi(n)} = \frac{1}{n} \prod_{p|n} \left(1 + \frac{1}{p} + \frac{1}{p^2} + \cdots \right) = \sum_{\substack{m \\ s(m)=n}} \frac{1}{m},$$

so that the sum in (3.18) is

$$\sum_{\substack{m \\ s(m) \le z}} \frac{1}{m}.$$

Since $s(m) \leq m$, this latter sum is

$$\geq \sum_{m \leq z} \frac{1}{m} > \log z.$$

Here the last inequality is obtained by the integral test. With more work one can derive an asymptotic formula for the the sum in (3.18) (recall Exercise 2.1.17).

By taking $z = y^{1/2}$ in Theorem 3.2, and appealing to (3.18), we obtain

Theorem 3.3 *Let* $P = \prod_{p \leq \sqrt{y}} p$. *Then for any* x *and any* $y \geq 2$,

$$S(x, y; P) \leq \frac{2y}{\log y} \left(1 + O\left(\frac{1}{\log y} \right) \right).$$

By combining the above with (3.3) we obtain an immediate application to the distribution of prime numbers.

Corollary 3.4 *For any* $x \geq 0$ *and any* $y \geq 2$,

$$\pi(x + y) - \pi(x) \leq \frac{2y}{\log y} \left(1 + O\left(\frac{1}{\log y} \right) \right).$$

In Theorem 3.3 we consider only a very special sort of P, but the following lemma enables us to obtain corresponding results for more general P.

Lemma 3.5 *Put* $M(y; P) = \max_x S(x, y; P)$. *If* $(P, q) = 1$, *then*

$$M(y; P) \leq \frac{q}{\varphi(q)} M(y; qP).$$

Proof It suffices to show that

$$\varphi(q) S(x, y; P) = \sum_{m=1}^{q} S(x + Pm, y; qP), \qquad (3.19)$$

since the right-hand side is bounded above by $q M(y; qP)$. Suppose that $x + Pm < n \leq x + Pm + y$ and that $(n, qP) = 1$. Put $r = n - Pm$. Then $x < r \leq x + y, (r, P) = 1$, and $(r + Pm, q) = 1$. Thus the right-hand side above is

$$\sum_{m} \sum_{r} 1 = \sum_{\substack{x < r \leq x+y \\ (r,P)=1}} \sum_{\substack{1 \leq m \leq q \\ (r+Pm,q)=1}} 1.$$

Since $(P, q) = 1$, the map $m \mapsto r + Pm$ permutes the residue classes (mod q). Hence the inner sum above is $\varphi(q)$, and we have (3.19). $\qquad\square$

Theorem 3.6 *For any real* x *and any* $y \geq 2$,

$$S(x, y; P) \leq e^{C_0} y \left(\prod_{\substack{p|P \\ p \leq \sqrt{y}}} \left(1 - \frac{1}{p} \right) \right) \left(1 + O\left(\frac{1}{\log y} \right) \right).$$

Proof   Let

$$P_1 = \prod_{\substack{p|P \\ p \le \sqrt{y}}} p, \quad q_1 = \prod_{\substack{p \nmid P \\ p \le \sqrt{y}}} p.$$

Theorem 3.3 provides an upper bound for $M(y; q_1 P_1)$, and hence by Lemma 3.5 we have an upper bound for $M(y; P_1)$. To complete the argument it suffices to note that $S(x, y; P) \le S(x, y; P_1) \le M(y; P_1)$, and to appeal to Mertens' formula (Theorem 2.7(e)).   □

We note that Theorem 3.3 is a special case of Theorem 3.6. Although we have taken great care to derive uniform estimates, for many purposes it is enough to know that

$$S(x, y; P) \ll y \prod_{\substack{p|P \\ p \le y}} \left(1 - \frac{1}{p}\right). \tag{3.20}$$

This follows from Theorem 3.6 since $\prod_{\sqrt{y} < p \le y}(1 - 1/p)^{-1} \ll 1$ by Mertens' formula. To obtain an estimate in the opposite direction, write $P = P_1 q_1$ where P_1 is composed entirely of primes $> y$, and q_1 is composed entirely of primes $\le y$. Since the integers in the interval $(0, y]$ have no prime factor $> y$, we see that $M(y; P_1) \ge [y]$. Hence by Lemma 3.5,

$$M(y; P) \ge [y] \prod_{\substack{p|P \\ p \le y}} \left(1 - \frac{1}{p}\right). \tag{3.21}$$

Thus the bound (3.20) is of the correct order of magnitude.

The advantage of Theorem 3.6 lies in its uniformity. On the other hand, the use of Lemma 3.5 is wasteful if the P in Theorem 3.6 is much smaller than in Theorem 3.3. For example, if $P = \prod_{p \le y^{1/4}} p$, then by Theorem 3.6 we find that

$$S(x, y; P) \le \frac{cy}{\log y} \left(1 + O\left(\frac{1}{\log y}\right)\right)$$

with $c = 4$, whereas by Theorem 3.2 with $z = y^{1/2}$ we obtain the above with the better constant

$$c = \frac{4}{3 - 2\log 2} = 2.4787668\ldots.$$

To see this, we note that

$$L_P(z) = \sum_{n \le z} \frac{\mu(n)^2}{\varphi(n)} - \sum_{z^{1/2} < p \le z} \frac{1}{p - 1} \sum_{n \le z/p} \frac{\mu(n)^2}{\varphi(n)}. \tag{3.22}$$

Then by Exercise 2.1.17 and Mertens' estimates (Theorem 2.7) it follows that this is $\frac{1}{4}(3 - 2 \log 2) \log y + O(1)$.

3.2.1   Exercises

1. Let Λ_d be defined as in the proof of Theorem 3.2.
   (a) Show that

   $$\Lambda_d \ll \frac{d}{L_P(z)\varphi(d)} \log \frac{2z}{d}$$

   for $d \leq z$.
   (b) Use the above to give a second proof of (3.17).
2. Show that for $y \geq 2$ the number of prime powers p^k in the interval $(x, x + y]$ is

   $$\leq \frac{2y}{\log y} \left(1 + O\left(\frac{1}{\log y} \right) \right).$$

3. (Chowla 1932) Let $f(n)$ be an arithmetic function, put

   $$g(n) = \sum_{[d,e]=n} f(d)\overline{f(e)},$$

   and let σ_c denote the abscissa of convergence of the Dirichlet series $\sum g(n)n^{-s}$.
   (a) Show that if $\sigma > \max(1, \sigma_c)$, then

   $$\zeta(s) \sum_{d,e} \frac{f(d)\overline{f(e)}}{[d, e]^s} = \sum_{n=1}^{\infty} \left| \sum_{d|n} f(d) \right|^2 n^{-s}.$$

   (b) Show that

   $$\sum_{d,e} \frac{\mu(d)\mu(e)}{[d, e]^2} = \frac{6}{\pi^2}.$$

   (c) Show that

   $$\sum_{\substack{d,e \\ [d,e]=n}} \mu(d)\mu(e) = \mu(n)$$

   for all positive integers n.
4. Let $f(n)$ be an arithmetic function such that $f(1) = 1$. Show that f is multiplicative if and only if $f(m)f(n) = f((m, n))f([m, n])$ for all pairs of positive integers m, n.

5. (Hensley 1978)

   (a) Let $P = \prod_{p \leq \sqrt{y}} p$. Show that the number of n, $x < n \leq x + y$, such that $\Omega(n) = 2$, is

$$\leq S(x, y; P) + \sum_{p \leq \sqrt{y}} \left(\pi \left(\frac{x+y}{p} \right) - \pi \left(\frac{x}{p} \right) \right).$$

   (b) By using Theorem 3.3 and Corollary 3.4, show that for $y \geq 2$,

$$\sum_{\substack{x < n \leq x+y \\ \Omega(n)=2}} 1 \leq \frac{2y \log \log y}{\log y} \left(1 + O \left(\frac{1}{\log \log y} \right) \right).$$

6. (H.-E. Richert, unpublished)

   (a) Show that

$$\sum_{x < n \leq x+y} \left(\sum_{d^2 | n} \Lambda_d \right)^2 = y \sum_{d, e} \frac{\Lambda_d \Lambda_e}{[d, e]^2} + O \left(\left(\sum_d |\Lambda_d| \right)^2 \right).$$

   (b) Let $f(n) = n^2 \prod_{p | n} (1 - p^{-2})$. Show that $\sum_{d | n} f(d) = n^2$.

   (c) For $1 \leq d \leq z$ let Λ_d be real numbers such that $\Lambda_1 = 1$. Show that the minimum of $\sum_{d, e} \Lambda_d \Lambda_e / [d, e]^2$ is $1/L$ where $L = \sum_{n \leq z} \mu(n)^2 / f(n)$. Show also that $\Lambda_d \ll 1$ for the extremal Λ_d.

   (d) Show that $\zeta(2) - 1/z \leq L \leq \zeta(2)$.

   (e) Let $Q(x)$ denote the number of square-free numbers not exceeding x. Show that for $x \geq 0$, $y \geq 1$,

$$Q(x + y) - Q(x) \leq \frac{y}{\zeta(2)} + O\left(y^{2/3}\right).$$

7. Let $m(y; P) = \min_x S(x, y; P)$. Show that if $(q, P) = 1$, then

$$m(y; P) \geq \frac{q}{\varphi(q)} m(y; qP).$$

8. (N. G. de Bruijn, unpublished; cf. van Lint & Richert 1964) Let $\mathcal{M}$ be an arbitrary set of natural numbers, and let $s(n)$ denote the largest square-free divisor of n. Show that

$$0 \leq \sum_{\substack{n \leq x \\ n \in \mathcal{M}}} \frac{\mu(n)^2}{\varphi(n)} - \sum_{\substack{n \leq x \\ s(n) \in \mathcal{M}}} \frac{1}{n} \leq \sum_{n \leq x} \frac{\mu(n)^2}{\varphi(n)} - \sum_{n \leq x} \frac{1}{n} \ll 1.$$

9. (van Lint & Richert 1965)

   (a) Show that

$$\sum_{n \leq z} \frac{\mu(n)^2}{\varphi(n)} \leq \left(\sum_{d | q} \frac{\mu(d)^2}{\varphi(d)} \right) \left(\sum_{\substack{m \leq z \\ (m, q)=1}} \frac{\mu(m)^2}{\varphi(m)} \right).$$

(b) Deduce that

$$\sum_{\substack{n \leq z \\ (n,q)=1}} \frac{\mu(n)^2}{\varphi(n)} \geq \frac{\varphi(q)}{q} \sum_{n \leq z} \frac{\mu(n)^2}{\varphi(n)}.$$

10. (Hooley 1972; Montgomery & Vaughan 1979)

    (a) Let λ_d^+ be an upper bound sifting function such that $\lambda_d^+ = 0$ for all $d > z$. Show that for any q,

    $$0 \leq \frac{\varphi(q)}{q} \sum_{\substack{d \\ (d,q)=1}} \frac{\lambda_d^+}{d} \leq \sum_d \frac{\lambda_d^+}{d}.$$

    (Hint: Multiply both sides by $P/\varphi(P) = \sum 1/m$ where m runs over all integers composed of the primes dividing P, and $P = \prod_{p \leq z} p$.)

    (b) Let Λ_d be real with $\Lambda_d = 0$ for $d > z$. Show that for any q,

    $$0 \leq \frac{\varphi(q)}{q} \sum_{\substack{d,e \\ (de,q)=1}} \frac{\Lambda_d \Lambda_e}{[d,e]} \leq \sum_{d,e} \frac{\Lambda_d \Lambda_e}{[d,e]}.$$

    (c) Let λ_d^- be a lower bound sifting function such that $\lambda_d^- = 0$ for $d > z$. Show that for any q,

    $$\frac{\varphi(q)}{q} \sum_{\substack{d \\ (d,q)=1}} \frac{\lambda_d^-}{d} \geq \sum_d \frac{\lambda_d^-}{d}.$$

3.3 Sifting an arithmetic progression

Thus far we have sifted only the zero residue class from a set of consecutive integers. We now widen the situation slightly.

Lemma 3.7 *Let P be a positive integer, and for each prime p dividing P suppose that one particular residue class a_p has been chosen. Let $S'(x, y; P)$ denote the number of integers m, $x < m \leq x + y$, such that for each $p|P$, $m \not\equiv a_p \pmod{p}$. Then*

$$\max_x S'(x, y; P) = \max_x S(x, y; P).$$

Since $S'(x, y; P)$ reduces to $S(x, y; P)$ when we take $a_p = 0$ for all $p|P$, we see that there is no loss of generality in sifting only the zero residue class, when the initial set of numbers consists of consecutive integers. Also, we note that the value of the maximum taken above is independent of the choice of the a_p.

Proof By the Chinese remainder theorem there is a number c such that $c \equiv a_p$ (mod p) for every $p|P$. Put $n = m - c$. Thus the inequality $x < m \le x + y$ is equivalent to $x - c < n \le x - c + y$, and the condition that $p|P$ implies $m \not\equiv a_p$ (mod p) is equivalent to $(n, P) = 1$. Hence $S'(x, y; P) = S(x - c, y; P)$, so that

$$\max_x S'(x, y; P) = \max_x S(x - c, y; P) = \max_x S(x, y; P),$$

and the proof is complete. $\qquad\square$

Theorem 3.8 *Suppose that $(a, q) = 1$, that $(P, q) = 1$, and that x and y are real numbers with $y \ge 2q$. The number of n, $x < n \le x + y$, such that $n \equiv a$ (mod q) and $(n, P) = 1$ is*

$$\le e^{C_0} \frac{y}{q} \left(\prod_{\substack{p|P \\ p \le \sqrt{y/q}}} \left(1 - \frac{1}{p}\right) \right) \left(1 + O\left(\frac{1}{\log y/q}\right)\right).$$

Proof Write $n = mq + a$, so that $x' < m \le x' + y'$ where $x' = (x - a)/q$ and $y' = y/q$. For each $p|P$ let a_p be the unique residue class (mod p) such that $a_p q + a \equiv 0$ (mod p). Thus $p|n$ if and only if $m \equiv a_p$ (mod p). Hence the number of n in question is $S'(x', y'; P)$, in the language of Lemma 3.7. The stated bound now follows from this lemma and Theorem 3.6. $\qquad\square$

Using the estimate above, we generalize Corollary 3.4 to arithmetic progressions. We let $\pi(x; q, a)$ denote the number of prime numbers $p \le x$ such that $p \equiv a \pmod{q}$.

Theorem 3.9 (Brun–Titchmarsh) *Let a and q be integers with $(a, q) = 1$, and let x and y be real numbers with $x \ge 0$ and $y \ge 2q$. Then*

$$\pi(x + y; q, a) - \pi(x; q, a) \le \frac{2y}{\varphi(q) \log y/q} \left(1 + O\left(\frac{1}{\log y/q}\right)\right). \qquad (3.23)$$

Proof Take P to be the product of those primes $p \le \sqrt{y/q}$ such that $p \nmid q$. Then

$$\prod_{p|P} \left(1 - \frac{1}{p}\right) = \prod_{\substack{p|q \\ p \le \sqrt{y/q}}} \left(1 - \frac{1}{p}\right)^{-1} \prod_{p \le \sqrt{y/q}} \left(1 - \frac{1}{p}\right)$$

$$\le \prod_{p|q} \left(1 - \frac{1}{p}\right)^{-1} \prod_{p \le \sqrt{y/q}} \left(1 - \frac{1}{p}\right).$$

By Mertens' estimate this is

$$= \frac{q}{\varphi(q)} \cdot \frac{2e^{-C_0}}{\log y/q} \left(1 + O\left(\frac{1}{\log y/q}\right)\right).$$

Thus by Theorem 3.8, the number of primes $p, x < p \leq x + y$, such that $p \equiv a$ (mod q) and $(p, P) = 1$ satisfies the bound (3.23). To complete the proof it remains to note that the number of primes $p, x < p \leq x + y$, such that $p \equiv a$ (mod q) and $p|P$ is at most $\omega(P) \leq \sqrt{y/q}$, which can be absorbed in the error term in (3.23). $\qquad\qquad\qquad\qquad\qquad\qquad\qquad\qquad\qquad\qquad\qquad\quad$ $\square$

3.4 Twin primes

Thus far we have removed at most one residue class per prime. More generally, we might wish to delete from an interval $(x, x + y]$ those numbers n that lie in a certain set $\mathcal{B}(p)$ of 'bad' residue classes modulo p. Let $b(p) = \text{card } \mathcal{B}(p)$ denote the number of residue classes to be removed, for $p|P$ where P is a given square-free number, and set

$$a(n) = \prod_{\substack{p|P \\ n \in \mathcal{B}(p) \,(\text{mod } p)}} p \,.$$

Thus the n that remain after sifting are precisely the n for which $(a(n), P) = 1$. By the sieve we obtain upper and lower bounds for the number of remaining n of the form

$$\sum_{x < n \leq x+y} \sum_{m|(a(n),P)} \lambda_m = \sum_{m|P} \lambda_m \sum_{\substack{x < n \leq x+y \\ m|a(n)}} 1 \,. \qquad (3.24)$$

Now $p|a(n)$ if and only if $n \in \mathcal{B}(p)$ (mod p). By the Chinese remainder theorem, this will be the case for all $p|m$ when n lies in one of precisely $\prod_{p|m} b(p)$ residue classes modulo m. The $b(p)$ are defined only for primes, but it is convenient now to extend the definition to all positive integers by putting

$$b(m) = \prod_{p^\alpha \| m} b(p)^\alpha \,.$$

Thus $b(m)$ is the totally multiplicative function generated by the $b(p)$. For square-free m, $b(m)$ represents the number of deleted residue classes modulo m. We are now in a position to estimate the inner sum above. We partition the interval $(x, x + y]$ into $[y/m]$ intervals of length m, and one interval of length $\{y/m\}m$. In each interval of length m there are precisely $b(m)$ values of n for which $m|a(n)$. In the final shorter interval, the number of such n lies between 0 and $b(m)$. Thus the inner sum on the right above is $= yb(m)/m + O(b(m))$, and hence the expression (3.24) is

$$= y \sum_{m|P} \frac{b(m)\lambda_m}{m} + O\left(\sum_{m|P} b(m)|\lambda_m|\right) \,. \qquad (3.25)$$

To continue from this point, one should specify the choice of λ_m, and then estimate the main term and error term. In the context of Selberg's Λ^2 method, we have real Λ_d with Λ_1 and $\Lambda_d = 0$ for $d > z$. The number of $n \in (x, x+y]$ that survive sifting is

$$\leq \sum_{x<n\leq x+y} \left(\sum_{d|(a(n),P)} \Lambda_d \right)^2 = \sum_{d|P} \sum_{e|P} \Lambda_d \Lambda_e \sum_{\substack{x<n\leq x+y \\ [d,e]|a(n)}} 1$$

$$= y \sum_{d|P} \sum_{e|P} \frac{b([d,e])}{[d,e]} \Lambda_d \Lambda_e + O\left(\sum_{d|P} \sum_{e|P} g([d,e]) |\Lambda_d \Lambda_e| \right). \qquad (3.26)$$

This is (3.25) with $\lambda_m = \sum_{[d,e]=m} \Lambda_d \Lambda_e$.

We consider first the main term above. Clearly $[d,e] = de/(d,e)$ and $b([d,e]) = b(d)b(e)/b((d,e))$. For square-free m put

$$g(m) = \prod_{p|m} \frac{b(p)}{p - b(p)}. \qquad (3.27)$$

Here we have 0 in the denominator if there is a prime p for which $b(p) = p$. However, in that case all residues modulo p are removed, and no integer survives sifting. Thus we may confine our attention to $b(p)$ such that $b(p) < p$ for all p. If m is square-free, then

$$\sum_{d|m} \frac{1}{g(d)} = \prod_{p|m} \left(1 + \frac{p - b(p)}{b(p)} \right) = \frac{m}{b(m)}.$$

By applying this with $m = (d, e)$ we see that the first sum in (3.26) is

$$\sum_{\substack{d|P \\ e|P}} \frac{b(d)\Lambda_d}{d} \cdot \frac{b(e)\Lambda_e}{e} \cdot \frac{(d,e)}{b((d,e))} = \sum_{\substack{d|P \\ e|P}} \frac{b(d)\Lambda_d}{d} \cdot \frac{b(e)\Lambda_e}{e} \sum_{\substack{f|d \\ f|e}} \frac{1}{g(f)}$$

$$= \sum_{f|P} \frac{1}{g(f)} \sum_{\substack{d \\ f|d|P}} \frac{b(d)}{d} \Lambda_d \sum_{\substack{e \\ f|e|P}} \frac{b(e)}{e} \Lambda_e$$

$$= \sum_{f|P} \frac{1}{g(f)} y_f^2 \qquad (3.28)$$

where

$$y_f = \sum_{\substack{d \\ f|d|P}} \frac{b(d)}{d} \Lambda_d. \qquad (3.29)$$

The linear change of variables from Λ_d to y_f is invertible:

$$\Lambda_d = \frac{d}{b(d)} \sum_{\substack{f \\ d|f|P}} y_f \mu(f/d). \qquad (3.30)$$

By the above formulæ we see that the condition that $\Lambda_d = 0$ for $d > z$ is equivalent to the condition that $y_f = 0$ for $f > z$. Also, the condition that $\Lambda_1 = 1$ is equivalent to

$$\sum_{f|P} y_f \mu(f) = 1. \qquad (3.31)$$

For such y_f we see that

$$\sum_{f|P} \frac{1}{g(f)} y_f^2 = \sum_{\substack{f|P \\ f \le z}} \frac{1}{g(f)} \left(y_f - \mu(f)g(f)/L\right)^2 + \frac{1}{L} \qquad (3.32)$$

where

$$L = \sum_{\substack{f \le z \\ f|P}} \mu(f)^2 g(f). \qquad (3.33)$$

Thus our main term is minimized by taking

$$y_f = \begin{cases} \mu(f)g(f)/L & (f \le z), \\ 0 & \text{(otherwise)}, \end{cases} \qquad (3.34)$$

and we note that these y_f satisfy (3.31). The size of L depends on P, z, and the $b(p)$. In the case of twin primes we obtain the following estimate.

Theorem 3.10 *Let $P = \prod_{p \le \sqrt{y}} p$ where $y \ge 4$. The number of integers $n \in (x, x + y]$, such that $(n, P) = (n + 2, P) = 1$ does not exceed*

$$\frac{8cy}{(\log y)^2} \left(1 + O\left(\frac{\log \log y}{\log y}\right)\right)$$

where

$$c = 2 \prod_{p>2} \left(1 - \frac{1}{(p-1)^2}\right).$$

The number of primes $p \in (x, x + y]$ for which $p|P$ is $\le \pi(\sqrt{y})$. Likewise, the number of primes $p \in (x, x + y]$ for which $p + 2$ is prime and $(p + 2)|P$ is $\le \pi(\sqrt{y})$. Otherwise, if $p \in (x, x + y]$ and $p + 2$ is prime, then $(p, P) = (p + 2, P) = 1$; the number of such p is bounded by the above. Since $\pi(\sqrt{y})$ is negligible by comparison, the above bound applies also to the number of primes $p \in (x, x + y]$ for which $p + 2$ is prime.

Proof   We first estimate L as given in (3.33). We have $b(2) = 1$ and $b(p) = 2$ for $p > 2$. Since $\mu(m)^2 g(m)$ is a multiplicative function that takes the value $2/(p-2)$ when $m = p > 2$, and since $d(n)/n$ is a multiplicative function that takes the value $2/p$ when $n = p$, we expect that $d(n)/n$ and $\mu(m)^2 g(m)$ are 'close' in the sense that we can obtain the latter function by convolving $d(n)/n$ with a fairly tame function $c(k)$. On comparing the Euler products of the respective Dirichlet series generating functions, we see that if the $c(k)$ are defined so that

$$\sum_{k=1}^{\infty} c(k) k^{-s} = (1 + 2^{-s})(1 - 2^{-s-1})^2 \prod_{p>2} \left(1 + \frac{2}{(p-2)p^s}\right) \left(1 - \frac{1}{p^{s+1}}\right)^2,$$

(3.35)

then

$$\mu(m)^2 g(m) = \sum_{\substack{k,n \\ kn=m}} c(k) d(n)/n.$$

Hence

$$L = \sum_{m \le z} \mu(m)^2 g(m) = \sum_{k \le z} c(k) \sum_{n \le z/k} d(n)/n.$$

By Theorem 2.3 and (Riemann–Stieltjes) integration by parts we see that

$$\sum_{n=1}^{N} \frac{d(n)}{n} = \frac{1}{2}(\log N)^2 + O(\log N).$$

Hence

$$L = \sum_{k \le z} c(k)((\log z/k)^2/2 + O(\log z))$$

$$= \frac{1}{2}(\log z)^2 \sum_{k \le z} c(k) + O\left((\log z) \sum_k |c(k)| \log 2k\right)$$

$$+ O\left(\sum_k |c(k)|(\log k)^2\right).$$

The Euler product in (3.35) is absolutely convergent for $\sigma > -1/2$. Hence $\sum |c(k)| k^{-\sigma} < \infty$ for $\sigma > -1/2$. Thus the two sums in the error terms above are convergent. Also,

$$\sum_{k>z} |c(k)| \le \frac{1}{\log z} \sum_{k=1}^{\infty} |c(k)| \log k \ll \frac{1}{\log z}.$$

Thus by taking $s = 0$ in (3.35) we find that

$$L = \frac{1}{2c}(\log z)^2 + O(\log z).$$

(3.36)

It remains to bound the error term in (3.26). Since $0 \le b([d, e]) \le b(d)b(e)$, the error term is

$$\ll \left(\sum_{d \le z} b(d)|\Lambda_d| \right)^2 .$$

From (3.30) and (3.34) we see that

$$\Lambda_d = \frac{d}{b(d)L} \sum_{\substack{f \le z \\ d|f}} \mu(f)g(f)\mu(f/d) = \frac{\mu(d)dg(d)}{b(d)L} \sum_{\substack{m \le z/d \\ (m,d)=1}} \mu(m)^2 g(m) .$$

Hence

$$\sum_{d \le z} b(d)|\Lambda_d| \ll \frac{1}{L} \sum_{d \le z} \mu(d)^2 dg(d) \sum_{m \le z/d} \mu(m)^2 g(m)$$

$$= \frac{1}{L} \sum_{m \le z} \mu(m)^2 g(m) \sum_{d \le z/m} \mu(d)^2 dg(d) .$$

By Corollary 2.15 we see that

$$\sum_{d \le D} \mu(d)^2 dg(d) \ll \frac{D}{\log D} \prod_{p \le D} (1 + g(p))$$

$$\ll \frac{D}{\log D} \prod_{p \le D} \left(1 - \frac{1}{p} \right)^{-2} \ll D \log D .$$

Since $L \asymp (\log z)^2$, it follows that

$$\sum_{d \le z} b(d)|\Lambda_d| \ll \frac{z}{\log z} \sum_{m \le z} \mu(m)^2 g(m)/m \ll \frac{z}{\log z} .$$

On combining our estimates, we see that the number of n, $x < n \le x + y$, such that $(a(n), P) = 1$ is

$$\le \frac{2cy}{(\log z)^2} + O\left(\frac{y}{(\log z)^3} \right) + O\left(\frac{z^2}{(\log z)^2} \right) .$$

In order that the last error term is majorized by the one before it, we take $z = (y/\log y)^{1/2}$. Then

$$\log z = \frac{1}{2} \log y + O(\log \log y),$$

so we obtain the stated result. $\qquad\qquad\qquad\qquad\qquad\qquad\qquad\square$

Corollary 3.11 (Brun) *Let $\sum_p^*$ denote a sum over those primes p for which $p + 2$ is prime. Then $\sum_p^* 1/p$ converges.*

Proof   The number of twin primes for which $2^{k-1} < p \le 2^k$ is $\ll 2^k/k^2$. Hence the contribution of such primes to the sum in question is $\ll 1/k^2$. But $\sum 1/k^2 < \infty$, so we obtain the stated result. $\square$

Let r be an even non-zero integer. To bound the number of primes p for which $p + r$ is also prime, it suffices to establish the following monotonicity principle, which is a natural generalization of Lemma 3.5.

Lemma 3.12   *For each prime p let $\mathcal{B}(p)$ be the union of $b(p)$ arithmetic progressions with common difference p. Put $\mathcal{B} = \bigcup_{p|P} \mathcal{B}(p)$, and set*

$$M(x, y; b) = \max_{\mathcal{B}} \sum_{\substack{x < n \le x+y \\ n \notin \mathcal{B}}} 1$$

where the maximum is over all choices of the $\mathcal{B}(p)$ with $b(p)$ fixed. If $0 \le b_1(p) \le b_2(p) < p$ for all p, then

$$M(x, y; b_1) \prod_{p|P} \left(1 - \frac{b_1(p)}{p} \right)^{-1} \le M(x, y; b_2) \prod_{p|P} \left(1 - \frac{b_2(p)}{p} \right)^{-1}.$$

Proof   We induct on $\sum_{p|P}(b_2(p) - b_1(p))$. If $b_1(p) = b_2(p)$ for all $p|P$, then we have equality in the above. Let $p'|P$ be a prime for which $b_1(p') < b_2(p')$. Suppose that the $\mathcal{B}_1(p)$ are chosen so that card $\mathcal{B}_1(p) = b_1(p)$ and

$$\sum_{\substack{x < n \le x+y \\ n \notin \mathcal{B}_1}} 1 = M(x, y; b_1).$$

We note that

$$\sum_{\substack{b=1 \\ b \notin \mathcal{B}_1(p')}}^{p'} \sum_{\substack{x < n \le x+y \\ n \notin \mathcal{B}_1 \\ n \not\equiv b \, (p')}} 1 = \sum_{\substack{x < n \le x+y \\ n \notin \mathcal{B}_1}} \sum_{\substack{b=1 \\ b \notin \mathcal{B}_1(p') \\ b \not\equiv n \, (p')}}^{p'} 1. \tag{3.37}$$

Consider the inner sum on the right. Since $n \notin \mathcal{B}_1(p')$, the variable b is restricted to lie in one of $p' - b_1(p') - 1$ residue classes. Hence the right-hand side above is

$$= (p' - b_1(p') - 1)M(x, y; b_1).$$

Since there are $p' - b_1(p')$ values of b in the outer sum on the left-hand side of (3.37), it follows that there is a choice of b such that $b \notin \mathcal{B}_1(p')$ and

$$\sum_{\substack{x < n \le x+y \\ n \notin \mathcal{B}_1 \\ n \not\equiv b \, (p')}} 1 \ge \frac{p' - b_1(p') - 1}{p' - b_1(p')} M(x, y; b_1).$$

Let $b_1'(p) = b_1(p)$ for $p \neq p'$, $b_1'(p') = b_1(p') + 1$. The left-hand side above is $\leq M(x, y; b_1')$, which by the inductive hypothesis is

$$\leq M(x, y; b_2) \frac{p - b_1(p') - 1}{p - b_2(p')} \prod_{\substack{p|P \\ p \neq p'}} \left(\frac{p - b_1(p)}{p - b_2(p)} \right).$$

Thus

$$M(x, y; b_1) \leq M(x, y; b_2) \prod_{p|P} \left(\frac{p - b_1(p)}{p - b_2(p)} \right),$$

and the induction is complete. $\qquad\qquad\qquad\qquad\qquad\qquad\qquad\square$

By combining Theorem 3.10 and Lemma 3.12, we obtain

Theorem 3.13 *Suppose that $y \geq 4$. Let $\mathcal{B}(p)$ be the union of $b(p)$ arithmetic progressions with common difference p, and put $\mathcal{B} = \bigcup_{p|P} \mathcal{B}(p)$. If $b(2) \leq 1$ and $b(p) \leq 2$ for $p > 2$, then the number of $n \in (x, x + y]$ such that $n \notin \mathcal{B}$ is*

$$\leq 8 \frac{y}{(\log y)^2} \left(\prod_{p|P} \left(1 - \frac{b(p)}{p} \right) \left(1 - \frac{1}{p} \right)^{-2} \right) \left(1 + O\left(\frac{\log \log y}{\log y} \right) \right).$$

Corollary 3.14 *Let r be an even non-zero integer, and suppose that $y \geq 4$. The number of primes $p \in (x, x + y]$ such that $p + r$ is also prime is*

$$\leq \frac{8c(r)y}{(\log y)^2} \left(1 + O\left(\frac{\log \log y}{\log y} \right) \right)$$

uniformly in r where

$$c(r) = \left(\prod_{p|r} \left(1 - \frac{1}{p} \right)^{-1} \right) \left(\prod_{p \nmid r} \left(1 - \frac{2}{p} \right) \left(1 - \frac{1}{p} \right)^{-2} \right) = \left(\prod_{\substack{p|r \\ p > 2}} \frac{p - 1}{p - 2} \right) c$$

and c is the constant in Theorem 3.10.

Suppose that r is a fixed even non-zero integer. It is conjectured that the number of primes $p \leq y$ such that $p + r$ is also prime is asymptotic to

$$\frac{c(r)y}{(\log y)^2}$$

as y tends to infinity. Thus the bound we have derived is larger than this by a factor of 8. We conclude with an application of the above.

Theorem 3.15 (Romanoff) *Let $N(x)$ denote the number of integers $n \leq x$ that can be expressed as a sum of a prime and a power of 2. Then $N(x) \gg x$ for $x \geq 4$.*

Proof   Let $r(n)$ denote the number of solutions of $n = p + 2^k$. By Cauchy's inequality,

$$\left(\sum_{n \leq x} r(n) \right)^2 \leq N(x) \sum_{n \leq x} r(n)^2 .$$

Thus to complete the proof it suffices to show that

$$\sum_{n \leq x} r(n) \gg x \qquad\qquad (x \geq 4), \qquad\qquad (3.38)$$

and that

$$\sum_{n \leq x} r(n)^2 \ll x . \qquad\qquad (3.39)$$

The first of these estimates is easy: Put $y = [(\log x)/ \log 2]$. If $0 \leq k \leq y - 1$, then $2^k \leq x/2$, and if also $p \leq x/2$, then $p + 2^k \leq x$. Thus the sum in (3.38) is

$$\geq \pi(x/2)y \gg \frac{x}{\log x} \log x \gg x$$

for $x \geq 4$.

To prove (3.39), we first observe that the sum on the left-hand side is

$$= \sum_{\substack{p_1, p_2, j, k \\ p_1 + 2^j \leq x \\ p_2 + 2^k \leq x \\ p_1 + 2^j = p_2 + 2^k}} 1 .$$

This sum includes 'diagonal' terms, in which $p_1 = p_2$ and $j = k$; there are $\ll x/\log x$ choices for p_1 and $\ll \log x$ choices for j, so there are $\ll x$ such terms. The remaining terms above contribute an amount that is

$$\ll \sum_{0 \leq j < k \leq y} \pi_2(x, 2^k - 2^j) \qquad\qquad (3.40)$$

where $\pi_2(x, r)$ denotes the number of primes $p \leq x$ for which $p + r$ is also prime. From Corollary 3.14 we know that if $r \neq 0$, then

$$\pi_2(x, r) \ll \frac{x}{(\log x)^2} \prod_{\substack{p|r \\ p>2}} \left(1 + \frac{1}{p} \right) \ll \frac{x}{(\log x)^2} \sum_{\substack{m|r \\ 2 \nmid m}} \frac{1}{m},$$

uniformly in r. Thus the expression (3.40) is

$$\ll \frac{x}{(\log x)^2} \sum_{0 \leq j < k \leq y} \sum_{\substack{m|(2^k - 2^j) \\ 2 \nmid m}} \frac{1}{m} .$$

Put $n = k - j$. Thus $0 < n \le y$. Let $h_2(m)$ denote the order of 2 modulo m, which is to say that $h_2(m)$ is the least positive integer h such that $2^h \equiv 1$ (mod m). We note that $m|(2^n - 1)$ if and only if $h_2(m)|n$. The number of such $n, 0 < n \le y,$ is $\le y/h_2(m)$. There are also $\le y$ choices of j. Thus to complete the proof of (3.39) it suffices to show that

$$\sum_{\substack{m \\ 2 \nmid m}} \frac{1}{mh_2(m)} < \infty. \tag{3.41}$$

To this end, let

$$a_n = \sum_{\substack{m \\ 2 \nmid m \\ h_2(m)=n}} \frac{1}{m},$$

and set $A(x) = \sum_{n \le x} a_n$. We shall show that

$$A(x) \ll \log x. \tag{3.42}$$

By summation by parts it follows that $\sum a_n/n$ converges. (Alternatively, we could appeal to Theorem 1.3, from which we see that $\sum a_n/n^s$ converges for $\sigma > 0$.) This suffices, since the sum in (3.41) is $\sum a_n/n$.

It remains to establish (3.42). Set

$$P = P(x) = \prod_{n \le x} (2^n - 1).$$

If $h_2(m) = n \le x$, then $m|P$. Hence

$$A(x) \le \sum_{m|P} \frac{1}{m} \le \prod_{p|P} \left(1 + \frac{1}{p} + \frac{1}{p^2} + \cdots\right) = \frac{P}{\varphi(P)} \ll \log\log P$$

by Theorem 2.9. But $P \le 2^{x^2}$, so we have (3.42), and the proof is complete. $\square$

3.4.1 Exercises

1. For each prime p let $\mathcal{B}(p)$ be the union of $b(p)$ 'bad' arithmetic progressions with common difference p. Put $\mathcal{B} = \bigcup_{p|P} \mathcal{B}(p)$, and let

$$m(x, y; b) = \min_{\mathcal{B}} \sum_{\substack{x < n \le x+y \\ n \notin \mathcal{B}}} 1$$

where the minimum is over all choices of the $\mathcal{B}(p)$ with $b(p)$ fixed. Show that if $b_1(p) \le b_2(p)$ for all p, then

$$m(x, y; b_1) \prod_p \left(1 - \frac{b_1(p)}{p}\right)^{-1} \ge m(x, y; b_2) \prod_p \left(1 - \frac{b_2(p)}{p}\right)^{-1}.$$

2. Show that the number of primes $p \leq 2n$ such that $2n - p$ is prime is

$$\leq 8c \left(\prod_{\substack{p|n \\ p>2}} \frac{p-1}{p-2} \right) \frac{2n}{(\log 2n)^2} \left(1 + O\left(\frac{\log\log 4n}{\log 2n} \right) \right)$$

where c is the constant in Theorem 3.10.

3. (Erdős 1940, Ricci 1954)

   (a) Show that

   $$\sum_{r \leq x} c(r) = x + O(\log x)$$

   where $c(r)$ is defined as in Corollary 3.14.

   (b) Let p' denote the least prime $> p$, and put $d(p) = p' - p$. Show that if a and b are fixed real numbers with $a < b$, then

   $$\sum_{\substack{p \leq x \\ a\log p \leq d(p) \leq b\log p}} \log p \lesssim 8(b-a)x.$$

   (c) Suppose that f is a non-negative, properly Riemann-integrable function on a finite interval $[a, b]$. Show that

   $$\sum_{p \leq x} f\left(\frac{d(p)}{\log p} \right) \log p \leq (8 + o(1))x \int_a^b f(u)\,du.$$

   (d) Show that if a and b are fixed real numbers with $a < b$, then

   $$\sum_{\substack{p \leq x \\ a\log p \leq d(p) \leq b\log p}} (b\log p - d(p)) \lesssim 4(b-a)^2 x.$$

   (e) Explain why

   $$\sum_{\substack{p \leq x \\ d(p)>b\log p}} (d(p) - b\log p) \geq 0.$$

   (f) Deduce that

   $$\sum_{\substack{p \leq x \\ d(p)\geq a\log p}} (b\log p - d(p)) \lesssim 4(b-a)^2 x.$$

   (g) Show that

   $$\sum_{p \leq x} d(p) \sim x.$$

   (h) Show that

   $$\sum_{p \leq x} (b\log p - d(p)) = (b - 1 + o(1))x.$$

(i) Take $b = a + 1/8$, and suppose that $d(p) \geq a \log p$ for all $p > p_0$. Show that the estimates of (f) and (h) are inconsistent if $a > 15/16$. Thus conclude that

$$\liminf_{p \to \infty} \frac{d(p)}{\log p} \leq \frac{15}{16}.$$

4. Let $r(n)$ be defined as in the proof of Theorem 3.15. Show that

$$\sum_{n \leq x} r(n) \sim \frac{x}{\log 2}.$$

5. Let $r(n)$ be defined as in the proof of Theorem 3.15. Show that

$$\sum_{\substack{n \leq x \\ 2 \mid n}} r(n) \ll \frac{x}{\log x}.$$

6. (Erdős 1950)
   (a) Show that if $n \equiv 1$ (mod 3) and $k \equiv 0$ (mod 2), then $3 \mid (n - 2^k)$.
   (b) Show that if $n \equiv 1$ (mod 7) and $k \equiv 0$ (mod 3), then $7 \mid (n - 2^k)$.
   (c) Show that if $n \equiv 2$ (mod 5) and $k \equiv 1$ (mod 4), then $5 \mid (n - 2^k)$.
   (d) Show that if $n \equiv 8$ (mod 17) and $k \equiv 3$ (mod 8), then $17 \mid (n - 2^k)$.
   (e) Show that if $n \equiv 11$ (mod 13) and $k \equiv 7$ (mod 12), then $13 \mid (n - 2^k)$.
   (f) Show that if $n \equiv 121$ (mod 241) and $k \equiv 23$ (mod 24), then $241 \mid (n - 2^k)$.
   (g) Show that every integer k satisfies at least one of the congruences $k \equiv 0$ (mod 2), $k \equiv 0$ (mod 3), $k \equiv 1$ (mod 4), $k \equiv 3$ (mod 8), $k \equiv 7$ (mod 12), $k \equiv 23$ (mod 24).
   (h) Show that if n satisfies all the congruences $n \equiv 1$ (mod 3), $n \equiv 1$ (mod 7), $n \equiv 2$ (mod 5), $n \equiv 8$ (mod 17), $n \equiv 11$ (mod 13), $n \equiv 121$ (mod 241), then $n - 2^k$ is divisible by at least one of the primes $3, 7, 5, 17, 13, 241$.
   (i) Show that these congruential conditions are equivalent to the single condition $n \equiv 172677$ (mod 3728270).
   (j) An integer n satisfying the above might still be representable in the form $p + 2^k$, but if it is, then the prime in question must be one of the six primes listed. Show that if in addition, $n \equiv 9$ or 11 or 15 (mod 16), then n cannot be expressed as a sum of a prime and a power of 2.

3.5 Notes

Sections 3.1, 3.2. The modern era of sieve methods began with the work of Brun (1915, 1919). Hardy & Littlewood (1922) used Brun's method to establish the estimate (3.9). The sharp form of this in Corollary 3.4 is due to Selberg (1952a,b). The Λ^2 method of Selberg (1947) provides only upper bounds, but lower bounds can also be derived from it by using ideas of Buchstab (1938).

In contrast to the elegance of the Selberg Λ^2 method, the further study of sieves leads us to construct asymptotic estimates for complicated sums over integers whose prime factors are distributed in certain ways. In this connection, the argument (3.22) is a simple foretaste of more complicated things to come. Hence further discussion of sieves is possible only after the appropriate technical tools are in place.

In this chapter we have applied the sieve only to arithmetic progressions, but it can be shown that the sieve is applicable to much more general sets. This makes sieves very versatile, but it also means that they are subject to certain unfortunate limitations. In order to estimate the number of elements of a set S that remain after sifting, it suffices to have a reasonably precise estimate of the number X_d of multiples of d in the set, say of the form $X_d = f(d)X/d + O(R_d)$ where X is an estimate for the cardinality of S, and f is a multiplicative function. Thus Theorem 3.3 can be generalized to much more general sets, and in that more general setting it is known that the constant 2 is best-possible. It may be true that the constant 2 can be improved in the special case that one is sifting an interval, but this has not been achieved thus far.

When sifting an interval, the error terms can be avoided by using Fourier analysis as in Selberg (1991, Sections 19–22), or by using the large sieve as in Montgomery & Vaughan (1973). In particular, the number of integers in $[M + 1, M + N]$ remaining after sifting is at most N/L where

$$L = \sum_{q \leq Q} \frac{\mu(q)^2}{1 + \frac{3}{2}qQ/N} \prod_{p|q} \frac{b(p)}{p - b(p)} . \tag{3.43}$$

Here $b(p)$ is the number of residue classes modulo p that are deleted. This is both a generalization and a sharpening of Theorem 3.2.

Section 3.3. Titchmarsh (1930) used Brun's method to obtain Theorem 3.9, but with a larger constant instead of 2. Montgomery & Vaughan (1973) have shown that Corollary 3.4 and Theorem 3.9 are still valid when the error terms are omitted. See also Selberg (1991, Section 22). The first significant improvement of Theorem 3.9 was obtained by Motohashi (1973). Other improvements of various kinds have been derived by Motohashi (1974), Hooley (1972, 1975), Goldfeld (1975), Iwaniec (1982), and Friedlander & Iwaniec (1997).

In Lemmas 3.5 and 3.12, and in Exercises 3.2.7, 3.2.9, 3.2.10, 3.4.1 we see evidence of a monotonicity principle that permeates sieve theory; cf. Selberg (1991, pp. 72–73).

Hooley (1994) has shown that quite sharp sieve bounds can be derived using the interrupted inclusion–exclusion idea that Brun started with. This approach has been developed further by Ford & Halberstam (2000). An exposition of sieves based on these ideas is given by Bateman & Diamond (2004, Chapters 12, 13). Still more extensive accounts of sieve methods have been given by Greaves (2001), Halberstam & Richert (1974), Iwaniec & Kowalski (2004, Chapter 6), Motohashi (1983), and Selberg (1971, 1991). In addition, a collection of applications of sieves to arithmetic problems has been given by Hooley (1976), and additional sieve ideas are found in Bombieri (1977), Bombieri, Friedlander & Iwaniec (1986, 1987, 1989), Fouvry & Iwaniec (1997), Friedlander & Iwaniec (1998a, b), and Iwaniec (1978, 1980a, b, 1981).

Section 3.4. The twin prime conjecture is a special case of the prime k-tuple conjecture. Suppose that $d_1, \ldots, d_k$ are distinct integers, and let $b(p)$ denote the number of distinct residue classes modulo p found among the d_i. The prime k-tuple conjecture asserts that if $b(p) < p$ for every prime number p, then there exist infinitely many positive integers n such that the k numbers $n + d_i$ are all prime. Hardy & Littlewood (1922) put this in a quantitative form: If $b(p) < p$ for all p, then the number of $n \leq N$ for which the k numbers $n + d_i$ are all prime is conjectured to be

$$\sim \mathfrak{S}(\boldsymbol{d}) \frac{N}{(\log N)^k} \tag{3.44}$$

as $N \to \infty$ where

$$\mathfrak{S}(\boldsymbol{d}) = \prod_p \left(1 - \frac{b(p)}{p}\right)\left(1 - \frac{1}{p}\right)^{-k}. \tag{3.45}$$

This product is absolutely convergent, since $b(p) = k$ for all sufficiently large primes p. Although this remains unproved, by sifting we can obtain an upper bound of the expected order of magnitude. In particular, from (3.43) it can be shown that the number of n, $M + 1 \leq n \leq M + N$, for which the numbers $n + d_i$ are all prime is

$$\lesssim 2^k k! \mathfrak{S}(\boldsymbol{d}) \frac{N}{(\log N)^k}. \tag{3.46}$$

Corollarys 3.4 and 3.14 are special cases of this.

Theorem 3.15 is due to Romanoff (1934). Once the bound for the number of twin primes is in place, the hardest part of the proof is to establish the estimate (3.41). Romanoff's original proof of this was rather difficult. Erdős & Turán (1935) gave a simpler proof, but the clever proof we have given is due to Erdős (1951). Let $r(n)$ be defined as in the proof of Theorem 3.15. Erdős (1950) showed that $r(n) = \Omega(\log \log n)$, and that $\sum_{n \leq x} r(n)^k \ll_k x$ for any positive k. Presumably $r(n) = o(\log n)$, but for all we know there could be, although it seems unlikely, infinitely many n such that $n - 2^k$ is prime whenever $0 < 2^k < n$. The number $n = 105$ has this property, and is probably the largest such number. The best upper bound we have for the number of such n not exceeding X is (Vaughan 1973),

$$X \exp \left(-\frac{c \log X \log \log \log X}{\log \log X} \right).$$

For generalizations of Romanoff's theorem, see Erdős (1950, 1951).

3.6   References

Ankeny, N. C. & Onishi, H. (1964/1965). The general sieve, *Acta Arith.* **10**, 31–62.

Bateman, P. T. & Diamond, H. (2004). *Analytic Number Theory*, Hackensack: World Scientific.

Behrend, F. A. (1948). Generalization of an inequality of Heilbronn and Rohrbach, *Bull. Amer. Math. Soc.* **54**, 681–684.

Bombieri, E. (1977). The asymptotic sieve, *Rend. Accad. Naz.* XL (5) **1/2** (1975/76), 243–269.

Bombieri, E., Friedlander, J. B., & Iwaniec, H. (1986). Primes in arithmetic progressions to large moduli, *Acta Math.* **156**, 203–251.

(1987). Primes in arithmetic progressions to large moduli, II, *Math. Ann.* **277**, 361–393.

(1989). Primes in arithmetic progressions to large moduli, III, *J. Amer. Math. Soc.* **2**, 215–224.

Brun, V. (1915). Über das Goldbachsche Gesetz und die Anzahl der Primzahlpaare, *Archiv for Math. og Naturvid.* B **34**, no. 8, 19 pp.

(1919). La série $1/5 + 1/7 + 1/11 + 1/13 + 1/17 + 1/19 + 1/29 + 1/31 + 1/41 + 1/43 + 1/59 + 1/61 + \cdots$ où les dénominateurs sont "nombres premiers jumeaus" est convergente ou finie, *Bull. Sci. Math.* (2) **43**, 100–104; 124–128.

(1967). Reflections on the sieve of Eratosthenes, *Norske Vid. Selsk. Skr.* Trondheim, no. 1, 9 pp.

Buchstab, A. A. (1938). New improvements in the method of the sieve of Eratosthenes, *Mat. Sb.* (N. S.) **4 (46)**, 375–387.

Chowla, S. (1932). Contributions to the analytic theory of numbers, *Math. Z.* **35**, 279–299.

Chung, K.-L. (1941). A generalization of an inequality in the elementary theory of numbers, *J. Reine Angew. Math.* **183**, 193–196.

van der Corput, J. G. (1958). Inequalities involving least common multiple and other arithmetical functions, *Nederl. Akad. Wetensch. Proc. Ser.* A **61** (= Indag. Math. **20**), 5–15.

Erdős, P. (1940). The difference of consecutive primes, *Duke Math. J.* **6**, 438–441.

(1946). On the coefficients of the cyclotomic polynomial, *Bull. Amer. Math. Soc.* **52**, 179–184.

(1950). On integers of the form $2^k + p$ and some related problems, *Summa Brasil. Math.* **2**, 113–123.

(1951). On some problems of Bellman and a theorem of Romanoff, *J. Chinese Math. Soc.* (N. S.) **1**, 409–421.

Erdős, P. & Turán, P. (1935). Ein zahlentheoretischer Satz, *Mitt. Forsch. Inst. Math. Mech. Univ. Tomsk* **1**, 101–103.

Ford, K. & Halberstam, H. (2000). The Brun–Hooley sieve, *J. Number Theory* **81**, 335–350.

Fouvry, E. & Iwaniec, H. (1997). Gaussian primes, *Acta Arith.* **79** (1997), 249–287.

Friedlander, J. B. & Iwaniec, H. (1997). The Brun–Titchmarsh theorem, *Analytic Number Theory* (Kyoto, 1996). London Math. Soc. Lecture Note Ser. **247**, Cambridge: Cambridge University Press, pp. 85–93.

(1998a). The polynomial $X^2 + Y^4$ captures its primes, *Ann. of Math.* (2) **148**, 945–1040.

(1998b). Asymptotic sieve for primes, *Ann. of Math.* (2) **148**, 1041–1065.

Goldfeld, D. M. (1975). A further improvement of the Brun–Titchmarsh theorem, *J. London Math. Soc.* (2) **11**, 434–444.

Greaves, G. (2001). *Sieves in Number Theory*. Berlin: Springer.

Halberstam, H. (1985). *Lectures on the linear sieve*, Topics in Analytic Number Theory (Austin, 1982). Austin: University of Texas Press, pp. 165–220.

Halberstam, H. & Richert, H.-E. (1973). Brun's method and the fundamental lemma, *Acta Arith.* **24**, 113–133.

(1974). *Sieve Methods*. London: Academic Press.

(1975). Brun's method and the fundamental lemma. II, *Acta Arith.* **27**, 51–59.

Hardy, G. H. & Littlewood, J. E. (1922). Some problems of 'Partitio Numerorum': III. On the expression of a number as a sum of primes, *Acta Math.* **44**, 1–70; *Collected Papers*, Vol. I, London: Oxford University Press, 1966, pp. 561–630.

Heilbronn, H. (1937). On an inequality in the elementary theory of numbers, *Proc. Cambridge Philos. Soc.* **33**, 207–209.

Hensley, D. (1978). An almost-prime sieve, *J. Number Theory* **10**, 250–262; *Corrigendum*, **12**, (1980), 437.

Hooley, C. (1972). On the Brun–Titchmarsh theorem, *J. Reine Angew. Math.* **255**, 60–79.

(1975). On the Brun–Titchmarsh theorem, II, *Proc. London Math. Soc.* (3) **30**, 114–128.

(1976). Applications of Sieve Methods to the Theory of Prime Numbers, *Cambridge Tract* 70. Cambridge: Cambridge University Press.

(1994). An almost pure sieve, *Acta Arith.* **66**, 359–368.

Iwaniec, H. (1978). Almost-primes represented by quadratic polynomials, *Invent. Math.* **47**, 171–188.

(1980a). Rosser's sieve, *Acta Arith.* **36**, 171–202.

(1980b). A new form of the error term in the linear sieve, *Acta Arith.* **37**, 307–320.

(1981). Rosser's sieve – bilinear forms of the remainder terms – some applications. *Recent Progress in Analytic Number Theory*, Vol. 1. New York: Academic Press, pp. 203–230.

(1982). On the Brun–Titchmarsh theorem, *J. Math. Soc. Japan* **34**, 95–123.

Iwaniec, H. & Kowalski, E. (2004). *Analytic Number Theory*, Colloquium Publications 53. Providence: Amer. Math. Soc.

Jurkat, W. B. & Richert, H.-E. (1965). An improvement in Selberg's sieve method, I, *Acta Arith.* **11**, 217–240.

Lehmer, D. H. (1955). The distribution of totatives, *Canad. J. Math.* **7**, 347–357.

van Lint, J. H. & Richert, H.-E. (1964). Über die Summe $\sum_{\substack{n \leq x \\ p(n) < y}} \frac{\mu^2(n)}{\varphi(n)}$ *Nederl. Akad. Wetensch. Proc. Ser.* A **67** (= Indag. Math. **26**), 582–587.

(1965). On primes in artihmetic progressions, *Acta Arith.* **11**, 209–216.

Montgomery, H. L. (1968). A note on the large sieve, *J. London Math. Soc.* **43**, 93–98.

Montgomery, H. L. & Vaughan, R. C. (1973). The large sieve, *Mathematika* **20**, 119–134.

(1979). Mean values of character sums, *Canad. J. Math.* **31**, 476–487.

Motohashi, Y. (1973). On some improvements of the Brun–Titchmarsh theorem, II, Research of analytic number theory (*Proc. Sympos., Res. Inst. Math. Sci.*, Kyoto, 1973), Søurikaisekikenkyøusho Kókyøuroku, No. 193, 97–109.

(1974). On some improvements of the Brun–Titchmarsh theorem, *J. Math. Soc. Japan* **26**, 306–323.

(1975). On some improvements of the Brun–Titchmarsh theorem, III, *J. Math. Soc. Japan* **27**, 444–453.

(1983). *Lectures on Sieve Methods and Prime Number theory.* Tata Institute of Fundamental Research (Bombay). Berlin: Springer-Verlag.

Ricci, G. (1954). Sull'andamento della differenza di numeri primi consecutivi, *Riv. Mat. Univ. Parma* **5**, 3–54.

Riesel, H. & Vaughan, R. C. (1983). On sums of primes, *Ark. Mat.* **21**, 46–74.

Rohrbach, H. (1937). Beweis einer zahlentheoretischen Ungleichung, *J. Reine Angew. Math.* **177**, 193–196.

Romanoff, N. P. (1934). Über einige Sätze der additiven Zahlentheorie, *Math. Ann.* **109**, 668–678.

Selberg, A. (1947). On an elementary method in the theory of primes, *Norske Vid. Selsk. Forh.*, Trondhjem **19**, no. 18, 64–67; *Collected Papers*, Vol. 1. Berlin: Springer-Verlag, 1989, pp. 363–366.

(1952a). On elementary methods in primenumber-theory and their limitations, *Den 11te Skandinaviske Matematikerkongress* (Trondheim, 1949), Oslo: Johan Grundt Tanums Forlag, pp. 13–22; *Collected Papers*, Vol. 1. Berlin: Springer-Verlag, 1989, pp. 388–397.

(1952b). The general sieve-method and its place in prime-number theory. Proceedings of the International Congress of Mathematicians (Cambridge MA, 1950), Vol. 1, Providence: Amer. Math. Soc., pp. 286–292; *Collected Papers*, Vol. 1. Berlin: Springer-Verlag, 1989, pp. 411–417.

(1971). *Sieve methods*, Proceedings of Symposium on Pure Mathematics (SUNY Stony Brook, 1969), Vol. XX. Providence: Amer. Math. Soc., 311–351; *Collected Papers*, Vol. 1. Berlin: Springer-Verlag, 1989, pp. 568–608.

(1972). *Remarks on sieves*, Proceedings of the Number Theory Conference (Boulder CO Aug. 14–18), pp. 205–216; *Collected Papers*, Vol. 1. Berlin: Springer-Verlag, 1989, pp. 609–615.

(1989). *Sifting problems, sifting density and sieves*, Number Theory, Trace Formulas, and Discrete Groups (Oslo, 1987), K. E. Aubert, E. Bombieri, D. Goldfeld, eds.

Boston: Academic Press, pp. 467–484; *Collected Papers*, Vol. 1. Berlin: Springer-Verlag, 1989, pp. 675–69.

(1991). Lectures on Sieves, *Collected Papers*, Vol. 2. Berlin: Springer-Verlag, pp. 65–247.

Titchmarsh, E. C. (1930). A divisor problem, *Rend. Circ. Math.* Palermo **54**, 414–429.

Tsang, K. M. (1989). *Remarks on the sieving limit of the Buchstab–Rosser sieve*, Number Theory, Trace Formulas and Discrete Groups (Oslo, 1987). Boston: Academic Press, pp. 485–502.

Vaughan, R. C. (1973). Some applications of Montgomery's sieve, *J. Number Theory* **5**, 64–79.

Vijayaraghavan, T. (1951). On a problem in elementary number theory, *J. Indian Math. Soc.* (N.S.) **15**, 51–56.

4

Primes in arithmetic progressions: I

4.1 Additive characters

If $f(z) = \sum_{n=0}^{\infty} c_n z^n$ is a power series, we can restrict our attention to terms for which n has prescribed parity by considering

$$\frac{1}{2}f(z) + \frac{1}{2}f(-z) = \sum_{\substack{n=0 \\ n \equiv 0\,(2)}}^{\infty} c_n z^n$$

or

$$\frac{1}{2}f(z) - \frac{1}{2}f(-z) = \sum_{\substack{n=0 \\ n \equiv 1\,(2)}}^{\infty} c_n z^n.$$

That is, we can express the characteristic function of an arithmetic progression (mod 2) as a linear combination $\frac{1}{2}1^n \pm \frac{1}{2}(-1)^n$ of 1^n and $(-1)^n$. Here 1 and -1 are the square-roots of 1, and we can similarly express the characteristic function of an arithmetic progression (mod q) as a linear combination of the sequences ζ^n where ζ runs over the q different q^{th} roots of unity. We write $e(\theta) = e^{2\pi i \theta}$, and then the qth roots of unity are the numbers $\zeta = e(a/q)$ for $1 \le a \le q$. If $(a, q) = 1$ then the least integer n such that $\zeta^n = 1$ is q, and we say that ζ is a *primitive q^{th} root of unity*. From the formula

$$\sum_{k=0}^{q-1} \zeta^k = \frac{1 - \zeta^q}{1 - \zeta}$$

for the sum of a geometric progression, we see that if ζ is a qth root of unity then

$$\sum_{k=1}^{q} \zeta^k = 0$$

unless $\zeta = 1$. Hence

$$\frac{1}{q} \sum_{k=1}^{q} e(-ka/q)e(kn/q) = \begin{cases} 1 & \text{if } n \equiv a \pmod{q}, \\ 0 & \text{otherwise,} \end{cases} \tag{4.1}$$

and thus the characteristic function of an arithmetic progression $(\bmod\ q)$ can be expressed as a linear combination of the sequences $e(kn/q)$. These functions are called the *additive characters* $(\bmod\ q)$ because they are the homomorphisms from the additive group $(\mathbb{Z}/q\mathbb{Z})^{+}$ of integers $(\bmod\ q)$ to the multiplicative group $\mathbb{C}^{\times}$ of non-zero complex numbers.

In the language of linear algebra we see that the arithmetic functions of period q form a vector space of dimension q. For any k, $1 \le k \le q$, the sequence $\{e(kn/q)\}_{n=-\infty}^{\infty}$ has period q, and these q sequences form a basis for the space of q-periodic arithmetic functions. Indeed, the formula (4.1) expresses the ath elementary vector as a linear combination of the vectors $[e(n/q), e(2n/q), \ldots, e((q-1)n/q), 1]$.

If $f(n)$ is an arithmetic function with period q then we define the *finite Fourier transform* of f to be the function

$$\widehat{f}(k) = \frac{1}{q} \sum_{n=1}^{q} f(n)e(-kn/q). \tag{4.2}$$

To obtain a Fourier representation of f we multiply both sides of (4.1) by $f(n)$ and sum over n to see that

$$\begin{aligned} f(a) &= \sum_{n=1}^{q} \frac{f(n)}{q} \sum_{k=1}^{q} e(-ka/q)e(kn/q) \\ &= \sum_{k=1}^{q} e(-ka/q)\frac{1}{q} \sum_{n=1}^{q} f(n)e(kn/q) \\ &= \sum_{k=1}^{q} e(-ka/q)\widehat{f}(-k). \end{aligned}$$

Here the exact values that k runs through are immaterial, as long as the set of these values forms a complete residue system modulo q. Hence we may replace k by $-k$ in the above, and so we see that

$$f(n) = \sum_{k=1}^{q} \widehat{f}(k)e(kn/q). \tag{4.3}$$

This includes (4.1) as a special case, for if we take f to be the characteristic function of the arithmetic progression $a \pmod{q}$ then by (4.2) we have $\widehat{f}(k) = e(-ka/q)/q$, and then (4.3) coincides with (4.1). The pair (4.2), (4.3) of inversion formulæ are analogous to the formula for the Fourier coefficients and Fourier expansion of a function $f \in L^1(\mathbb{T})$, but the situation here is simpler because our sums have only finitely many terms.

Let $v(h)$ be the vector $v(h) = [e(h/q), e(2h/q), \ldots, e((q-1)h/q), 1]$. From (4.1) we see that two such vectors $v(h_1)$ and $v(h_2)$ are orthogonal unless $h_1 \equiv h_2 \pmod{q}$. These vectors are not normalized, but they all have the same length $\sqrt{q}$, so apart from some rescaling, the transformation from f to $\widehat{f}$ is an isometry. More precisely, if f has period q and $\widehat{f}$ is given by (4.2), then by (4.3),

$$\sum_{n=1}^{q} |f(n)|^2 = \sum_{n=1}^{q} \left| \sum_{k=1}^{q} \widehat{f}(k) e(kn/q) \right|^2.$$

By expanding and taking the sum over n inside, we see that this is

$$= \sum_{j=1}^{q} \sum_{k=1}^{q} \widehat{f}(j) \overline{\widehat{f}(k)} \sum_{n=1}^{q} e(jn/q) e(-kn/q).$$

By (4.1) the innermost sum is q if $j = k$ and is 0 otherwise. Hence

$$\sum_{n=1}^{q} |f(n)|^2 = q \sum_{k=1}^{q} |\widehat{f}(k)|^2. \tag{4.4}$$

This is analogous to Parseval's identity for functions $f \in L^2(\mathbb{T})$, or to Plancherel's identity for functions $f \in L^2(\mathbb{R})$.

Among the exponential sums that we shall have occasion to consider is *Ramanujan's sum*

$$c_q(n) = \sum_{\substack{a=1 \\ (a,q)=1}}^{q} e(an/q). \tag{4.5}$$

We now establish some of the interesting properties of this quantity.

Theorem 4.1 *As a function of n, $c_q(n)$ has period q. For any given n, $c_q(n)$ is a multiplicative function of q. Also,*

$$\sum_{d|q} c_d(n) = \begin{cases} q & \text{if } q \mid n, \\ 0 & \text{otherwise.} \end{cases} \tag{4.6}$$

Finally,

$$c_q(n) = \sum_{d|(q,n)} d\mu(q/d) = \frac{\mu(q/(q,n))}{\varphi(q/(q,n))} \varphi(q). \tag{4.7}$$

The case $n = 1$ of this last formula is especially memorable:

$$\sum_{\substack{a=1 \\ (a,q)=1}}^{q} e(a/q) = \mu(q).$$

Proof The first assertion is evident, as each term in the sum (4.5) has period q. As for the second, suppose that $q = q_1 q_2$ where $(q_1, q_2) = 1$. By the Chinese Remainder Theorem, for each $a \pmod{q}$ there is a unique pair a_1, a_2 with a_i determined $\pmod{q_i}$, so that $a \equiv a_1 q_2 + a_2 q_1 \pmod{q}$. Moreover, under this correspondence we see that $(a, q) = 1$ if and only if $(a_i, q_i) = 1$ for $i = 1, 2$. Then

$$
\begin{aligned}
c_q(n) &= \sum_{\substack{a_1=1 \\ (a_1,q_1)=1}}^{q_1} \sum_{\substack{a_2=1 \\ (a_2,q_2)=1}}^{q_2} e((a_1 q_2 + a_2 q_1)n/(q_1 q_2)) \\
&= \left(\sum_{\substack{a_1=1 \\ (a_1,q_1)=1}}^{q_1} e(a_1 n/q_1) \right) \left(\sum_{\substack{a_2=1 \\ (a_2,q_2)=1}}^{q_2} e(a_2 n/q_2) \right) \\
&= c_{q_1}(n) c_{q_2}(n).
\end{aligned}
$$

To establish (4.6), suppose that $d|q$, and consider those a, $1 \le a \le q$, such that $(a, q) = d$. Put $b = a/d$. Then the numbers a are in one-to-one correspondence with those b, $1 \le b \le q/d$, for which $(b, q/d) = 1$. Hence

$$
\begin{aligned}
\sum_{a=1}^{q} e(na/q) &= \sum_{d|q} \sum_{\substack{a=1 \\ (a,q)=d}}^{q} e(na/q) \\
&= \sum_{d|q} \sum_{\substack{b=1 \\ (b,q/d)=1}}^{q/d} e(nb/(q/d)) \\
&= \sum_{d|q} c_{q/d}(n).
\end{aligned}
$$

By (4.1), the left-hand side above is q when $q|n$, and is 0 otherwise. Thus we have (4.6).

The first formula in (4.7) is merely the Möbius inverse of (4.6). To obtain the second formula in (4.7), we begin by considering the special case in which q is a prime power, $q = p^k$.

$$
\begin{aligned}
c_{p^k}(n) &= \sum_{\substack{a=1 \\ p\nmid a}}^{p^k} e(na/p^k) \\
&= \sum_{a=1}^{p^k} e(na/p^k) - \sum_{a=1}^{p^{k-1}} e(na/p^{k-1}).
\end{aligned}
$$

Here the first sum is p^k if $p^k | n$, and is 0 otherwise. Similarly, the second sum is p^{k-1} if $p^{k-1} | n$, and is 0 otherwise. Hence the above is

$$= \begin{cases} 0 & \text{if } p^{k-1} \nmid n, \\ -p^{k-1} & \text{if } p^{k-1} \| n, \\ p^k - p^{k-1} & \text{if } p^k | n \end{cases}$$

$$= \frac{\mu\left(p^k / (n, p^k)\right)}{\varphi\left(p^k / (n, p^k)\right)} \varphi(p^k).$$

The general case of (4.7) now follows because $c_q(n)$ is a multiplicative function of q. □

4.1.1 Exercises

1. Let $U = [u_{kn}]$ be the $q \times q$ matrix with elements $u_{kn} = e(kn/q)/\sqrt{q}$. Show that $UU^* = U^*U = I$, i.e., that U is unitary.
2. (Friedman 1957; cf. Reznick 1995)
   (a) Show that
   $$\int_0^1 \left(ue(\theta/2) + ve(-\theta/2)\right)^{2r} d\theta = \binom{2r}{r} u^r v^r$$
   for any non-negative integer r and arbitrary complex numbers u, v.
   (b) Show that if $u = (x - iy)/2$, $v = (x + iy)/2$, then
   $$x \cos \pi\theta + y \sin \pi\theta = ue(\theta/2) + ve(-\theta/2)$$
   for all θ.
   (c) Show that
   $$\int_0^1 \left(x \cos \pi\theta + y \sin \pi\theta\right)^{2r} d\theta = \binom{2r}{r} 2^{-2r} (x^2 + y^2)^r$$
   for any non-negative integer r and arbitrary real or complex numbers x, y.
   (d) Show that
   $$\sum_{a=1}^q \left(ue^{\pi i a/q} + ve^{-\pi i a/q}\right)^{2r} = q \binom{2r}{r} u^r v^r$$
   if r is an integer, $0 \le r < q$.
   (e) Show that
   $$\sum_{a=1}^q (x \cos \pi a/q + y \sin \pi a/q)^{2r} = q \binom{2r}{r} 2^{-2r} (x^2 + y^2)^r$$
   if r is an integer, $0 \le r < q$.

3. Show that $|c_q(n)| \le (q, n)$.
4. (Carmichael 1932)
   (a) Show that if $q > 1$, then
   $$\sum_{n=1}^{q} c_q(n) = 0.$$
   (b) Show that if $q_1 \ne q_2$ and $[q_1, q_2] | N$, then
   $$\sum_{n=1}^{N} c_{q_1}(n) c_{q_2}(n) = 0.$$
   (c) Show that if $q | N$, then
   $$\sum_{n=1}^{N} c_q(n)^2 = N\varphi(q).$$
5. (Grytczuk 1981; cf. Redmond 1983) Show that
   $$\sum_{d|q} |c_d(n)| = 2^{\omega(q/(q,n))}(q, n).$$
6. (Ramanujan 1918) Show that
   $$\frac{\varphi(n)}{n} = \sum_{d=1}^{\infty} \frac{\mu(d)}{d^2} \sum_{q|d} c_q(n) = \sum_{q=1}^{\infty} a_q c_q(n)$$
   where
   $$a_q = \frac{6\mu(q)}{\pi^2 q^2} \prod_{p|q} \left(1 - \frac{1}{p^2}\right)^{-1}.$$
7. (Wintner 1943, Sections 33–35) The orthogonality relations of Exercise 4 give us hope that it might be possible to represent an arithmetic function $F(n)$ in the form
   $$F(n) = \sum_{q=1}^{\infty} a_q c_q(n) \tag{4.8}$$
   by taking
   $$a_q = \frac{1}{\varphi(q)} \lim_{x \to \infty} \frac{1}{x} \sum_{n \le x} F(n) c_q(n). \tag{4.9}$$
   In the following, suppose that $f(r)$ is chosen so that $F(n) = \sum_{r|n} f(r)$ for all n.
   (a) Suppose that
   $$\sum_{r=1}^{\infty} \frac{|f(r)|}{r} < \infty. \tag{4.10}$$

Let d be a fixed positive integer. Show that

$$\sum_{\substack{n \le x \\ d|n}} F(n) = \frac{x}{d} \sum_{r=1}^{\infty} \frac{f(r)}{r}(d, r) + o(x)$$

as $x \to \infty$.

(b) Suppose that (4.10) holds. Show that

$$\lim_{x \to \infty} \frac{1}{x} \sum_{n \le x} F(n)c_q(n) = \varphi(q) \sum_{\substack{r=1 \\ q|r}}^{\infty} \frac{f(r)}{r}.$$

(c) Put

$$a_q = \sum_{\substack{r=1 \\ q|r}}^{\infty} \frac{f(r)}{r}.$$

Show that if

$$\sum_{r=1}^{\infty} \frac{|f(r)|d(r)}{r} < \infty \tag{4.11}$$

then (4.8) and (4.9) hold, and moreover that $\sum_{q=1}^{\infty} |a_q c_q(n)| < \infty$.

8. (Ramanujan 1918) Show that if $q > 1$, then $\sum_{n=1}^{\infty} c_q(n)/n = -\Lambda(q)$. (See also Exercise 8.3.4.)

9. Let $\Phi_q(z)$ denote the qth cyclotomic polynomial, i.e., the monic polynomial whose roots are precisely the primitive q^{th} roots of unity, so that

$$\Phi_q(z) = \prod_{\substack{n=1 \\ (n,q)=1}}^{q} (z - e(n/q)).$$

(a) Show that

$$\Phi_q(z) = \prod_{d|q}(z^d - 1)^{\mu(q/d)}$$

and that $(z^d - 1)^{\mu(q/d)}$ has a power series expansion, valid when $|z| < 1$, with integer coefficients. Deduce that $\Phi_q(z) \in \mathbb{Z}[z]$.

(b) Suppose that $z \in \mathbb{Z}$ and $p \mid \Phi_q(z)$ and let e denote the order of z modulo p. Show that $e \mid q$ and that if $p \mid (z^d - 1)$ then $e \mid d$.

(c) Choose t so that $p^t \| (z^e - 1)$. Show that for $m \in \mathbb{N}$ with $p \nmid m$ one has $p^t \| (z^{me} - 1)$.

(d) Show that if $p \nmid q$, then $p^{ht} \| \Phi_q(z)$ where $h = \sum_{e|d|q} \mu(q/d)$. Deduce that $e = q$ and that $q \mid (p - 1)$.

(e) By taking z to be a suitable multiple of q, or otherwise, show that there are infinitely many primes p with $p \equiv 1 \pmod{q}$.

4.2 Dirichlet characters

In the preceding section we expressed the characteristic function of an arithmetic progression as a linear combination of additive characters. For purposes of multiplicative number theory we shall similarly represent the characteristic function of a reduced residue class (mod q) as a linear combination of totally multiplicative functions $\chi(n)$ each one supported on the reduced residue classes and having period q. These are the *Dirichlet characters*. Since $\chi(n)$ has period q we may think of it as mapping from residue classes, and since $\chi(n) \neq 0$ if and only if $(n, q) = 1$, we may think of χ as mapping from the multiplicative group of reduced residue classes to the multiplicative group $\mathbb{C}^\times$ of non-zero complex numbers. As χ is totally multiplicative, $\chi(mn) = \chi(m)\chi(n)$ for all m, n, we see that the map $\chi : (\mathbb{Z}/q\mathbb{Z})^\times \longrightarrow \mathbb{C}^\times$ is a homomorphism. The method we use to describe these characters applies when $(\mathbb{Z}/q\mathbb{Z})^\times$ is replaced by an arbitrary finite abelian group G, so we consider the slightly more general problem of finding all homomorphisms $\chi : G \to \mathbb{C}^\times$ from such a group G to $\mathbb{C}^\times$. We call these homomorphisms the characters of G, and let $\widehat{G}$ denote the set of all characters of G. We let χ_0 denote the *principal character*, whose value is identically 1. We note that if $\chi \in \widehat{G}$, then $\chi(e) = 1$ where e denotes the identity in G. Let n denote the order of G. If $g \in G$ and $\chi \in \widehat{G}$, then $g^n = e$, and hence $\chi(g^n) = 1$. Consequently $\chi(g)^n = 1$, and so we see that all values taken by characters are n^{th} roots of unity. In particular, this implies that $\widehat{G}$ is finite, since there can be at most n^n such maps. If χ_1 and χ_2 are two characters of G, then we can define a product character $\chi_1\chi_2$ by $\chi_1\chi_2(g) = \chi_1(g)\chi_2(g)$. For $\chi \in \widehat{G}$, let $\overline{\chi}$ be the character $\overline{\chi(g)}$. Then $\chi \cdot \overline{\chi} = \chi_0$, and we see that $\widehat{G}$ is a finite abelian group with identity χ_0. The following lemmas prepare for a full description of $\widehat{G}$ in Theorem 4.4.

Lemma 4.2 *Suppose that G is cyclic of order n, say $G = (a)$. Then there are exactly n characters of G, namely $\chi_k(a^m) = e(km/n)$ for $1 \leq k \leq n$. Moreover,*

$$\sum_{g \in G} \chi(g) = \begin{cases} n & \text{if } \chi = \chi_0, \\ 0 & \text{otherwise,} \end{cases} \tag{4.12}$$

and

$$\sum_{\chi \in \widehat{G}} \chi(g) = \begin{cases} n & \text{if } g = e, \\ 0 & \text{otherwise.} \end{cases} \tag{4.13}$$

In this situation, $\widehat{G}$ is cyclic, $\widehat{G} = (\chi_1)$.

Proof Suppose that $\chi \in \widehat{G}$. As we have observed, $\chi(a)$ is an nth root of unity, say $\chi(a) = e(k/n)$ for some k, $1 \leq k \leq n$. Hence $\chi(a^m) = \chi(a)^m = e(km/n)$.

Since the characters are now known explicitly, the remaining assertions are easily verified. □

Next we describe the characters of the direct product of two groups in terms of the characters of the factors.

Lemma 4.3 *Suppose that G_1 and G_2 are finite abelian groups, and that $G = G_1 \otimes G_2$. If χ_i is a character of G_i, $i = 1, 2$, and $g \in G$ is written $g = (g_1, g_2)$, $g_i \in G_i$, then $\chi(g) = \chi_1(g_1)\chi_2(g_2)$ is a character of G. Conversely, if $\chi \in \widehat{G}$, then there exist unique $\chi_i \in \widehat{G_i}$ such that $\chi(g) = \chi_1(g_1)\chi_2(g_2)$. The identities (4.12) and (4.13) hold for G if they hold for both G_1 and G_2.*

We see here that each $\chi \in \widehat{G}$ corresponds to a pair $(\chi_1, \chi_2) \in \widehat{G_1} \times \widehat{G_2}$. Thus $G \cong \widehat{G_1} \otimes \widehat{G_2}$.

Proof The first assertion is clear. As for the second, put $\chi_1(g_1) = \chi((g_1, e_2))$, $\chi_2(g_2) = \chi((e_1, g_2))$. Then $\chi_i \in \widehat{G_i}$ for $i = 1, 2$, and $\chi_1(g_1)\chi_2(g_2) = \chi(g)$. The χ_i are unique, for if $g = (g_1, e_2)$, then

$$\chi(g) = \chi((g_1, e_2)) = \chi_1(g_1)\chi_2(e_2) = \chi_1(g_1),$$

and similarly for χ_2. If $\chi(g) = \chi_1(g_1)\chi_2(g_2)$, then

$$\sum_{g \in G} \chi(g) = \left(\sum_{g_1 \in G_1} \chi_1(g_1)\right)\left(\sum_{g_2 \in G_2} \chi_2(g_2)\right),$$

so that (4.12) holds for G if it holds for G_1 and for G_2. Similarly, if $g = (g_1, g_2)$, then

$$\sum_{\chi \in \widehat{G}} \chi(g) = \left(\sum_{\chi_1 \in \widehat{G_1}} \chi_1(g_1)\right)\left(\sum_{\chi_1 \in \widehat{G_2}} \chi_2(g_2)\right),$$

so that (4.13) holds for G if it holds for G_1 and G_2. □

Theorem 4.4 *Let G be a finite abelian group. Then $\widehat{G}$ is isomorphic to G, and (4.12) and (4.13) both hold.*

Proof Any finite abelian group is isomorphic to a direct product of cyclic groups, say

$$G \cong C_{n_1} \otimes C_{n_2} \otimes \cdots \otimes C_{n_r}.$$

The result then follows immediately from the lemmas. □

Though G and $\widehat{G}$ are isomorphic, the isomorphism is not canonical. That is, no particular one-to-one correspondence between the elements of G and those of $\widehat{G}$ is naturally distinguished.

Corollary 4.5 *The multiplicative group* $(\mathbb{Z}/q\mathbb{Z})^\times$ *of reduced residue classes* (mod q) *has* $\varphi(q)$ *Dirichlet characters. If* χ *is such a character, then*

$$\sum_{\substack{n=1 \\ (n,q)=1}}^{q} \chi(n) = \begin{cases} \varphi(q) & \text{if } \chi = \chi_0, \\ 0 & \text{otherwise.} \end{cases} \tag{4.14}$$

If $(n, q) = 1$, *then*

$$\sum_{\chi} \chi(n) = \begin{cases} \varphi(q) & \text{if } n \equiv 1 \ (\text{mod } q), \\ 0 & \text{otherwise,} \end{cases} \tag{4.15}$$

where the sum is extended over the $\varphi(q)$ *Dirichlet characters* χ (mod q).

As we remarked at the outset, for our purposes it is convenient to define the Dirichlet characters (mod q) on all integers; we do this by setting $\chi(n) = 0$ when $(n, q) > 1$. Thus χ is a totally multiplicative function with period q that vanishes whenever $(n, q) > 1$, and any such function is a Dirichlet character (mod q). In this book a character is understood to be a Dirichlet character unless the contrary is indicated.

Corollary 4.6 *If* χ_i *is a character* (mod q_i) *for* $i = 1, 2$, *then* $\chi_1(n)\chi_2(n)$ *is a character* (mod $[q_1, q_2]$). *If* $q = q_1 q_2$, $(q_1, q_2) = 1$, *and* χ *is a character* (mod q), *then there exist unique characters* χ_i (mod q), $i = 1, 2$, *such that* $\chi(n) = \chi_1(n)\chi_2(n)$ *for all* n.

Proof The first assertion follows immediately from the observations that $\chi_1(n)\chi_2(n)$ is totally multiplicative, that it vanishes if $(n, [q_1, q_2]) > 1$, and that it has period $[q_1, q_2]$. As for the second assertion, we may suppose that $(n, q) = 1$. By the Chinese Remainder Theorem we see that

$$(\mathbb{Z}/q\mathbb{Z})^\times \cong (\mathbb{Z}/q_1\mathbb{Z})^\times \otimes (\mathbb{Z}/q_2\mathbb{Z})^\times$$

if $(q_1, q_2) = 1$. Thus the result follows from Lemma 4.2. $\qquad\square$

Our proof of Theorem 4.4 depends on Abel's theorem that any finite abelian group is isomorphic to the direct product of cyclic groups, but we can prove Corollary 4.5 without appealing to this result, as follows. By the Chinese Remainder Theorem we see that

$$(\mathbb{Z}/q\mathbb{Z})^\times \cong \bigotimes_{p^\alpha \| q} (\mathbb{Z}/p^\alpha\mathbb{Z})^\times.$$

If p is odd, then the reduced residue classes (mod p^α) form a cyclic group; in classical language we say there is a primitive root g. Thus if $(n, p) = 1$, then there is a unique v (mod $\varphi(p^\alpha)$) such that $g^v \equiv n$ (mod p^α). The number v is called the index of n, and is denoted $\nu = \text{ind}_g n$. From Lemma 4.2 it follows that the characters (mod p^α), $p > 2$, are given by

$$\chi_k(n) = e\left(\frac{k \, \text{ind}_g n}{\varphi(p^\alpha)}\right) \tag{4.16}$$

for $(n, p) = 1$. We obtain $\varphi(p^\alpha)$ different characters by allowing k to assume integral values in the range $1 \le k \le \varphi(p^\alpha)$. By Lemma 4.3 it follows that if q is odd, then the general character (mod q) is given by

$$\chi(n) = e\left(\sum_{p^\alpha \| q} \frac{k \, \text{ind}_g n}{\varphi(p^\alpha)}\right) \tag{4.17}$$

for $(n, q) = 1$, where it is understood that $k = k(p^\alpha)$ is determined (mod $\varphi(p^\alpha)$) and that $g = g(p^\alpha)$ is a primitive root (mod p^α).

The multiplicative structure of the reduced residues (mod 2^α) is more complicated. For $\alpha = 1$ or $\alpha = 2$ the group is cyclic (of order 1 or 2, respectively), and (4.16) holds as before. For $\alpha \ge 3$ the group is not cyclic, but if n is odd, then there exist unique μ (mod 2) and ν (mod $2^{\alpha-2}$) such that $n \equiv (-1)^\mu 5^\nu$ (mod 2^α). In group-theoretic terms this means that

$$(\mathbb{Z}/2^\alpha\mathbb{Z})^\times \cong C_2 \otimes C_{2^{\alpha-2}}$$

when $\alpha \ge 3$. By Lemma 4.3 the characters in this case take the form

$$\chi(n) = e\left(\frac{j\mu}{2} + \frac{k\nu}{2^{\alpha-2}}\right) \tag{4.18}$$

for odd n where $j = 0$ or 1 and $1 \le k \le 2^{\alpha-2}$. Thus (4.17) holds if $8 \nmid q$, but if $8 | q$, then the general character takes the form

$$\chi(n) = e\left(\frac{j\mu}{2} + \frac{k\nu}{2^{\alpha-2}} + \sum_{\substack{p^\alpha \| q \\ p>2}} \frac{\ell \, \text{ind}_g n}{\varphi(p^\alpha)}\right) \tag{4.19}$$

when $(n, q) = 1$.

By definition, if $f(n)$ is totally multiplicative, $f(n) = 0$ whenever $(n, q) > 1$, and $f(n)$ has period q, then f is a Dirichlet character (mod q). It is useful to note that the first condition can be relaxed.

Theorem 4.7 *If f is multiplicative, $f(n) = 0$ whenever $(n, q) > 1$, and f has period q, then f is a Dirichlet character modulo q.*

Proof It suffices to show that f is totally multiplicative. If $(mn, q) > 1$, then $f(mn) = f(m)f(n)$ since $0 = 0$. Suppose that $(mn, q) = 1$. Hence in particular $(m, q) = 1$, so that the map $k \mapsto n + kq$ (mod m) permutes the residue classes (mod m). Thus there is a k for which $n + kq \equiv 1$ (mod m), and consequently $(m, n + kq) = 1$. Then

$$
\begin{aligned}
f(mn) &= f(m(n+kq)) && \text{(by periodicity)} \\
&= f(m)f(n+kq) && \text{(by multiplicativity)} \\
&= f(m)f(n) && \text{(by periodicity)},
\end{aligned}
$$

and the proof is complete. $\qquad\square$

We shall discuss further properties of Dirichlet characters in Chapter 9.

4.2.1 Exercises

1. Let G be a finite abelian group of order n. Let $g_1, g_2, \ldots, g_n$ denote the elements of G, and let $\chi_1(g), \chi_2(g), \ldots, \chi_n(g)$ denote the characters of G. Let $U = [u_{ij}]$ be the $n \times n$ matrix with elements $u_{ij} = \chi_i(g_j)/\sqrt{n}$. Show that $UU^* = U^*U = I$, i.e., that U is unitary.
2. Show that for arbitrary real or complex numbers $c_1, \ldots, c_q$,

$$
\sum_{\chi} \left| \sum_{n=1}^{q} c_n \chi(n) \right|^2 = \varphi(q) \sum_{\substack{n=1 \\ (n,q)=1}}^{q} |c_n|^2
$$

where the sum on the left-hand side runs over all Dirichlet characters $\chi \pmod q$.
3. Show that for arbitrary real or complex numbers c_χ,

$$
\sum_{n=1}^{q} \left| \sum_{\chi} c_\chi \chi(n) \right|^2 = \varphi(q) \sum_{\chi} |c_\chi|^2
$$

where the sum over χ is extended over all Dirichlet characters $\pmod q$.
4. Let $(a, q) = 1$, and suppose that k is the order of a in the multiplicative group of reduced residue classes $\pmod q$.
   (a) Show that if χ is a Dirichlet character $\pmod q$, then $\chi(a)$ is a kth root of unity.
   (b) Show that if z is a kth root of unity, then

$$
1 + z + \cdots + z^{k-1} = \begin{cases} k & \text{if } z = 1, \\ 0 & \text{otherwise.} \end{cases}
$$

   (c) Let ζ be a kth root of unity. By taking $z = \chi(a)/\zeta$, show that each kth root of unity occurs precisely $\varphi(q)/k$ times among the numbers $\chi(a)$ as χ runs over the $\varphi(q)$ Dirichlet characters $\pmod q$.
5. Let χ be a Dirichlet character $\pmod q$, and let k denote the order of χ in the character group.
   (a) Show that if $(a, q) = 1$, then $\chi(a)$ is a kth root of unity.

(b) Show that each kth root of unity occurs precisely $\varphi(q)/k$ times among the numbers $\chi(a)$ as a runs over the $\varphi(q)$ reduced residue classes (mod q).

6. Let χ be a character (mod q) such that $\chi(a) = \pm 1$ whenever $(a, q) = 1$, and put $S(\chi) = \sum_{n=1}^{q} n\chi(n)$. Thus $S(\chi)$ is an integer.
   (a) Show that if $(a, q) = 1$ then $a\chi(a)S(\chi) \equiv S(\chi)$ (mod q).
   (b) Show that there is an a such that $(a, q) = 1$ and $(a\chi(a) - 1, q)|12$.
   (c) Deduce that $12S(\chi) \equiv 0$ (mod q).

In algebraic number fields we encounter not only Dirichlet characters, but also characters of ideal class groups and of Galois groups. In addition, algebraic number fields possessing one or more complex embeddings also have a further kind of character, Hecke's *Grössencharaktere*. In a sequence of exercises, beginning with the one below, we develop the basic properties of these characters for the Gaussian field $\mathbb{Q}(\sqrt{-1})$.

7. Let K be the Gaussian field,

$$K = \mathbb{Q}\left(\sqrt{-1}\right) = \{a + bi : a, b \in \mathbb{Q}\},$$

and let $\mathcal{O}_K$ be the ring of algebraic integers in K,

$$\mathcal{O}_K = \{a + bi : a, b \in \mathbb{Z}\}.$$

Elements $\alpha = a + bi \in K$ have a *norm*, $N(\alpha) = a^2 + b^2$, and we observe that $N(\alpha\beta) = N(\alpha)N(\beta)$. An element α of a ring is a *unit* if α has an inverse in the ring. The ring $\mathcal{O}_K$ has precisely four units, namely i^k for $k = 0, 1, 2, 3$. Two elements $\alpha, \beta \in \mathcal{O}_K$ are *associates* if $\alpha = u\beta$ for some unit u. For each integer m we define the Hecke *Grössencharakter*

$$\chi_m(\alpha) = \begin{cases} e^{4mi\,\arg\alpha} & \text{if } \alpha \neq 0, \\ 0 & \text{if } \alpha = 0. \end{cases}$$

(a) Show that if α and β are associates then $\chi_m(\alpha) = \chi_m(\beta)$.
(b) Show that $\chi_m(\alpha\beta) = \chi_m(\alpha)\chi_m(\beta)$ for all α and β in $\mathcal{O}_K$.

4.3 Dirichlet *L*-functions

Let χ be a character (mod q). For $\sigma > 1$ we put

$$L(s, \chi) = \sum_{n=1}^{\infty} \chi(n)n^{-s}. \tag{4.20}$$

Since χ is totally multiplicative, by Theorem 1.9 we have

$$L(s, \chi) = \prod_{p}(1 - \chi(p)p^{-s})^{-1} \tag{4.21}$$

for $\sigma > 1$. Thus we see that

$$L(s, \chi_0) = \sum_{\substack{n=1 \\ (n,q)=1}}^{\infty} n^{-s} = \zeta(s) \prod_{p|q} \left(1 - p^{-s}\right) \tag{4.22}$$

for $\sigma > 1$. By (4.14) we see that if $\chi \neq \chi_0$, then

$$\sum_{1 \leq n \leq kq} \chi(n) = 0$$

for $k = 1, 2, 3, \ldots$. Hence

$$\left| \sum_{n \leq x} \chi(n) \right| \leq q \tag{4.23}$$

for any x, so that by Theorem 1.3, the series (4.20) converges for $\sigma > 0$. This result is best possible since the terms in (4.20) do not tend to 0 when $\sigma = 0$. On the other hand, we shall show in Chapter 10 that the function $L(s, \chi)$ is entire if $\chi \neq \chi_0$. For $\sigma > 1$ we can take logarithms in (4.21), and differentiate, as in Corollary 1.11, and thus we obtain

Theorem 4.8 *If $\chi \neq \chi_0$, then $L(s, \chi)$ is analytic for $\sigma > 0$. On the other hand, the function $L(s, \chi_0)$ is analytic in this half-plane except for a simple pole at $s = 1$ with residue $\varphi(q)/q$. In either case,*

$$\log L(s, \chi) = \sum_{n=2}^{\infty} \frac{\Lambda(n)}{\log n} \chi(n) n^{-s} \tag{4.24}$$

for $\sigma > 1$, and

$$-\frac{L'}{L}(s, \chi) = \sum_{n=1}^{\infty} \Lambda(n) \chi(n) n^{-s}. \tag{4.25}$$

In these last formulæ we see how relations for L-functions parallel those for the zeta functions. Indeed, when manipulating Dirichlet series formally, the only property of n^{-s} that is used is that it is totally multiplicative. Hence all such calculations can be made with n^{-s} replaced by $\chi(n)n^{-s}$. For example, we know that $\sum \mu(n)^2 n^{-s} = \zeta(s)/\zeta(2s)$ for $\sigma > 1$. Hence formally

$$\sum_{n=1}^{\infty} \mu(n)^2 \chi(n) n^{-s} = L(s, \chi)/L(2s, \chi^2). \tag{4.26}$$

Since $|\chi(n)n^{-s}| \leq n^{-\sigma}$, this latter series is absolutely convergent whenever the former one is, and by (4.21) we see that (4.26) holds for $\sigma > 1$. In fact, by a theorem of Stieltjes (see Exercise 1.3.2), the identity (4.26) holds for $\sigma > 1/2$ if $\chi \neq \chi_0$.

We now use the identity (4.15) to capture a prescribed residue class. If $(a, q) = 1$, then

$$\frac{1}{\varphi(q)} \sum_{\chi} \overline{\chi}(a)\chi(n) = \begin{cases} 1 & \text{if } n \equiv a \pmod{q}, \\ 0 & \text{otherwise} \end{cases} \tag{4.27}$$

where the sum is extended over all characters $\chi \pmod{q}$. This is the multiplicative analogue of (4.1). Hence if $(a, q) = 1$ then

$$\sum_{\substack{n=1 \\ n \equiv a\,(q)}}^{\infty} \Lambda(n)n^{-s} = \frac{1}{\varphi(q)} \sum_{n=1}^{\infty} \Lambda(n)n^{-s} \sum_{\chi} \overline{\chi}(a)\chi(n)$$

$$= \frac{-1}{\varphi(q)} \sum_{\chi} \overline{\chi}(a)\frac{L'}{L}(s, \chi) \tag{4.28}$$

for $\sigma > 1$. As $L(s, \chi_0)$ has a simple pole at $s = 1$, the function $\frac{L'}{L}(s, \chi)$ has a simple pole at 1 with residue -1. Thus the term arising from χ_0 on the right-hand side above is

$$\frac{1}{\varphi(q)(s - 1)} + O_q(1) \tag{4.29}$$

as $s \to 1^+$. This enables us to prove that there are infinitely many primes $p \equiv a \pmod{q}$, provided that we can show that the terms from $\chi \neq \chi_0$ on the right-hand side of (4.28) do not interfere with the main term (4.29). But $L(s, \chi)$ is analytic for $\sigma > 0$, so that $\frac{L'}{L}(s, \chi)$ is analytic except at zeros of $L(s, \chi)$. Hence

$$\lim_{s \to 1^+} \frac{L'}{L}(s, \chi) = \frac{L'}{L}(1, \chi) \tag{4.30}$$

for $\chi \neq \chi_0$, provided that $L(1, \chi) \neq 0$. Thus the following result lies at the heart of the matter.

Theorem 4.9   (Dirichlet) *If χ is a character (mod q) with $\chi \neq \chi_0$, then $L(1, \chi) \neq 0$.*

Suppose that $(a, q) = 1$. Then the above, with (4.28), (4.29), and (4.30) give the estimate

$$\sum_{\substack{n=1 \\ n \equiv a\,(q)}}^{\infty} \Lambda(n)n^{-s} = \frac{1}{\varphi(q)(s - 1)} + O_q(1)$$

as $s \to 1^+$. Consequently

$$\sum_{\substack{n=1 \\ n\equiv a\,(q)}}^{\infty} \frac{\Lambda(n)}{n} = \infty.$$

Here the contribution of the proper prime powers is

$$\sum_{\substack{p^k\equiv a\,(q) \\ k\geq 2}} \frac{\log p}{p^k} \leq \sum_p \log p \sum_{k=2}^{\infty} p^{-k} = \sum_p \frac{\log p}{p(p-1)} < \infty, \qquad (4.31)$$

and thus we have

Corollary 4.10 (Dirichlet's theorem) *If $(a,q) = 1$, then there are infinitely many primes $p \equiv a \pmod q$, and indeed*

$$\sum_{p\equiv a\,(q)} \frac{\log p}{p} = \infty.$$

We call a character *real* if all its values are real (i.e., $\chi(n) = 0$ or ± 1 for all n). Otherwise a character is *complex*. A character is *quadratic* if it has order 2 in the character group: $\chi^2 = \chi_0$ but $\chi \neq \chi_0$. Thus a quadratic character is real, and a real character is either principal or quadratic. In Chapter 9 we shall express quadratic characters in terms of the Kronecker symbol $\left(\frac{d}{n}\right)$.

Proof of Theorem 4.9 We treat quadratic and complex characters separately.
Case 1: Complex χ. From (4.24) we have

$$\prod_\chi L(s,\chi) = \exp\left(\sum_\chi \sum_{n=2}^{\infty} \frac{\Lambda(n)}{\log n}\chi(n)n^{-s}\right)$$

for $\sigma > 1$. By (4.15) this is

$$= \exp\left(\varphi(q) \sum_{\substack{n=2 \\ n\equiv 1\,(q)}}^{\infty} \frac{\Lambda(n)}{\log n}n^{-s}\right).$$

If we take $s = \sigma > 1$, then the sum above is a non-negative real number, and hence we see that

$$\prod_\chi L(\sigma,\chi) \geq 1 \qquad (4.32)$$

for $\sigma > 1$. Now $L(s,\chi_0)$ has a simple pole at $s = 1$, but the other $L(s,\chi)$ are analytic at $s = 1$. Thus $L(1,\chi) = 0$ can hold for at most one χ, since otherwise the product in (4.32) would tend to 0 as $\sigma \to 1^+$. If χ is a character $\pmod q$, then $\overline{\chi}$ is a character $\pmod q$, and $\chi \neq \overline{\chi}$ if χ is complex. Moreover

$\overline{L(s,\chi)} = L(\overline{s}, \overline{\chi})$ by the Schwarz reflection principle, so that $L(1, \overline{\chi}) = 0$ if $L(1, \chi) = 0$. Consequently $L(1, \chi) \neq 0$ for complex χ.

Case 2: Quadratic χ. Let $r(n) = \sum_{d|n} \chi(d)$. Thus $\sum_{n=1}^{\infty} r(n)n^{-s} = \zeta(s)L(s, \chi)$ for $\sigma > 1, r(n)$ is multiplicative, and

$$r(p^{\alpha}) = \begin{cases} 1 & \text{if } p \mid q, \\ \alpha + 1 & \text{if } \chi(p) = 1, \\ 1 & \text{if } \chi(p) = -1 \text{ and } 2 \mid \alpha, \\ 0 & \text{if } \chi(p) = -1 \text{ and } 2 \nmid \alpha. \end{cases}$$

Hence $r(n) \geq 0$ for all n, and $r(n^2) \geq 1$ for all n. Suppose that $L(1, \chi) = 0$. Then $\zeta(s)L(s, \chi)$ is analytic for $\sigma > 0$, and by Landau's theorem (Theorem 1.7) the series $\sum r(n)n^{-s}$ converges for $\sigma > 0$. But this is false, since

$$\sum_{n=1}^{\infty} r(n)n^{-1/2} \geq \sum_{n=1}^{\infty} r(n^2)n^{-1} \geq \sum_{n=1}^{\infty} n^{-1} = +\infty.$$

Hence $L(1, \chi) \neq 0$. Since $L(\sigma, \chi) > 0$ for $\sigma > 1$ when χ is quadratic, we see in fact that $L(1, \chi) > 0$ in this case. $\square$

By using the techniques of Chapter 2 we can prove more than the mere divergence of the series in Corollary 4.10.

Theorem 4.11 *Suppose that χ is a non-principal Dirichlet character. Then for $x \geq 2$,*

(a)
$$\sum_{n \leq x} \frac{\chi(n)\Lambda(n)}{n} \ll_{\chi} 1,$$

(b)
$$\sum_{p \leq x} \frac{\chi(p)\log p}{p} \ll_{\chi} 1,$$

(c)
$$\sum_{p \leq x} \frac{\chi(p)}{p} = b(\chi) + O_{\chi}\left(\frac{1}{\log x}\right),$$

(d)
$$\prod_{p \leq x} \left(1 - \frac{\chi(p)}{p}\right)^{-1} = L(1, \chi) + O_{\chi}\left(\frac{1}{\log x}\right)$$

where

$$b(\chi) = \log L(1, \chi) - \sum_{\substack{p^k \\ k>1}} \frac{\chi(p^k)}{kp^k}.$$

Proof   We show first that

$$\sum_{n \leq x} \frac{\chi(n)\log n}{n} = -L'(1, \chi) + O_q\left(\frac{\log x}{x}\right). \tag{4.33}$$

To this end we put $S(x) = \sum_{n \leq x} \chi(n)$. Then from (4.23) we see that $S(x) \ll_{\chi} 1$.

Thus the error term above is

$$\sum_{n>x} \frac{\chi(n)\log n}{n} = \int_x^\infty \frac{\log u}{u}\, dS(u)$$

$$= -\frac{S(x)\log x}{x} - \int_x^\infty S(u)(1-\log u)u^{-2}\, du$$

$$\ll_\chi \frac{\log x}{x}.$$

As $\log n = \sum_{d|n} \Lambda(d)$, the left-hand side of (4.33) is

$$\sum_{md\le x} \frac{\Lambda(d)\chi(md)}{md} = \sum_{d\le x} \frac{\Lambda(d)\chi(d)}{d} \sum_{m\le x/d} \frac{\chi(m)}{m}. \qquad (4.34)$$

Here the inner sum is of the form

$$\sum_{m\le y} \frac{\chi(m)}{m} = L(1,\chi) - \sum_{m>y} \frac{\chi(m)}{m},$$

and this last sum is

$$\int_y^\infty u^{-1}\, dS(u) = -\frac{S(y)}{y} + \int_y^\infty S(u)u^{-2}\, du \ll_\chi y^{-1}.$$

Hence the right-hand side of (4.34) is

$$L(1,\chi)\sum_{d\le x} \frac{\Lambda(d)\chi(d)}{d} + O_\chi\left(\frac{1}{x}\sum_{d\le x}\Lambda(d)\right).$$

This last error term is $\ll_\chi 1$, and then (a) follows from (4.33) and the fact that $L(1,\chi)\ne 0$. The derivation of (b) from (a), and of (c) from (b) proceeds as in the proof of Theorem 2.7. Continuing as in that proof, we see from (c) that

$$\sum_{1<n\le x} \frac{\Lambda(n)\chi(n)}{n\log n} = c(\chi) + O_\chi\left(\frac{1}{\log x}\right)$$

where

$$c(\chi) = b(\chi) + \sum_{\substack{p^k \\ k>1}} \frac{\chi(p^k)}{kp^k}.$$

We let $s \to 1^+$ in (4.24), and deduce by Theorem 1.1 that $c(\chi) = \log L(1,\chi)$. To complete the derivation of (d) it suffices to argue as in the proof of Theorem 2.7. $\square$

By forming a linear combination of these estimates as in (4.27) we obtain

Corollary 4.12 *If $(a,q)=1$ and $x \ge 2$, then*

(a)
$$\sum_{\substack{n\le x \\ n\equiv a\,(q)}} \frac{\Lambda(n)}{n} = \frac{1}{\varphi(q)}\log x + O_q(1),$$

(b)
$$\sum_{\substack{p \leq x \\ n \equiv a\,(q)}} \frac{\log p}{p} = \frac{1}{\varphi(q)} \log x + O_q(1),$$

(c)
$$\sum_{\substack{p \leq x \\ n \equiv a\,(q)}} \frac{1}{p} = \frac{1}{\varphi(q)} \log \log x + b(q, a) + O_q\left(\frac{1}{\log x}\right),$$

(d)
$$\prod_{\substack{p \leq x \\ n \equiv a\,(q)}} \left(1 - \frac{1}{p}\right)^{-1} = c(q, a)(\log x)^{1/\varphi(q)}\left(1 + O_q\left(\frac{1}{\log x}\right)\right)$$

where

$$b(q, a) = \frac{1}{\varphi(q)}\left(C_0 + \sum_{p|q}\log\left(1 - \frac{1}{p}\right) + \sum_{\chi \neq \chi_0}\overline{\chi}(a)\log L(1, \chi)\right) - \sum_{\substack{p^k \equiv a\,(q) \\ k > 1}}\frac{1}{kp^k}$$

and

$$c(q, a) = \left(e^{C_0}\frac{\varphi(q)}{q}\prod_{\chi \neq \chi_0}\left(L(1, \chi)^{\overline{\chi}(a)}\prod_p\left(1 - \frac{1}{p}\right)^{-\chi(p)}\left(1 - \frac{\chi(p)}{p}\right)\right)\right)^{1/\varphi(q)}.$$

Proof   To derive (a) from Theorem 4.11(a) we use (4.27) and the estimate

$$\sum_{n \leq x}\frac{\Lambda(n)\chi_0(n)}{n} = \log x + O_q(1),$$

which follows from Theorem 2.7(a) since

$$\sum_{\substack{p^k \\ p|q}}\frac{\log p}{p^k} = \sum_{p|q}\frac{\log p}{p - 1} \ll_q 1.$$

We derive (b) and (c) similarly from the corresponding parts of Theorem 4.11. In the latter case we use the estimate

$$\sum_{p \leq x}\frac{\chi_0(p)}{p} = \log \log x + b(\chi_0) + O_q\left(\frac{1}{\log x}\right)$$

where

$$b(\chi_0) = C_0 + \sum_{p|q}\log\left(1 - \frac{1}{p}\right) - \sum_{\substack{p^k \\ k > 1}}\frac{\chi_0(p^k)}{kp^k}.$$

To derive (d) we observe first that

$$\prod_{p \leq x}\left(1 - \frac{\chi_0(p)}{p}\right)^{-1} = \prod_{\substack{p \leq x \\ p|q}}\left(1 - \frac{1}{p}\right)\prod_{p \leq x}\left(1 - \frac{1}{p}\right)^{-1},$$

which by Theorem 2.7(e) is

$$= \frac{\varphi(q)}{q}\left(\prod_{\substack{p|q \\ p > x}}\left(1 - \frac{1}{p}\right)\right)^{-1}e^{-C_0}(\log x)\left(1 + O\left(\frac{1}{\log x}\right)\right).$$

Here each term in the product is $1 + O(1/x)$, and the number of factors is $\leq \omega(q)$, so the product is $1 + O_q(1/x)$, and hence the above is

$$= e^{C_0} \frac{\varphi(q)}{q} (\log x) \left(1 + O_q \left(\frac{1}{\log x}\right)\right).$$

To complete the proof it suffices to combine this with Theorem 4.11(d) in (4.27). $\qquad\square$

4.3.1 Exercises

1. Let χ be a Dirichlet character (mod q). Show that if $\sigma > 1$, then

(a) $$\sum_{n=1}^{\infty} (-1)^{n-1} \chi(n) n^{-s} = (1 - \chi(2) 2^{1-s}) L(s, \chi);$$

(b) $$\sum_{n=1}^{\infty} d(n)^2 \chi(n) n^{-s} = \frac{L(s, \chi)^4}{L(2s, \chi^2)}.$$

2. (Mertens 1895a,b) Let $r(n) = \sum_{d|n} \chi(d)$.
   (a) Show that if χ is a non-principal character (mod q), then

   $$\sum_{n>x} \frac{\chi(n)}{\sqrt{n}} \ll_\chi \frac{1}{\sqrt{x}}.$$

   (b) Show that if χ is a non-principal character (mod q), then

   $$\sum_{n \leq x} \frac{r(n)}{n^{1/2}} = 2x^{1/2} L(1, \chi) + O_\chi(1).$$

   (c) Recall that if χ is quadratic then $r(n) \geq 0$ for all n, and that $r(n^2) \geq 1$. Deduce that if χ is a quadratic character, then the left-hand side above is $\gg \log x$.
   (d) Conclude that if χ is a quadratic character, then $L(1, \chi) > 0$.

3. (Mertens 1897, 1899) For $u \geq 0$, put $f(u) = \sum_{m \leq u} (1 - m/u)$.
   (a) Show that $f(u) \geq 0$, that $f(u)$ is continuous, and that if u is not an integer, then

   $$f'(u) = \frac{[u]([u] + 1)}{2u^2};$$

   deduce that f is increasing.
   (b) Show also that

   $$f(u) = \frac{u}{2} - \frac{1}{u} \int_0^u \{v\} \, dv = \frac{u}{2} - \frac{1}{2} + O(1/u).$$

   (c) Let $r(n) = \sum_{d|n} \chi(d)$, and assume that χ is non-principal. Show that

   $$\sum_{n \leq x} r(n)(1 - n/x) = \sum_{d \leq x} \chi(d) f(x/d).$$

(d) Write $\sum_{d\le x} = \sum_{d\le y} + \sum_{y<d\le x} = S_1 + S_2$ where $1 \le y \le x$. Use part (b) to show that $S_1 = \frac{1}{2}xL(1, \chi) + O_\chi(x/y) + O(y^2/x)$.

(e) Use the results of part (a) to show that $S_2 \ll_\chi f(x/y)$.

(f) By making an appropriate choice of y, deduce that if χ is a non-principal character, then

$$\sum_{n\le x} r(n)(1 - n/x) = \frac{x}{2}L(1, \chi) + O_\chi\left(x^{1/3}\right).$$

(g) Argue that if χ is a quadratic character, then the left-hand side above is $\gg x^{1/2}$; deduce that $L(1, \chi) > 0$.

4. (Ingham 1929) Let $f_1(n)$ and $f_2(n)$ be totally multiplicative functions, and suppose that $|f_i(n)| \le 1$ for all n.

(a) Show that if $\sigma > 1$, then

$$\sum_{n=1}^{\infty}\left(\sum_{d|n} f_1(d)\right)\left(\sum_{d|n} f_2(d)\right) n^{-s}$$

$$= \frac{\zeta(s)\left(\sum_{n=1}^{\infty} f_1(n)n^{-s}\right)\left(\sum_{n=1}^{\infty} f_2(n)n^{-s}\right)\left(\sum_{n=1}^{\infty} f_1(n)f_2(n)n^{-s}\right)}{\sum_{n=1}^{\infty} f_1(n)f_2(n)n^{-2s}}$$

$$= \frac{\prod_p\left(1 - \frac{f_1(p)f_2(p)}{p^{2s}}\right)}{\prod_p\left(1 - \frac{1}{p^s}\right)\left(1 - \frac{f_1(p)}{p^s}\right)\left(1 - \frac{f_2(p)}{p^s}\right)\left(1 - \frac{f_1(p)f_2(p)}{p^s}\right)}.$$

(b) By considering

$$F(s) = \sum_{n=1}^{\infty}\left|\sum_{d|n} \chi(d)d^{-iu}\right|^2 n^{-s},$$

show that $L(1 + iu, \chi) \ne 0$.

5. Let $\pi(x; q, a)$ denote the number of primes $p \equiv a \pmod{q}$ with p not exceeding x. Similarly, let

$$\vartheta(x; q, a) = \sum_{\substack{p\le x \\ p\equiv a\,(q)}} \log p, \qquad \psi(x; q, a) = \sum_{\substack{n\le x \\ n\equiv a\,(q)}} \Lambda(n).$$

(a) Show that

$$\vartheta(x; q, a) = \psi(x; q, a) + O\left(x^{1/2}\right).$$

(b) Show that

$$\pi(x; q, a) = \frac{\vartheta(x; q, a)}{\log x} + O\left(\frac{x}{(\log x)^2}\right).$$

(c) Show that if $x \geq C, C \geq 2$, and $(a, q) = 1$, then

$$\sum_{\substack{x/C < p \leq x \\ p \equiv a \,(q)}} \frac{\log p}{p} = \frac{\log C}{\varphi(q)} + O_q(1).$$

(d) Show that for any positive integer q there is a small number c_q and a large number C_q such that if $x \geq 2C_q$ and $(a, q) = 1$, then

$$\sum_{\substack{x/C_q < p \leq x \\ p \equiv a \,(q)}} \frac{\log p}{p} > c_q.$$

(e) Show that for any positive integer q there is a C_q such that if $(a, q) = 1$, then

$$\pi(x; q, a) \gg_q \frac{x}{\log x}$$

uniformly for $x \geq C_q$.

(f) Show that if $(a, q) = 1$, then

$$\liminf_{x \to \infty} \frac{\pi(x; q, a)}{x/\log x} \leq \frac{1}{\varphi(q)}, \qquad \limsup_{x \to \infty} \frac{\pi(x; q, a)}{x/\log x} \geq \frac{1}{\varphi(q)}.$$

6. (a) Show that

$$\vartheta(x) \leq \pi(x) \log x \leq \vartheta(x) + O\left(\frac{x}{\log x}\right)$$

for $x \geq 2$.

(b) Let $\mathcal{P}$ denote a set of prime numbers, and put

$$\pi_{\mathcal{P}}(x) = \sum_{\substack{p \leq x \\ p \in \mathcal{P}}} 1, \qquad \vartheta_{\mathcal{P}}(x) = \sum_{\substack{p \leq x \\ p \in \mathcal{P}}} \log p.$$

Show that

$$\vartheta_{\mathcal{P}}(x) = \pi_{\mathcal{P}}(x) \log x + O\left(\frac{x}{\log x}\right)$$

for $x \geq 2$, where the implicit constant is absolute.

(c) Let

$$n = \prod_{\substack{p \leq y \\ p \in \mathcal{P}}} p.$$

Show that $\log n = \omega(n) \log y + O(y/\log y)$ for $y \geq 2$.

(d) From now on, assume that $\vartheta_{\mathcal{P}}(x) \gg x$ for all sufficiently large x, where the implicit constant may depend on $\mathcal{P}$. Show that $\log \log n = \log y + O_{\mathcal{P}}(1)$.

(e) Deduce that

$$d(n) = n^{(\log 2 + o(1))/\log\log n}$$

as $y \to \infty$.

7. Let $R(n)$ denote the number of ordered pairs a, b such that $a^2 + b^2 = n$ with $a \geq 0$ and $b > 0$. Also, let $r(n)$ denote the number of such pairs for which $(a, b) = 1$. Finally, let $\chi_{-4} = \left(\frac{-4}{n}\right)$ be the non-principal character (mod 4). We recall that if the prime factorization of n is written in the form

$$n = 2^{\alpha} \prod_{\substack{p^{\beta} \| n \\ p \equiv 1\,(4)}} p^{\beta} \prod_{\substack{q^{\gamma} \| n \\ q \equiv 3\,(4)}} q^{\gamma},$$

then $r(n) > 0$ if and only if $\gamma = 0$ for all primes q and $\alpha \leq 1$. We also recall that

$$R(n) = \sum_{d^2 \mid n} r(n/d^2) = \sum_{d\mid n} \chi_{-4}(d) = \begin{cases} \prod_p (\beta + 1) & \text{if } 2\mid\gamma \text{ for all } q, \\ 0 & \text{otherwise.} \end{cases}$$

(a) Show that $\sum_{n=1}^{\infty} R(n)n^{-s} = \zeta(s)L(s, \chi_{-4})$ for $\sigma > 1$.
(b) Show that $\sum_{n=1}^{\infty} r(n)n^{-s} = \zeta(s)L(s, \chi_{-4})/\zeta(2s)$ for $\sigma > 1$.
(c) Show that if $x \geq 0$ and $y \geq 2$, then

$$\text{card}\{n \in (x, x+y] : r(n) > 0\} \ll \frac{y}{\sqrt{\log y}}.$$

(d) Show that

$$\text{card}\{n \leq x : R(n) > 0\} \ll \frac{x}{\sqrt{\log x}}$$

for $x \geq 2$.

(e) Suppose that n is of the form

$$n = \prod_{\substack{p \leq y \\ p \equiv 1\,(4)}} p.$$

Thus $\log n = \vartheta(y; , 4, 1) \asymp y$ for $y \geq 5$, and hence $\log y = \log\log n + O(1)$. Show that for such n,

$$R(n) = n^{(\log 2 + o(1))/\log\log n}.$$

In the above it is noteworthy that although $R(n) \leq d(n)$ for all n, that $R(n)$ is usually 0 and has a smaller average value (cf. Exercise 2.1.9) than $d(n)$ (cf. Theorem 2.3), the maximum order of magnitude of $R(n)$ is the same as for $d(n)$.

8. Let $K = \mathbb{Q}(\sqrt{-1})$ be the Gaussian field, $\mathcal{O}_K = \{a + ib : a, b \in \mathbb{Z}\}$ the ring of integers in K. Ideals $\mathfrak{a}$ in $\mathcal{O}_K$ are principal, $\mathfrak{a} = (a + ib)$, and have norm $N(\mathfrak{a}) = a^2 + b^2$.

(a) Explain why the number of ideals $\mathfrak{a}$ with $N(\mathfrak{a}) \leq x$ is $\frac{\pi}{4}x + O(x^{1/2})$.

(b) For $\sigma > 1$, let $\zeta_K(s) = \sum_{\mathfrak{a}} N(\mathfrak{a})^{-s}$ be the Dedekind zeta function of K. Show that $\zeta_K(s) = \zeta(s)L(s, \chi_{-4})$.

(c) For the Gaussian field K, show that $N(\mathfrak{a}\mathfrak{b}) = N(\mathfrak{a})N(\mathfrak{b})$. (This is true in any algebraic number field.)

(d) Assume that ideals in K factor uniquely into prime ideals. (This is true in any algebraic number field, and is particularly easy to establish for the Gaussian field since it has a division algorithm.) Deduce that if $\sigma > 1$, then

$$\zeta_K(s) = \prod_{\mathfrak{p}} \left(1 - \frac{1}{N(\mathfrak{p})}\right)^{-1}$$

where the product runs over all prime ideals $\mathfrak{p}$ in $\mathcal{O}_K$.

(e) Define a function $\mu(\mathfrak{a}) = \mu_K(\mathfrak{a})$ in such a way that

$$\frac{1}{\zeta_K(s)} = \sum_{\mathfrak{a}} \frac{\mu(\mathfrak{a})}{N(\mathfrak{a})^s}$$

for $\sigma > 1$.

(f) Let $\mathfrak{a}$ and $\mathfrak{b}$ be given ideals. Show that

$$\sum_{\substack{\mathfrak{d}|\mathfrak{a} \\ \mathfrak{d}|\mathfrak{b}}} \mu(\mathfrak{d}) = \begin{cases} 1 & \text{if } \gcd(\mathfrak{a}, \mathfrak{b}) = 1, \\ 0 & \text{otherwise.} \end{cases}$$

(g) Among pairs $\mathfrak{a}$, $\mathfrak{b}$ of ideals with $N(\mathfrak{a}) \leq x$, $N(\mathfrak{b}) \leq x$, show that the probability that $\gcd(\mathfrak{a}, \mathfrak{b}) = 1$ is

$$\frac{1}{\zeta_K(2)} + O\left(x^{-1/2}\right) = \frac{6}{\pi^2 L(2, \chi_{-4})} + O\left(x^{1/2}\right).$$

9. (Erdős 1946, 1949, 1957, Vaughan 1974, Saffari, unpublished, but see Bateman, Pomerance & Vaughan 1981; cf. Exercise 2.3.7) Let $\Phi_q(z) = \prod_{d|q}(z^d - 1)^{\mu(q/d)}$ denote the qth cyclotomic polynomial. Suppose that

$$q = \prod_{\substack{p \leq y \\ p \equiv \pm 2 \ (5)}} p$$

where y is chosen so that $\omega(q)$ is odd.

(a) Show that if $d|q$ and $\omega(d)$ is even, then $|e(d/5) - 1| = |e(1/5) - 1|$.

(b) Show that if $d|q$ and $\omega(d)$ is odd, then $|e(d/5) - 1| = |e(2/5) - 1|$.

(c) Deduce that $|\Phi_q(e(1/5))| = |e(1/5) + 1|^{d(q)/2}$.

(d) Deduce that $\Phi_q(z)$ has a coefficient whose absolute value is at least

$$\exp\left(q^{(\log 2 - \varepsilon)/\log\log q}\right)$$

if $y > y_0(\varepsilon)$.

10. *Grössencharaktere* for $\mathbb{Q}(\sqrt{-1})$, continued from Exercise 4.2.7.

(a) For $\sigma > 1$ put

$$L(s, \chi_m) = \sum_{\alpha \in \mathcal{O}_K}' \chi_m(\alpha) N(\alpha)^{-s} = \frac{1}{4} \sum_{\substack{a,b \in \mathbb{Z} \\ (a,b)\neq(0,0)}} \chi_m(a + bi)(a^2 + b^2)^{-s}$$

where $\sum_\alpha'$ denotes a sum over unassociated members of $\mathcal{O}_K$. Show that the above sum is absolutely convergent in this half-plane.

(b) We recall that members of $\mathcal{O}_K$ factor uniquely into Gaussian primes. Also, the Gaussian primes are obtained by factoring the rational primes: The prime 2 ramifies, $2 = i^3(1 + i)^2$, the rational primes $p \equiv 1 \pmod 4$ split into two distinct Gaussian primes, $p = (a + bi)(a - bi)$, and the rational primes $q \equiv 3 \pmod 4$ are inert. Show that

$$L(s, \chi_m) = \prod_{\mathfrak{p}}(1 - \chi_m(\mathfrak{p})N(\mathfrak{p})^{-s})^{-1}$$

for $\sigma > 1$ where the product is over an unassociated family of Gaussian primes $\mathfrak{p}$.

(c) By grouping associates together, show that if $4 \nmid m$, then the sum

$$\sum_{\substack{a,b \in \mathbb{Z} \\ (a,b)\neq(0,0)}} e^{mi \arg(a+bi)}(a^2 + b^2)^{-s}$$

vanishes identically for $\sigma > 1$.

(d) For $0 \le \theta \le 2\pi$, put $N(x;\theta) = \text{card}\{(a, b) \in \mathbb{Z}^2 : a^2 + b^2 \le x, 0 < \arg(a + bi) \le \theta\}$. Show that for $x \ge 1$,

$$N(x;\theta) = \frac{\theta}{2}x + O\left(x^{1/2}\right)$$

uniformly in θ.

(e) Show that if $m \neq 0$, then

$$\sum_{\substack{a^2+b^2 \le x \\ a>0, b\ge 0}} \chi_m(a + bi) = \int_0^{\pi/2} e^{4mi\theta}\, dN(x;\theta) \ll |m|x^{1/2}.$$

(f) Show that if $m \neq 0$, then the Dirichlet series $L(s, \chi_m)$ is convergent for $\sigma > 1/2$.

(g) Show that $L(s, \chi_m)$ and $L(s, \chi_{-m})$ are identically equal, and hence that $L(\sigma, \chi_m) \in \mathbb{R}$ for $\sigma > 1/2$.

4.4 Notes

Section 4.1. Ramanujan's sum was introduced by Ramanujan (1918). Incredibly, both Hardy and Ramanujan missed the fact that $c_q(n)$ be written in closed form: The formula on the extreme right of (4.7) is due to Hölder (1936). Normally one would say that a function f is even if $f(x) = f(-x)$. However, in the present context, an arithmetic function f with period q is said to be *even* if $f(n)$ is a function only of (n, q). Thus $c_q(n)$ is an even function. The space of almost-even functions is rather small, but includes several arithmetic functions of interest. For such functions one may hope for a representation in the form $f(n) = \sum_{q=1}^{\infty} a_q c_q(n)$, called a *Ramanujan expansion*. For a survey of the theory of such expansions, see Schwarz (1988). Hildebrand (1984) established definitive results concerning the pointwise convergence of Ramanujan expansions. An appropriate Parseval identity has been established for mean-square summable almost-even functions; see Hildebrand, Schwarz & Spilker (1988).

Section 4.2. The first instance of characters of a non-cyclic group occurs in Gauss's analysis of the genus structure of the class group of binary quadratic forms. The quotient of the class group by the principal genus is isomorphic to $C_2 \otimes C_2 \otimes \cdots \otimes C_2$, and the associated characters are given by Kronecker's symbol. Dirichlet (1839) defined the Dirichlet characters for the multiplicative group $(\mathbb{Z}/q\mathbb{Z})^\times$ of reduced residues modulo q, and the same technique suffices to construct the characters for any finite Abelian group. More generally, if G is a group, then a homomorphism $h : G \longrightarrow GL(n, \mathbb{C})$ is called a *group representation*, and the trace of $h(g)$ is a *group character*. Note that if a and b are conjugate elements of G, say $a = gbg^{-1}$, then $h(a)$ and $h(b)$ are similar matrices. Hence they have the same eigenvalues, and in particular tr $h(a) =$ tr $h(b)$. Thus a group character is constant on conjugacy classes. In the case of a finite Abelian group it suffices to take $n = 1$, and in this case the representation and its trace are essentially the same. For an introduction to characters in a wider setting, see Serre (1977).

Section 4.3. Dirichlet (1837a,b,c) first proved Corollary 4.10 in the case that q is prime. The definition of the Dirichlet characters is not difficult in that case, since the multiplicative group $(\mathbb{Z}/p\mathbb{Z})^\times$ of reduced residues is cyclic. The most challenging part of the proof is to show that $L(1, \chi)$ when χ is the Legendre symbol (mod p). If $p \equiv 3$ (mod 4), then

$$\sum_{a=1}^{p-1} a \left(\frac{a}{p} \right) \equiv \sum_{a=1}^{p-1} a = \frac{p(p-1)}{2} \equiv 1 \pmod 2,$$

and hence the sum on the left is non-zero. It follows by (9.9) that $L(1, \chi_p) \neq 0$ in this case. If $p \equiv 1$ (mod 4), then one has the identity of Exercise 9.3.7(c), and thus to show that $L(1, \chi_p) \neq 0$ it suffices to show that $Q \neq 1$. Dirichlet established this by means of Gauss's theory of cyclotomy. Accounts of this are found in Davenport (2000, Sections 1–3), and in Narkiewicz (2000, pp. 64–65). An alternative proof that $Q \neq 1$ was given more recently by Chowla & Mordell (1961) (cf. Exercise 9.3.8). In order to prove that $L(1, \chi) \neq 0$ when χ is quadratic, Dirichlet related $L(1, \chi)$ to the class number of binary quadratic forms. Suppose that d is a fundamental quadratic discriminant, and put $\chi_d(n) = \left(\frac{d}{n}\right)$, the Kronecker symbol (as discussed in Section 9.3). Suppose first that $d > 0$. Among the solutions of Pell's equation $x^2 - dy^2 = 4$, let (x_0, y_0) be the solution with $x_0 > 0$, $y_0 > 0$, and y_0 minimal, and put $\eta = \frac{1}{2}(x_0 + y_0\sqrt{d})$. Dirichlet showed that

$$L(1, \chi_d) = \frac{h \log \eta}{\sqrt{d}} \qquad (4.35)$$

where h is the number of equivalence classes of binary quadratic forms with discriminant d. Since $h \geq 1$ and $y_0 \geq 1$, it follows that $L(1, \chi_d) \gg (\log d)/\sqrt{d}$ in this case. Now suppose that $d < 0$ and that w denotes the number of automorphs of the positive definite binary quadratic forms of discriminant d (i.e., $w = 6$ if $d = -3$, $w = 4$ if $d = -4$, and $w = 2$ if $d < -4$). Dirichlet showed that

$$L(1, \chi_d) = \frac{2\pi h}{w\sqrt{-d}} . \qquad (4.36)$$

Thus $L(1, \chi_d) \geq \pi/\sqrt{-d}$ when $d < -4$.

Our treatment of quadratic characters in the proof of Theorem 4.9 is due to Landau (1906). Mertens (1895a,b, 1897, 1899) gave two elementary proofs that $L(1, \chi) > 0$ when χ is quadratic; cf. Exercises 2.4.2 and 2.4.3. For a definitive account of Mertens' methods, see Bateman (1959). Other proofs have been given by Teege (1901), Gel'fond & Linnik (1962, Chapter 3 Section 2), Bateman (1966, 1997), Pintz (1971), and Monsky (1993). See also Baker, Birch & Wirsing (1973).

4.5 References

Baker, A., Birch, B. J., & Wirsing, E. A. (1973). On a problem of Chowla, *J. Number Theory* **5**, 224–236.

Bateman, P. T. (1959). Theorems implying the non-vanishing of $\sum \chi(m)m^{-1}$ for real residue-characters, *J. Indian Math. Soc.* **23**, 101–115.

(1966). Lower bounds for $\sum h(m)/m$ for arithmetical function h similar to real residue characters, *J. Math. Anal. Appl.* **15**, 2–20.

(1997). A theorem of Ingham implying that Dirichlet's L-functions have no zeros with real part one, *Enseignement Math.* (2) **43**, 281–284.

Bateman, P. T., Pomerance, C., & Vaughan, R. C. (1981). On the size of the coefficients of the cyclotomic polynomial, *Coll. Math. Soc. J. Bolyai*, pp. 171–202.

Carmichael, R. (1932). Expansions of arithmetical functions in infinite series, *Proc. London Math. Soc.* (2) **34**, 1–26.

Chowla, S. & Mordell, L. J. (1961). Note on the nonvanishing of $L(1)$, *Proc. Amer. Math. Soc.* **12**, 283–284.

Davenport, H. (2000). *Multiplicative Number Theory*, Graduate Texts Math. 74. New York: Springer-Verlag.

Delange, H. (1976). On Ramanujan expansions of certain arithmetical functions, *Acta Arith.* **31**, 259–270.

Dirichlet, P. G. L. (1839a). Sur l'usage des intétrales définies dans la sommation des séries finies ou infinies, *J. Reine Angew. Math.* **17**, 57–67; *Werke*, Vol. 1, Berlin: Reimer, 1889, pp. 237–256.

(1837b). Beweis eines Satzes ueber die arithmetische Progression, *Ber. Verhandl. Kgl. Preuss. Akad. Wiss.*, 108–110; *Werke*, Vol. 1, Berlin: Reimer, 1889, pp. 307–312.

(1837c). Beweis des Satzes, dass jede unbegrenzte arithmetische Progression, deren erstes Glied und Differenz ganze Zahlen ohne gemeinschaftlichen Factor sind, unendlich viele Primzahlen enthält, *Abhandl. Kgl. Preuss. Akad. Wiss.* 45–81; *Werke*, Vol. 1, Berlin: Reimer, 1889, pp. 313–342.

(1839). Recherches sur diverses applications de l'analyse infinitésimale a la théorie des nombres, *J. Reine Angew. Math.* **19**, 324–369; *Werke*, Vol. 1, Berlin: Reimer, 1889, pp. 411–496.

Erdős, P. (1946). On the coefficients of the cyclotomic polynomial, *Bull. Amer. Math. Soc.* **52**, 179–184.

(1949). On the coefficients of the cyclotomic polynomial, *Portugal. Math.* **8**, 63–71.

(1957). On the growth of the cyclotomic polynomial in the interval (O, 1). *Proc. Glasgow Math. Assoc.* **3**, 102–104.

Friedman, A. (1957). Mean-values and polyharmonic polynomials, *Michigan Math. J.* **4**, 67–74.

Gel'fond, A. O. & Linnik, Ju. V. (1962). *Elementary Methods in Analytic Number Theory*. Moscow: Gosudarstv. Izdat. Fiz.-Mat. Lit.; English translation, Chicago: Rand McNally, 1965; English translation, Cambridge: M. I. T. Press, 1966.

Grytczuk, A. (1981). An identity involving Ramanujan's sum, *Elem. Math.* **36**, 16–17.

Hildebrand, A. (1984). Über die punkweise Konvergenz von Ramanujan-Entwicklungen zahlentheoretischer Funktionen, *Acta Arith.* **44**, 108–140.

Hildebrand, A., Schwarz, W., & Spilker, J. (1988). Still another proof of Parseval's equation for almost-even arithmetical functions, *Aequationes Math.* **35**, 132–139.

Hölder, O. (1936). Zur Theorie der Kreisteilungsgleichung, *Prace Mat.–Fiz.* **43**, 13–23.

Ingham, A. E. (1929). Note on Riemann's ζ-function and Dirichlet's L-functions, *J. London Math. Soc.* **5**, 107–112.

Landau, E. (1906). Über das Nichtverschwinden einer Dirichletschen Reihe, *Sitzungsber. Akad. Wiss.* Berlin **11**, 314–320; *Collected Works*, Vol. 2. Essen: Thales, 1986, pp. 230–236.

Mertens, F. (1895a). Über Dirichletsche Reihen, *Sitzungsber. Kais. Akad. Wiss. Wien* **104**, 2a, 1093–1153.

(1895b). Über das Nichtverschwinden Dirichletscher Reihen mit reelen Gliedern, *Sitzber. Kais. Akad. Wiss. Wien* **104**, 2a, 1158–1166.

(1897). Über Multiplikation und Nichtverschwinden Dirichlet'scher Reihen, *J. Reine Angew. Math.* **117**, 169–184.

(1899). Eine asymptotische Aufgabe, *Sitzber. Kais. Akad. Wiss. Wien* **108**, 2a, 32–37.

Monsky, P. (1993). Simplifying the proof of Dirichlet's theorem, *Amer. Math. Monthly* **100**, 861–862.

Narkiewicz, W. (2000). *The Development of Prime Number Theory*, Berlin: Springer-Verlag.

Pintz, J. (1971). On a certain point in the theory of Dirichlet's *L*-functions, I,II, *Mat. Lapok* **22**, 143–148; 331–335.

Ramanujan, S. (1918). On certain trigonometrical sums and their applications in the theory of numbers, Trans. Cambridge Philos. Soc. **22**, 259–276; *Collected papers*. Cambridge: Cambridge University Press, 1927, pp. 179–199.

Redmond, D. (1983). A remark on a paper: "An identity involving Ramanujan's sum" by A. Grytczuk, *Elem. Math.* **38**, 17–20.

Reznick, B. (1995). Some constructions of spherical 5-designs, *Linear Algebra Appl.*, 226/228, 163–196.

Schwarz, W. (1988). Ramanujan expansions of arithmetical functions, *Ramanujan revisited*, Proc. Centenary Conference (Urbana, June 1987). Boston: Academic Press, pp. 187–214.

Serre, J.–P. (1977). *Linear representation of finite groups*, Graduate Texts Math. 42. New York: Springer-Verlag.

Teege, H. (1901). Beweis, daß die unendliche Reihe $\sum_{n=1}^{n=\infty} \left(\frac{p}{n} \right) \frac{1}{n}$ einen positiven von Null verschiedenen Wert hat, *Mitt. Math. Ges. Hamburg* **4**, 1–11.

Vaughan, R. C. (1974). Bounds for the coefficients of cyclotomic polynomials, *Michigan Math. J.* **21**, 289–295.

Wintner, A. (1943). *Eratosthenian averages*. Baltimore: Waverly Press.

5

Dirichlet series: II

5.1 The inverse Mellin transform

In Chapter 1 we saw that we can express a Dirichlet series $\alpha(s) = \sum_{n=1}^{\infty} a_n n^{-s}$ in terms of the coefficient sum $A(x) = \sum_{n \leq x} a_n$, by means of the formula

$$\alpha(s) = s \int_{1}^{\infty} A(x) x^{-s-1} \, dx, \tag{5.1}$$

which holds for $\sigma > \max(0, \sigma_c)$. This is an example of a Mellin transform. In the reverse direction, Perron's formula asserts that

$$A(x) = \frac{1}{2\pi i} \int_{\sigma_0 - i\infty}^{\sigma_0 + i\infty} \alpha(s) \frac{x^s}{s} \, ds \tag{5.2}$$

for $\sigma_0 > \max(0, \sigma_c)$. This is an example of an inverse Mellin transform.

To understand why we might expect that (2) should be true, note that if $\sigma_0 > 0$, then by the calculus of residues

$$\frac{1}{2\pi i} \int_{\sigma_0 - i\infty}^{\sigma_0 + i\infty} y^s \frac{ds}{s} = \begin{cases} 1 & \text{if } y > 1, \\ 0 & \text{if } 0 < y < 1. \end{cases} \tag{5.3}$$

Thus we would expect that

$$\frac{1}{2\pi i} \int_{\sigma_0 - i\infty}^{\sigma_0 + i\infty} \alpha(s) \frac{x^s}{s} \, ds = \sum_{n} \frac{a_n}{2\pi i} \int_{\sigma_0 - i\infty}^{\sigma_0 + i\infty} \left(\frac{x}{n}\right)^s \frac{ds}{s} = \sum_{n \leq x} a_n. \tag{5.4}$$

The interchange of limits here is difficult to justify, since $\alpha(s)$ may not be uniformly convergent, and because the integral in (5.3) is neither uniformly nor absolutely convergent. Moreover, if x is an integer, then the term $n = x$ in (5.4) gives rise to the integral (5.3) with $y = 1$, and this integral does not converge, although its Cauchy principal value exists:

$$\lim_{T \to \infty} \frac{1}{2\pi i} \int_{\sigma_0 - iT}^{\sigma_0 + iT} \frac{ds}{s} = \frac{1}{2} \tag{5.5}$$

for $\sigma_0 > 0$. We now give a rigorous form of Perron's formula.

137

Theorem 5.1   (Perron's formula) *If $\sigma_0 > \max(0, \sigma_c)$ and $x > 0$, then*

$$\sideset{}{'}\sum_{n \leq x} a_n = \lim_{T \to \infty} \frac{1}{2\pi i} \int_{\sigma_0 - iT}^{\sigma_0 + iT} \alpha(s) \frac{x^s}{s}\, ds.$$

Here $\sum'$ indicates that if x is an integer, then the last term is to be counted with weight $1/2$.

Proof   Choose N so large that $N > 2x + 2$, and write

$$\alpha(s) = \sum_{n \leq N} a_n n^{-s} + \sum_{n > N} a_n n^{-s} = \alpha_1(s) + \alpha_2(s),$$

say. By (5.4), modified in recognition of (5.5), we see that

$$\sideset{}{'}\sum_{n \leq x} a_n = \lim_{T \to \infty} \frac{1}{2\pi i} \int_{\sigma_0 - iT}^{\sigma_0 + iT} \alpha_1(s) \frac{x^s}{s}\, ds;$$

here the justification is trivial since there are only finitely many terms. As for $\alpha_2(s)$, we observe that

$$\alpha_2(s) = \int_N^\infty u^{-s}\, d(A(u) - A(N)) = s \int_N^\infty (A(u) - A(N)) u^{-s-1}\, du.$$

But $A(u) - A(N) \ll u^\theta$ for $\theta > \max(0, \sigma_c)$, and hence

$$\alpha_2(s) \ll \left(1 + \frac{|s|}{\sigma - \theta}\right) N^{\theta - \sigma}$$

for $\sigma > \theta > \max(0, \sigma_c)$. Implicit constants here and in the rest of this proof may depend on the a_n. Hence

$$\int_{\sigma_0 \pm iT}^{T \pm iT} \alpha_2(s) \frac{x^s}{s}\, ds \ll \frac{N^\theta}{\sigma_0 - \theta} \int_{\sigma_0}^\infty \left(\frac{x}{N}\right)^\sigma d\sigma \ll \frac{N^\theta}{\sigma_0 - \theta} \frac{(x/N)^{\sigma_0}}{\log N/x},$$

and

$$\int_{T-iT}^{T+iT} \alpha_2(s) \frac{x^s}{s}\, ds \ll N^\theta (x/N)^{\sigma_0}$$

for large T. We take θ so that $\sigma_0 > \theta > \max(0, \sigma_c)$. Hence by Cauchy's theorem

$$\int_{\sigma_0 - iT}^{\sigma_0 + iT} = \int_{\sigma_0 - iT}^{T-iT} + \int_{T-iT}^{T+iT} + \int_{T+iT}^{\sigma_0 + iT} \ll x^{\sigma_0} N^{\theta - \sigma_0}.$$

On combining our estimates, we see that

$$\limsup_{T \to \infty} \left| \sideset{}{'}\sum_{n \leq x} a_n - \frac{1}{2\pi i} \int_{\sigma_0 - iT}^{\sigma_0 + iT} \alpha(s) \frac{x^s}{s}\, ds \right| \ll x_0^\sigma N^{\theta - \sigma_0}.$$

Since this holds for arbitrarily large N, it follows that the lim sup is 0, and the proof is complete.   □

We have now established a precise relationship between (5.1) and (5.2), but Theorem 5.1 is not sufficiently quantitative to be useful in practice. We express the error term more explicitly in terms of the *sine integral*

$$\operatorname{si}(x) = - \int_x^\infty \frac{\sin u}{u}\, du.$$

By integration by parts we see that $\operatorname{si}(x) \ll 1/x$ for $x \geq 1$, and hence that

$$\operatorname{si}(x) \ll \min(1, 1/x) \tag{5.6}$$

for $x > 0$. We also note that

$$\operatorname{si}(x) + \operatorname{si}(-x) = - \int_{-\infty}^{+\infty} \frac{\sin u}{u}\, du = -\pi. \tag{5.7}$$

Theorem 5.2 *If $\sigma_0 > \max(0, \sigma_a)$ and $x > 0$, then*

$$\sideset{}{'}\sum_{n \leq x} a_n = \frac{1}{2\pi i} \int_{\sigma_0 - iT}^{\sigma_0 + iT} \alpha(s) \frac{x^s}{s}\, ds + R \tag{5.8}$$

where

$$R = \frac{1}{\pi} \sum_{x/2 < n < x} a_n \operatorname{si}\left(T \log \frac{x}{n}\right)$$

$$- \frac{1}{\pi} \sum_{x < n < 2x} a_n \operatorname{si}\left(T \log \frac{n}{x}\right) + O\left(\frac{4^{\sigma_0} + x^{\sigma_0}}{T} \sum_n \frac{|a_n|}{n^{\sigma_0}}\right).$$

Proof Since the series $\alpha(s)$ is absolutely convergent on the interval $[\sigma_0 - iT, \sigma_0 + iT]$, we see that

$$\frac{1}{2\pi i} \int_{\sigma_0 - iT}^{\sigma_0 + iT} \alpha(s) \frac{x^s}{s}\, ds = \sum_n a_n \frac{1}{2\pi i} \int_{\sigma_0 - iT}^{\sigma_0 + iT} \left(\frac{x}{n}\right)^s \frac{ds}{s}.$$

Thus it suffices to show that

$$\frac{1}{2\pi i} \int_{\sigma_0 - iT}^{\sigma_0 + iT} y^s \frac{ds}{s} = \begin{cases} 1 + O(y^{\sigma_0}/T) & \text{if } y \geq 2, \\ 1 + \frac{1}{\pi}\operatorname{si}(T \log y) + O(2^{\sigma_0}/T) & \text{if } 1 \leq y \leq 2, \\ -\frac{1}{\pi}\operatorname{si}(T \log 1/y) + O(2^{\sigma_0}/T) & \text{if } 1/2 \leq y \leq 1, \\ O(y^{\sigma_0}/T) & \text{if } y \leq 1/2 \end{cases} \tag{5.9}$$

for $\sigma_0 > 0$.

To establish the first part of this formula, suppose that $y \geq 2$, and let $\mathcal{C}$ be the piecewise linear path from $-\infty - iT$ to $\sigma_0 - iT$ to $\sigma_0 + iT$ to $-\infty + iT$. Then by the calculus of residues we see that

$$\frac{1}{2\pi i} \int_{\mathcal{C}} y^s \frac{ds}{s} = 1,$$

since the integrand has a pole with residue 1 at $s = 0$. In addition,

$$\int_{-\infty \pm iT}^{\sigma_0 \pm iT} y^s \frac{ds}{s} = \int_{-\infty}^{\sigma_0} \frac{y^{\sigma \pm iT}}{\sigma \pm iT} d\sigma \ll \frac{1}{T} \int_{-\infty}^{\sigma_0} y^\sigma \, d\sigma = \frac{y^{\sigma_0}}{T \log y} \ll \frac{y^{\sigma_0}}{T},$$

so we have (5.9) in the case $y \geq 2$. The case $y \leq 1/2$ is treated similarly, but the contour is taken to the right, and there is no residue.

Suppose now that $1 \leq y \leq 2$, and take $\mathcal{C}$ to be the closed rectangular path from $\sigma_0 - iT$ to $\sigma_0 + iT$ to iT to $-iT$ to $\sigma_0 - iT$, with a semicircular indentation of radius ε at $s = 0$. Then by Cauchy's theorem

$$\frac{1}{2\pi i} \int_{\mathcal{C}} y^s \frac{ds}{s} = 0.$$

We note that

$$\int_{\pm iT}^{\sigma_0 \pm iT} y^s \frac{ds}{s} \ll \frac{1}{T} \int_0^{\sigma_0} y^\sigma \, d\sigma \leq \frac{1}{T} \int_0^{\sigma_0} 2^\sigma \, d\sigma \ll \frac{2^{\sigma_0}}{T}.$$

The integral around the semicircle tends to $1/2$ as $\varepsilon \to 0$, and the remaining integral is

$$\frac{1}{2\pi i} \lim_{\varepsilon \to 0} \left(\int_{i\varepsilon}^{iT} + \int_{-iT}^{-i\varepsilon} \right) y^s \frac{ds}{s} = \frac{1}{2\pi i} \lim_{\varepsilon \to 0} \int_\varepsilon^T \left(y^{it} - y^{-it} \right) \frac{dt}{t}$$

$$= \frac{1}{\pi} \int_0^{T \log y} \sin v \, \frac{dv}{v}$$

$$= \frac{1}{2} + \frac{1}{\pi} \mathrm{si}(T \log y)$$

by (5.7). This gives (5.9) when $1 \leq y \leq 2$ and the case $1/2 \leq y \leq 1$ is treated similarly. $\square$

In many situations, Theorem 5.2 contains more information than is really needed – it is often more convenient to appeal to the following less precise result.

Corollary 5.3   *In the situation of* Theorem 5.2,

$$R \ll \sum_{\substack{x/2 < n < 2x \\ n \neq x}} |a_n| \min\left(1, \frac{x}{T|x - n|}\right) + \frac{4^{\sigma_0} + x^{\sigma_0}}{T} \sum_{n=1}^\infty \frac{|a_n|}{n^{\sigma_0}}.$$

Proof   From (5.6) we see that

$$\mathrm{si}(T|\log n/x|) \ll \min\left(1, \frac{1}{T|\log n/x|}\right).$$

But $n/x = 1 + (n - x)/x$ and $|\log(1 + \delta)| \asymp |\delta|$ uniformly for $-1/2 \leq \delta \leq 1$, so the above is

$$\asymp \min\left(1, \frac{x}{T|x - n|}\right)$$

if $x/2 \leq n \leq 2x$. Thus the stated bound follows from Theorem 5.2. $\square$

In classical harmonic analysis, for $f \in L^1(\mathbb{T})$ we define Fourier coefficients $\widehat{f}(k) = \int_0^1 f(x)e(-k\alpha)\,d\alpha$, and we expect that the Fourier series $\sum \widehat{f}(k)e(k\alpha)$ provides a useful formula for $f(\alpha)$. As it happens, the Fourier series may diverge, or converge to a value other than $f(\alpha)$, but for most f a satisfactory alternative can be found. For example, if f is of bounded variation, then

$$\frac{f(\alpha^-) + f(\alpha^+)}{2} = \lim_{K \to \infty} \sum_{-K}^{K} \widehat{f}(k)e(k\alpha).$$

A sharp quantitative form of this is established in Appendix D.1. Analogously, if $f \in L^1(\mathbb{R})$, then we can define the Fourier transform of f,

$$\widehat{f}(t) = \int_{-\infty}^{+\infty} f(x)e(-tx)\,dx, \tag{5.10}$$

and we expect that

$$f(x) = \int_{-\infty}^{+\infty} \widehat{f}(t)e(tx)\,dt. \tag{5.11}$$

As in the case of Fourier series, this may fail, but it is not difficult to show that if f is of bounded variation on $[-A, A]$ for every A, then

$$\frac{f(\alpha^-) + f(\alpha^+)}{2} = \lim_{T \to \infty} \int_{-T}^{T} \widehat{f}(t)e(tx)\,dt. \tag{5.12}$$

The relationship between (5.1) and (5.2) is precisely the same as between (5.10) and (5.11). Indeed, if we take $f(x) = A(e^{2\pi x})e^{-2\pi\sigma x}$, then $f \in L^1(\mathbb{R})$ by Theorem 1.3, and by changing variables in (5.1) we find that

$$\widehat{f}(t) = \frac{\alpha(\sigma + it)}{2\pi(\sigma + it)}.$$

Thus (5.2) is equivalent to (5.11), and an appeal to (5.12) provides a second (real variable) proof of Theorem 5.1.

In general, if

$$F(s) = \int_0^\infty f(x)x^{s-1}\,dx, \tag{5.13}$$

then we say that $F(s)$ is the *Mellin transform* of $f(x)$. By (5.10) and (5.11) we expect that

$$f(x) = \frac{1}{2\pi i} \int_{\sigma_0 - i\infty}^{\sigma_0 + i\infty} F(s)x^{-s}\,ds, \tag{5.14}$$

and when this latter formula holds we say that f is the *inverse Mellin transform* of F. Thus if $A(x)$ is the summatory function of a Dirichlet series $\alpha(s)$, then $\alpha(s)/s$ is the Mellin transform of $A(1/x)$ for $\sigma > \max(0, \sigma_c)$, and Perron's formula (Theorem 5.1) asserts that if $\sigma_0 > \max(0, \sigma_c)$, then $A(1/x)$ is the inverse

Mellin transform of $\alpha(s)/s$. Further instances of this pairing arise if we take a *weight function* $w(x)$, and form a *weighted summatory function*

$$A_w(x) = \sum_{n=1}^{\infty} a_n w(n/x).$$

Let $K(s)$ denote the Mellin transform of $w(x)$,

$$K(s) = \int_0^{\infty} w(x) x^{s-1}\, dx.$$

Then we expect that

$$\alpha(s)K(s) = \int_0^{\infty} A_w(x) x^{-s-1}\, dx, \tag{5.15}$$

and that

$$A_w(x) = \frac{1}{2\pi i} \int_{\sigma_0 - i\infty}^{\sigma_0 + i\infty} \alpha(s)K(s)x^s\, ds. \tag{5.16}$$

Alternatively, we may start with a *kernel* $K(s)$, and define the weight $w(x)$ to be its inverse Mellin transform. The precise conditions under which these identities hold depends on the weight or kernel; we mention several important examples.

1. Cesàro weights. For a positive integer k, put

$$C_k(x) = \frac{1}{k!} \sum_{n \le x} a_n (x-n)^k. \tag{5.17}$$

Then $C_k(x) = \int_0^x C_{k-1}(u)\, du$ for $k \ge 1$ where $C_0(x) = A(x)$, and hence $C_k(x) \ll x^\theta$ for $\theta > k + \max(0, \sigma_c)$. (The implicit constant here may depend on k, on θ, and on the a_n.) By integrating (5.1) by parts repeatedly, we see that

$$\alpha(s) = s(s+1)\cdots(s+k) \int_1^{\infty} C_k(x) x^{-s-k-1}\, dx \tag{5.18}$$

for $\sigma > \max(0, \sigma_c)$. By following the method used to prove Theorem 5.1, it may also be shown that

$$C_k(x) = \frac{1}{2\pi i} \int_{\sigma_0 - i\infty}^{\sigma_0 + i\infty} \alpha(s) \frac{x^{s+k}}{s(s+1)\cdots(s+k)}\, ds \tag{5.19}$$

when $x > 0$ and $\sigma_0 > \max(0, \sigma_c)$. Here the critical step is to show that if $y \ge 1$ and $\sigma_0 > 0$, then

$$\frac{1}{2\pi i} \int_{\sigma_0 - i\infty}^{\sigma_0 + i\infty} \frac{y^s}{s(s+1)\cdots(s+k)}\, ds = \sum_{j=0}^{k} \operatorname{Res}\left(\frac{y^s}{s(s+1)\cdots(s+k)}\right)\Big|_{s=-j}$$

by the calculus of residues; this is

$$= \sum_{j=0}^{k} \frac{(-1)^j y^{-j}}{j!(k-j)!} = \frac{1}{k!}(1-1/y)^k$$

by the binomial theorem.

2. Riesz typical means. For positive integers k and positive real x put

$$R_k(x) = \frac{1}{k!} \sum_{n \leq x} a_n (\log x/n)^k. \qquad (5.20)$$

Then $R_k(x) = \int_0^x R_{k-1}(u)/u \, du$ where $R_0(x) = A(x)$, so that $R_k(x) \ll x^\theta$ for $\theta > \max(0, \sigma_c)$. (The implicit constant here may depend on k, on θ, and on the a_n.) By integrating (5.1) by parts repeatedly we see that

$$\alpha(s) = s^{k+1} \int_1^\infty R_k(x) x^{-s-1} \, dx \qquad (5.21)$$

for $\sigma > \max(0, \sigma_c)$. By following the method used to prove Theorem 5.1 we also find that

$$R_k(x) = \frac{1}{2\pi i} \int_{\sigma_0 - i\infty}^{\sigma_0 + i\infty} \alpha(s) \frac{x^s}{s^{k+1}} \, ds \qquad (5.22)$$

when $x > 0$ and $\sigma_0 > \max(0, \sigma_c)$. Here the critical observation is that if $y \geq 1$ and $\sigma_0 > 0$, then

$$\frac{1}{2\pi i} \int_{\sigma_0 - i\infty}^{\sigma_0 + i\infty} \frac{y^s}{s^{k+1}} \, ds = \text{Res}\left(\frac{y^s}{s^{k+1}} \right)\bigg|_{s=0} = \frac{1}{k!}(\log y)^k.$$

3. Abelian weights. For $\sigma > 0$ we have

$$\Gamma(s) = \int_0^\infty e^{-u} u^{s-1} \, du = n^s \int_0^\infty e^{-nx} x^{s-1} \, dx.$$

We multiply by $a_n n^{-s}$ and sum, to find that

$$\alpha(s)\Gamma(s) = \int_0^\infty P(x) x^{s-1} \, dx \qquad (5.23)$$

where

$$P(x) = \sum_{n=1}^\infty a_n e^{-nx}. \qquad (5.24)$$

These operations are valid for $\sigma > \max(0, \sigma_a)$, but by partial summation $P(x) \ll x^{-\theta}$ as $x \to 0^+$ for $\theta > \max(0, \sigma_c)$, so that the integral in (5.23) is absolutely convergent in the half-plane $\sigma > \max(0, \sigma_c)$. Hence the integral is an analytic function in this half-plane, so that by the principle of uniqueness of analytic continuation it follows that (5.23) holds for $\sigma > \max(0, \sigma_c)$. In the opposite direction,

$$P(x) = \frac{1}{2\pi i} \int_{\sigma_0 - i\infty}^{\sigma_0 + i\infty} \alpha(s)\Gamma(s)x^{-s}\, ds \tag{5.25}$$

for $x > 0$, $\sigma > \max(0, \sigma_c)$. To prove this we recall from Theorem 1.5 that $\alpha(s) \ll \tau$ uniformly for $\sigma \geq \varepsilon + \max(0, \sigma_c)$, and from Stirling's formula (Theorem C.1) we see that $|\Gamma(s)| \asymp e^{-\frac{\pi}{2}|t|}|t|^{\sigma - 1/2}$ as $|t| \to \infty$ with σ bounded. Thus the value of the integral is independent of σ_0, and in particular we may assume that $\sigma_0 > \max(0, \sigma_a)$. Consequently the terms in $\alpha(s)$ can be integrated individually, and it suffices to appeal to Theorem C.4.

The formulæ (5.23) and (5.25) provide an important link between the Dirichlet series $\alpha(s)$ and the power series generating function $P(x)$. Indeed, these formulæ hold for complex x, provided that $\Re x > 0$. In particular, by taking $x = \delta - 2\pi i \alpha$ we find that

$$\sum_{n=1}^{\infty} a_n e(n\alpha)e^{-n\delta} = \frac{1}{2\pi i} \int_{\sigma_0 - i\infty}^{\sigma_0 + i\infty} \alpha(s)\Gamma(s)(\delta - 2\pi i\alpha)^{-s}\, ds.$$

It may be noted in the above examples that smoother weights $w(x)$ give rise to kernels $K(s)$ that tend to 0 rapidly as $|t| \to \infty$. Further useful kernels can be constructed as linear combinations of the above kernels.

Since the Mellin transform is a Fourier transform with altered variables, all results pertaining to Fourier transforms can be reformulated in terms of Mellin transforms. Particularly useful is Plancherel's identity, which asserts that if $f \in L^1(\mathbb{R}) \cap L^2(\mathbb{R})$, then $\|f\|_2 = \|\widehat{f}\|_2$. This is the analogue for Fourier transforms of Parseval's identity for Fourier series, which asserts that $\sum_k |\widehat{f}(k)|^2 = \|f\|_2^2$. By the changes of variables we noted before, we obtain

Theorem 5.4   (Plancherel's identity) *Suppose that $\int_0^\infty |w(x)|x^{-\sigma-1}\, dx < \infty$, and also that $\int_0^\infty |w(x)|^2 x^{-2\sigma - 1}\, dx < \infty$. Put $K(s) = \int_0^\infty w(x)x^{-s-1}\, dx$. Then*

$$2\pi \int_0^\infty |w(x)|^2 x^{-2\sigma - 1}\, dx = \int_{-\infty}^{+\infty} |K(\sigma + it)|^2\, dt.$$

Among the many possible applications of this theorem, we note in particular that

$$2\pi \int_0^\infty |A(x)|^2 x^{-2\sigma - 1}\, dx = \int_{-\infty}^{+\infty} \left| \frac{\alpha(\sigma + it)}{\sigma + it} \right|^2\, dt \tag{5.26}$$

for $\sigma > \max(0, \sigma_c)$.

5.1.1 Exercises

1. Show that if $\sigma_c < \sigma_0 < 0$, then

$$\lim_{T \to \infty} \frac{1}{2\pi i} \int_{\sigma_0 - iT}^{\sigma_0 + iT} \alpha(s) \frac{x^s}{s} \, ds = {\sum_{n > x}}' a_n.$$

2. (a) Show that if $y \geq 0$, then

$$-\frac{\pi}{2} = \operatorname{si}(0) \leq \operatorname{si}(y) \leq \operatorname{si}(\pi) = 0.28114 \ldots.$$

(b) Show that if $y \geq 0$, then

$$\Im \int_y^\infty \frac{e^{iu}}{u} \, du = \Im \int_y^{y + i\infty} \frac{e^{iz}}{z} \, dz.$$

(c) Deduce that if $y \geq 0$, then $|\operatorname{si}(y)| < 1/y$.

3. (a) Let $\beta > 0$ be fixed. Show that if $\sigma_0 > 0$, then

$$\frac{1}{2\pi i} \int_{\sigma_0 - i\infty}^{\sigma_0 + i\infty} \Gamma(s/\beta) y^s \, ds = \beta e^{-y^{-\beta}}.$$

(b) Let $\beta > 0$ be fixed. Show that if $x > 0$ and $\sigma_0 > \max(0, \sigma_c)$, then

$$\frac{1}{2\pi i} \int_{\sigma_0 - i\infty}^{\sigma_0 + i\infty} \alpha(s) \Gamma(s/\beta) x^s \, ds = \beta \sum_{n=1}^\infty a_n e^{-(n/x)^\beta}.$$

4. (a) Suppose that $a > 0$ and that b is real. Explain why

$$\frac{1}{2\pi i} \int_{\sigma_0 - i\infty}^{\sigma_0 + i\infty} e^{a^2 s^2/2 + bs} \, ds = \frac{e^{-b^2/(2a^2)}}{2\pi i} \int_{\sigma_0 - i\infty}^{\sigma_0 + i\infty} e^{a^2(s + b/a^2)^2/2} \, ds.$$

(b) Explain why the values of the integrals above are independent of the value of σ_0. Hence show that if $\sigma_0 = -b/a^2$, then the above is

$$= \frac{e^{-b^2/(2a^2)}}{2\pi} \int_{-\infty}^{+\infty} e^{-a^2 t^2/2} \, dt = \frac{1}{\sqrt{2\pi} a} e^{-b^2/a^2}.$$

(c) Show that if $a > 0$, $x > 0$ and $\sigma_0 > \sigma_c$, then

$$\frac{1}{2\pi i} \int_{\sigma_0 - i\infty}^{\sigma_0 + i\infty} \alpha(s) e^{a^2 s^2/2} x^s \, ds = \frac{1}{\sqrt{2\pi} a} \sum_{n=1}^\infty a_n \exp\left(-\frac{(\log x/n)^2}{2a^2}\right).$$

5. Take $k = 1$ in (5.22) for several different values of x, and form a suitable linear combination, to show that if $x \geq 0$ and and $\sigma_c < 0$, then

$$\frac{2}{\pi} \int_{-\infty}^{+\infty} \alpha(it) \left(\frac{\sin \frac{1}{2} t \log x}{t}\right)^2 \, dt = \sum_{n \leq x} a_n \log x/n.$$

6. Let $w(x) \nearrow$, and suppose that $w(x) \ll x^\sigma$ as $x \to \infty$ for some fixed σ. Let σ_w be the infimum of those σ such that $\int_0^\infty w(x)x^{-\sigma-1}\,dx < \infty$, and put

$$K(s) = \int_0^\infty w(x)x^{-s-1}\,dx$$

for $\sigma > \sigma_w$.

(a) Show that $A_w(x) = \sum_{n=1}^\infty a_n w(x/n)$ satisfies $A_w(x) \ll x^\theta$ for $\theta > \max(\sigma_w, \sigma_c)$.

(b) Show that

$$K(s)\alpha(s) = \int_0^\infty A_w(x)x^{-s-1}\,dx$$

for $\sigma > \max(\sigma_w, \sigma_c)$.

(c) Show that

$$\tfrac{1}{2}(A_w(x^-) + A_w(x^+)) = \frac{1}{2\pi i}\lim_{T\to\infty}\int_{\sigma_0-iT}^{\sigma_0+iT}\alpha(s)K(s)x^s\,ds$$

for $\sigma_0 > \max(\sigma_w, \sigma_c)$, $x > 0$.

7. Show that

$$\zeta(s) = -s\int_0^\infty \frac{\{x\}}{x^{s+1}}\,dx$$

for $0 < \sigma < 1$, and that

$$2\pi\int_0^\infty \{x\}^2 x^{-2\sigma-1}\,dx = \int_{-\infty}^{+\infty}\left|\frac{\zeta(\sigma+it)}{\sigma+it}\right|^2\,dt$$

for $0 < \sigma < 1$.

8. (a) Show that if $f \in L^1(\mathbb{R})$ and $f' \in L^1(\mathbb{R})$, then $\widehat{f'}(t) = 2\pi i t\,\widehat{f}(t)$.

(b) Suppose that f is a function such that $f \in L^1(\mathbb{R})$, that $xf(x) \in L^2(\mathbb{R})$, and that $f' \in L^1(\mathbb{R}) \cap L^2(\mathbb{R})$. Show that

$$\int_{-\infty}^{+\infty}|f(x)|^2\,dx = -\int_{-\infty}^{+\infty} x\left(f'(x)\overline{f}(x) + f(x)\overline{f'}(x)\right)\,dx.$$

The Cauchy–Schwarz inequality asserts that

$$\left|\int_{-\infty}^{+\infty} a(x)b(x)\,dx\right|^2 \leq \left(\int_{-\infty}^{+\infty}|a(x)|^2\,dx\right)\left(\int_{-\infty}^{+\infty}|b(x)|^2\,dx\right).$$

By means of this inequality, or otherwise, show that

$$\left(\int_{-\infty}^{+\infty}|xf(x)|^2\,dx\right)\left(\int_{-\infty}^{+\infty}|t\widehat{f}(t)|^2\,dt\right) \geq \frac{1}{16\pi^2}\left(\int_{-\infty}^{+\infty}|f(x)|^2\,dx\right)^2.$$

This is a form of the Heisenberg uncertainty principle. From it we see that if f tends to 0 rapidly outside $[-A, A]$, and if $\widehat{f}$ tends to 0 rapidly outside $[-B, B]$, then $AB \gg 1$.

9. (a) Note the identity

$$f\bar{g} = \tfrac{1}{2}|f+g|^2 - \tfrac{1}{2}|f-g|^2 + \tfrac{i}{2}|f+ig|^2 - \tfrac{i}{2}|f-ig|^2.$$

   (b) Show that if $f \in L^1(\mathbb{R}) \cap L^2(\mathbb{R})$ and if $g \in L^1(\mathbb{R}) \cap L^2(\mathbb{R})$, then

$$\int_{-\infty}^{+\infty} f(x)\overline{g(x)}\, dx = \int_{-\infty}^{+\infty} \widehat{f}(t)\overline{\widehat{g}(t)}\, dt.$$

10. Suppose that F is strictly increasing, and that for $i = 1, 2$ the functions f_i are real-valued with $f_i \in L^1(\mathbb{R}) \cap L^2(\mathbb{R})$ and $F(f_i) \in L^1(\mathbb{R}) \cap L^2(\mathbb{R})$.

   (a) Show that

$$\int_{-\infty}^{+\infty} (f_1(x) - f_2(x))(F(f_1(x)) - F(f_2(x)))\, dx$$
$$= \int_{-\infty}^{+\infty} \big(\widehat{f_1}(t) - \widehat{f_2}(t)\big)\big(\widehat{F(f_1)}(t) - \widehat{F(f_2)}(t)\big)\, dt.$$

   (b) Suppose additionally that $\widehat{f_i}(t) = 0$ for $|t| \geq T$, and that $\widehat{F(f_1)}(t) = \widehat{F(f_2)}(t)$ for $-T \leq t \leq T$. Show that $f_1 = f_2$ a.e.

5.2 Summability

We say that an infinite series $\sum a_n$ is Abel summable to a, and write $\sum a_n = a$ (A) if

$$\lim_{r \to 1^-} \sum_{n=0}^{\infty} a_n r^n = a.$$

Abel proved that if a series converges, then it is A-summable to the same value. Because of this historical antecedent, we call a theorem 'Abelian' if it states that one kind of summability implies another. Perhaps the simplest Abelian theorem asserts that if $\sum_{n=1}^{\infty} a_n$ converges to a, then

$$\lim_{N \to \infty} \sum_{n=1}^{N} \left(1 - \frac{n}{N}\right) a_n = a. \qquad (5.27)$$

This is the Cesàro method of summability of order 1, and so we abbreviate the relation above as $\sum a_n = a$ (C, 1). On putting $s_N = \sum_{n=1}^{N} a_n$, we reformulate the above by saying that if $\lim_{N \to \infty} s_N = a$, then

$$\lim_{N \to \infty} \frac{1}{N} \sum_{n=1}^{N} s_n = a. \qquad (5.28)$$

Here, as in Abel summability and in most other summabilities, each term in the second limit is a linear function of the terms in the first limit. Following Toeplitz and Schur, we characterize those linear transformations $T = [t_{mn}]$ that preserves limits of sequences. We call T *regular* if the following three conditions are satisfied:

$$\text{There is a } C = C(T) \text{ such that } \sum_{n=1}^{\infty} |t_{mn}| \leq C \text{ for all } m; \qquad (5.29)$$

$$\lim_{m \to \infty} t_{mn} = 0 \text{ for all } n; \qquad (5.30)$$

$$\lim_{m \to \infty} \sum_{n=1}^{\infty} t_{mn} = 1. \qquad (5.31)$$

We now show that regular transformations preserve limits, and relegate the verification of the converse to exercises.

Theorem 5.5 *Suppose that T satisfies (5.29) above. If $\{a_n\}$ is a bounded sequence, then the sequence*

$$b_m = \sum_{n=1}^{\infty} t_{mn} a_n \qquad (5.32)$$

is also bounded. If T satisfies (5.29) and (5.30), and if $\lim_{n \to \infty} a_n = 0$, then $\lim_{m \to \infty} b_m = 0$. Finally, if T is regular and $\lim_{n \to \infty} a_n = a$, then $\lim_{m \to \infty} b_m = a$.

The important special case (5.28) is obtained by noting that the (semi-infinite) matrix $[t_{mn}]$ with

$$t_{mn} = \begin{cases} 1/m & \text{if } 1 \leq n \leq m, \\ 0 & \text{if } n > m \end{cases}$$

is regular. Moreover, the proof of Theorem 5.5 requires only a straightforward elaboration of the usual proof of (5.28).

Proof If $|a_n| \leq A$ and (5.29) holds, then

$$|b_m| \leq \sum_{n=1}^{\infty} |t_{mn} a_n| \leq A \sum_{n=1}^{\infty} |t_{mn}| \leq CA.$$

To establish the second assertion, suppose that $\varepsilon > 0$ and that $|a_n| < \varepsilon$ for $n > N = N(\varepsilon)$. Now

$$|b_m| \le \sum_{n=1}^{N} |t_{mn}a_n| + \sum_{n>N} |t_{mn}a_n| = \Sigma_1 + \Sigma_2,$$

say. From (5.29) and the argument above with $A = \varepsilon$ we see that $\Sigma_2 \le C\varepsilon$. From (5.30) we see that $\lim_{m\to\infty} \Sigma_1 = 0$. Hence $\limsup_{m\to\infty} |b_m| \le C\varepsilon$, and we have the desired conclusion since ε is arbitrary. Finally, suppose that T is regular and that $\lim_{n\to\infty} a_n = a$. We write $a_n = a + \alpha_n$, so that

$$b_m = a \sum_{n=1}^{\infty} t_{mn} + \sum_{n=1}^{\infty} t_{mn}\alpha_n.$$

Since $\lim_{n\to\infty} \alpha_n = 0$, we may appeal to the preceding case to see that the second sum tends to 0 as $m \to \infty$. Hence by (5.31) we conclude that $\lim_{m\to\infty} b_m = a$, and the proof is complete. □

In Chapter 1 we used Theorem 1.1 to show that if $\mathcal{S}$ is a sector of the form $\mathcal{S} = \{s : \sigma > \sigma_0,\ |t - t_0| \le H(\sigma - \sigma_0)\}$ where H is an arbitrary positive constant, and if the Dirichlet series $\alpha(s)$ converges at the point s_0, then

$$\lim_{\substack{s \to s_0 \\ s \in \mathcal{S}}} \alpha(s) = \alpha(s_0).$$

To see how this may also be derived from Theorem 5.5, let $\{s_m\}$ be an arbitrary sequence of points of $\mathcal{S}$ for which $\lim_{m\to\infty} s_m = s_0$. It suffices to show that $\lim_{m\to\infty} \alpha(s_m) = \alpha(s_0)$. Take

$$t_{mn} = n^{s_0-s_m} - (n+1)^{s_0-s_m},$$

so that

$$\alpha(s_m) = \sum_{n=1}^{\infty} t_{mn} \left(\sum_{k=1}^{n} a_k k^{-s_0} \right).$$

In view of Theorem 5.5, it suffices to show that $[t_{mn}]$ is regular. The conditions (5.30) and (5.31) are clearly satisfied, and (5.29) follows on observing that if $s \in \mathcal{S}$, then $s - s_0 \ll_H \sigma - \sigma_0$, so that

$$\left| n^{s_0-s} - (n+1)^{s_0-s} \right| = \left| (s - s_0) \int_n^{n+1} u^{s_0-s-1}\, du \right|$$

$$\ll_H (\sigma - \sigma_0) \int_n^{n+1} u^{\sigma_0-\sigma-1}\, du$$

$$= n^{\sigma_0-\sigma} - (n+1)^{\sigma_0-\sigma}.$$

Thus we have the result. Abel's analogous theorem on the convergence of power series can be derived similarly from Theorem 5.5.

The converse of Abel's theorem on power series is false, but Tauber (1897) proved a partial converse: If $a_n = o(1/n)$ and $\sum a_n = a$ (A), then $\sum a_n = a$. Following Hardy and Littlewood, we call a theorem 'Tauberian' if it provides a partial converse of an Abelian theorem. The qualifying hypothesis ('$a_n = o(1/n)$' in the above) is the 'Tauberian hypothesis'. For simplicity we begin with partial converses of (5.27).

Theorem 5.6 *If $\sum_{n=1}^{\infty} a_n = a$ (C, 1), then $\sum a_n = a$ provided that one of the following hypotheses holds:*
(a) $a_n \geq 0$ *for* $n \geq 1$;
(b) $a_n = O(1/n)$ *for* $n \geq 1$;
(c) *There is a constant A such that* $a_n \geq -A/n$ *for all* $n \geq 1$.

Proof Clearly (a) implies (c). If (b) holds, then both $\Re a_n$ and $\Im a_n$ satisfy (c). Thus it suffices to prove that $\sum a_n = a$ when (c) holds. We observe that if H is a positive integer, then

$$\sum_{n=1}^{N} a_n = \frac{N+H}{H} \sum_{n=1}^{N+H} a_n \left(1 - \frac{n}{N+H}\right) - \frac{N}{H} \sum_{n=1}^{N} a_n \left(1 - \frac{n}{N}\right)$$
$$- \frac{1}{H} \sum_{N<n<N+H} a_n(N + H - n) \tag{5.33}$$
$$= T_1 - T_2 - T_3,$$

say. Take $H = [\varepsilon N]$ for some $\varepsilon > 0$. By hypothesis, $\lim_{N\to\infty} T_1 = a(1+\varepsilon)/\varepsilon$, and $\lim_{N\to\infty} T_2 = a/\varepsilon$. From (c) we see that

$$T_3 \geq -A \sum_{N<n<N+H} \frac{1}{n} \geq -\frac{AH}{N} \geq -A\varepsilon.$$

Hence on combining these estimates in (5.33) we see that

$$\limsup_{N\to\infty} \sum_{n=1}^{N} a_n \leq a + A\varepsilon.$$

Since ε can be taken arbitrarily small, it follows that

$$\limsup_{N\to\infty} \sum_{n=1}^{N} a_n \leq a.$$

To obtain a corresponding lower bound we note that

$$\sum_{n=1}^{N} a_n = \frac{N}{H} \sum_{n=1}^{N} a_n \left(1 - \frac{n}{N}\right) - \frac{N-H}{H} \sum_{n=1}^{N-H} a_n \left(1 - \frac{n}{N-H}\right)$$
$$+ \frac{1}{H} \sum_{N-H<n<N} a_n(n + H - N). \tag{5.34}$$

Arguing as we did before, we find that

$$\liminf_{N\to\infty} \sum_{n=1}^{N} a_n \geq a - A\varepsilon/(1-\varepsilon),$$

so that

$$\liminf_{N\to\infty} \sum_{n=1}^{N} a_n \geq a,$$

and the proof is complete. □

If we had argued from (a) or (b), then the treatment of the term T_3 above would have been simpler, since from (a) it follows that $T_3 \geq 0$, while from (b) we have $T_3 \ll \varepsilon$.

Our next objective is to generalize and strengthen Theorem 5.6. The type of generalization we have in mind is exhibited in the following result, which can be established by adapting the above proof: Let β be fixed, $\beta \geq 0$. If

$$\sum_{n=1}^{N} a_n \left(1 - \frac{n}{N}\right) = (a + o(1))N^{\beta},$$

and if $a_n \geq -An^{\beta-1}$, then

$$\sum_{n=1}^{N} a_n = (a(\beta+1) + o(1))N^{\beta}.$$

Concerning the possibility of strengthening Theorem 5.6, we note that by an Abelian argument (or by an application of Theorem 5.5) it may be shown that $\sum a_n = a$ (C, 1) implies that $\sum a_n = a$ (A). Thus if we replace (C, 1) by (A) in Theorem 5.6, then we have weakened the hypothesis, and the result would therefore be stronger. Indeed, Hardy (1910) conjectured and Littlewood (1911) proved that if $\sum a_n = a$ (A) and $a_n = O(1/n)$, then $\sum a_n = a$. That is, the condition '$a_n = o(1/n)$' in Tauber's theorem can be replaced by the condition (b) above. In fact the still weaker condition (c) suffices, as will be seen by taking $\beta = 0$ in Corollary 5.9 below. We now formulate a general result for the Laplace transform, from which the analogues for power series and Dirichlet series follow easily.

Theorem 5.7 (Hardy–Littlewood) *Suppose that $a(u)$ is Riemann-integrable over $[0, U]$ for every $U > 0$, and that the integral*

$$I(\delta) = \int_0^\infty a(u)e^{-u\delta}\, du$$

converges for every $\delta > 0$. Let β be fixed, $\beta \geq 0$, and suppose that

$$I(\delta) = (\alpha + o(1))\delta^{-\beta} \tag{5.35}$$

as $\delta \to 0^+$. If, moreover, there is a constant $A \geq 0$ such that

$$a(u) \geq -A(u+1)^{\beta-1} \tag{5.36}$$

for all $u \geq 0$, then

$$\int_0^U a(u)\,du = \left(\frac{\alpha}{\Gamma(\beta+1)} + o(1)\right) U^\beta. \tag{5.37}$$

The basic properties of the gamma function are developed in Appendix C, but for our present purposes it suffices to put

$$\Gamma(\beta) = \int_0^\infty u^{\beta-1} e^{-u}\,du$$

for $\beta > 0$. From this it follows by integration by parts that

$$\beta\Gamma(\beta) = \Gamma(\beta+1) \tag{5.38}$$

when $\beta > 0$.

The amount of unsmoothing required in deriving (5.37) from (5.35) is now much greater than it was in the proof of Theorem 5.6. Nevertheless we follow the same line of attack. To obtain the proper perspective we review the preceding proof. Let $\mathcal{J} = [0, 1]$, let $\chi_{\mathcal{J}}(u)$ be its characteristic function, and put $K(u) = \max(0, 1-u)$ for $u \geq 0$. Thus $\sum_{n=1}^N a_n = \sum_n a_n \chi_{\mathcal{J}}(n/N)$, and $\sum_{n=1}^N a_n(1 - n/N) = \sum_n a_n K(n/N)$. Our strategy was to approximate to $\chi_{\mathcal{J}}(u)$ by linear combinations of $K(\kappa u)$ for various values of κ, $\kappa > 0$. The relation underlying (5.33) and (5.34) is both simple and explicit:

$$\frac{1}{\varepsilon}\left(K(u) - (1-\varepsilon)K(u/(1-\varepsilon))\right) \leq \chi_{\mathcal{J}}(u) \leq \frac{1}{\varepsilon}((1+\varepsilon)K(u/(1+\varepsilon)) - K(u));$$
$$\tag{5.39}$$

we took $\varepsilon = H/N$. In the present situation we wish to approximate to $\chi_{\mathcal{J}}(u)$ by linear combinations of $e^{-\kappa u}$, $\kappa > 0$. We make the change of variable $x = e^{-u}$, so that $0 \leq x \leq 1$, and we put $\mathcal{J} = [1/e, 1]$. Then we want to approximate to $\chi_{\mathcal{J}}(x)$ by a linear combination $P(x)$ of the functions x^κ, $\kappa > 0$. In fact it suffices to use only integral values of κ, so that $P(x)$ is a polynomial that vanishes at the origin. In place of (5.33), (5.34) and (5.39) we shall substitute

Lemma 5.8 *Let ε be given, $0 < \varepsilon < 1/4$, and put $\mathcal{J} = [1/e, 1]$, $\mathcal{K} = [e^{-1-\varepsilon}, e^{-1+\varepsilon}]$. There exist polynomials $P_\pm(x)$ such that for $0 \leq x \leq 1$ we have*

$$P_-(x) \leq \chi_{\mathcal{J}}(x) \leq P_+(x) \tag{5.40}$$

and

$$|P_\pm(x) - \chi_{\mathcal{J}}(x)| \leq \varepsilon x(1-x) + 5\chi_{\mathcal{K}}(x). \tag{5.41}$$

Proof Let $g(x) = (\chi_{\mathcal{J}}(x) - x)/(x(1-x))$. Then g is continuous in $[0, 1]$ apart from a jump discontinuity at $x = 1/e$ of height $e^2/(e-1) < 5$. Hence by Weierstrass's theorem on the uniform approximation of continuous functions by polynomials we see that there are polynomials $Q_\pm(x)$ such that $Q_-(x) \leq g(x) \leq Q_+(x)$ for $0 \leq x \leq 1$, and for which

$$|g(x) - Q_\pm(x)| \leq \varepsilon + 5\chi_{\mathcal{K}}(x) \tag{5.42}$$

for $0 \leq x \leq 1$. Then the polynomials $P_\pm(x) = x + x(1-x)Q_\pm(x)$ have the desired properties. $\qquad\qquad\square$

Proof of Theorem 5.7 We suppose first that $\alpha = 0$. We note that if $P(x)$ is a polynomial such that $P(0) = 0$, say $P(x) = \sum_{r=1}^{R} c_r x^r$, then by (5.35) we see that

$$\int_0^\infty a(u)P(e^{-u\delta})\,du = \sum_{r=1}^{R} c_r I(r\delta) = o(\delta^{-\beta}) \tag{5.43}$$

as $\delta \to 0^+$. In the notation of the above lemma,

$$\int_0^U a(u)\,du = \int_0^\infty a(u)\chi_{\mathcal{J}}(e^{-u/U})\,du.$$

If (5.40) holds, then by (5.36) we see that

$$\int_0^\infty a(u)\big(P_+\big(e^{-u/U}\big) - \chi_{\mathcal{J}}\big(e^{-u/U}\big)\big)\,du$$
$$\geq -A \int_0^\infty (u+1)^{\beta-1}\big(P_+\big(e^{-u/U}\big) - \chi_{\mathcal{J}}\big(e^{-u/U}\big)\big)\,du.$$

By (5.41) this latter integral is

$$\ll \varepsilon \int_0^\infty (u+1)^{\beta-1} e^{-u/U}(1 - e^{-u/U})\,du + \int_{(1-\varepsilon)U}^{(1+\varepsilon)U} (u+1)^{\beta-1}\,du.$$

In the first term, the integrand is $\ll (u+1)^\beta U^{-1}$ for $0 \leq u \leq U$; it is $\ll u^{\beta-1}e^{-u/U}$ for $u \geq U$. Hence the first integral is $\ll U^\beta$. The second integral is $\ll \varepsilon U^\beta$. On taking $\delta = 1/U$, $P = P_+$ in (5.43) and combining our results, we find that

$$\int_0^U a(u)\,du \leq A_1\varepsilon U^\beta + o(U^\beta).$$

Since ε can be arbitrarily small, we deduce that

$$\limsup_{U\to\infty} U^{-\beta} \int_0^U a(u)\, du \le 0.$$

By arguing similarly with P_- instead of P_+, we see that the corresponding liminf is ≥ 0, and so we have (5.37) in the case $\alpha = 0$.

Suppose now that $\alpha \ne 0$, $\beta > 0$. We note first that

$$\int_0^\infty (u+1)^{\beta-1} e^{-u\delta}\, du = e^\delta \int_1^\infty v^{\beta-1} e^{-v\delta}\, dv = e^\delta \int_0^\infty v^{\beta-1} e^{-v\delta}\, dv + O(e^\delta),$$

and that

$$\int_0^\infty v^{\beta-1} e^{-v\delta}\, dv = \delta^{-\beta} \int_0^\infty w^{\beta-1} e^{-w}\, dw = \delta^{-\beta}\Gamma(\beta).$$

Hence if $b(u) = a(u) - \alpha(u+1)^{\beta-1}/\Gamma(\beta)$, then $b(u) \ge -B(u+1)^{\beta-1}$, and

$$\int_0^\infty b(u) e^{-u\delta}\, du = o(\delta^{-\beta}).$$

Thus $\int_0^U b(u)\, du = o(U^\beta)$, so that

$$\int_0^U a(u)\, du = \frac{\alpha}{\beta\Gamma(\beta)} U^\beta + o(U^\beta),$$

and we have (5.37), in view of (5.38).

For the remaining case, $\beta = 0$, it suffices to consider $b(u) = a(u) - \alpha \chi_{[0,1]}(u)$. $\quad\square$

Corollary 5.9 *Suppose that $p(z) = \sum_{n=0}^\infty a_n z^n$ converges for $|z| < 1$, and that $\beta \ge 0$. If $p(x) = (\alpha + o(1))(1-x)^{-\beta}$ as $x \to 1^-$, and if $a_n \ge -An^{\beta-1}$ for $n \ge 1$, then*

$$\sum_{n=0}^N a_n = \left(\frac{\alpha}{\Gamma(\beta+1)} + o(1)\right) N^\beta.$$

Proof Put $a(u) = a_n$ for $n \le u < n+1$. Then (5.36) holds, and

$$I(\delta) = \sum_{n=0}^\infty a_n \int_n^{n+1} e^{-u\delta}\, du = \frac{1-e^{-\delta}}{\delta} p(e^{-\delta}).$$

But $1 - e^{-\delta} \sim \delta$ as $\delta \to 0^+$, so that (5.35) holds. The result now follows by taking $U = N+1$ in (5.37). $\quad\square$

Corollary 5.10 *If $\sum a_n = \alpha$ (A), and if the sequence $s_N = \sum_{n=0}^N a_n$ is bounded, then $\sum a_n = \alpha$ (C, 1).*

Proof Take $\beta = 1$, $p(z) = \sum_{n=0}^{\infty} s_n z^n = (1-z)^{-1} \sum_{n=0}^{\infty} a_n z^n$ in Corollary 5.9. Then $\sum_{n=0}^{N} s_n = (\alpha + o(1))N$, which is the desired result. □

For Dirichlet series we have similarly

Theorem 5.11 *Suppose that* $\alpha(s) = \sum_{n=1}^{\infty} a_n n^{-s}$ *converges for* $\sigma > 1$, *and that* $\beta \geq 0$. *If* $\alpha(\sigma) = (\alpha + o(1))(\sigma - 1)^{-\beta}$ *as* $\sigma \to 1^{+}$, *and if* $a_n \geq -A(1 + \log n)^{\beta-1}$, *then*

$$\sum_{n=1}^{N} \frac{a_n}{n} = \left(\frac{\alpha}{\Gamma(\beta+1)} + o(1)\right)(\log N)^{\beta}.$$

Proof Take $a(u) = \sum_{u-1 \leq \log n < u} a_n/n$. Then $I(\delta)$ converges for $\delta > 0$, and moreover

$$I(\delta) = \sum_{n=1}^{\infty} \frac{a_n}{n} \int_{\log n}^{1+\log n} e^{-u\delta}\, du = \frac{1-e^{-\delta}}{\delta} \alpha(1+\delta),$$

so that (5.37) follows. To obtain the desired conclusion we require a further appeal to our Tauberian hypothesis. We note that

$$\int_{0}^{\log N} a(u)\, du = \sum_{n \leq N} \frac{a_n}{n} - \sum_{N/e < n \leq N} \frac{a_n}{n} \log \frac{ne}{N}.$$

By our Tauberian hypothesis this is

$$\leq \sum_{n \leq N} \frac{a_n}{n} + A_1 (\log N)^{\beta-1},$$

so that

$$\sum_{n \leq N} \frac{a_n}{n} \geq \left(\frac{\alpha}{\Gamma(\beta+1)} + o(1)\right)(\log N)^{\beta} - A_1(\log N)^{\beta-1}.$$

On taking $U = 1 + \log N$ in (5.37) we may derive a corresponding upper bound to complete the proof. □

The qualitative arguments we have given can be put in quantitative form as the need arises. For example, it is easy to see that if

$$\sum_{n=1}^{N} a_n = N + O(\sqrt{N}), \tag{5.44}$$

then

$$\sum_{n=1}^{N} a_n(N-n) = \frac{1}{2}N^2 + O(N^{3/2}). \tag{5.45}$$

This is best possible (take $a_n = 1 + n^{-1/2}$), but if the error term is oscillatory, then smoothing may reduce its size (consider $a_n = \cos\sqrt{n}$). Conversely if (5.45) holds and if the sequence a_n is bounded, then the method used to prove Theorem 5.6 can be used to show that

$$\sum_{n=1}^{N} a_n = N + O\left(N^{3/4}\right). \tag{5.46}$$

This conclusion, though it falls short of (5.44), is best possible (take $a_n = 1 + \cos n^{1/4}$). We can also put Theorem 5.7 in quantitative form, but here the loss in precision is much greater, and in general the importance of Theorem 5.7 and its corollaries lies in its versatility. For example, it can be shown that if $\sum_{n=0}^{\infty} a_n r^n = (1-r)^{-1} + O(1)$ as $r \to 1^-$, and if $a_n = O(1)$, then

$$\sum_{n=0}^{N} a_n = N + O\left(\frac{N}{\log N}\right).$$

This error term, though weak, is best possible (take $a_n = 1 + \cos(\log n)^2$).
    For Dirichlet series it can be shown that if

$$\alpha(s) = \sum_{n=1}^{\infty} a_n n^{-s} = \frac{1}{s-1} + O(1)$$

as $s \to 1^+$, and if the sequence a_n is bounded, then

$$\sum_{n=1}^{N} \frac{a_n}{n} = \log N + O\left(\frac{\log N}{\log\log N}\right).$$

This is also best possible (take $a_n = 1 + \cos(\log\log n)^2$), but we can obtain a sharper result by strengthening our analytic hypothesis. For example, it can be shown that if $\alpha(s)$ is analytic in a neighbourhood of 1 and if the sequence a_n is bounded, then

$$\sum_{n=1}^{N} \frac{a_n}{n} = O(1).$$

However, even this stronger assumption does not allow us to deduce that

$$\sum_{n=1}^{N} a_n = o(N),$$

as we see by considering $a_n = \cos\log n$. In Chapter 8 we shall encounter further Tauberian theorems in which the above conclusion is derived from hypotheses concerning the behaviour of $\alpha(s)$ throughout the half-plane $\sigma \geq 1$.

5.2.1 Exercises

1. Let T be a regular matrix such that $t_{mn} \geq 0$ for all m, n. Show that if $\lim_{n\to\infty} a_n = +\infty$, then $\lim_{m\to\infty} b_m = +\infty$.

2. Show that if $T = [t_{mn}]$ and $U = [u_{mn}]$ are regular matrices, then so is $TU = V = [v_{mn}]$ where

$$v_{mn} = \sum_{k=1}^{\infty} t_{mk} u_{kn}.$$

3. Show that if $b = Ta$ and $\lim_{m\to\infty} b_m = a$ whenever $\lim_{n\to\infty} a_n = a$, then T is regular.

4. For $n = 0, 1, 2, \ldots$ let $t_n(x)$ be defined on $[0, 1)$, and suppose that the t_n satisfy the following conditions:
   (i) There is a constant C such that if $x \in [0, 1)$, then $\sum_{n=0}^{\infty} |t_n(x)| \leq C$.
   (ii) For all n, $\lim_{x\to 1^-} t_n(x) = 0$.
   (iii) $\lim_{x\to 1^-} \sum_{n=0}^{\infty} t_n(x) = 1$.
   Show that if $\lim_{n\to\infty} a_n = a$ and if $b(x) = \sum_{n=0}^{\infty} a_n t_n(x)$, then $\lim_{x\to 1^-} b(x) = a$.

5. (Kojima 1917) Suppose that the numbers t_{mn} satisfy the following conditions:
   (i) There is a constant C such that $\sum_{n=1}^{\infty} |t_{mn}| \leq C$ for all m.
   (ii) For all n, $\lim_{m\to\infty} t_{mn}$ exists.
   (iii) $\lim_{m\to\infty} \sum_{n=1}^{\infty} t_{mn}$ exists.
   Show that if $\lim_{n\to\infty} a_n$ exists and if $b_m = \sum_{n=1}^{\infty} t_{mn} a_n$, then $\lim_{m\to\infty} b_m$ exists.

6. For positive integers n let $K_n(x)$ be a function defined on $[0, \infty)$ such that
   (i) $\int_0^{\infty} K_n(x)\,dx \to 1$ as $n \to \infty$;
   (ii) $\int_0^{\infty} |K_n(x)|\,dx \leq C$ for all n;
   (iii) $\lim_{n\to\infty} K_n(x) = 0$ uniformly for $0 \leq x \leq X$.
   Suppose that $a(x)$ is a bounded function, and that $b_n = \int_0^{\infty} a(x) K_n(x)\,dx$. Show that if $\lim_{x\to\infty} a(x) = a$, then $\lim_{n\to\infty} b_n = a$.

7. Let r_m be a sequence of positive real numbers with $r_m \to 1^-$ as $m \to \infty$. For $m \geq 1$, $n \geq 1$, put $t_{mn} = n r_m^{n-1}(1 - r_m)^2$.
   (a) Show that $[t_{mn}]$ is regular.
   (b) Show that if $a_n = \sum_{k=0}^{n-1} c_k(1 - k/n)$ and b_m is defined by (5.32), then
   $$b_m = \sum_{k=0}^{\infty} c_k r_m^k.$$
   (c) Show that if $\sum c_n = c$ (C, 1), then $\sum c_n = c$ (A).

8. Suppose that $T = [t_{mn}]$ is given by

$$t_{mn} = \begin{cases} 0 & \text{if } n = 0, \\ \dfrac{m!n}{m^{n+1}(m-n)!} & \text{if } m \geq n > 0, \\ 0 & \text{if } m < n. \end{cases}$$

(a) Show that

$$\sum_{n=k}^{m} t_{mn} = \frac{m!}{m^k(m-k)!}$$

for $1 \le k \le m$.

(b) Verify that T is regular.

(c) Show that if $a_n = \sum_{k=0}^{n} x^k/k!$ for $n \ge 0$, then $b_m = (1+x/m)^m$ for $m \ge 1$.

9. (Mercer's theorem) Suppose that

$$b_m = \frac{1}{2}a_m + \frac{1}{2} \cdot \frac{a_1 + a_2 + \cdots + a_m}{m}$$

for $m \ge 1$. Show that

$$a_n = \frac{2n}{n+1}b_n - \frac{2}{n(n+1)}\sum_{m=1}^{n-1} m b_m.$$

Conclude that $\lim_{n\to\infty} a_n = a$ if and only if $\lim_{m\to\infty} b_m = a$.

10. For a non-negative integer k we say that $\sum a_n = a$ (C, k) if

$$\lim_{x\to\infty} \sum_{n\le x} a_n \left(1 - \frac{n}{x}\right)^k = a.$$

This is *Cesàro summability of order k*.

(a) Show that if $\sum a_n = a$ (C, j), then $\sum a_n = a$ (C, k) for all $k \ge j$.

(b) Show that if $\sum a_n = a$ (C, k) for some k, then $\sum a_n = a$ (A).

11. Show that if $\sum a_n = a$ (A), then $\lim_{s\to 0+} \sum a_n n^{-s} = a$. (See Wintner 1943 for Tauberian converses.)

12. For a non-negative integer k we say that $\sum a_n = a$ (R, k) if

$$\lim_{x\to\infty} \sum_{n\le x} a_n \left(1 - \frac{\log n}{\log x}\right)^k = a.$$

This is *Riesz summability of order k*.

(a) Show that if $\sum a_n = a$ (R, j), then $\sum a_n = a$ (R, k) for all $k \ge j$.

(b) Show that if $\sum a_n = a$ (R, k) for some k, then $\sum_{s\to 0+} \alpha(s) = a$.

13. Put $t_{mn} = 0$ for $n > m$, set

$$t_{mm} = \frac{m+1}{\log(m+1)}(\log(m+1) - \log m),$$

while for $1 \le n < m$ put

$$t_{mn} = \frac{n+1}{\log(m+1)}(-\log n + 2\log(n+1) - \log(n+2)).$$

(a) Show that if

$$a_n = \sum_{k=1}^{n} c_k \left(1 - \frac{k}{n+1} \right)$$

for $n \geq 1$, then the b_m given in (5.32) satisfies

$$b_m = \sum_{k=1}^{m} c_k \left(1 - \frac{\log k}{\log(n+1)} \right).$$

(b) Show that $t_{mn} \geq 0$ for all m, n.

(c) Show that

$$\sum_{n=1}^{\infty} t_{mn} = 1 + \frac{\log 2}{\log(m+1)}.$$

(d) Show that $\lim_{m \to \infty} t_{mn} = 0$.

(e) Conclude that if $\sum c_k = c$ (C, 1), then $\sum c_k = c$ (R, 1).

14. Let $A(x) = \sum_{0 < n \leq x} a_n$.

(a) Show that

$$\sum_{n=1}^{N} a_n \left(1 - \frac{n}{N} \right) = \frac{1}{N} \int_0^N A(x)\, dx.$$

(b) Show that

$$\sum_{n=1}^{N} a_n \left(1 - \frac{\log n}{\log N} \right) = \frac{1}{\log N} \int_1^N \frac{A(x)}{x}\, dx.$$

(c) Suppose that t is a fixed non-zero real number. By Corollary 1.15, or otherwise, show that

$$\sum_{n=1}^{N} n^{-1-it} \left(1 - \frac{n}{N} \right) = \frac{N^{-it}}{(1-it)^2} + \zeta(1+it) + O\left(\frac{\log N}{N} \right).$$

(d) Similarly, show that

$$\sum_{n=1}^{N} n^{-1-it} \left(1 - \frac{\log n}{\log N} \right) = \zeta(1+it) + O\left(\frac{1}{\log N} \right).$$

(e) Conclude that $\sum_{n=1}^{\infty} n^{-1-it}$ is not summable (C, 1), but that it is summable (R, 1) to $\zeta(1+it)$.

15. We say that a series is *Lambert summable*, and write $\sum a_n = a$ (L), if

$$\lim_{r \to 1^-} (1-r) \sum_{n=1}^{\infty} \frac{n a_n r^n}{1 - r^n} = a.$$

(a) Show that if $\sum a_n = a$, then $\sum a_n = a$ (L).

(b) Show that if a_n is a bounded sequence and $|z| < 1$, then

$$\sum_{n=1}^{\infty} \frac{n a_n z^n}{1 - z^n} = \sum_{n=1}^{\infty} \left(\sum_{d|n} d a_d \right) z^n.$$

(c) Show that $\sum_{n=1}^{\infty} \mu(n)/n = 0$ (L).

(d) Deduce that if $\sum_{n=1}^{\infty} \mu(n)/n$ converges, then its value is 0. (See (6.18) and (8.6).)

(e) Show that $\sum_{n=1}^{\infty} (\Lambda(n) - 1)/n = -2C_0$ (L).

(f) Deduce that if $\sum_{n \leq x} \Lambda(n)/n = \log x + c + o(1)$ then $c = -C_0$. (See Exercise 8.1.1.)

16. (Bohr 1909; Riesz 1909; Phragmén (cf. Landau 1909, pp. 762, 904)) Let $\alpha(s) = \sum a_n n^{-s}$, $\beta(s) = \sum b_n n^{-s}$, and $\gamma(s) = \alpha(s)\beta(s) = \sum c_n n^{-s}$ where $c_n = \sum_{d|n} a_d b_{n/d}$. Further, put $A(x) = \sum_{n \leq x} a_n$ and $B(x) = \sum_{n \leq x} b_n$.

(a) Show that

$$\int_1^x A(y) B(x/y) \frac{dy}{y} = \sum_{n \leq x} c_n \log x/n.$$

(b) Show that if $\sum a_n$ converges and $\sum b_n$ converges, then $\sum c_n = \alpha(0)\beta(0)$ (R, 1).

(c) (Landau 1907) By taking $j = 0$ in Exercise 12(a), or otherwise, show that if the three series $\sum a_n$, $\sum b_n$, $\sum c_n$ all converge, then $\sum c_n = \left(\sum a_n \right) \left(\sum b_n \right)$.

17. Suppose that $f(n) \nearrow \infty$. Construct a_n so that $|a_n| \leq f(n)/n$ for all n,

$$\limsup_{N \to \infty} \sum_{n=1}^{N} a_n = 1, \qquad \liminf_{N \to \infty} \sum_{n=1}^{N} a_n = -1,$$

but

$$\lim_{N \to \infty} \sum_{n=1}^{N} a_n (1 - n/N) = 0.$$

18. (Landau 1908) Show that if $f(x) \sim x$ as $x \to \infty$ and $x f'(x)$ is increasing, then $\lim_{x \to \infty} f'(x) = 1$.

19. (Landau (1913); cf. Littlewood (1986, p. 54–55); Schoenberg 1973) Show that if $f(x) \to 0$ as $x \to \infty$, and if $f''(x) = O(1)$, then $f'(x) \to 0$ as $x \to \infty$.

20. (Tauber's 'second theorem') Suppose that $P(\delta) = \sum_{n=0}^{\infty} a_n e^{-n\delta}$ for $\delta > 0$, and put $s_N = \sum_{n=0}^{N} a_n$.

(a) Show that if $a_n = O(1/n)$, then $s_N = P(1/N) + O(1)$.

(b) Show that if $a_n = o(1/n)$, then $s_N = P(1/N) + o(1)$.

(c) Let $B(N) = \sum_{n=1}^{N} n a_n$. Show that if $\sum a_n$ converges, then $B(N) = o(N)$ as $N \to \infty$.

(d) Show that if $P(\delta)$ converges for $\delta > 0$, then

$$s_N - P(1/N) = \frac{B(N)}{N} + \int_1^N B(u) \left(\frac{1}{u^2} - \frac{e^{-u/N}}{u^2} - \frac{e^{-u/N}}{uN} \right) du$$

$$+ \int_N^{\infty} B(u) e^{-u/N} \left(\frac{u}{N} - 1 \right) \frac{du}{u^2}.$$

(e) Show that if $B(N) = o(N)$, then $s_N - P(1/N) = o(1)$.

(f) Show that if $\sum a_n = a$ (A), then $\sum a_n = a$ if and only if $B(N) = o(N)$.

21. (a) Using Ramanujan's identity $\sum_{n=1}^{\infty} d(n)^2 n^{-s} = \zeta(s)^4 / \zeta(2s)$ and Theorem 5.11, show that $\sum_{n \le x} d(n)^2 / n \sim (4\pi^2)^{-1} (\log x)^4$.

(b) Show that if $\sum_{n \le x} d(n)^2 \sim cx(\log x)^3$ as $x \to \infty$, then $c = 1/\pi^2$.

22. Show that $\sum_{n=1}^{\infty} 1/(d(n) n^s) \sim c(s-1)^{-1/2}$ as $s \to 1^+$ where

$$c = \prod_p \left((p^2 - p)^{1/2} \log \left(\frac{p}{p-1} \right) \right).$$

Deduce that

$$\sum_{n \le x} \frac{1}{n d(n)} \sim \frac{2c}{\sqrt{\pi}} (\log x)^{1/2}$$

as $x \to \infty$.

23. Show that if $\sum_{n \le N} a_n / n = O(1)$ and $\lim_{s \to 1^+} \sum_{n=1}^{\infty} a_n n^{-s} = a$, then

$$\lim_{x \to \infty} \sum_{n \le x} \frac{a_n}{n} \left(1 - \frac{\log n}{\log x} \right) = a.$$

24. Show that

$$\int_0^{\infty} \frac{\sin x}{x} e^{-sx} \, dx = \arctan 1/s$$

for $s > 0$. Using Theorem 5.7, deduce that

$$\int_0^{\infty} \frac{\sin x}{x} \, dx = \frac{\pi}{2}.$$

25. Suppose that $f(u) \ge 0$, that $\int_0^{\infty} f(u) \, du < \infty$, and that $\int_0^{\infty} (1 - e^{-\delta u}) \, du \sim \delta^{1/2}$ as $\delta \to 0^+$. Show that $\int_U^{\infty} f(u) \, du \sim (\pi U)^{-1/2}$ as $U \to \infty$.

26. Show that $\sum_{n=1}^{\infty} a_n = a$ if and only if

$$\lim_{r \to 1^-} \sum_{n=0}^{\infty} a_n r^{2^n} = a.$$

27. Suppose that for every $\varepsilon > 0$ there is an $\eta > 0$ such that $\sum_{N < n \leq (1+\eta)N} |a_n| < \varepsilon$ whenever $N > 1/\eta$. Show that if $\sum a_n = a$ (A), then $\sum a_n = a$.

28. Show that if $\sum a_n = a$ (C, 1) and if $a_{n+1} - a_n = O(|a_n|/n)$, then $\sum a_n = a$.

29. (Hardy & Littlewood 1913, Theorem 27) Show that if $\sum a_n = a$ (A) and if $a_{n+1} - a_n = O(|a_n|/n)$, then $\sum a_n = a$.

30. (Hardy 1907) Show that

$$\lim_{x \to 1^-} \sum_{k=0}^{\infty} (-1)^k x^{2^k}$$

does not exist.

5.3 Notes

Section 5.1. Theorem 5.1 and the more general (5.22) were first proved rigorously by Perron (1908). Although the Mellin transform had been used by Riemann and Cahen, it was Mellin (1902) who first described a general class of functions for which the inversion succeeds. Hjalmar Mellin was Finnish, but his family name is of Swedish origin, so it is properly pronounced mĕ · lēn'. However, in English-speaking countries the uncultured pronunciation mĕl' · ĭn is universal.

In connection with Theorem 5.4, it should be noted that Plancherel's formula $\|f\|_2 = \|\hat{f}\|_2$ holds not just for all $f \in L^1(\mathbb{R}) \cap L^2(\mathbb{R})$ but actually for all $f \in L^2(\mathbb{R})$. However, in this wider setting one must adopt a new definition for $\hat{f}$, since the definition we have taken is valid only for $f \in L^1(\mathbb{R})$. See Goldberg (1961, pp. 46–47) for a resolution of this issue.

For further material concerning properties of Dirichlet series, one should consult Hardy & Riesz (1915), Titchmarsh (1939, Chapter 9), or Widder (1971, Chapter 2). Beyond the theory developed in these sources, we call attention to two further topics of importance in number theory. Wiener (1932, p. 91) proved that if the Fourier series of $f \in L^1(\mathbb{T})$ is absolutely convergent and is never zero, then the Fourier series of $1/f$ is also absolutely convergent. Wiener's proof was rather difficult, but Gel'fand (1941) devised a simpler proof depending on his theory of normed rings. Lévy (1934) proved more generally that the Fourier series of $F(f)$ is absolutely convergent provided that F is analytic at all points in the range of f. Elementary proofs of these theorems have been given by Zygmund (1968, pp. 245–246) and Newman (1975). These theorems were generalized to absolutely convergent Dirichlet series by Hewitt & Williamson (1957), who showed that if $\alpha(s) = \sum a_n n^{-s}$ is absolutely convergent for $\sigma \geq \sigma_0$, then $1/\alpha(s)$ is represented by an absolutely convergent Dirichlet series in the same half-plane, if and only if the values taken by $\alpha(s)$ in this half-plane are bounded away from 0. Ingham (1962) noted a fallacy in Zygmund's account of Lévy's theorem, corrected it, and gave an elementary proof of the generalization to absolutely convergent Dirichlet series. See also Goodman & Newman (1984). Secondly, Bohr (1919) developed a theory concerning the values taken on by an absolutely convergent Dirichlet series. This is described by Titchmarsh (1986, Chapter 11), and in greater detail by Apostol (1976, Chapter 8). For a small footnote to this theory, see Montgomery & Schinzel (1977).

Section 5.2. That conditions (5.29)–(5.31) are necessary and sufficient for the transformation T to preserve limits was proved by Toeplitz (1911) for upper triangular matrices, and by Steinhaus (1911) in general. See also Kojima (1917) and Schur (1921). For more on the Toeplitz matrix theorem and various aspects of Tauberian theorems, see Peyerimhoff (1969).

Theorem 5.6 under the hypothesis (a) is trivial by dominated convergence. Theorem 5.6(b) is a special case of a theorem of Hardy (1910), who considered the more general (C,k) convergence, and Theorem 5.6(c) is similarly a special case of a theorem of Landau (1910, pp. 103–113).

Tauber (1897) proved two theorems, the second of which is found in Exercise 5.2.18. Littlewood (1911) derived his strengthening of Tauber's first theorem by using high-order derivatives. Subsequently Hardy & Littlewood (1913, 1914a, b, 1926, 1930) used the same technique to obtain Theorem 5.8 and its corollaries. Karamata (1930, 1931a, b) introduced the use of Weierstrass's approximation theorem. Karamata also considered a more general situation, in which the right-hand sides of (5.35) and (5.36) are multiplied by a slowly oscillating function $L(1/\delta)$, and the right-hand side of (5.37) is multiplied by $L(U)$. Our exposition employs a further simplification due to Wielandt (1952). Other proofs of Littlewood's theorem have been given by Delange (1952) and by Eggleston (1951). Ingham (1965) observed that a peak function similar to Littlewood's can be constructed by using high-order differencing instead of differentiation. Since many proofs of the Weierstrass theorem involve constructing a peak function, the two methods are not materially different. Sharp quantitative Tauberian theorems have been given by Postnikov (1951), Korevaar (1951, 1953, 1954a–d), Freud (1952, 1953, 1954), Ingham (1965), and Ganelius (1971).

For other accounts of the Hardy–Littlewood theorem, see Hardy (1949) or Widder (1946, 1971). For a brief survey of applications of summability to classical analysis, see Rubel (1989).

Wiener (1932, 1933) invented a general Tauberian theory that contains the Hardy–Littlewood theorems for power series (Theorem 5.8 and its corollaries)

164   *Dirichlet series: II*

as a special case. Wiener's theory is discussed by Hardy (1949), Pitt (1958), and Widder (1946). Among the longer expositions of Tauberian theory, the recent accounts of Korevaar (2002, 2004) are especially recommended.

5.4   References

Apostol, T. (1976). *Modular Functions and Dirichlet Series in Number Theory*, Graduate Texts Math. 41. New York: Springer-Verlag.

Bohr, H. (1909). Über die Summabilität Dirichletscher Reihen, *Nachr. König. Gesell. Wiss. Göttingen Math.-Phys. Kl.*, 247–262; *Collected Mathematical Works*, Vol. I. København: Dansk Mat. Forening, 1952, A2.

(1919). Zur Theorie algemeinen Dirichletschen Reihen, *Math. Ann.* **79**, 136–156; *Collected Mathematical Works*, Vol. I. København: Dansk Mat. Forening, 1952, A13.

Delange, H. (1952). Encore une nouvelle démonstration du théorème taubérien de Littlewood, *Bull. Sci. Math.* (2) **76**, 179–189.

Edwards, D. A. (1957). On absolutely convergent Dirichlet series, *Proc. Amer. Math. Soc.* **8**, 1067–1074.

Eggleston, H. G. (1951). A Tauberian lemma, *Proc. London Math. Soc.* (3) **1**, 28–45.

Freud, G. (1952). Restglied eines Tauberschen Satzes, I, *Acta Math. Acad. Sci. Hungar.* **2**, 299–308.

(1953). Restglied eines Tauberschen Satzes, II, *Acta Math. Acad. Sci. Hungar.* **3**, 299–307.

(1954). Restglied eines Tauberschen Satzes, III, *Acta Math. Acad. Sci. Hungar.* **5**, 275–289.

Ganelius, T. (1971). *Tauberian Remainder Theorems*, Lecture Notes Math. 232. Berlin: Springer-Verlag.

Gel'fand, I. M. (1941). Über absolut konvergente trigonometrische Reihen und Integrale, *Mat. Sb. N. S.* **9**, 51–66.

Goldberg R. R. (1961). *Fourier Transforms*, Cambridge Tract 52. Cambridge: Cambridge University Press.

Goodman, A. & Newman, D. J. (1984). A Wiener type theorem for Dirichlet series, *Proc. Amer. Math. Soc.* **92**, 521–527.

Hardy, G. H. (1907). On certain oscillating series, *Quart. J. Math.* **38**, 269–288; *Collected Papers*, Vol. 6. Oxford: Clarendon Press, 1974, pp. 146–167.

(1910). Theorems relating to the summability and convergence of slowly oscillating series, *Proc. London Math. Soc.* (2) **8**, 301–320; *Collected Papers*, Vol. 6. Oxford: Clarendon Press, 1974, pp. 291–310.

(1949). *Divergent Series*, Oxford: Oxford University Press.

Hardy, G. H. & Littlewood, J. E. (1913). Contributions to the arithmetic theory of series, *Proc. London Math. Soc.* (2) **11**, 411–478; *Collected Papers*, Vol. 6. Oxford: Clarendon Press, 1974, pp. 428–495.

(1914a). Tauberian theorems concerning power series and Dirichlet series whose coefficients are positive, *Proc. London Math. Soc.* (2) **13**, 174–191; *Collected Papers*, Vol. 6. Oxford: Clarendon Press, 1974, pp. 510–527.

(1914b). Some theorems concerning Dirichlet's series, *Messenger Math.* **43**, 134–147; *Collected Papers*, Vol. 6. Oxford: Clarendon Press, 1974, pp. 542–555.

(1926). A further note on the converse of Abel's theorem, *Proc. London Math. Soc.* (2) **25**, 219–236; *Collected Papers*, Vol. 6. Oxford: Clarendon Press, 1974, pp. 699–716.

(1930). Notes on the theory of series XI: On Tauberian theorems, *Proc. London Math. Soc.* (2) **30**, 23–37; *Collected Papers*, Vol. 6. Oxford: Clarendon Press, 1974, pp. 745–759.

Hardy, G. H. & Riesz, M. (1915). *The General Theory of Dirichlet's Series*, Cambridge Tract No. 18. Cambridge: Cambridge University Press. Reprint: Stechert–Hafner (1964).

Hewitt, E. & Williamson, H. (1957). Note on absolutely convergent Dirichlet series, *Proc. Amer. Math. Soc.* **8**, 863–868.

Ingham, A. E. (1962). *On absolutely convergent Dirichlet series.* Studies in Mathematical Analysis and Related Topics. Stanford: Stanford University Press, pp. 156–164.

(1965). On tauberian theorems, *Proc. London Math. Soc.* (3) **14A**, 157–173.

Karamata, J. (1930). Über die Hardy–Littlewoodschen Umkehrungen des Abelschen Stetigkeitssatzes, *Math. Z.* **32**, 319–320.

(1931a). Neuer Beweis und Verallgemeinerung einiger Tauberian-Sätze, *Math. Z.* **33**, 294–300.

(1931b). Neuer Beweis und Verallgemeinerung der Tauberschen Sätze, welche die Laplacesche und Stieltjessche Transformation betreffen, *J. Reine Angew. Math.* **164**, 27–40.

Kojima, T. (1917). On generalized Toeplitz's theorems on limit and their application, *Tôhoku Math. J.* **12**, 291–326.

Korevaar, J. (1951). An estimate of the error in Tauberian theorems for power series, *Duke Math. J.* **18**, 723–734.

(1953). Best L_1 approximation and the remainder in Littlewood's theorem, *Proc. Nederl. Akad. Wetensch. Ser. A* **56** (= Indagationes Math **15**), 281–293.

(1954a). A very general form of Littlewood's theorem, *Proc. Nederl. Akad. Wetensch. Ser. A* **57** (= Indagationes Math. **16**), 36–45.

(1954b). Another numerical Tauberian theorem for power series, *Proc. Nederl. Akad. Wetensch. Ser. A* **57** (= Indagationes Math. **16**), 46–56.

(1954c). Numerical Tauberian theorems for Dirichlet and Lambert series, *Proc. Nederl. Akad. Wetensch. Ser. A* **57** (= Indagationes Math. **16**), 152–160.

(1954d). Numerical Tauberian theorems for power series and Dirichlet series, I, II, *Proc. Nederl. Akad. Wetensch. Ser. A* **57** (= Indagationes Math. **16**), 432–443, 444–455.

(2001). Tauberian theory, approximation, and lacunary series of powers, *Trends in approximation theory* (Nashville, 2000), Innov. Appl. Math. Nashville: Vanderbilt University Press, pp. 169–189.

(2002). A century of complex Tauberian theory, *Bull. Amer. Math. Soc.* (N.S.) **39**, 475–531.

(2004). Tauberian Theory. A Century of Developments. *Grundl. Math. Wiss.* 329. Berlin: Springer-Verlag.

Landau, E. (1907). Über die Multiplikation Dirichletscher Reihen, *Rend. Circ. Mat. Palermo* **24**, 81–160.

(1908). Zwei neue Herleitungen für die asymptotische Anzahl der Primzahlen unter einer gegebenen Grenze, *Sitzungsberichte Akad. Wiss.* Berlin 746–764; *Collected Works*, Vol.4. Essen: Thales Verlag, 1986, pp. 21–39.

(1909). *Handbuch der Lehre von der Verteilung der Primzahlen*, Leipzig: Teubner. Reprint: Chelsea (New York), 1953.

(1910). Über die Bedeutung einiger neuerer Grenzwertsätze der Herren Hardy und Axer, *Prace mat.-fiz.* (Warsaw) **21**, 97–177; *Collected Works*, Vol. 4. Essen: Thales Verlag, 1986, pp. 267–347.

(1913). Einige Ungleichungen für zweimal differentiierbare Funktionen, *Proc. London Math. Soc.* (2) **13**, 43–49; *Collected Works*, Vol. 6. Essen: Thales Verlag, 1986, pp. 49–55.

Lévy, P. (1934). Sur la convergence absolue des séries de Fourier, *Compositio Math.* **1**, 1–14.

Littlewood, J. E. (1911). The converse of Abel's theorem on power series, *Proc. London Math. Soc.* (2) **9**, 434–448; *Collected Papers*, Vol. 1. Oxford: Oxford University Press, 1982, pp. 757–773.

(1986). *Littlewood's Miscellany*, Bollobas, B. Ed., Cambridge: Cambridge University Press.

van de Lune, J. (1986). *An Introduction to Tauberian Theory: From Tauber to Wiener.* CWI Syllabus 12. Amsterdam: Mathematisch Centrum.

Mellin, H. (1902). Über den Zusammenhang zwischen den linearen Differential- und Differenzengleichungen, *Acta Math.* **25**, 139–164.

Montgomery, H. L. & Schinzel, A. (1977). Some arithmetic properties of polynomials in several variables. *Transcendence Theory: Advances and Applications* (Cambridge, 1976). London: Academic Press, pp. 195–203.

Newman, D. J. (1975). A simple proof of Wiener's $1/f$ theorem, *Proc. Amer. Math. Soc.* **48**, 264–265.

Perron, O. (1908). Zur Theorie der Dirichletschen Reihen, *J. Reine Angew. Math.* **134**, 95–143.

Peyerimhoff, A. (1969). *Lectures on summability*, Lecture Notes Math. 107. Berlin: Springer-Verlag.

Pitt, H. R. (1958). *Tauberian Theorems*. Tata Monographs. London: Oxford University Press.

Postnikov, A. G. (1951). The remainder term in the Tauberian theorem of Hardy and Littlewood, *Dokl. Akad. Nauk SSSR N. S.* **77**, 193–196.

Riesz, M. (1909). Sur la sommation des séries de Dirichlet, *C. R. Acad. Sci.* Paris **149**, 18–21.

Rubel, L. (1989). Summability theory: a neglected tool of analysis, *Amer. Math. Monthly* **96**, 421–423.

Schoenberg, I. J. (1973). The elementary cases of Landau's problem of inequalities between derivatives, *Amer. Math. Monthly* **80**, 121–158.

Schur, I. (1921). Über lineare Transformationen in der Theorie der unendlichen Reihen, *J. Reine Angew. Math.* **151**, 79–111.

Steinhaus, H. (1911). *Kilka słów o uogólnieniu pojęcia granicy*, Warsaw: Prace mat-fiz **22**, 121–134.

Tauber, A. (1897). Ein Satz aus der Theorie der unendlichen Reihen, *Monat. Math.* **8**, 273–277.

Titchmarsh, E. C. (1939). *The Theory of Functions*, Second Edition. Oxford: Oxford University Press.

(1986). *The Theory of the Riemann Zeta-function*, Second Edition. Oxford: Oxford University Press.

Toeplitz, O. (1911). *Über algemeine lineare Mittelbildungen*, Warsaw: Prace mat–fiz **22**, 113–119.

Widder, D. V. (1946). *The Laplace transform*, Princeton: Princeton University Press.

(1971). *An Introduction to Transform Theory*. New York: Academic Press.

Wielandt, H. (1952). Zur Umkehrung des Abelschen Stetigkeitssatzes, *Math Z.* **56**, 206–207.

Wiener, N. (1932). Tauberian theorems, *Ann. of Math.* (2) **33**, 1–100.

(1933). *The Fourier Integral, and Certain of its Applications*. Cambridge: Cambridge University Press.

Wintner, A. (1943). *Eratosthenian averages*. Baltimore: Waverly Press.

Zygmund, A. (1968). *Trigonometric series*, Vol. 1, Second Edition. Cambridge: Cambridge University Press.

6

The Prime Number Theorem

6.1 A zero-free region

The Prime Number Theorem (PNT) asserts that

$$\pi(x) \sim \frac{x}{\log x}$$

as x tends to infinity. We shall prove this by using Perron's formula, but in the course of our arguments it will be important to know that $\zeta(s) \neq 0$ for $\sigma \geq 1$. In Chapter 1 we saw that $\zeta(s) \neq 0$ for $\sigma > 1$, but it remains to show that $\zeta(1 + it) \neq 0$. To obtain a quantitative form of the Prime Number Theorem we take some care to show that $\zeta(s) \neq 0$ for $\sigma \geq 1 - \delta(t)$ where $\delta(t)$ is some function of t. We would like the width $\delta(t)$ of the zero-free region to be as large as possible, as the rate at which $\delta(t)$ tends to 0 determines the size of the estimate we can derive for the error term in the Prime Number Theorem.

We begin by reviewing some basic facts concerning functions of a complex variable. If $P(z)$ is a polynomial, then the rate of growth of $|P(z)|$ as $|z| \to \infty$ reflects the number of zeros of $P(z)$. This is generalized to other analytic functions by Jensen's formula. For our purposes we are content to establish the following simple consequence of Jensen's formula.

Lemma 6.1 (Jensen's inequality) *If $f(z)$ is analytic in a domain containing the disc $|z| \leq R$, if $|f(z)| \leq M$ in this disc, and if $f(0) \neq 0$, then for $r < R$ the number of zeros of f in the disc $|z| \leq r$ does not exceed*

$$\frac{\log M/|f(0)|}{\log R/r}.$$

Proof   Let $z_1, z_2, \ldots, z_K$ denote the zeros of f in the disc $|z| \leq R$, and

168

put

$$g(z) = f(z) \prod_{k=1}^{K} \frac{R^2 - z\overline{z}_k}{R(z - z_k)}.$$

The kth factor of the product has been constructed so that it has a pole at z_k, and so that it has modulus 1 on the circle $|z| = R$. Hence g is an analytic function in the disc $|z| \leq R$, and if $|z| = R$, then $|g(z)| = |f(z)| \leq M$. Hence by the maximum modulus principle, $|g(0)| \leq M$. But

$$|g(0)| = |f(0)| \prod_{k=1}^{K} \frac{R}{|z_k|}.$$

Each factor in the product is ≥ 1, and if $|z_k| \leq r$, then the factor is $\geq R/r$. If there are L such zeros, then the above is $\geq |f(0)|(R/r)^L$, which gives the stated upper bound for L. $\qquad\square$

We now show that a bound for the modulus of an analytic function can be derived from a one-sided bound for its real part in a slightly larger region.

Lemma 6.2 (The Borel–Carathéodory Lemma) *Suppose that $h(z)$ is analytic in a domain containing the disc $|z| \leq R$, that $h(0) = 0$, and that $\Re h(z) \leq M$ for $|z| \leq R$. If $|z| \leq r < R$, then*

$$|h(z)| \leq \frac{2Mr}{R - r}$$

and

$$|h'(z)| \leq \frac{2MR}{(R - r)^2}.$$

Proof   It suffices to show that

$$\left| \frac{h^{(k)}(0)}{k!} \right| \leq \frac{2M}{R^k} \tag{6.1}$$

for all $k \geq 1$, for then

$$|h(z)| \leq \sum_{k=1}^{\infty} \left| \frac{h^{(k)}(0)}{k!} \right| r^k \leq 2M \sum_{k=1}^{\infty} \left(\frac{r}{R} \right)^k = \frac{2Mr}{R - r},$$

and

$$|h'(z)| \leq \sum_{k=1}^{\infty} \frac{|h^{(k)}(0)|kr^{k-1}}{k!} \leq \frac{2M}{R} \sum_{k=1}^{\infty} k \left(\frac{r}{R} \right)^{k-1} = \frac{2MR}{(R - r)^2}.$$

To prove (6.1) we first note that

$$\int_0^1 h(Re(\theta)) \, d\theta = \frac{1}{2\pi i} \oint_{|z|=R} h(z) \frac{dz}{z} = h(0) = 0.$$

Moreover, if $k > 0$, then

$$\int_0^1 h(Re(\theta))e(k\theta)\,d\theta = \frac{R^{-k}}{2\pi i}\oint_{|z|=R} h(z)z^{k-1}\,dz = 0,$$

and

$$\int_0^1 h(Re(\theta))e(-k\theta)\,d\theta = \frac{R^k}{2\pi i}\oint_{|z|=R} h(z)z^{-k-1}\,dz = \frac{R^k h^{(k)}(0)}{k!}.$$

By forming a linear combination of these identities we see that if $k > 0$, then

$$\int_0^1 h(Re(\theta))(1 + \cos 2\pi(k\theta + \phi))\,d\theta = \frac{R^k e(-\phi)h^{(k)}(0)}{2\cdot k!}.$$

By taking real parts it follows that

$$\Re\left(\frac{1}{2}R^k e(-\phi)h^{(k)}(0)/k!\right) \leq M\int_0^1 (1 + \cos 2\pi(k\theta + \phi))\,d\theta = M$$

for $k > 0$. Since this holds for any real ϕ, we are free to choose ϕ so that $e(-\phi)h^{(k)}(0) = |h^{(k)}(0)|$. Then the above inequality gives (6.1), and the proof is complete. □

If $P(z) = c\prod_{k=1}^K (z - z_k)$, then

$$\frac{P'}{P}(z) = \sum_{k=1}^K \frac{1}{z - z_k}.$$

We now generalize this to analytic functions $f(z)$, to the extent that f'/f can be approximated by a sum over its nearby zeros.

Lemma 6.3 *Suppose that $f(z)$ is analytic in a domain containing the disc $|z| \leq 1$, that $|f(z)| \leq M$ in this disc, and that $f(0) \neq 0$. Let r and R be fixed, $0 < r < R < 1$. Then for $|z| \leq r$ we have*

$$\frac{f'}{f}(z) = \sum_{k=1}^K \frac{1}{z - z_k} + O\left(\log\frac{M}{|f(0)|}\right)$$

where the sum is extended over all zeros z_k of f for which $|z_k| \leq R$. (The implicit constant depends on r and R, but is otherwise absolute.)

Proof If $f(z)$ has zeros on the circle $|z| = R$, then we replace R by a very slightly larger value. Thus we may assume that $f(z) \neq 0$ for $|z| = R$. Set

$$g(z) = f(z)\prod_{k=1}^K \frac{R^2 - z\overline{z_k}}{R(z - z_k)}.$$

By Lemma 6.1 we know that

$$K \leq \frac{\log M/|f(0)|}{\log 1/R} \ll \log \frac{M}{|f(0)|}. \tag{6.2}$$

If $|z| = R$, then each factor in the product has modulus 1. Consequently $|g(z)| \leq M$ when $|z| = R$, and by the maximum modulus principle $|g(z)| \leq M$ for $|z| \leq R$. We also note that

$$|g(0)| = |f(0)| \prod_{k=1}^{K} \frac{R}{|z_k|} \geq |f(0)|.$$

Since $g(z)$ has no zeros in the disc $|z| \leq R$, we may put $h(z) = \log(g(z)/g(0))$. Then $h(0) = 0$, and

$$\Re h(z) = \log|g(z)| - \log|g(0)| \leq \log M - \log|f(0)|$$

for $|z| \leq R$. Hence by the Borel–Carathéodory lemma we see that

$$h'(z) \ll \log \frac{M}{|f(0)|} \tag{6.3}$$

for $|z| \leq r$. But

$$h'(z) = \frac{g'}{g}(z) = \frac{f'}{f}(z) - \sum_{k=1}^{K} \frac{1}{z - z_k} + \sum_{k=1}^{K} \frac{1}{z - R^2/\overline{z_k}}. \tag{6.4}$$

Now $|R^2/\overline{z_k}| \geq R$, so that if $|z| \leq r$ then $|z - R^2/\overline{z_k}| \geq R - r$. Hence for $|z| \leq r$ the last sum above has modulus

$$\leq \frac{K}{R - r} \ll \log \frac{M}{|f(0)|}$$

by (6.2). To obtain the stated result it suffices to combine this estimate and (6.3) in (6.4). □

We now apply these general principles to the zeta function.

Lemma 6.4 *If* $|t| \geq 7/8$ *and* $5/6 \leq \sigma \leq 2$, *then*

$$\frac{\zeta'}{\zeta}(s) = \sum_{\rho} \frac{1}{s - \rho} + O(\log \tau)$$

where $\tau = |t| + 4$ *and the sum is extended over all zeros* ρ *of* $\zeta(s)$ *for which* $|\rho - (3/2 + it)| \leq 5/6$.

Proof We apply Lemma 6.3 to the function $f(z) = \zeta(z + (3/2 + it))$, with $R = 5/6$ and $r = 2/3$. To complete the proof it suffices to note that $|f(0)| \gg 1$ by the (absolutely convergent) Euler product formula (1.17), and that $f(z) \ll \tau$ for $|z| \leq 1$ by Corollary 1.17. □

If the zeta function were to have a zero of multiplicity m at $1 + i\gamma$, then we would have

$$\frac{\zeta'}{\zeta}(1 + \delta + i\gamma) \sim \frac{m}{\delta}$$

as $\delta \to 0^+$. But

$$\Re\frac{\zeta'}{\zeta}(1 + \delta + i\gamma) = -\sum_{n=1}^{\infty} \Lambda(n)n^{-1-\delta}\cos(\gamma \log n),$$

and in the very worst case this could be no larger than

$$\sum_{n=1}^{\infty} \Lambda(n)n^{-1-\delta} = -\frac{\zeta'}{\zeta}(1 + \delta) \sim \frac{1}{\delta}.$$

Thus m is at most 1, and even in this case ζ'/ζ would be essentially as large as it could possibly be. Roughly speaking, this would imply that $p^{i\gamma}$ is near -1 for most primes. But then it would follow that $p^{2i\gamma}$ is near 1 for most primes, so that

$$\frac{\zeta'}{\zeta}(1 + \delta + 2i\gamma) \sim -\frac{1}{\delta}$$

as $\delta \to 0^+$. Then $\zeta(s)$ would have a pole at $1 + 2i\gamma$, contrary to Corollary 1.13. The essence of this informal argument is captured very effectively by the following elementary inequality.

Lemma 6.5 *If $\sigma > 1$, then*

$$\Re\left(-3\frac{\zeta'}{\zeta}(\sigma) - 4\frac{\zeta'}{\zeta}(\sigma + it) - \frac{\zeta'}{\zeta}(\sigma + 2it)\right) \geq 0.$$

Proof From Corollary 1.11 we see that the left-hand side above is

$$\sum_{n=1}^{\infty} \Lambda(n)n^{-1-\delta}\big(3 + 4\cos(t \log n) + \cos(2t \log n)\big).$$

It now suffices to note that $3 + 4\cos\theta + \cos 2\theta = 2(1 + \cos\theta)^2 \geq 0$ for all θ. □

We now use Lemmas 6.4 and 6.5 to establish the existence of a zero-free region for the zeta function.

Theorem 6.6 *There is an absolute constant $c > 0$ such that $\zeta(s) \neq 0$ for $\sigma \geq 1 - c/\log\tau$.*

This is the classical zero-free region for the zeta function.

Proof Since $\zeta(s)$ is given by the absolutely convergent product (1.17) for $\sigma > 1$, it suffices to consider $\sigma \leq 1$. From (1.24) we see that

$$\left| \zeta(s) - \frac{s}{s-1} \right| \leq |s| \int_1^\infty u^{-\sigma - 1}\, du = \frac{|s|}{\sigma} \qquad (6.5)$$

for $\sigma > 0$. From this we see that $\zeta(s) \neq 0$ when $\sigma > |s-1|$, i.e., in the parabolic region $\sigma > (1 + t^2)/2$. In particular, $\zeta(s) \neq 0$ in the rectangle $8/9 \leq \sigma \leq 1$, $|t| \leq 7/8$. Now suppose that $\rho_0 = \beta_0 + i\gamma_0$ is a zero of the zeta function with $5/6 \leq \beta_0 \leq 1$, $|\gamma_0| \geq 7/8$. Since $\Re\rho \leq 1$ for all zeros ρ of $\zeta(s)$, it follows that $\Re 1/(s - \rho) > 0$ whenever $\sigma > 1$. Hence by Lemma 6.4 with $s = 1 + \delta + i\gamma_0$ we see that

$$-\Re\frac{\zeta'}{\zeta}(1 + \delta + i\gamma_0) \leq -\frac{1}{1 + \delta - \beta_0} + c_1 \log(|\gamma_0| + 4).$$

Similarly, by Lemma 6.4 with $s = 1 + \delta + 2i\gamma_0$ we find that

$$\Re - \frac{\zeta'}{\zeta}(1 + \delta + 2i\gamma_0) \leq c_1 \log(|2\gamma_0| + 4).$$

From Corollary 1.13 we see that

$$-\frac{\zeta'}{\zeta}(1 + \delta) = \frac{1}{\delta} + O(1).$$

On combining these estimates in Lemma 6.5 we conclude that

$$\frac{3}{\delta} - \frac{4}{1 + \delta - \beta_0} + c_2 \log(|\gamma_0| + 4) \geq 0.$$

We take $\delta = 1/(2c_2 \log(|\gamma_0| + 4))$. Thus the above gives

$$7c_2 \log(|\gamma_0| + 4) \geq \frac{4}{1 + \delta - \beta_0},$$

which is to say that

$$1 + \frac{1}{2c_2 \log(|\gamma_0| + 4)} - \beta_0 \geq \frac{4}{7c_2 \log(|\gamma_0| + 4)}.$$

Hence

$$1 - \beta_0 \geq \frac{1}{14c_2 \log(|\gamma_0| + 4)},$$

so the proof is complete. $\qquad\qquad\square$

In the above argument it is essential that the coefficient of $\zeta(s)$ is larger than the coefficient of $\zeta(\sigma)$. Among non-negative cosine polynomials $T(\theta) =$

$a_0 + a_1 \cos 2\pi\theta + \cdots + a_N \cos 2\pi N\theta$, the ratio a_1/a_0 can be arbitrarily close to 2, as we see in the Fejér kernel

$$\Delta_N(\theta) = 1 + 2\sum_{n=1}^{N-1}\left(1 - \frac{n}{N}\right)\cos 2n\pi\theta = \frac{1}{N}\left(\frac{\sin \pi N\theta}{\sin \pi\theta}\right)^2 \geq 0,$$

but it must be strictly less than 2 since

$$a_0 - \tfrac{1}{2}a_1 = \int_0^1 T(\theta)(1 - \cos 2\pi\theta)\, d\theta > 0.$$

It is useful to have bounds for the zeta function and its logarithmic derivative in the zero-free region.

Theorem 6.7 *Let c be the constant in* Theorem 6.6. *If $\sigma > 1 - c/(2\log\tau)$ and $|t| \geq 7/8$, then*

$$\frac{\zeta'}{\zeta}(s) \ll \log\tau, \tag{6.6}$$

$$|\log\zeta(s)| \leq \log\log\tau + \cdot O(1), \tag{6.7}$$

and

$$\frac{1}{\zeta(s)} \ll \log\tau. \tag{6.8}$$

On the other hand, if $1 - c/(2\log\tau) < \sigma \leq 2$ and $|t| \leq 7/8$, then $\frac{\zeta'}{\zeta}(s) = -1/(s-1) + O(1)$, $\log\big(\zeta(s)(s-1)\big) \ll 1$, and $1/\zeta(s) \ll |s-1|$.

Proof If $\sigma > 1$, then by Corollary 1.11 and the triangle inequality we see that

$$\left|\frac{\zeta'}{\zeta}(s)\right| \leq \sum_{n=1}^{\infty}\Lambda(n)n^{-\sigma} = -\frac{\zeta'}{\zeta}(\sigma) \ll \frac{1}{\sigma - 1}.$$

Hence (6.6) is obvious if $\sigma \geq 1 + 1/\log\tau$. Let $s_1 = 1 + 1/\log\tau + it$. In particular we have

$$\frac{\zeta'}{\zeta}(s_1) \ll \log\tau. \tag{6.9}$$

From this estimate and Lemma 6.4 we deduce that

$$\sum_{\rho}\Re\frac{1}{s_1 - \rho} \ll \log\tau \tag{6.10}$$

where the sum is over those zeros ρ for which $|\rho - (3/2 + it)| \leq 5/6$. Suppose that $1 - c/(2\log\tau) \leq \sigma \leq 1 + 1/\log\tau$. Then by Lemma 6.4 we see that

$$\frac{\zeta'}{\zeta}(s) - \frac{\zeta'}{\zeta}(s_1) = \sum_{\rho}\left(\frac{1}{s - \rho} - \frac{1}{s_1 - \rho}\right) + O(\log\tau). \tag{6.11}$$

Since $|s - \rho| \asymp |s_1 - \rho|$ for all zeros ρ in the sum, it follows that

$$\frac{1}{s - \rho} - \frac{1}{s_1 - \rho} \ll \frac{1}{|s_1 - \rho|^2 \log \tau} \ll \Re \frac{1}{s_1 - \rho}.$$

Now (6.6) follows on combining this with (6.9) and (6.10) in (6.11).

To derive (6.7) we begin as in our proof of (6.6). From Corollary 1.11 and the triangle inequality we see that if $\sigma > 1$, then

$$|\log \zeta(s)| \leq \sum_{n=2}^{\infty} \frac{\Lambda(n)}{\log n} n^{-\sigma} = \log \zeta(\sigma).$$

But by Theorem 1.14 we know that $\zeta(\sigma) < 1 + 1/(\sigma - 1)$, so that (6.7) holds when $\sigma \geq 1 + 1/\log \tau$. In particular (6.7) holds at the point $s_1 = 1 + 1/\log \tau + it$, so that to treat the remaining s it suffices to bound the difference

$$\log \zeta(s) - \log \zeta(s_1) = \int_{s_1}^{s} \frac{\zeta'}{\zeta}(w) \, dw.$$

We take the path of integration to be the line segment joining the endpoints. Then the length of this interval multiplied by the bound (6.6) gives the error term $O(1)$ in (6.7).

The estimate (6.8) follows directly from (6.7), since $\log 1/|\zeta| = -\Re \log \zeta$. The remaining estimates follow trivially from (6.5). □

The ideas we have used enable us not only to derive a zero-free region but also to place a bound on the number of zeros ρ that might lie near the point $1 + it$.

Theorem 6.8 *Let $n(r;t)$ denote the number of zeros ρ of $\zeta(s)$ in the disc $|\rho - (1 + it)| \leq r$. Then $n(r;t) \ll r \log \tau$, uniformly for $r \leq 3/4$.*

Proof If c_1 is a small positive constant and $r < c_1/\log \tau$, then $n(r;t) = 0$ by Theorem 6.6. Suppose that $c_1/\log \tau \leq r \leq 1/6$, $|t| \geq 7/8$. As in the proof of Theorem 6.7, the estimate (6.10) holds when we take $s_1 = 1 + r + it$. In the sum over ρ, each term is non-negative, and those zeros ρ counted in $n(r;t)$ contribute at least $1/(2r)$ apiece. Hence their number is $\ll r \log \tau$. If $1/6 < r \leq 3/4$ and $|t| \geq 3$, then the desired bound follows at once by applying Jensen's inequality (Lemma 6.1 above) to the function $f(z) = \zeta(z + 2 + it)$, with $R = 11/6$, in view of the bounds provided by Corollary 1.17. Note that $|f(0)| \gg 1$ because of the absolute convergence of the Euler product. If $1/6 < r \leq 3/4$ and $|t| \leq 3$, then we apply Jensen's inequality to the function $f(z) = (z + 1 + it)\zeta(z + 2 + it)$. □

6.1.1   Exercises

1. (a) Show that if $|z| < R$, $|w| \le R$, and $z \ne w$, then

$$\left| \frac{z\overline{w} - R^2}{(z - w)R} \right| \ge 1.$$

   (b) Show that if $|w| \le \rho < R$, $|z| = r < R$, and $z \ne w$, then

$$\left| \frac{z\overline{w} - R^2}{(z - w)R} \right| \ge \frac{r\rho + R^2}{(r + \rho)R}.$$

   (c) Suppose that f is analytic in the disc $|z| \le R$. For $r \le R$ put $M(r) = \max_{|z| \le r} |f(z)|$. Show that if $0 < r < R$ and $0 < \rho < R$, then the number of zeros of f in the disc $|z| \le \rho$ does not exceed

$$\frac{\log \dfrac{M(R)}{M(r)}}{\log \dfrac{r\rho + R^2}{(r + \rho)R}}.$$

2. Suppose that R, M, and ε are positive real numbers, and set $h(z) = 2Mz/(z + R + \varepsilon)$.

   (a) Show that $h(0) = 0$, that $h(z)$ is analytic for $|z| < R + \varepsilon$, and that $\Re h(z) \le M$ for $|z| \le R + \varepsilon$.

   (b) Show that if $0 < r < R$, then

$$\max_{|z| \le r} |h(z)| = -h(-r) = \frac{2Mr}{R + \varepsilon - r}.$$

   (c) Show that if $0 < r < R$, then

$$\max_{|z| \le r} |h'(z)| = h'(-r) = \frac{2M(R + \varepsilon)}{(R + \varepsilon - r)^2}.$$

3. Show that, in the situation of the Borel–Carathéodory lemma (Lemma 6.2), if $|z| \le r < R$, then

$$|h''(z)| \le \frac{4MR}{(R - r)^3}.$$

4. (Mertens 1898) Use the Dirichlet series expansion of $\log \zeta(s)$ to show that if $\sigma > 1$, then

$$|\zeta(\sigma)^3 \zeta(\sigma + it)^4 \zeta(\sigma + 2it)| \ge 1.$$

The method used to establish a zero-free region for the zeta function can be applied to any particular Dirichlet L-function, though the constants involved may depend on the function. We shall pursue this systematically in Chapter 11, but in the exercise below we treat one interesting example.

5. Let χ_0 denote the principal character (mod 4), and χ_1 the non-principal character (mod 4).
   (a) Show that $L(1, \chi_1) = \pi/4$, and hence that there is a neighbourhood of 1 in which $L(s, \chi_1) \neq 0$.
   (b) Show that if $\sigma > 1$, then

   $$\Re\left(-3\frac{L'}{L}(\sigma, \chi_0) - 4\frac{L'}{L}(\sigma + it, \chi_1) - \frac{L'}{L}(\sigma + 2it, \chi_0)\right) \geq 0.$$

   (c) Show that there is a constant $c > 0$ such that $L(s, \chi_1) \neq 0$ for $\sigma > 1 - c/\log \tau$.
   (d) Show that there is a constant $c > 0$ such that if $\sigma > 1 - c/\log \tau$, then

   $$\frac{L'}{L}(s, \chi_1) \ll \log \tau,$$

   $$|\log L(s, \chi_1)| \leq \log \log \tau + O(1),$$

   $$\frac{1}{L(s, \chi_1)} \ll \log \tau.$$

6. (a) Show that if $1 < \sigma_1 \leq \sigma_2$, then

   $$\frac{\zeta(\sigma_2)}{\zeta(\sigma_1)} \leq \left|\frac{\zeta(\sigma_2 + it)}{\zeta(\sigma_1 + it)}\right| \leq \frac{\zeta(\sigma_1)}{\zeta(\sigma_2)}$$

   for all real t.
   (b) Show that if $1 < \sigma_1 \leq \sigma_2 \leq 2$, then

   $$\frac{\sigma_1 - 1}{\sigma_2 - 1} \ll \left|\frac{\zeta(\sigma_2 + it)}{\zeta(\sigma_1 + it)}\right| \ll \frac{\sigma_2 - 1}{\sigma_1 - 1}$$

   uniformly in t.

7. (Montgomery & Vaughan 2001)
   (a) Show that if $\sigma > 1$, then

   $$\left|\frac{\zeta(\sigma + i(t + 1))}{\zeta(\sigma + it)}\right| \leq \exp\left(2\sum_{n=1}^{\infty} \frac{\Lambda(n)}{n^\sigma \log n} |\sin\left(\tfrac{1}{2}\log n\right)|\right)$$

   uniformly for all real t.
   (b) Put $f(\theta) = |\sin \pi\theta|$, and for integers k set $\widehat{f}(k) = \int_0^1 f(\theta)e(-k\theta)\,d\theta$ where $e(\theta) = e^{2\pi i\theta}$. Show that $\widehat{f}(k) = -2/(\pi(4k^2 - 1))$.
   (c) By Corollary D.3, or otherwise, show that

   $$|\sin \pi\theta| = \sum_{k=-\infty}^{\infty} \widehat{f}(k)e(k\theta).$$

(d) Show that if $1 < \sigma \le 2$, then

$$\left| \frac{\zeta(\sigma + i(t+1))}{\zeta(\sigma + it)} \right| \le \prod_{k=-\infty}^{\infty} |\zeta(\sigma + ik)|^{2\widehat{f}(k)}$$

uniformly for all real t.

(e) Show that if $\sigma > 1$, then

$$(\sigma - 1)^{4/\pi} \ll \left| \frac{\zeta(\sigma + i(t+1))}{\zeta(\sigma + it)} \right| \ll (\sigma - 1)^{-4/\pi}$$

uniformly in t.

(f) Show that

$$(\log t)^{-4/\pi} \ll \left| \frac{\zeta(1 + i(t+1))}{\zeta(1 + it)} \right| \ll (\log t)^{4/\pi}$$

uniformly for $t \ge 2$.

8. Suppose that a and b are fixed, $0 < a < b < 1$. Suppose that f is analytic in a domain containing the disc $|z| \le R$, that $f(0) \ne 0$, and that $|f(z)| \le M$ for $|z| \le R$. Show that

$$\frac{f'}{f}(z) = \sum_{k=1}^{K} \frac{1}{z - z_k} + O\left(\frac{1}{R} \log \frac{M}{|f(0)|} \right)$$

for $|z| \le aR$ where the sum is over those zeros z_k of $f(z)$ for which $|z_k| \le bR$.

9. (Landau 1924a) Suppose that $\theta(t)$ and $\phi(t)$ are functions with the following properties: $\phi(t) > 0$, $\phi(t) \nearrow$, $e^{-\phi(t)} \le \theta(t) \le 1/2$, $\theta(t) \searrow$. Suppose also that

$$\zeta(s) \ll e^{\phi(t)}$$

for $\sigma \ge 1 - \theta(t)$, $t \ge 2$.

(a) Show that

$$\frac{\zeta'}{\zeta}(s) = \sum_{\rho} \frac{1}{s - \rho} + O\left(\frac{\phi(t+1)}{\theta(t+1)} \right)$$

for $\sigma \ge 1 - \theta(t+1)/3$ where the sum is over zeros ρ for which $|\rho - (1 + \theta(t+1) + it)| \le 5\theta(t+1)/3$.

(b) Show that there is an absolute constant $c > 0$ such that $\zeta(s) \ne 0$ for

$$\sigma \ge 1 - c\frac{\theta(2t+1)}{\phi(2t+1)}.$$

(c) Show that the zero-free region (6.26) follows from the estimate (6.25).

(d) By mimicking the proof of Theorem 6.7, but with $s_1 = 1 + \theta(2t+1)/\phi(2t+1) + it$, show that

$$\frac{\zeta'}{\zeta}(s) \ll \frac{\phi(2t+2)}{\theta(2t+2)},$$

$$|\log \zeta(s)| \leq \log \frac{\phi(2t+2)}{\theta(2t+2)} + O(1),$$

$$\frac{1}{\zeta(s)} \ll \frac{\phi(2t+2)}{\theta(2t+2)}$$

for $\sigma \geq 1 - \frac{1}{2}c\theta(2t+2)/\phi(2t+2)$.

10. Suppose that $\zeta(s) \neq 0$ for $\sigma \geq \eta(t)$, $t \geq 2$, where $\eta(t) \searrow$, $\eta(t) \gg 1/\log t$. Show that

$$\frac{\zeta'}{\zeta}(s) \ll \log t$$

for $\sigma \geq 1 - \frac{1}{2}\eta(t+1)$, $t \geq 2$.

6.2   The Prime Number Theorem

We are now in a position to prove the Prime Number Theorem in a quantitative form. We apply Perron's formula to $\frac{\zeta'}{\zeta}(s)$ to obtain an asymptotic estimate for

$$\psi(x) - \sum_{n \leq x} \Lambda(n),$$

and then use partial summation to derive an estimate for $\pi(x)$. It would be more direct to apply Perron's formula to $\log \zeta(s)$, but our approach is technically simpler since $\log \zeta(s)$ has a logarithmic singularity at $s = 1$ while $\frac{\zeta'}{\zeta}(s)$ has only a simple pole there.

Theorem 6.9 *There is a constant $c > 0$ such that*

$$\psi(x) = x + O\left(\frac{x}{\exp(c\sqrt{\log x})}\right), \tag{6.12}$$

$$\vartheta(x) = x + O\left(\frac{x}{\exp(c\sqrt{\log x})}\right), \tag{6.13}$$

and

$$\pi(x) = \mathrm{li}(x) + O\left(\frac{x}{\exp(c\sqrt{\log x})}\right) \tag{6.14}$$

uniformly for $x \geq 2$.

Here li(x) is the *logarithmic integral*,

$$\text{li}(x) = \int_2^x \frac{1}{\log u}\, du.$$

By integrating this integral by parts K times we see that

$$\text{li}(x) = x \sum_{k=1}^{K-1} \frac{(k-1)!}{(\log x)^k} + O_K\left(\frac{x}{(\log x)^K}\right). \tag{6.15}$$

On combining this with (6.14) we see that

$$\pi(x) = \frac{x}{\log x} + O\left(\frac{x}{(\log x)^2}\right).$$

This is a quantitative form of the Prime Number Theorem. When this main term is used, the error term is genuinely of the indicated size, since by (6.14) and (6.15) again we see that

$$\pi(x) = \frac{x}{\log x} + \frac{x}{(\log x)^2} + O\left(\frac{x}{(\log x)^3}\right).$$

Thus we see that in order to obtain a precise estimate of $\pi(x)$, it is essential to use the logarithmic integral (or some similar function) to express the main term.

Proof From Corollary 1.11 and Theorem 5.2 we see that

$$\psi(x) = \frac{-1}{2\pi i} \int_{\sigma_0 - iT}^{\sigma_0 + iT} \frac{\zeta'}{\zeta}(s)\frac{x^s}{s}\, ds + R \tag{6.16}$$

for $\sigma_0 > 1$, where by Corollary 5.3 we see that

$$R \ll \sum_{x/2 < n < 2x} \Lambda(n) \min\left(1, \frac{x}{T|x-n|}\right) + \frac{(4x)^{\sigma_0}}{T} \sum_{n=1}^{\infty} \frac{\Lambda(n)}{n^{\sigma_0}}.$$

Here the second sum is $-\frac{\zeta'}{\zeta}(\sigma_0)$, which is $\asymp 1/(\sigma_0 - 1)$ for $1 < \sigma_0 \leq 2$. To estimate the first sum we note that $\Lambda(n) \leq \log n \ll \log x$. For the n that is nearest to x we replace the minimum by its first member, and for all other values of n we replace it by its second member. Thus the first sum is

$$\ll (\log x)\left(1 + \frac{x}{T} \sum_{1 \leq k \leq x} \frac{1}{k}\right) \ll \log x + \frac{x}{T}(\log x)^2.$$

Suppose that $2 \leq T \leq x$ and that $\sigma_0 = 1 + 1/\log x$. Then

$$R \ll \frac{x}{T}(\log x)^2.$$

Put $\sigma_1 = 1 - c/\log T$ where c is a small positive constant, and let $\mathcal{C}$ denote the closed contour that consists of line segments joining the points $\sigma_0 - iT$, $\sigma_0 + iT$, $\sigma_1 + iT$, $\sigma_1 - iT$. From Theorem 6.6 we know that $\frac{\zeta'}{\zeta}(s)$ has a simple pole with residue -1 at $s = 1$, but that it is otherwise analytic within $\mathcal{C}$. Hence by the calculus of residues,

$$\frac{-1}{2\pi i} \int_{\mathcal{C}} \frac{\zeta'}{\zeta}(s) \frac{x^s}{s}\, ds = x.$$

If c is small, then the estimate (6.6) of Theorem 6.7 applies on this contour. Hence

$$-\int_{\sigma_0+iT}^{\sigma_1+iT} \frac{\zeta'}{\zeta}(s) \frac{x^s}{s}\, ds \ll \frac{\log T}{T} x^{\sigma_0}(\sigma_0 - \sigma_1) \ll \frac{x}{T},$$

and similarly for the integral from $\sigma_1 - iT$ to $\sigma_0 - iT$. Using (6.6) again, we also see that

$$-\int_{\sigma_1+iT}^{\sigma_1-iT} \frac{\zeta'}{\zeta}(s) \frac{x^s}{s}\, ds \ll x^{\sigma_1}(\log T) \int_{-T}^{T} \frac{dt}{1+|t|} + x^{\sigma_1} \int_{-1}^{1} \frac{dt}{|\sigma_1 + it - 1|}$$

$$\ll x^{\sigma_1}(\log T)^2 + \frac{x^{\sigma_1}}{1-\sigma_1} \ll x^{\sigma_1}(\log T)^2.$$

On combining these estimates we conclude that

$$\psi(x) = x + O\left(x(\log x)^2 \left(\frac{1}{T} + x^{-c/\log T}\right)\right).$$

We choose T so that the two terms in the last factor of the error term are equal, i.e., $T = \exp\left(\sqrt{c \log x}\right)$. With this choice of T, the error term above is

$$\ll x(\log x)^2 \exp\left(-\sqrt{c \log x}\right) \ll x \exp\left(-c\sqrt{\log x}\right)$$

since we may suppose that $0 < c < 1$. Thus the proof of (6.12) is complete.

To derive (6.13) it suffices to combine (6.12) with the first estimate of Corollary 2.5. As for (6.14), we note that

$$\pi(x) = \int_{2-}^{x} \frac{1}{\log u}\, d\vartheta(u) = \mathrm{li}(x) + \int_{2-}^{x} \frac{1}{\log u}\, d(\vartheta(u) - u).$$

By integrating by parts we see that this last integral is

$$\frac{\vartheta(u) - u}{\log u}\bigg|_{2-}^{x} + \int_{2}^{x} \frac{\vartheta(u) - u}{u(\log u)^2}\, du,$$

and by (6.13) it follows that this is $\ll x \exp(-c\sqrt{\log x})$. Thus we have (6.14), and the proof is complete. $\qquad\square$

The method we used to derive Theorem 6.9 is very flexible, and can be applied to many other situations. For example, the summatory function

$$M(x) = \sum_{n \le x} \mu(n)$$

can be estimated by applying the above method with ζ'/ζ replaced by $1/\zeta$. Thus it may be shown that

$$M(x) \ll x \exp\left(-c\sqrt{\log x}\right) \tag{6.17}$$

for $x \ge 2$. If instead we were to apply the method to the function $1/\zeta(s+1)$, we would find that

$$\sum_{n \le x} \frac{\mu(n)}{n} \ll \exp\left(-c\sqrt{\log x}\right), \tag{6.18}$$

since $1/(s\zeta(s+1))$ is analytic at $s = 0$. Hence in particular,

$$\sum_{n=1}^{\infty} \frac{\mu(n)}{n} = 0. \tag{6.19}$$

6.2.1　Exercises

1. (Landau 1901b; cf. Rosser & Schoenfeld 1962) Use Theorem 6.9 to show that

$$\pi(2x) - 2\pi(x) = -2(\log 2)x(\log x)^{-2} + O(x(\log x)^{-3}).$$

Deduce that for all large x, the interval $(x, 2x]$ contains fewer prime numbers than the interval $(0, x]$.

2. Use Theorem 6.9 to show that if n is of the form $n = \prod_{p \le y} p$ where y is sufficiently large, then $d(n) > n^{(\log 2)/\log\log n}$.

3. (a) Use Theorem 6.9 to show that

$$\sum_{x < p \le y} \frac{1}{p} = \log\frac{\log y}{\log x} + O\left(\exp\left(-c\sqrt{\log x}\right)\right).$$

(b) Use the above and Theorem 2.7 to show that

$$\sum_{p \le x} \frac{1}{p} = \log\log x + b + O\left(\exp\left(-c\sqrt{\log x}\right)\right)$$

where $b = C_0 - \sum_p \sum_{k=2}^{\infty} 1/(kp^k)$.

4. Show that for $x \ge 2$,

$$\sum_{n \le x} \frac{\Lambda(n)}{n} = \log x - C_0 + O\left(\exp\left(-c\sqrt{\log x}\right)\right).$$

5. (cf. Cipolla 1902; Rosser 1939) Let $p_1 < p_2 < \cdots$ denote the prime numbers. Show that

$$p_n = n\left(\log n + \log\log n - 1 + \frac{\log\log n}{\log n} - \frac{2}{\log n} + O\left(\frac{(\log\log n)^2}{(\log n)^2}\right)\right).$$

6. (Landau 1900) Let $\pi_k(x)$ denote the number of integers not exceeding x that are composed of exactly k distinct primes.
   (a) Show that

   $$\pi_2(x) = \sum_{p\leq\sqrt{x}} \pi(x/p) + O\left(x(\log x)^{-2}\right).$$

   (b) Show that the sum above is

   $$\sum_{p\leq\sqrt{x}} \frac{x}{p\log x/p} + O\left(x(\log\log x)(\log x)^{-2}\right).$$

   (c) Using Theorem 6.9 and integration by parts, show that the sum above is

   $$x\int_2^{\sqrt{x}} \frac{du}{u(\log x/u)\log u} + O(x/\log x).$$

   (d) Conclude that $\pi_2(x) = x(\log\log x)/\log x + O(x/\log x)$.

7. (D. E. Knutson) Let d_n denote the least common multiple of the numbers $1, 2, \ldots, n$.
   (a) Show that $d_n = \exp(\psi(n))$.
   (b) Let $E(z) = \sum_{n=1}^{\infty} z^n/d_n$. Show that this power series has radius of convergence e.
   (c) Show that $E(1)$ is irrational.

8. (Landau 1905) Let $Q(x)$ denote the number of square-free integers not exceeding x, and define $R(x)$ by the relation $Q(x) = (6/\pi^2)x + R(x)$.
   (a) Show that

   $$R(x) = M(y)\{x/y^2\} - \sum_{d\leq y}\mu(d)\{x/d^2\}$$
   $$+ \sum_{m\leq x/y^2} M\left(\sqrt{x/m}\right) - 2x\int_y^{\infty} M(u)u^{-3}\,du.$$

   (b) Taking $y = x^{1/2}\exp(-c\sqrt{\log x})$ where c is sufficiently small, show that $R(x) \ll x^{1/2}\exp(-c\sqrt{\log x})$.

9. Let $N = N(Q) = 1 + \sum_{q\leq Q}\varphi(q)$ be the number of Farey points of order Q, and for $0 \leq \alpha \leq 1$ write

   $$\text{card}\{(a, q) : q \leq Q, \ (a, q) = 1, \ a/q \leq \alpha\} = N\alpha + R.$$

where $R = R(Q, \alpha)$.

(a) Show that if $\alpha = (1/Q)^-$, then $R = -N/Q \asymp -Q$.

(b) Show that if $\alpha = 1 - 1/Q$, then $R = N/Q - 1 \asymp Q$.

(c) Show that

$$R = -\sum_{r \leq Q} \{r\alpha\} M(Q/r)$$

for $0 \leq \alpha \leq 1$.

(d) Show that $R \ll Q$ uniformly for $0 \leq \alpha \leq 1$.

10. (Landau 1903b; Massias, Nicolas & Robin 1988, 1989) Let $f(n)$ denote the maximal order of any element of the symmetric group S_n.

(a) Show that $f(n) = \max \operatorname{lcm}(n_1, n_2, \ldots, n_k)$ where the maximum is extended over all sets $\{n_1, n_2, \ldots, n_k\}$ of natural numbers for which $n_1 + n_2 + \cdots + n_k \leq n$.

(b) Choose y as large as possible so that $\sum_{p \leq y} p \leq n$. Show that

$$\log f(n) \geq \sum_{p \leq y} \log p = (1 + o(1))(n \log n)^{1/2}.$$

(c) Show that $f(n) = \max q_1 q_2 \cdots q_k$ where $q_i = p_i^{a(i)}$, $p_i \neq p_j$ for $i \neq j$, and $\sum q_i \leq n$.

(d) Use the arithmetic–geometric mean inequality to show that $\prod q_i \leq (n/k)^k$.

(e) Show that if k is the number of q_i's in (c), then $k \leq (2 + o(1))(n/\log n)^{1/2}$.

(f) Conclude that $\log f(n) \asymp (n \log n)^{1/2}$.

11. Let $\lambda(n) = (-1)^{\Omega(n)}$ be Liouville's lambda function.

(a) Show that $\sum_{n=1}^{\infty} \lambda(n) n^{-s} = \zeta(2s)/\zeta(s)$ for $\sigma > 1$.

(b) Using the method of proof of Theorem 6.9, show that

$$\sum_{n \leq x} \lambda(n) \ll x \exp\left(-c\sqrt{\log x}\right).$$

(c) Use (6.17) and the fact that $\lambda(n) = \sum_{d^2 | n} \mu(n/d^2)$ to give a second proof of the above estimate.

12. (Landau 1907, Section 14) Let $c_n = 1$ if n is a prime or a prime power, $c_n = 0$ otherwise.

(a) Show that $\mu(n)\omega(n) = -\sum_{d | n} c_d \mu(n/d)$.

(b) Use (6.18) and the method of the hyperbola to show that

$$\sum_{n=1}^{\infty} \frac{\mu(n)\omega(n)}{n} = 0.$$

13. Use the method of proof of Theorem 6.9 to show that

$$\sum_{n \le x} \Lambda(n) n^{-it} = \frac{x^{1-it}}{1-it} + O(x \exp(-c\sqrt{\log x}))$$

$$+ O\left(x(\log x)^2 \exp\left(-c\frac{\log x}{\log \tau}\right)\right)$$

uniformly for $|t| \le x$.

14. Use the method of proof of Theorem 6.9 to show that for any fixed real t,

$$\sum_{n=1}^{\infty} \mu(n) n^{-1-it} = \frac{1}{\zeta(1+it)}.$$

15. (a) Use the method of proof of Theorem 6.9 to show that for any fixed $t \ne 0$,

$$\sum_{n=1}^{\infty} \frac{\Lambda(n)}{\log n} n^{-1-it} = \log \zeta(1+it).$$

(b) Deduce that for any $t \ne 0$,

$$\prod_{p} (1 - p^{-1-it})^{-1} = \zeta(1+it).$$

16. (Landau 1899b, 1901a, 1903c) Use the method of proof of Theorem 6.9 to show that

(a)
$$\sum_{n=1}^{\infty} \frac{\mu(n) \log n}{n} = -1;$$

(b)
$$\sum_{n=1}^{\infty} \frac{\mu(n)(\log n)^2}{n} = -2C_0;$$

(c)
$$\sum_{n=1}^{\infty} \frac{\lambda(n) \log n}{n} = -\zeta(2).$$

17. Taking (6.18) and a quantitative form of the first part of the preceding exercise for granted, use elementary reasoning to show that if $q \le x$ then

(a)
$$\sum_{\substack{n \le x \\ (n,q)=1}} \frac{\mu(n)}{n} \ll \exp(-c\sqrt{\log x}),$$

(b)
$$\sum_{\substack{n \le x \\ (n,q)=1}} \frac{\mu(n) \log n}{n} = -\frac{q}{\varphi(q)} + O(\exp(-c\sqrt{\log x})).$$

18. (Hardy 1921) Use the method of proof of Theorem 6.9 to show that

(a)
$$\sum_{n=1}^{\infty} \frac{\mu(n)}{\varphi(n)} = 0;$$

(b)
$$\sum_{n=1}^{\infty} \frac{\mu(n) \log n}{\varphi(n)} = 0;$$

(c)
$$\sum_{n=1}^{\infty} \frac{\mu(n)(\log n)^2}{\varphi(n)} = 4A \log 2$$

where $A = \prod_{p>2} \left(1 - \frac{1}{(p-1)^2}\right)$.

19. Let $Q(x)$ denote the number of square-free integers not exceeding x, and recall Theorem 2.2.

   (a) Show that
   $$Q(x) = \frac{6}{\pi^2}x - x \sum_{n>\sqrt{x}} \frac{\mu(n)}{n^2} - \sum_{n\leq\sqrt{x}} \mu(n)\{x/n^2\}$$

   where $\{\theta\} = x - [x]$ is the fractional part of θ.

   (b) Show that $\sum_{n>y} \mu(n)/n^2 \ll y^{-1} \exp(-c\sqrt{\log y})$ for $y \geq 2$.

   (c) Note that if k is a positive integer, then $\{x/n^2\}$ is monotonic for n in the interval $\sqrt{x/(k+1)} < n \leq \sqrt{x/k}$. Deduce that if $x \geq 2k^2$, then
   $$\sum_{\sqrt{x/(k+1)}<n\leq\sqrt{x/k}} \mu(n)\{x/n^2\} \ll \sqrt{x/k} \exp\left(-c\sqrt{\log x}\right).$$

   (d) By using the above for $1 \leq k \leq K = \exp(-b\sqrt{\log x})$ where b is suitably chosen in terms of c, show that
   $$Q(x) = \frac{6}{\pi^2}x + O\left(x^{1/2} \exp\left(-\frac{c}{2}\sqrt{\log x}\right)\right).$$

20. (Ingham 1945) Let $F(n) = \sum_{d|n} f(d)$ for all n. From our remarks at the beginning of Chapter 2 we see that it is natural to expect a connection between

   (i) $S(x) := \sum_{n\leq x} F(n) = cx + o(x)$;
   (ii) $\sum_{n=1}^{\infty} f(n)/n = c$.

   Neither of these implies the other, but we show now that (i) implies that the series (ii) is (C,1) summable to c.

   (a) Show that $S(x) = \sum_{n\leq x} f(n)[x/n]$.

   (b) Show that
   $$\sum_{n\leq x} \frac{f(n)}{n}\left(1 - \frac{n}{x}\right) = \int_1^x S(v)\left(\sum_{d\leq x/v} \mu(d)/d\right) \frac{dv}{v^2}.$$

   (c) Show that
   $$\int_1^x \sum_{d\leq x/v} \frac{\mu(d)}{d} \frac{dv}{v} \to 1$$

   as $x \to \infty$.

(d) Use the estimate $\sum_{d \leq y} \mu(d)/d \ll (\log 2y)^{-2}$ to show that

$$\int_1^x \left| \sum_{d \leq x/v} \frac{\mu(d)}{d} \right| \frac{dv}{v} \ll 1.$$

(e) Mimic the proof of Theorem 5.5, or use Exercise 5.2.6 to show that if (i) holds, then

$$\lim_{x \to \infty} \sum_{n \leq x} \frac{f(n)}{n} \left(1 - \frac{n}{x}\right) = c.$$

(f) Use Theorem 5.6 to show that if (i) holds and $f(n) = O(1)$, then (ii) follows.

(g) Take $f(n) = \mu(n)$ to deduce that $\sum_{n=1}^{\infty} \mu(n)/n = 0$. (Of course we used much more above in (d). For a result in the converse direction, see Exercise 8.1.5.)

21. (Landau 1908b) Let $\mathcal{R}$ be the set of positive integers that can be expressed as a sum of two squares, let $R(x)$ denote the number of such integers not exceeding x, and let χ_1 denote the non-principal character (mod 4), as in Exercise 6.1.5.

(a) Show that

$$\sum_{n \in \mathcal{R}} n^{-s} = (1 - 2^{-s})^{-1} \prod_{p \equiv 1\,(4)} (1 - p^{-s})^{-1} \prod_{p \equiv 3\,(4)} (1 - p^{-2s})^{-1}$$

for $\sigma > 1$.

(b) Show that the Dirichlet series above is $f(s)\sqrt{\zeta(s)L(s, \chi_1)}$ where

$$f(s) = (1 - 2^{-s})^{-1/2} \prod_{p \equiv 3\,(4)} (1 - p^{-2s})^{-1/2}$$

is a Dirichlet series with abscissa of convergence $\sigma_c = 1/2$.

(c) Deduce that the Dirichlet series generating function for $\mathcal{R}$ has a quadratic singularity at $s = 1$.

(d) Show that

$$R(x) = \frac{1}{2\pi i} \int_{\mathcal{C}} f(s)\sqrt{\zeta(s)L(s, \chi_1)} \frac{x^s}{s}\, ds + O\left(x \exp\left(-c\sqrt{\log x}\right)\right)$$

where $\mathcal{C}$ is the contour running from $1 - c - i\delta$ along a straight line to $1 - i\delta$, then along the semicircle $1 + \delta e^{i\theta}$, $-\pi/2 \leq \theta \leq \pi/2$, and finally along a straight line to $1 - c + i\delta$. Here c should be sufficiently small and $\delta = 1/\log x$.

(e) Show that the integral above is

$$= \frac{1}{2\pi i} \int_{\mathcal{C}} \frac{g(s)x^s}{\sqrt{s-1}}\, ds$$

where

$$g(s) = \frac{f(s)}{s}\sqrt{(s-1)\zeta(s)L(s,\chi_1)}$$

is analytic in a neighbourhood of 1.

(f) Show that

$$g(1) = \sqrt{\frac{\pi}{2}} \prod_{p \equiv 3\,(4)} (1 - p^{-2})^{-1/2}.$$

(g) Show that $g(s) = g(1) + O(|s-1|)$ when s is near 1.

(h) By means of Theorem C.3 with $s = 1/2$, or otherwise, show that

$$\frac{1}{2\pi i} \int_C \frac{x^s}{\sqrt{s-1}}\,ds = \frac{x}{\sqrt{\pi \log x}} + O(x^{1-c}).$$

(i) Show that if $\delta = 1/\log x$, then

$$\int_C |s-1|^{1/2} x^\sigma\,|ds| \ll \frac{x}{(\log x)^{3/2}}.$$

(j) Show that

$$R(x) = \frac{bx}{\sqrt{\log x}} + O\big(x(\log x)^{-3/2}\big)$$

where

$$b = 2^{-1/2} \prod_{p \equiv 3\,(4)} (1 - p^{-2})^{-1/2}.$$

22. Let $\mathcal{A}$ denote the set of those positive integers that are composed entirely of the prime 2 and primes $\equiv 1 \pmod 4$, and let $\mathcal{B}$ be the the set of those positive integers that are composed entirely of primes $\equiv 3 \pmod 4$.

(a) Explain why any positive integer n has a unique representation in the form $n = a(n)b(n)$ where $a(n) \in \mathcal{A}$ and $b(n) \in \mathcal{B}$.

(b) Let $A(x)$ denote the number of $a \in \mathcal{A}$, $a \le x$. Show that

$$A(x) = \frac{\alpha x}{\sqrt{\log x}} + O\left(\frac{x}{(\log x)^{3/2}}\right)$$

where $\alpha = 1/\sqrt{2}$.

(c) Let $B(x)$ denote the number of $b \in \mathcal{B}$, $b \le x$. Show that

$$B(x) = \frac{\beta x}{\sqrt{\log x}} + O\left(\frac{x}{(\log x)^{3/2}}\right)$$

where $\beta = \sqrt{2}/\pi$.

(d) For $0 \leq \kappa \leq 1$ let $N_\kappa(x)$ denote the number of $n \leq x$ such that $a(n) \leq n^\kappa$. Show that

$$N_\kappa(x) = \sum_{\substack{a \leq x^\kappa \\ a \in \mathcal{A}}} \sum_{\substack{a^{1/\kappa - 1} \leq b \leq x/a \\ b \in \mathcal{B}}} 1.$$

(e) Show that if κ is fixed, $0 \leq \kappa \leq 1$, then

$$N_k(x) = c(\kappa)x + O\left(\frac{x}{\sqrt{\log x}}\right)$$

where

$$c(\kappa) = \frac{1}{\pi} \int_0^\kappa \frac{du}{\sqrt{u(1-u)}}.$$

23. The definition of li(x) is somewhat arbitrary because of the casual choice of the lower endpoint of integration. A more intrinsic logarithmic integral is Li(x), which is defined to be

$$\text{Li}(x) = \lim_{\varepsilon \to 0^+} \left(\int_0^{1-\varepsilon} + \int_{1+\varepsilon}^x \right) \frac{dt}{\log t} \qquad (6.20)$$

for $x > 1$. (Note that li(x) = Li(x) − Li(2).)

(a) Show that

$$\int_0^{1-\varepsilon} \frac{dt}{\log t} = -\int_{-\log(1-\varepsilon)}^\infty e^{-v} \frac{dv}{v}.$$

(b) Show that

$$\int_0^{1-\varepsilon} \frac{dt}{\log t} = \log \varepsilon - \int_0^\infty (\log v) e^{-v} \, dv + O(\varepsilon \log 1/\varepsilon),$$

and explain why the integral on the right is $\Gamma'(1) = -C_0$.

(c) Show that if $x > 1$, then

$$\int_{1+\varepsilon}^x \frac{dt}{\log t} = \int_{\log(1+\varepsilon)}^{\log x} e^v \frac{dv}{v}.$$

(d) Show that if $x > 1$, then

$$\int_{1+\varepsilon}^x \frac{dt}{\log t} = \log \log x - \log \varepsilon + \int_1^{\log x} \frac{e^v - 1}{v} \, dv + O(\varepsilon).$$

(e) Show that if $x > 1$, then

$$\text{Li}(x) = \log \log x + C_0 + \int_0^{\log x} \frac{e^v - 1}{v} \, dv.$$

(f) Expand e^v as a power series, and integrate term-by-term, to show that if $x > 1$, then

$$\text{Li}(x) = \log\log x + C_0 + \sum_{n=1}^{\infty} \frac{(\log x)^n}{n!n}. \qquad (6.21)$$

24. For $0 < x < 1$ let

$$\text{Li}(x) = \int_0^x \frac{dt}{\log t}.$$

(a) Show that if $0 < x < 1$, then

$$\text{Li}(x) = x\log\log 1/x - \int_{-\log x}^{\infty} e^{-v}\log v\, dv.$$

(b) Show that if $0 < x < 1$, then

$$\text{Li}(x) = x\log\log 1/x + C_0 + \int_0^{-\log x} e^{-v}\log v\, dv.$$

(c) Show that if $0 < x < 1$, then

$$\text{Li}(x) = \log\log 1/x + C_0 - \int_0^{-\log x} \frac{1 - e^{-v}}{v}\, dv.$$

(d) Show that if $0 < x < 1$, then

$$\text{Li}(x) = \log\log 1/x + C_0 + \sum_{n=1}^{\infty} \frac{(\log x)^n}{n!n}.$$

(e) (Pólya & Szegö 1972, p. 8) Show that

$$\sum_{n=1}^{\infty} \frac{z^n}{n!n} = -e^z \sum_{n=1}^{\infty} \left(\sum_{k=1}^{n} \frac{1}{k}\right) \frac{(-z)^n}{n!}.$$

(f) Show that if $0 < x < 1$, then

$$\text{Li}(x) = \log\log 1/x + C_0 - x\sum_{n=1}^{\infty} \left(\sum_{k=1}^{n} \frac{1}{k}\right) \frac{(\log 1/x)^n}{n!}. \qquad (6.22)$$

25. By repeated integration by parts we know that

$$\text{Li}(x) = x\sum_{k=1}^{K} \frac{(k-1)!}{(\log x)^k} + O_K\left(\frac{x}{(\log x)^{K+1}}\right).$$

Our object is to determine how closely one can approximate to $\text{Li}(x)$ by partial sums of the formal asymptotic expansion

$$\text{Li}(x) \sim x \sum_{k=1}^{\infty} \frac{(k-1)!}{(\log x)^k}.$$

(a) Show that the least term in the sum above occurs when $k = [\log x] + 1$.

(b) Show that if $x \geq e^K$, then

$$\text{Li}(x) = x \sum_{k=1}^{K} \frac{(k-1)!}{(\log x)^k} + \text{Li}(e)$$

$$+ \sum_{k=1}^{K-1} \left(k! \int_{e^k}^{e^{k+1}} \frac{dt}{(\log t)^{k+1}} - \frac{(k-1)!e^k}{k^k} \right)$$

$$- \frac{(K-1)!e^K}{K^K} + K! \int_{e^K}^{x} \frac{dt}{(\log t)^{K+1}}.$$

(c) Define $R(x)$ by the relation

$$\text{Li}(x) = x \sum_{k=1}^{[\log x]} \frac{(k-1)!}{(\log x)^k} + R(x).$$

Show that $R(x)$ is increasing, continuous, and convex downward for $x \in [e^K, e^{K+1})$. Let $\alpha_K = R(e^K)$, and let β_K be the limit of $R(x)$ as x tends to e^{K+1} from below.

(d) Show that

$$\int_{e^K}^{e^{K+1}} \frac{dt}{(\log t)^{K+1}} = \frac{e^K}{K^K} \int_0^{1/K} \frac{e^{Kw}}{(1+w)^{K+1}} \, dw.$$

(e) Show that the integrand on the right above is ≤ 1 in the range of integration.

(f) Show that the minimum of $e^{Kw}/(1+w)^{K+1}$ for $w > 0$ occurs when $w = 1/K$.

(g) Show that

$$\frac{e^{K+1}}{(K+1)^{K+1}} < \int_{e^K}^{e^{K+1}} \frac{dt}{(\log t)^{K+1}} < \frac{e^K}{K^{K+1}}.$$

(h) Show that $\alpha_K \nearrow$ and that $\beta_K \searrow$.

(i) Show that $\beta_K - \alpha_K \ll K^{-1/2}$.

(j) Show that $R(x) = c + O((\log x)^{-1/2})$ where

$$c = \text{Li}(e) + \sum_{k=1}^{\infty} \left(k! \int_{e^k}^{e^{k+1}} \frac{dt}{(\log t)^{k+1}} - \frac{(k-1)!e^k}{k^k} \right).$$

(k) Show that if $x \geq e$, then

$$\alpha_1 \leq \mathrm{Li}(x) - x \sum_{k=1}^{[\log x]} \frac{(k-1)!}{(\log x)^k} \leq \beta_1 \tag{6.23}$$

where $\alpha_1 = -0.82316\ldots$ and $\beta_1 = 1.259706\ldots\ldots$

26. (Ingham 1932, pp. 60–63) Suppose that $\eta(t)$ is defined for $t \geq 2$, that $\eta'(t)$ is continuous, $\eta'(t) \to 0$ as $t \to \infty$, that $\eta(t) \searrow$, that $1/\log t \ll \eta(t) \leq 1/2$, and that $\zeta(s) \neq 0$ for $\sigma \geq 1 - \eta(t)$, $t \geq 2$. For $x \geq 2$, put

$$\omega(x) = \min_{2 \leq t < \infty} \eta(t)\log x + \log t .$$

(a) Show that there is an absolute constant $c > 0$ such that

$$\pi(x) = \mathrm{li}(x) + O(x \exp(-c\omega(x))).$$

(b) Show that if $a > 0$ is fixed and (6.24) below holds, then (6.27) below holds with $b = 1/(1+a)$.

(c) Show that (6.28) follows from (6.26).

6.3   Notes

Section 6.1. Jensen (1899) proved that if f satisfies the hypotheses of Lemma 6.1, then

$$|f(0)| \prod_{k=1}^{n} \frac{R}{|z_k|} = \exp\left(\frac{1}{2\pi} \int_0^{2\pi} \log |f(Re^{i\theta})| \, d\theta \right)$$

where $z_1, \ldots, z_n$ are the zeros of f in the disc $|z| \leq R$. Here the right-hand side may be regarded as being the geometric mean of $|f(z)|$ for z on the circle $|z| = R$. Each factor of the product above is ≥ 1, and if $|z_k| \leq r$, then $R/|z_k| \geq R/r$. Thus Lemma 6.1 follows easily from the above. The products used in the proofs of Lemmas 6.1 and 6.3 are known as Blaschke products. Their use (usually with infinitely many factors) is an important tool of complex analysis. Lemma 6.2 is due to Borel (1897); it refines an earlier estimate of Hadamard. Carathéodory's contributions on this subject are recounted by Landau (1906; Section 4).

Lemma 6.4 is implicit in Landau (1909, p. 372), and may have been known earlier. It can also be easily derived from the identity (10.29) that arises by applying Hadamard's theory of entire functions to the zeta function.

The Prime Number Theorem was first proved, in the qualitative form $\pi(x) \sim x/\log x$, independently by Hadamard (1896) and de la Vallée Poussin (1896). In these papers, it was shown that $\zeta(1 + it) \neq 0$, but no specific zero-free region was established. The first proof that $\zeta(1 + it) \neq 0$ given by de la Vallée Poussin was rather complicated, but later in his long paper he gave a second proof depending on the inequality $1 - \cos 2\theta \leq 4(1 + \cos\theta)$. This is equivalent to the non-negativity of the cosine polynomial $3 + 4\cos\theta + \cos 2\theta$, which Mertens (1898) used to obtain the result of Exercise 6.4. Our Lemma 6.5 is derived by the same method. The classical zero-free region of Theorem 6.6 was established first by de la Vallée Poussin (1899). The estimates (6.6) and (6.8) of Theorem 6.7 were first proved by Gronwall (1913).

Wider zero-free regions have been established by using exponential sum estimates to obtain better upper bounds for $|\zeta(s)|$ when σ is near 1. The first such improvement was derived by Hardy & Littlewood. Their paper on this was never published, but accounts of their approach have been given by Landau (1924b) and Titchmarsh (1986, Chapter 5). Littlewood (1922) announced that from these estimates he had deduced that $\zeta(s) \neq 0$ for $\sigma \geq 1 - c(\log\log\tau)/\log\tau$. As explained by Ingham (1932, p. 66), Littlewood never published his complicated proof, because the simpler method of Landau (1924a) had become available.

In 1935, Vinogradov introduced a new method for estimating Weyl sums. A *Weyl sum* is a sum of the form $\sum_{n=1}^{N} e(f(n))$ where $f \in \mathbb{R}[x]$. The quality of Vinogradov's estimate depends on rational approximations to the coefficients of f, and on the degree of f. The function $f(x) = t\log x$ is not a polynomial, but by approximating to it by polynomials one can make Vinogradov's method apply. This was first done by Chudakov (1936 a, b, c), who derived estimates for $\zeta(s)$ for σ near 1 that allowed him to deduce that $\zeta(s) \neq 0$ for

$$\sigma > 1 - c(\log\tau)^{-a} \qquad (6.24)$$

for $a > 10/11$. Vinogradov (1936b) gave stronger exponential sum estimates, which Titchmarsh (1938) used to obtain a zero-free region of the above form for $a > 4/5$. Hua (1949) introduced a further refinement of Vinogradov's method, from which Titchmarsh (1951, Chapter 6) and Tatuzawa (1952) derived the zero-free region

$$\sigma > 1 - c(\log\tau)^{-3/4}(\log\log\tau)^{-3/4}.$$

By refining the passage from Weyl sums to the zeta function, Korobov (1958a) obtained (6.24) for $a > 5/7$, and then Korobov (1958b, c) and Vinogradov (1958) obtained $a > 2/3$. In fact, Vinogradov claimed that one can take $a = 2/3$, but this seems to be still out of reach. Richert's polished exposition of Vinogradov's method is reproduced in Walfisz (1963). Other expositions have since been given by Karatsuba & Voronin (1992, Chapter 4), Montgomery (1994, Chapter 4), and Vaughan (1997). Richert (1967) used Vinogradov's method to show that

$$\zeta(s) \ll t^{100(1-\sigma)^{3/2}}(\log t)^{2/3} \tag{6.25}$$

for $\sigma \le 1, t \ge 2$. From this it follows that $\zeta(s) \ne 0$ for

$$\sigma \ge 1 - c(\log \tau)^{-2/3}(\log \log \tau)^{-1/3}. \tag{6.26}$$

The methods of Hadamard and de la Vallée Poussin depended on the analytic continuation of $\zeta(s)$, on bounds for the size of $\zeta(s)$ in the complex plane, and on Hadamard's theory of entire functions. The first two of these are achieved most easily by Riemann's functional equation (see Corollaries 10.3–10.5). An abbreviated account of the third is found in Lemma 10.11. Landau (1903a) showed that one can obtain a zero-free region using only the local analytic properties of the zeta function. This enabled Landau to prove the Prime Ideal Theorem, which is the natural extension of the Prime Number Theorem to algebraic number fields: If K is an algebraic number field, then the number of prime ideals $\mathfrak{p}$ in K with $N(\mathfrak{p}) \le x$ is asymptotic to $x/\log x$ as $x \to \infty$. This could not have been done at that time by the methods of Hadamard and de la Vallée Poussin, since the analytic continuation and functional equation of the Dedekind zeta function $\zeta_K(s)$ was established only later, by Hecke (1917). Landau did not achieve Theorem 6.6 at the first attempt, but he refined his approach in a series of papers culminating in the polished exposition of Landau (1924a).

   Section 6.2. Ingham (1932, pp. 60–65; cf. Titchmarsh 1986, pp. 56–60) developed a general system by which any given zero-free region of the zeta function can be used to derive an associated bound for the error term in the Prime Number Theorem. In particular, he showed that if $\zeta(s) \ne 0$ for s in the region (6.24), then

$$\psi(x) = x + O(x \exp(-c(\log x)^b)) \tag{6.27}$$

where $b = 1/(1 + a)$. Similarly, from the zero-free region (6.26) it follows that

$$\pi(x) = \text{li}(x) + O\left(x \exp\left(-c(\log x)^{3/5}(\log \log x)^{-1/5}\right)\right). \tag{6.28}$$

Turán (1950) used his method of power sums to show conversely that (6.27) implies (6.24). More general converse theorems have since been established by Stás (1961) and Pintz (1980, 1983, 1984). A similar converse theorem in which an upper bound for $M(x) = \sum_{n \le x} \mu(n)$ is used to produce a zero-free region has been given by Allison (1970).

   That $M(x) = o(x)$ was first proved by von Mangoldt (1897). The quantitative estimate (6.17) is due to Landau (1908a). The relation (6.19), asserted by Euler

(1748; Chapter 15, no. 277), was first proved by von Mangoldt (1897). Landau
(1899a) and de la Vallée Poussin (1899) shortly gave simpler proofs.

6.4 References

Allison, D. (1970). On obtaining zero-free regions for the zeta-function from estimates
of $M(x)$, *Proc. Cambridge Philos. Soc.* **67**, 333–337.

Borel, E. (1897). Sur les zéros des fonctions entièrs, *Acta Math.* **20**, 357–396.

Chudakov, N. G. (1936a). Sur les zéros de la fonction $\zeta(s)$, *C. R. Acad. Sci.* Paris **202**,
191–193.

(1936b). On zeros of the function $\zeta(s)$, *Dokl. Akad. Nauk SSSR* **1**, 201–204.

(1936c). On zeros of Dirichlet's L-functions, *Mat. Sb.* (1) **43**, 591–602.

(1937). On Weyl's sums, *Mat. Sb.* (2) **44**, 17–35.

(1938). On the functions $\zeta(s)$ and $\pi(x)$, *Dokl. Akad. Nauk SSSR* **21**, 421–422.

Cipolla, M. (1902). La determinazione assintotica dell' n^{imo} numero primo, *Rend. Accad.
Sci. Fis-Mat. Napoli* (3) **8**, 132–166.

Euler, L. (1748). *Introductio in analysin infinitorum*, I, Lausanne; *Opera omnia* Ser 1,
Vol. 8, Teubner, 1922.

Gronwall, T. H. (1913). Sur la fonction $\zeta(s)$ de Riemann au voisinage de $\sigma = 1$, *Rend.
Mat. Cir. Palermo* **35**, 95–102.

Hadamard, J. (1896). Sur la distribution des zéros de la fonction $\zeta(s)$ et ses conséquences
arithmétiques, *Bull. Soc. Math.* France **24**, 199–220.

Hardy, G. H. (1921). Note on Ramanujan's trigonometrical function $c_q(n)$, and certain
series of arithmetical functions, *Proc. Cambridge Philos. Soc.* **20**, 263–271.

Hecke, E. (1917). Über die Zetafunktion beliebiger algebraischer Zahlkörper, *Nachr.
Akad. Wiss. Göttingen*, 77–89; *Mathematische Werke*, Göttingen: Vandenhoeck &
Ruprecht, 1959, pp. 159–171.

Hua, L. K. (1949). An improvement of Vinogradov's mean-value theorem and several
applications, *Quart. J. Math. Oxford Ser.* **20**, 48–61.

Ingham, A. E. (1932). The Distribution of Prime Numbers, *Cambridge Tracts Math.* 30.
Cambridge: Cambridge University Press.

(1945). Some Tauberian theorems connected with the Prime Number Theorem, *J.
London Math. Soc.* **20**, 171–180.

Jensen, J. L. W. V. (1899). Sur un nouvel et important théorème de la théorie des
fonctions, *Acta Math.* **22**, 359–364.

Karatsuba, A. A. & Voronin, S. M. (1992). *The Riemann Zeta-function*. Berlin: de
Gruyter.

Korobov, N. M. (1958a). On the zeros of the function $\zeta(s)$, *Dokl. Akad. Nauk SSSR* **118**,
231–232.

(1958b). Weyl's estimates of sums and the distribution of primes, *Dokl. Akad. Nauk
SSSR* **123**, 28–31.

(1958c). Evaluation of trigonometric sums and their applications, *Usp. Mat. Nauk* **13**,
no. 4, 185–192.

Landau, E. (1899a). *Neuer Beweis der Gleichung* $\sum_{k=1}^{\infty} \frac{\mu(k)}{k} = 0$, Inaugural Dissertation,
Berlin; *Collected Works*, Vol. 1. Essen: Thales Verlag, pp. 69–83.

(1899b). Contribution à la théorie de la fonction $\zeta(s)$ de Riemann, *C. R. Acad. Sci. Paris*, **129**, 812–815; *Collected Works*, Vol. 1. Essen: Thales Verlag, 1985, pp. 84–88.

(1900). Sur quelques problèmes rélatifs à la distribution des nombres premiers, *Bull. Soc. Math. France* **28**, 25–38; *Collected Works*, Vol. 1. Essen: Thales Verlag, 1985, pp. 92–105.

(1901a). Über die asymptotischen Werthe einiger zahlentheoretischer Functionen, *Math. Ann.* **54**, 570–591; *Collected Works*, Vol. 1. Essen: Thales Verlag, 1985, pp. 141–162.

(1901b). Solutions de questions proposées, *Nouv. Ann. de Math.* (4) **1**, 281–283; *Collected Works*, Vol. 1. Essen: Thales Verlag, 1985, pp. 181–182.

(1903a). Neuer Beweis des Primzahlsatzes und Beweis des Primidealsatzes, *Math. Ann.* **56**, 645–670; *Collected Works*, Vol. 1. Essen: Thales Verlag, 1985, pp. 327–353.

(1903b). Über die Maximalordnung der Permutationen gegebenen Grades, *Arch. Math. Phys.* (3) **5**, 92–103; *Collected Works*, Vol. 1. Essen: Thales Verlag, 1985, pp. 384–396.

(1903c). Über die zahlentheoretische Funktion $\mu(k)$, *Sitzungsber. Kaiserl. Akad. Wiss. Wien math-natur. Kl.* **112**, 537–570; *Collected Works*, Vol. 2. Essen: Thales Verlag, 1986, pp. 60–93.

(1905). Sur quelques inégalités dans la théorie de la fonction $\zeta(s)$ de Riemann, *Bull. Soc. Math. France* **33**, 229–241; *Collected Works*, Vol. 2. Essen: Thales Verlag, 1986, pp. 167–179.

(1906). Über den Picardschen Satz, *Vierteljahrschr. der Naturf. Ges. Zürich* **51**, 252–318; *Collected Works*, Vol. 3. Essen: Thales Verlag, 1986, pp. 113–179.

(1907). Über die Multiplikation Dirichlet'scher Reihen, *Rend. Circ. Mat. Palermo* **24**, 81–160; *Collected Works*, Vol. 3. Essen: Thales Verlag, 1986, pp. 323–401.

(1908a). Beiträge zur analytischen Zahlentheorie, *Rend. Mat. Circ. Palermo* **26**, 169–302; *Collected Works*, Vol. 3. Essen: Thales Verlag, 1986, pp. 411–544.

(1908b). Über die Einteilung der positiven ganzen Zahlen in vier Klassen nach der Mindestzahl der zu ihrer additiven Zusammensetzung erforderlichen Quadrate, *Arch. Math Phys.* (3) **13**, 305–312; *Collected Works*, Vol. 4. Essen: Thales Verlag, 1986, 59–66.

(1909). *Handbuch der Lehre von der Verteilung der Primzahlen*, Leipzig: Teubner.

(1924a). Über die Wurzeln der Zetafunktion, *Math. Z.* **20**, 98–104; *Collected Works*, Vol. 8. Essen: Thales Verlag, 1987, pp. 70–76.

(1024b). Über die ζ-funktion und die L-funktionen, *Math. Z.* **20**, 105–125; *Collected Works*, Vol. 8. Essen: Thales Verlag, 1987, pp. 77–98.

Littlewood, J. E. (1922). Researches in the theory of the Riemann ζ-function, *Proc. London Math. Soc.* (2), **20**, xxii–xxvii; *Collected papers*, Vol. 2. Oxford: Oxford University Press, 1982, pp. 844–850.

von Mangoldt, H. (1897). Beweis der Gleichung $\sum_{k=1}^{\infty} \frac{\mu(k)}{k} = 0$, *Sitzungsber. Königl. Preuß. Akad. Wiss.* Berlin, 835–852.

Massias, J.-P., Nicolas, J.-L., & Robin, G. (1988). Évaluation asymptotique de l'ordre maximum d'un élément du groupe symétrique, *Acta Arith.* **50**, 221–242.

(1989). Effective bounds for the maximal order of an element in the symmetric group, *Math. Comp.* **53**, 665–678.

Mertens, F. (1897). Ueber eine Zahlentheoretische Function, *Sitzungsber. Akad. Wiss. Wien Abt.* 2a **106**.

(1898). Über eine Eigenschaft der Riemannscher ζ-Funktion, *Sitzungsber. Kais. Akad. Wiss. Wien Abt.* 2a **107**, 1429–1434.

Montgomery, H. L. (1994). *Ten Lectures on the Interface Between Analytic Number Theory and Harmonic Analysis*, CBMS Regional Conf. Series in Math. 84. Providence: Amer. Math. Soc.

Montgomery, H. L. & Vaughan, R. C. (2001). Mean values of multiplicative functions, *Period. Math. Hungar.* **43**, 199–214.

Pintz, J. (1980). On the remainder term of the prime number formula, II. On a theorem of Ingham, *Acta Arith.* **37**, 209–220.

(1983). Oscillatory Properties of the Remainder Term of the Prime Number Formula, *Studies in Pure Math.* Basel: Birkhäuser, pp. 551–560.

(1984). On the remainder term of the prime number formula and the zeros of Riemann's zeta-function, *Number Theory* (Noordwijkerhout, 1983). Lecture notes in math. 1068. Berlin: Springer-Verlag, pp. 186–197.

Pólya, G. & Szegö, G. (1972). Problems and Theorems in Analysis, Vol. 1. *Grundl. math. Wiss.* 193. New York: Springer-Verlag.

Richert, H.-E. (1967). Zur Abschätzung der Riemannschen Zetakunktion in der Nähe der Vertikalen $\sigma = 1$, *Math. Ann.* **169**, 97–101.

Rosser, J. B. (1939). The n-th prime is greater than $n \log n$, *Proc. London Math. Soc.* (2) **45**, 21–44.

Rosser, J. B. & Schoenfeld, L. (1962). Approximate formulas for some functions of prime numbers, *Illinois J. Math.* **6**, 64–94.

Stás, W. (1961). Über die Umkehrung eines Satzes von Ingham, *Acta Arith.* **6**, 435–446.

Tatuzawa, T. (1952). On the number of primes in an arithmetic progression, *Jap. J. Math.* **21**, 93–111.

Titchmarsh, E. C. (1938). On $\zeta(s)$ and $\pi(x)$, *Quart. J. Math. Oxford Ser.* **9**, 97–108.

(1951). *The Theory of the Riemann Zeta-function*, Oxford: Oxford University Press.

(1986). *The Theory of the Riemann Zeta-function*, Second Ed. Oxford: Oxford University Press.

Turán, P. (1950). On the remainder-term in the prime-number formula, II, *Acta. Math. Acad. Sci. Hungar.* **1**, 155–166; *Collected Papers*, Vol. 1. Budapest: Akadémiai Kiado, 1990, pp. 541–551.

de la Vallée Poussin, C. J. (1896). Recherches analytiques sur la théorie des nombres premiers, I–III, *Ann. Soc. Sci. Bruxelles* **20**, 183–256, 281–362, 363–397.

(1899). Sur la fonction $\zeta(s)$ et le nombre des nombres premiers inférieurs à une limite donnée, *Mem. Couronnés de l'Acad. Roy. Sci. Bruxelles* **59**.

Vaughan, R. C. (1997). *The Hardy–Littlewood Method*, Second Edition, Cambridge Tracts in Math. 125, Cambridge: Cambridge University Press.

Vinogradov, I. M. (1935). On Weyl's sums, *Mat. Sb.* **42**, 521–530.

(1936a). A new method for resolving certain general questions in the theory of numbers, *Mat. Sb.* (1) **43**, 9–19.

(1936b). A new method of estimation of trigonometrical sums, *Mat. Sb.* (1) **43**, 175–188.

(1947). *The Method of Trigonometrical Sums in the Theory of Numbers*, Trav. Inst. Math. Stecklov **23**; English translation, London: Interscience Publishers, 1954.

(1958). A new evaluation of $\zeta(1 + it)$, *Izv. Akad. Nauk SSSR* **22**, 161–164.

Walfisz, A. (1963). *Weylsche Exponentialsummen in der neuren Zahlentheorie*, Math. Forschungsberichte 15. Berlin: Deutscher Verlag Wiss.

7

Applications of the Prime Number Theorem

We now use the Prime Number Theorem, and other estimates obtained by similar methods, to estimate the number of integers whose multiplicative structure is of a specified type.

7.1   Numbers composed of small primes

Let $\psi(x, y)$ denote the number of integers n, $1 \le n \le x$, all of whose prime factors are $\le y$. Obviously, if $y \ge x$, then

$$\psi(x, y) = [x] = x + O(1). \tag{7.1}$$

Also, if $n \le x$, then n can have at most one prime factor $p > \sqrt{x}$, and hence if $x^{1/2} \le y \le x$, then

$$\psi(x, y) = [x] - \sum_{y < p \le x} \sum_{\substack{n \le x \\ p|n}} 1$$

$$= [x] - \sum_{y < p \le x} [x/p]$$

$$= x - x \sum_{y < p \le x} \frac{1}{p} + O(\pi(x)).$$

By the estimates of Chebyshev and Mertens (Corollary 2.6 and Theorem 2.7(d)), this is

$$= x \left(1 - \log \frac{\log x}{\log y}\right) + O\left(\frac{x}{\log x}\right).$$

Thus if we take $u = (\log x)/(\log y)$, so that $y = x^{1/u}$, then we see that

$$\psi\left(x, x^{1/u}\right) = (1 - \log u)x + O\left(\frac{x}{\log x}\right) \tag{7.2}$$

199

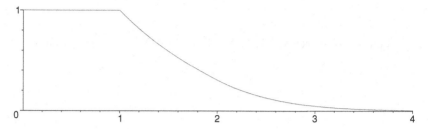

Figure 7.1 The Dickman function $\rho(u)$ for $0 \le u \le 4$.

uniformly for $1 \le u \le 2$. We shall show more generally that there is a function $\rho(u) > 0$ such that

$$\psi\left(x, x^{1/u}\right) \sim \rho(u)x \tag{7.3}$$

as $x \to \infty$ with u bounded. The function $\rho(u)$ that arises here is known as the *Dickman function*; it may be defined to be the unique continuous function on $[0, \infty)$ satisfying the differential–delay equation

$$u\rho'(u) = -\rho(u - 1) \tag{7.4}$$

for $u > 1$ together with the initial condition that

$$\rho(u) = 1 \tag{7.5}$$

for $0 \le u \le 1$. Before proceeding further we note some simple properties of this function. By dividing both sides of (7.4) by u and then integrating, we find that

$$\rho(v) = \rho(u) - \int_u^v \rho(t - 1)\,\frac{dt}{t} \tag{7.6}$$

for $1 \le u \le v$. Also, from (7.4) we see that $(u\rho(u))' = \rho(u) - \rho(u - 1)$, so that by integrating it follows that

$$u\rho(u) = \int_{u-1}^u \rho(v)\,dv + C$$

for $u \ge 1$, where C is a constant of integration. On taking $u = 1$ we deduce that $C = 0$, and hence that

$$u\rho(u) = \int_{u-1}^u \rho(v)\,dv \tag{7.7}$$

for $u \ge 1$.

As might be surmised from Figure 7.1, $\rho(u)$ is positive and decreasing. To prove this, let u_0 be the infimum of the set of all solutions of the equation $\rho(u) = 0$. By the continuity of ρ it follows that $\rho(u_0) = 0$. But $\rho(u) > 0$ for

$0 \le u < u_0$, and hence if we take $u = u_0$ in (7.7), then the left-hand side is 0 while the right-hand side is positive, a contradiction. Thus $\rho(u) > 0$ for all $u \ge 0$, and by (7.4) it follows that $\rho'(u) < 0$ for all $u > 1$. Figure 7.1 also suggests that $\rho(u)$ tends to 0 rapidly as $u \to \infty$. We now establish a crude estimate in this direction.

Lemma 7.1 *The function $\rho(u)$ is positive and decreasing for $u \ge 0$, and satisfies the inequalities*

$$\frac{1}{2\Gamma(2u+1)} \le \rho(u) \le \frac{1}{\Gamma(u+1)}.$$

Proof For positive integers U we prove by induction that the upper bound holds for $0 \le u \le U$. To provide the basis of the induction we need to show that $\Gamma(s) \le 1$ for $1 \le s \le 2$. This is immediate from the relations

$$\Gamma(1) = \Gamma(2) = 1, \quad \Gamma''(s) = \int_0^\infty e^{-x} x^{s-1} (\log x)^2 \, dx > 0 \quad (0 < s < \infty).$$
(7.8)

Since $\rho(u)$ is decreasing, we see by (7.7) that $u\rho(u) \le \rho(u-1)$. Thus if the desired upper bound holds for $u \le U$ and if $U \le u \le U+1$, then

$$\rho(u) \le \frac{\rho(u-1)}{u} \le \frac{1}{u\Gamma(u)} = \frac{1}{\Gamma(u+1)}$$

by (C.4).

After making the change of variables $u = v/2$, the desired lower bound asserts that $\rho(v/2) \ge 1/(2\Gamma(v+1))$. We let V run through positive integral values, and prove by induction on V that the lower bound holds for $0 \le v \le V$. To establish the lower bound for $0 \le v \le 2$ it suffices to show that $\Gamma(s) \ge 1/2$ for all $s > 0$. From (7.8) we see that $\Gamma(s) \ge 1$ for $0 < s \le 1$ and for $s \ge 2$; thus it remains to note that if $1 \le s \le 2$, then

$$\Gamma(s) = \int_0^\infty e^{-x} x^{s-1} \, dx \ge \int_0^1 e^{-x} x \, dx + \int_1^\infty e^{-x} \, dx = 1 - \frac{1}{e} > \frac{1}{2}.$$

(The actual fact of the matter is that $\min_{s>0} \Gamma(s) = \Gamma(1.4616\ldots) = 0.8856\ldots$.) Since $\rho(u)$ is decreasing, we see by (7.7) that $u\rho(u) \ge \rho(u-1/2)/2$. Thus if the lower bound holds for $0 \le v \le V$ and if $V \le v \le V+1$, then

$$\rho(v/2) \ge \frac{\rho((v-1)/2)}{v} \ge \frac{1}{2v\Gamma(v)} = \frac{1}{2\Gamma(v+1)}$$

by (C.4). This completes the inductive step, so the proof is complete. □

We now use elementary reasoning to show that (7.3) holds uniformly for u in bounded intervals.

Theorem 7.2   (Dickman) *Let $\psi(x, y)$ be the number of positive integers not exceeding x composed entirely of prime numbers not exceeding y, and let $\rho(u)$ be defined as above. Then for any $U \geq 0$ we have*

$$\psi\left(x, x^{1/u}\right) = \rho(u)x + O\left(\frac{x}{\log x}\right) \qquad (7.9)$$

uniformly for $0 \leq u \leq U$ and all $x \geq 2$.

Proof   We restrict U to integral values, and induct on U. The basis of the induction is provided by (7.1) and (7.5). Also, (7.2) gives (7.9) for $1 \leq u \leq 2$ since from (7.6) we see that

$$\rho(u) = 1 - \log u \qquad (7.10)$$

for $1 \leq u \leq 2$. Suppose now that U is an integer, $U \geq 2$, and that (7.9) holds uniformly for $0 \leq u \leq U$. We show that (7.9) holds uniformly for $U \leq u \leq U + 1$. To this end we classify n according to the size of the largest prime factor $P(n)$ of n. Thus we see that

$$\psi(x, y) = 1 + \sum_{p \leq y} \operatorname{card}\{n \leq x : P(n) = p\}.$$

Here the first term on the right reflects the fact that if $x \geq 1$, then $\psi(x, y)$ counts the number $n = 1$ for which $P(1)$ is undefined. In the sum on the right, the summand is $\psi(x/p, p)$, and hence we see that

$$\psi(x, y) = 1 + \sum_{p \leq y} \psi(x/p, p). \qquad (7.11)$$

On differencing, it follows that if $y \leq z$, then

$$\psi(x, y) = \psi(x, z) - \sum_{y < p \leq z} \psi(x/p, p). \qquad (7.12)$$

Suppose that $z = x^{1/U}$ and that $y = x^{1/u}$ with $U \leq u \leq U + 1$. Define u_p by the relation $p = (x/p)^{1/u_p}$. That is,

$$u_p = \frac{\log x}{\log p} - 1,$$

which is $\leq u - 1 \leq U$ if $p \geq y$. Hence by the inductive hypothesis the right-hand side of (7.12) is

$$\rho(U)x + O\left(\frac{x}{\log x}\right) - x \sum_{y < p \leq z} \frac{\rho((\log x)/(\log p) - 1)}{p}$$

$$+ O\left(x \sum_{y < p \leq z} \frac{1}{p \log x/p}\right). \qquad (7.13)$$

Let $s(w) = \sum_{p \le w} 1/p$, and write Mertens' estimate (Theorem 2.7(d)) in the form $s(w) = \log \log w + c + r(w)$. Then the sum in the main term above is

$$\int_y^z \rho((\log x)/(\log w) - 1) \, ds(w) = \int_y^z \rho((\log x)/(\log w) - 1) \, d \log \log w$$
$$+ \int_y^z \rho((\log x)/(\log w) - 1) \, dr(w).$$
$$(7.14)$$

We put $t = (\log x)/(\log w)$. Since

$$d \log \log w = \frac{dw}{w \log w} = -\frac{dt}{t},$$

the first integral on the right-hand side of (7.14) is

$$\int_U^u \rho(t - 1) \frac{dt}{t}. \qquad (7.15)$$

By integrating by parts and the estimate $r(w) \ll 1/\log w$ we see that the second integral on the right-hand side of (7.14) is

$$\rho((\log x)/(\log w) - 1) r(w) \Big|_y^z - \int_y^z r(w) \, d\rho((\log x)/(\log w) - 1)$$

$$\ll \frac{1}{\log x} \left(1 + \int_y^z 1 \, |d\rho((\log x)/(\log w) - 1)| \right)$$

$$\ll \frac{1}{\log x}$$

since ρ is monotonic and bounded. By Mertens' estimate (Theorem 2.7(d)) we also see that the error term in (7.13) is

$$\ll \frac{x}{\log x} \sum_{y < p \le z} \frac{1}{p} \ll \frac{x}{\log x}$$

since $\log \log z = \log \log y + O(1)$. On combining our estimates in (7.12) we find that

$$\psi(x, x^{1/u}) = x \left(\rho(U) - \int_U^u \rho(t - 1) \frac{dt}{t} \right) + O \left(\frac{x}{\log x} \right).$$

Thus by (7.6) we have the desired estimate for $U \le u \le U + 1$, and the proof is complete. $\qquad \square$

As for $\psi(x, y)$ when $y < x^\varepsilon$, we show next that

$$\psi(x, (\log x)^a) = x^{1 - 1/a + o(1)} \qquad (7.16)$$

for any fixed $a \ge 1$. The upper bound portion of this is obtained by means of bounds for an associated Dirichlet series, while the lower bound is derived by combinatorial reasoning.

An upper bound for $\psi(x, y)$ can be constructed by observing that if $\sigma > 0$, then

$$\psi(x, y) \leq \sum_{\substack{n \leq x \\ p|n \Rightarrow p \leq y}} \left(\frac{x}{n}\right)^{\sigma} \leq x^{\sigma} \sum_{p|n \Rightarrow p \leq y} \frac{1}{n^{\sigma}} = x^{\sigma} \prod_{p \leq y} \left(1 - \frac{1}{p^{\sigma}}\right)^{-1}. \quad (7.17)$$

Rankin used this chain of inequalities to derive an upper bound for $\psi(x, y)$. This approach is fruitful in a variety of settings, and has become known as 'Rankin's method'.

To use the above, we must establish an upper bound for the product on the right-hand side. The size of this product is a little difficult to describe, because its behaviour depends on the size of σ. If σ is near 0, then most of the factors are approximately $(1 - y^{-\sigma})^{-1}$, and hence we expect the product to be approximately $(1 - y^{-\sigma})^{-y/\log y}$. If σ is larger (but still < 1), then the general factor is approximately $\exp(p^{-\sigma})$, and hence the product is approximately the exponential of

$$\sum_{p \leq y} p^{-\sigma} \sim \int_2^y \frac{dt}{t^{\sigma} \log t} \sim \frac{y^{1-\sigma}}{(1 - \sigma) \log y}.$$

We begin by making these relations precise.

Lemma 7.3 *If $0 \leq \sigma \leq 1$, then*

$$\sum_{p \leq y} p^{-\sigma} = \int_2^y \frac{du}{u^{\sigma} \log u} + O\left(y^{1-\sigma} \exp\left(- c\sqrt{\log y}\right)\right) + O(1). \quad (7.18)$$

Proof We write the left-hand side as

$$\int_{2-}^y u^{-\sigma} \, d\pi(u) = \int_{2-}^y u^{-\sigma} \, d\,\mathrm{li}(u) + \int_{2-}^y u^{-\sigma} \, dr(u)$$

where $r(u) = \pi(u) - \mathrm{li}(u)$. The first integral on the right is $\int_2^y u^{-\sigma} (\log u)^{-1} \, du$. By integrating by parts we find that the second integral is

$$y^{-\sigma} r(y) - 2^{-\sigma} r(2^-) + \sigma \int_2^y r(u) u^{-\sigma-1} \, du.$$

Suppose that b is a positive constant chosen so that $r(u) \ll u \exp(-b\sqrt{\log u})$. Then the first two terms above can be absorbed into the error terms in (7.18) if $c < b$. To complete the proof it suffices to show that

$$\int_2^y u^{-\sigma} \exp(-b\sqrt{\log u}) \, du \ll 1 + y^{1-\sigma} \exp\left(- \tfrac{b}{3}\sqrt{\log y}\right), \quad (7.19)$$

for then we have (7.18) with $c = b/3$.

To prove (7.19) we note that if $\sigma \geq 1 - b/(2\sqrt{\log y})$, then

$$u^{1-\sigma} \exp\left(- \tfrac{b}{2}\sqrt{\log u}\right) = \exp\left((1 - \sigma) \log u - \tfrac{b}{2}\sqrt{\log u}\right)$$
$$\leq \exp\left(\tfrac{b}{2}(\log u)/\sqrt{\log y} - \tfrac{b}{2}\sqrt{\log u}\right)$$
$$\leq 1$$

for $2 \le u \le y$. Hence for σ in this range the integral in (7.19) is

$$\le \int_2^y \frac{du}{u \exp\left(\frac{b}{2}\sqrt{\log u}\right)} < \int_2^\infty \frac{du}{u \exp\left(\frac{b}{2}\sqrt{\log u}\right)} \ll 1.$$

Now suppose that

$$\sigma \le 1 - \frac{b}{2\sqrt{\log y}}. \tag{7.20}$$

We write the integral in (7.19) as $\int_2^{y^{1/4}} + \int_{y^{1/4}}^y = I_1 + I_2$, say. Then

$$I_1 \le \int_2^{y^{1/4}} u^{-\sigma}\, du < \frac{y^{(1-\sigma)/4}}{1-\sigma},$$

which by (7.20) is

$$\ll y^{1-\sigma}\sqrt{\log y}\exp\left(-\tfrac{3}{4}(1-\sigma)\log y\right) \ll y^{1-\sigma}\exp\left(-\tfrac{b}{3}\sqrt{\log y}\right).$$

As for I_2, we note that if $u \ge y^{1/4}$, then $\log u \ge \tfrac{1}{4}\log y$. Hence

$$I_2 \le \exp\left(-\tfrac{b}{2}\sqrt{\log y}\right)\int_2^y u^{-\sigma}\, du \le \exp\left(-\tfrac{b}{2}\sqrt{\log y}\right)\frac{y^{1-\sigma}}{1-\sigma}$$

$$\ll \exp\left(-\tfrac{b}{2}\sqrt{\log y}\right)y^{1-\sigma}\sqrt{\log y} \ll y^{1-\sigma}\exp\left(-\tfrac{b}{3}\sqrt{\log y}\right).$$

These estimates combine to give (7.19), so the proof is complete. $\qquad\square$

Lemma 7.4 *If $y \ge 2$ and $1 - 4/\log y \le \sigma \le 1$, then*

$$\sum_{p\le y} p^{-\sigma} = \log\log y + O(1). \tag{7.21}$$

If $y \ge 2$ and $0 \le \sigma \le 1 - 4/\log y$, then

$$\sum_{p\le y} p^{-\sigma} = \frac{y^{1-\sigma}}{(1-\sigma)\log y} + \log\frac{1}{1-\sigma} + O\left(\frac{y^{1-\sigma}}{(1-\sigma)^2(\log y)^2}\right). \tag{7.22}$$

Proof Suppose that $1 - 4/\log y \le \sigma \le 1$. If $u \le y$, then

$$u^{-\sigma} = u^{-1}u^{1-\sigma} = u^{-1}\exp\left((1-\sigma)\log u\right) = u^{-1}\left(1 + O((1-\sigma)\log u)\right)$$
$$= u^{-1} + O\left(u^{-1}(1-\sigma)\log u\right).$$

Hence

$$\int_2^y \frac{du}{u^\sigma\log u} = \int_2^y \frac{du}{u\log u} + O\left((1-\sigma)\int_2^y \frac{du}{u}\right) = \log\log y + O(1).$$

Thus (7.21) follows from Lemma 7.3.

To prove (7.22) we let $v = \exp(4/(1-\sigma))$, and observe that $v \le y$. We write the integral in Lemma 7.3 as $\int_2^v + \int_v^y = I_1 + I_2$, say. By the above we see that $I_1 = \log\log v + O(1) = \log 1/(1-\sigma) + O(1)$. By integration by parts we see that

$$I_2 = \frac{y^{1-\sigma}}{(1-\sigma)\log y} - \frac{v^{1-\sigma}}{(1-\sigma)\log v} + \frac{1}{1-\sigma}\int_v^y \frac{du}{u^\sigma(\log u)^2}.$$

Here the first term on the right is one of the main terms in (7.22), and the second term is $O(1)$. Let J denote the integral on the right. To complete the proof it suffices to show that

$$J \ll \frac{y^{1-\sigma}}{(1-\sigma)(\log y)^2}. \tag{7.23}$$

To this end we integrate by parts again:

$$J = \frac{y^{1-\sigma}}{(1-\sigma)(\log y)^2} - \frac{v^{1-\sigma}}{(1-\sigma)(\log v)^2} + \frac{2}{1-\sigma}\int_v^y \frac{dw}{w^\sigma(\log w)^3}.$$

Here the second term on the right-hand side is $e^4 2^{-4}(1-\sigma) \ll 1-\sigma$, while the first term on the right-hand side is larger. As for the integral on the right, we observe that if $w \geq v$, then $(\log w)^3 \geq 4(\log w)^2/(1-\sigma)$. Hence the last term on the right above has absolute value not exceeding $J/2$. Thus we have (7.23), and the proof is complete. ☐

Lemma 7.5 *Suppose that $y \geq 2$. If $\max\left(2/\log y, 1 - 4/\log y\right) \leq \sigma \leq 1$, then*

$$\prod_{p\leq y}\left(1 - p^{-\sigma}\right)^{-1} \asymp \log y. \tag{7.24}$$

If $2/\log y \leq \sigma \leq 1 - 4/\log y$, then

$$\prod_{p\leq y}(1 - p^{-\sigma})^{-1} = \frac{1}{1-\sigma}$$

$$\times \exp\left(\frac{y^{1-\sigma}}{(1-\sigma)\log y}\left(1 + O\left(\frac{1}{(1-\sigma)\log y}\right) + O(y^{-\sigma})\right)\right). \tag{7.25}$$

Proof The bound (7.24) is trivial when $\sigma \leq 2/3$ since then $y \leq e^{12}$. The estimate $(1-\delta)^{-1} = \exp\left(\delta + O(\delta^2)\right)$ holds uniformly for $|\delta| \leq 1/2$. We take $\delta = p^{-\sigma}$ for $p > v = e^{1/\sigma}$ to deduce that

$$\prod_{v<p\leq y}\left(1 - p^{-\sigma}\right)^{-1} = \exp\left(\sum_{v<p\leq y} p^{-\sigma} + O\left(\sum_{v<p\leq y} p^{-2\sigma}\right)\right).$$

Now (7.24) follows at once from Lemma 7.4 when $\sigma \geq 2/3$. Thus it remains to establish (7.25). The sum in the error term above is $\ll 1$ for $\sigma > 5/8$. If $3/8 \leq \sigma \leq 5/8$, then by Lemma 7.4 it is $\ll y^{1/4}/\log y$. If $2/\log y \leq \sigma \leq 3/8$, then by Lemma 7.4 the sum is $\ll y^{1-2\sigma}/\log y$. Thus in any case this error term is majorized by the error terms on the right-hand side of (7.25). By Lemma 7.4, the main term is

$$\sum_{v < p \le y} p^{-\sigma} = \frac{y^{1-\sigma}}{(1-\sigma)\log y} + \log \frac{1}{1-\sigma}$$

$$+ O\left(\frac{y^{1-\sigma}}{(1-\sigma)^2(\log y)^2}\right) + O\left(\frac{v}{\log v}\right).$$

Since $2/\log y \le \sigma \le 1 - 4/\log y$, y satisfies $y \ge e^6$, and $\sigma(1-\sigma)\log y \ge 2(1 - 2/\log y) \ge 4/3$. Hence $(y^{1-\sigma})^{3/4} \ge v$ and the second error term above is dominated by the first.

It remains to consider the contribution of the primes $p \le v$. If $\sigma > 1/3$, then the contribution of these primes is $\ll 1$, so we may suppose that $2/\log y \le \sigma \le 1/3$. In this range

$$1 - p^{-\sigma} \asymp \sigma \log p = \frac{\log p}{\log v}.$$

Since

$$\sum_{p \le v} \log\left(C\frac{\log v}{\log p}\right) \ll v,$$

it follows that

$$\prod_{p \le v}(1 - p^{-\sigma})^{-1} < \exp(Cv) = \exp\left(Ce^{1/\sigma}\right) \le \exp\left(Cy^{1/2}\right),$$

which suffices. Thus the proof is complete. □

We now bound $\psi(x, y)$ by combining Lemma 7.5 with the inequalities (7.17).

Theorem 7.6 *If* $y = x^{1/u}$ *and* $\log x \le y \le x^{1/9}$, *then*

$$\psi(x, y) < x(\log y)\exp\left(-u\log u - u\log\log u + u - \frac{u\log\log u}{\log u}\right.$$

$$\left. + O\left(\frac{u}{\log u}\right) + O\left(\frac{u^2\log u}{y}\right)\right).$$

Here the first error term is larger than the second if $y \ge (\log x)\log\log x$, while if y is smaller, then the second error term dominates.

Proof We first note that we may suppose that $y \ge 9\log x$, since the bound for smaller y follows by taking $y = 9\log x$. To motivate the choice of σ in (7.17) we note that the expression to be minimized is approximately

$$x^\sigma \exp\left(\int_2^y \frac{u^{-\sigma}}{\log u}\,du\right).$$

On taking logarithmic derivatives, this suggests that we should take σ to be the root of the equation

$$\log x = \frac{y^{1-\sigma}}{1-\sigma}. \tag{7.26}$$

In actual fact we take

$$\sigma = 1 - \frac{\log u + \log \log u}{\log y}. \tag{7.27}$$

It is easy to see that for this σ the right-hand side of (7.26) is

$$\log x \frac{\log u}{\log u + \log \log u},$$

so it is reasonable to expect that the simple choice (7.27) is close enough to the root of (7.26) for our present purposes.

From the inequalities $9 \log x \le y \le x^{1/9}$ it follows that the σ given by (7.27) satisfies $2/\log y \le \sigma \le 1 - 1/\log y$. Hence the stated upper bound follows by combining (7.17) with the estimates of Lemma 7.5. $\qquad\qquad\square$

To obtain companion lower bounds we observe that if k is chosen so that $y^k \le x$, then $\psi(x, y)$ certainly counts all integers n composed of primes $p \le y$ such that $\Omega(n) \le k$. Put $r = \pi(y)$, and suppose that $p_1, p_2, \ldots, p_r$ are the primes not exceeding y. Then n is of the form $n = p_1^{a_1} p_2^{a_2} \cdots p_r^{a_r}$, and $\psi(x, y)$ is at least as large as the number of solutions of the inequality $a_1 + a_2 + \cdots + a_r \le k$ in non-negative integers a_i. For this quantity we have an exact formula, as follows.

Lemma 7.7 *Let $A(r, k)$ denote the number of solutions of the inequality $a_1 + a_2 + \cdots + a_r \le k$ in non-negative integers a_i. Then $A(r, k) = \binom{r+k}{k}$.*

Analytic Proof Let $a_{r+1} = k - \sum_{i=1}^{r} a_i$. Then $A(r, k)$ is the number of ways of writing $k = a_1 + a_2 + \cdots + a_{r+1}$, which is the coefficient of x^k in the power series

$$\left(\sum_{a=0}^{\infty} x^a \right)^{r+1} = (1-x)^{-r-1} = \sum_{k=0}^{\infty} \binom{r+k}{k} x^k$$

by the 'negative' binomial theorem. $\qquad\qquad\square$

Combinatorial Proof Suppose that we have k circles $\circ$ and r bars $|$ arranged in a line. Let a_1 be the number of circles to the left of the first bar, let a_2 be the number of circles between the first and second bar, and so on, so that a_r is the number of circles between the last two bars. (The number of circles to the right of the last bar is $k - \sum a_i$.) Thus a configuration of circles and bars determines a choice of non-negative a_i with $a_1 + a_2 + \cdots + a_r \le k$. But conversely, a choice of such a_i determines a configuration of circles and bars. The number of ways of choosing the positions of the k circles in the $r + k$ available places is $\binom{r+k}{k}$. $\qquad\qquad\qquad\qquad\qquad\qquad\qquad\qquad\qquad\qquad\qquad\qquad\qquad$ $\square$

Theorem 7.8 *If* $\log x \leq y \leq x$, *then*

$$\psi(x, y) \gg \frac{x}{y} \exp(-u \log \log x + u/2).$$

Proof Let $r = \pi(y)$ and let k be the largest integer such that $y^k \leq x$. That is, $k = [u]$. Then by Lemma 7.7 and Stirling's formula we see that

$$\psi(x, y) \geq \binom{r+k}{k} \asymp \left(\frac{r+k}{k}\right)^k \left(\frac{r+k}{r}\right)^r \frac{1}{\sqrt{k}}. \qquad (7.28)$$

The identity

$$k \log(1 + r/k) + r \log(1 + k/r) = \int_0^r \log(1 + k/t)\, dt$$

shows that the left-hand side is an increasing function of r. It can be supposed that x is sufficiently large. Let $z = y/(k \log y)$. Then the expression (7.28) is

$$\gg \left(1 + \frac{y}{k \log y}\right)^k \left(1 + \frac{k \log y}{y}\right)^{y/\log y} \frac{1}{\sqrt{k}} \geq (z(1 + 1/z)^z)^k,$$

Moreover $u - 1 < k \leq u \leq y/\log y$ and $z(1 + 1/z)^z$ is increasing for $z \geq 1$. Thus the above is $\geq (z'(1 + 1/z')^{z'})^k \geq (z'(1 + 1/z')^{z'})^{u-1}$ where $z' = y/(u \log y)$. As $z' \leq y/\sqrt{k}$ this is

$$\geq \frac{1}{y} \left(\frac{y}{u \log y}\right)^u \left(1 + \frac{u \log y}{y}\right)^{y/\log y}$$

$$= \frac{x}{y} \exp\left(-u \log \log x + \frac{y}{\log y} \log(1 + (\log x)/y)\right).$$

The stated inequality now follows on noting that $\log(1 + \delta) \geq \delta/2$ for $0 \leq \delta \leq 1$. $\qquad\qquad\qquad\qquad\qquad\qquad\qquad\qquad\qquad\qquad\qquad\qquad\qquad$ $\square$

When y is of the form $y = (\log x)^a$ with a not too large, the upper bound of Theorem 7.6 and the lower bound of Theorem 7.8 are quite close, and we have

Corollary 7.9 *If* $y = (\log x)^a$ *and* $1 \leq a \leq (\log x)^{1/2}/(2 \log \log x)$, *then*

$$x^{1-1/a} \exp\left(\frac{\log x}{5a \log \log x}\right) < \psi(x, y) < x^{1-1/a} \exp\left(\frac{(\log a + O(1)) \log x}{a \log \log x}\right).$$

Proof The lower bound follows from Theorem 7.8 since $\log y \leq (\log x)/(4a \log \log x)$ in the range under consideration. As for the upper bound, we note that $\log u \asymp \log \log x$, so that $\log \log u = \log \log \log x + O(1)$. Hence

$\log u + \log \log u = \log \log x - \log a + O(1)$, and the result follows from Theorem 7.6. $\qquad\Box$

For $1 \le u \le 4$ we may use the differential equation (7.4) and the initial condition (7.5) to derive formulæ for $\rho(u)$ (see Exercise 7.1.6 below), but for larger u we take a different approach.

Theorem 7.10 *For any real or complex number s we have*

$$\int_0^\infty \rho(u)e^{-us}\,du = \exp\left(C_0 + \int_0^s \frac{e^{-z}-1}{z}\,dz\right) \qquad (7.29)$$

where C_0 is Euler's constant. Conversely, for any $u > 0$ and any real σ_0 we have

$$\rho(u) = \frac{e^{C_0}}{2\pi i}\int_{\sigma_0-i\infty}^{\sigma_0+i\infty}\exp\left(\int_0^s \frac{e^{-z}-1}{z}\,dz\right)e^{us}\,ds. \qquad (7.30)$$

Proof Let $F(s)$ denote the integral on the left-hand side of (7.29); this is the Laplace transform of $\rho(u)$. In view of the rapid decay of $\rho(u)$ established in Lemma 7.1, we see that the integral converges for all s, and hence that $F(s)$ is an entire function. On integrating by parts we see that

$$F(s) = \frac{1}{s} + \frac{1}{s}\int_1^\infty \rho'(u)e^{-us}\,du,$$

and hence that

$$(sF(s))' = -\int_1^\infty u\rho'(u)e^{-us}\,du.$$

The differential–delay identity (7.4) for $\rho(u)$ thus yields a differential equation for $F(s)$,

$$(sF(s))' = e^{-s}F(s).$$

By separation of variables it follows that

$$F(s) = F(0)\exp\left(\int_0^s \frac{e^{-z}-1}{z}\,dz\right).$$

To determine the value of $F(0)$ we note that

$$1 = \lim_{s\to+\infty} sF(s) = F(0)\exp\left(\int_0^1 \frac{e^{-z}-1}{z}\,dz + \int_1^\infty \frac{e^{-z}}{z}\,dz\right).$$

By integration by parts we see that

$$\int_0^1 \frac{e^{-z}-1}{z}\,dz + \int_1^\infty \frac{e^{-z}}{z}\,dz = \int_0^\infty e^{-z}\log z\,dz = \Gamma'(1) = -C_0 \qquad (7.31)$$

by (C.12) and Theorem C.2. Hence $F(0) = e^{C_0}$. An arithmetic proof of this is found in Exercise 7.1.7 below. Thus we have the identity (7.29), and (7.30) follows by applying the inverse Laplace transform to both sides.   □

7.1.1   Exercises

1. (Chowla & Vijayaraghavan 1947) Show that if $f(x)$ is a function that tends to infinity in such a way that $\log f(x) = o(\log x)$ then almost all integers n have a prime factor larger than $f(n)$. That is

$$\lim_{x \to \infty} \frac{1}{x} \operatorname{card}\{n \le x : P(n) > f(n)\} = 1$$

where $P(n)$ denotes the largest prime factor of n.

2. (de Bruijn 1951b) Let $P(n)$ denote the largest prime factor of n. Show that

$$\sum_{n \le x} \log P(n) \sim Dx \log x$$

where $D = \int_0^\infty \rho(u)(u+1)^{-2}\, du$ is called *Dickman's constant*.

3. (cf. Alladi & Erdős 1977) Let $P(n)$ denote the largest prime factor of n.

(a) Show that

$$\sum_{n \le x} P(n) = \sum_{\sqrt{x} < p \le x} p\left[\frac{x}{p}\right] + O(x^{3/2}).$$

(b) Show that the sum on the right above is

$$= \sum_{1 \le k \le \sqrt{x}} k \sum_{x/(k+1) < p \le x/k} p + O(x^{3/2}).$$

(c) Show that

$$\sum_{p \le y} p = \frac{y^2}{2 \log y} + O\left(\frac{y^2}{(\log y)^2}\right).$$

(d) Show that

$$\sum_{k=1}^\infty k\left(\frac{1}{k^2} - \frac{1}{(k+1)^2}\right) = \frac{\pi^2}{6}.$$

(e) Conclude that

$$\sum_{n \le x} P(n) = \frac{\pi^2}{12} \frac{x^2}{\log x} + O\left(\frac{x^2}{(\log x)^2}\right).$$

4. Show that $\rho^{(k)}(u)$ has a jump discontinuity at $u = k$, and is continuous for $u > k$.

5. (a) Show that $\rho(u)$ is convex upwards for all $u \ge 1$.
   (b) Show that if $u \ge 2$, then $u\rho(u) \ge \rho(u - 1/2)$.

    (c) Show that if $u \geq 2$, then $(2u - 1)\rho(u) \leq \rho(u - 1)$.

6. (a) Show that if $1 \leq u \leq 2$, then $\rho(u) = 1 - \log u$.

    (b) Show that if $2 \leq u \leq 3$, then

$$\rho(u) = 1 - \log u + \int_2^u \frac{\log(t - 1)}{t}\, dt.$$

    (c) Show that if $3 \leq u \leq 4$, then

$$\rho(u) = 1 - \log u + \int_2^u \frac{\log(t - 1)}{t}\, dt - \int_3^u \frac{(\log u/t)\log(t - 2)}{t - 1}\, dt.$$

7. Let $P(\sigma) = \prod_{p \leq y}(1 - p^{-\sigma})^{-1}$.

    (a) Explain why

$$P(1) = \sum_{p\mid n \Rightarrow p \leq y} \frac{1}{n} = e^{C_0} \log y + O(1).$$

    (b) Show that if $\sigma \geq 1$, then $\frac{P'}{P}(\sigma) \ll \log y$.

    (c) Deduce that

$$-P'(1) = \sum_{\substack{n \\ p\mid n \Rightarrow p \leq y}} \frac{\log n}{n} \ll (\log y)^2.$$

    (d) Conclude that

$$\sum_{\substack{n > x \\ p\mid n \Rightarrow p \leq y}} \frac{1}{n} \ll \frac{(\log y)^2}{\log x}.$$

    (e) Show that

$$\sum_{\substack{n \leq x \\ p\mid n \Rightarrow p \leq y}} \frac{1}{n} = (\log y) \int_0^u \frac{\psi(y^v, y)}{y^v}\, dv + O(1)$$

    where $u = (\log x)/\log y$.

    (f) Deduce that

$$\int_0^\infty \rho(u)\, du = e^{C_0}.$$

    (g) Show that $\sum_{n=1}^\infty n\rho(n) = e^{C_0}$.

8. (Erdős & Nicolas 1981) Let α be fixed, $0 < \alpha < 1$.

    (a) Let k be the least integer $> \alpha(\log x)/\log \log x$, put $y = x^{1/k}$, and set $r = \pi(y)$. Show that there are at least $\binom{r}{k}$ integers $n \leq x$ such that $\omega(n) > \alpha(\log x)/\log \log x$.

    (b) Show that the number of integers $n \leq x$ such that $\omega(n) > \alpha(\log x)/\log \log x$ is at least $x^{1-\alpha+o(1)}$.

(c) Show that if $\sigma > 1$ and $A \geq 1$, then the number of integers $n \leq x$ such that $\omega(n) > \alpha(\log x)/\log\log x$ is at most

$$x^\sigma A^{-k} \sum_{n=1}^\infty \frac{A^{\omega(n)}}{n^\sigma}.$$

(d) Show that if $A = \log x$ and $\sigma = 1 + (\log\log\log x)/\log\log x$, then the above is $x^{1-\alpha+o(1)}$.

9. (de Bruijn 1966) Assume that $0 < \sigma \leq 3/\log y$, and note that this interval covers a range that is not treated in Lemma 7.5.

   (a) Show that $1 - p^{-\sigma} \asymp \sigma\log p$, and hence deduce that

   $$\prod_{p \leq y}(1 - p^{-\sigma})^{-1} \leq \exp\left(\sum_{p \leq y} \log \frac{C}{\sigma\log p}\right)$$

   $$\leq \exp\left(\frac{Cy}{\log y} \log \frac{4}{\sigma\log y}\right) \qquad (7.32)$$

   for a suitable constant C.

   (b) Write

   $$\prod_{p \leq y}(1 - p^{-\sigma})^{-1} = (1 - y^{-\sigma})^{-\pi(y)} \prod_{p \leq y} \frac{1 - y^{-\sigma}}{1 - p^{-\sigma}} = F_1 \cdot F_2,$$

   say. Show that

   $$F_1 \leq (1 - y^{-\sigma})^{-y/\log y} \exp\left(\frac{Cy}{(\log y)^2} \log \frac{4}{\sigma\log y}\right).$$

   (c) Note that

   $$\frac{1 - p^{-\sigma}}{1 - y^{-\sigma}} = 1 - \frac{(y/p)^\sigma - 1}{y^\sigma - 1}, \qquad (7.33)$$

   and hence deduce that the above is $\geq 1 - c\frac{\log y/p}{\log y}$, so that

   $$F_2 \leq \exp\left(\frac{C}{\log y} \sum_{p \leq y} \log y/p\right) \leq \exp\left(Cy/(\log y)^2\right).$$

   (d) Conclude that

   $$\prod_{p \leq y}(1 - p^{-\sigma})^{-1} \leq (1 - y^{-\sigma})^{-y/\log y} \exp\left(\frac{Cy}{(\log y)^2} \log \frac{4}{\sigma\log y}\right)$$

   for $0 < \sigma \leq 3/\log y$.

10. (de Bruijn 1966) Lemma 7.5 suffers from a loss of precision when $3/\log y \leq \sigma \leq (\log\log y)/\log y$. To obtain a refined estimate in this range, write

$$\prod_{p \leq y}(1 - p^{-\sigma})^{-1} = F_1 \cdot F_2 \cdot F_3$$

where the F_i are products over the intervals $p \leq \exp(1/\sigma)$, $\exp(1/\sigma) < p \leq y/\exp(1/\sigma)$, and $y/\exp(1/\sigma) < p \leq y$, respectively.

(a) Use (7.32) to show that $F_1 \leq \exp\left(C\sigma e^{1/\sigma}\right)$.

(b) Use Lemma 7.5 to show that
$$F_2 \leq \exp\left(\frac{Cy^{1-\sigma}}{e^{1/\sigma}\log y}\right).$$

(c) Use the identity (7.33) to show that
$$\frac{1-p^{-\sigma}}{1-y^{-\sigma}} \geq 1 - \frac{c\sigma \log y/p}{y^{\sigma}},$$

and hence deduce that
$$F_3 \leq (1-y^{-\sigma})^{-\pi(y)} \exp\left(C\sigma \sum_{p\leq y} \frac{\log y/p}{y^{\sigma}}\right)$$
$$\leq (1-y^{-\sigma})^{-y/\log y} \exp\left(\frac{y^{1-\sigma}}{(\log y)^2} + \frac{C\sigma y^{1-\sigma}}{\log y}\right).$$

(d) Conclude that
$$\prod_{p\leq y}(1-p^{-\sigma})^{-1} \leq (1-y^{-\sigma})^{-y/\log y} \exp\left(\frac{C\sigma y^{1-\sigma}}{\log y}\right)$$

when $3/\log y \leq \sigma \leq (\log \log y)/\log y$.

11. (de Bruijn 1966)

(a) For $\sigma > 0$ let $f(\sigma) = x^{\sigma}(1-y^{-\sigma})^{-y/\log y}$. Show that $f(\sigma)$ is minimized precisely when
$$\sigma = \frac{\log(1 + y/\log x)}{\log y}.$$

(b) Show that for the above σ,
$$f(\sigma) = \exp\left(\frac{\log x}{\log y}\log\left(\frac{y+\log x}{\log x}\right) + \frac{y}{\log y}\log\left(\frac{y+\log x}{y}\right)\right).$$

(c) Show that if $y \leq \log x$, then
$$\psi(x,y) \leq \exp\left(\frac{\log x}{\log y}\log\left(\frac{y+\log x}{\log x}\right)\right.$$
$$\left. + \frac{y}{\log y}\left(1 + O\left(\frac{1}{\log y}\right)\right)\log\left(\frac{y+\log x}{y}\right)\right).$$

(d) Show that if $\log x \leq y \leq (\log x)^2$, then
$$\psi(x,y) \leq \exp\left(\frac{\log x}{\log y}\left(1 + O\left(\frac{1}{\log y}\right)\right)\log\left(\frac{y+\log x}{\log x}\right)\right.$$
$$\left. + \frac{y}{\log y}\log\left(\frac{y+\log x}{y}\right)\right).$$

12. (Erdős 1963) Show that

$$\psi(x, \log x) = \exp\left((2\log 2 + o(1))\frac{\log x}{\log\log x}\right).$$

13. (de Bruijn 1966) Show that if a is fixed, $0 < a < 1$, then

$$\psi(x, (\log x)^a) = \exp((1/a - 1 + o(1))(\log x)^a).$$

14. Let $\psi_2(x, y)$ denote the number of square-free integers $n \le x$ composed entirely of primes $p \le y$.
    (a) Show that

$$\psi_2(x, y) = \sum_{\substack{d \le x \\ p|d \Rightarrow p \le y}} \mu(d)\psi(x/d^2, y).$$

    (b) (Ivić) Let $\delta > 0$ be fixed. Then

$$\psi_2(x, y) \sim \frac{6}{\pi^2}\psi(x, y)$$

    uniformly for $x^\delta \le y \le x$.
    (c) Show that $\psi_2(x, \log x) = \psi(x, \log x)^{1/2 + o(1)}$.
    (d) Show that if $a > 1$ and $y \ge (\log x)^a$, then $\psi_2(x, y) = \psi(x, y)^{1 + o(1)}$.
    (e) Show that if $0 < a < 1$ and $y \le (\log x)^a$, then $\psi_2(x, y) = \psi(x, y)^{o(1)}$.
    (f) Show that $\psi_2(x, c \log x) = \psi(x, c \log x)^{\phi(c) + o(1)}$ for any fixed $c > 0$, where

$$\phi(c) = \begin{cases} \dfrac{c\log 2}{(c+1)\log(c+1) - c\log c} & (0 < c \le 2), \\[3mm] \dfrac{c\log c - (c-1)\log(c-1)}{(c+1)\log(c+1) - c\log c} & (c \ge 2). \end{cases}$$

7.2 Numbers composed of large primes

Let $\Phi(x, y)$ denote the number of integers $n \le x$ composed entirely of primes $p \ge y$. The number 1 is such a number as it is an empty product. Thus it is clear that if $y > x$, then

$$\Phi(x, y) = 1 \tag{7.34}$$

Also, if $x^{1/2} \le y \le x$, then

$$\Phi(x, y) = \pi(x) - \pi(y^-) + O(1) = \frac{x}{\log x} - \frac{y}{\log y} + O\left(\frac{x}{(\log x)^2}\right) \tag{7.35}$$

For smaller values of y we show that

$$\Phi(x, y) \sim \frac{w(u)x}{\log y} \tag{7.36}$$

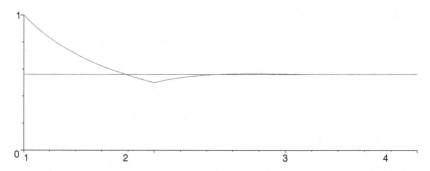

Figure 7.2 Buchstab's function $w(u)$ and its horizontal asymptote e^{-C_0} for $1 \leq u \leq 4$.

where $u = (\log x)/\log y$ and $w(u)$ is a function determined by the initial condition

$$w(u) = 1/u \tag{7.37}$$

for $1 < u \leq 2$ and for $u > 2$ by the differential–delay equation

$$(uw(u))' = w(u-1). \tag{7.38}$$

Before proceeding further we first derive some of the simplest properties of the function $w(u)$ depicted in Figure 7.2. By integrating (7.38) we deduce that $uw(u) = \int_1^{u-1} w(v)\,dv + C$ for $u > 2$, and by letting u tend to 2 we find that $C = 1$ so that

$$uw(u) = \int_1^{u-1} w(v)\,dv + 1 \tag{7.39}$$

for $u \geq 2$. From this it is evident that if $w(v) \leq 1$ for $v \leq u-1$, then $w(v) \leq 1$ for $v \leq u$, and that if $w(v) \geq 1/2$ for $v \leq u-1$, then $w(v) \geq 1/2$ for $v \leq u$. Thus we conclude that $1/2 \leq w(u) \leq 1$ for all $u > 1$. From the identity $uw'(u) = w(u-1) - w(u)$ we deduce that $|w'(u)| \leq 1/(2u)$ for all $u > 2$. Let $M(u) = \max_{v \geq u} |w'(v)|$. Since $w(u-1) - w(u) = -w'(\xi)$ for some ξ, $u - 1 < \xi < u$, we know that

$$M(u) \leq M(u-1)/u.$$

Let k be chosen so that $1 < u - k \leq 2$. By using the above inequality k times we find that

$$M(u) \leq \frac{M(u-k)}{u(u-1)\cdots(u-k+1)} \ll \frac{1}{\Gamma(u+1)}.$$

That is,

$$w'(u) \ll \frac{1}{\Gamma(u+1)} \tag{7.40}$$

for $u > 2$. Since $w'(u)$ tends to 0 rapidly, it follows that the integral $\int_2^\infty w'(v)\,dv$ converges absolutely, and hence we see that $\lim_{u\to\infty} w(u)$ exists. Since it is to be expected that $\Phi(x, y)$ is approximately $x \prod_{p<y}(1 - 1/p)$ when y is small, it is not surprising that

$$\lim_{u\to\infty} w(u) = e^{-C_0}. \tag{7.41}$$

We shall prove this later, as a consequence of Theorem 7.12. First we establish the basic asymptotic estimate (7.36).

Theorem 7.11   (Buchstab) *Let* $\Phi(x, y)$ *denote the number of positive integers* $n \leq x$ *composed entirely of prime numbers* $p \geq y$, *and let* $w(u)$ *be defined as above. Then*

$$\Phi(x, y) = \frac{w(u)x}{\log y} - \frac{y}{\log y} + O\left(\frac{x}{(\log x)^2}\right) \tag{7.42}$$

uniformly for $1 \leq u \leq U$ *and all* $y \geq 2$. *Here* $u = (\log x)/\log y$, *which is to say that* $y = x^{1/u}$.

The term $-y/\log y$ can be included in the error term when $y \ll x/\log x$ but, in view of (7.35), has to be present when y is close to x. It might be difficult to prove that the above holds uniformly for all $u \geq 1$ because of the precise form of the error term, but the weaker assertion (7.36) can be shown to hold for $u \geq 1 + \varepsilon$, since sieve methods can be used when u is large.

Proof   The number of positive integers $n \leq x$ whose least prime factor is p is exactly $\Phi(x/p, p)$. Hence by classifying integers according to their least prime factor we see that

$$\Phi(x, y) = 1 + \sum_{y \leq p \leq x} \Phi(x/p, p). \tag{7.43}$$

This is an identity of Buchstab; similar 'Buchstab identities' are important in sieve theory. We show by induction on U that

$$\Phi(x, y) = \frac{w(u)x}{\log y} - \frac{y}{\log y} + O\left(\frac{x}{(\log x)^2}\right) \tag{7.44}$$

for $U \leq u \leq U + 1$. When $U = 1$ this is (7.35), and it is only in this first range that the second main term is significant. For the inductive step we apply (7.43) with $y = x^{1/u}$ and with $y = x^{1/U}$ and subtract to see that

$$\Phi\left(x, x^{1/u}\right) = \Phi\left(x, x^{1/U}\right) + \sum_{x^{1/u} \leq p < x^{1/U}} \Phi(x/p, p).$$

Choose u_p so that $p = (x/p)^{1/u_p}$. Then the above is

$$\Phi\left(x, x^{1/U}\right) + \sum_{x^{1/u} \le p < x^{1/U}} \Phi\left(x/p, (x/p)^{1/u_p}\right).$$

But $u_p = (\log x)/\log p - 1 \in [U - 1, U]$, so by the inductive hypothesis, when $U \ge 2$, the above is

$$\frac{U w(U) x}{\log x} + O\left(\frac{x}{(\log x)^2}\right)$$

$$+ \sum_{x^{1/u} \le p < x^{1/U}} \left(\frac{u_p w(u_p) x}{p \log x/p} + O\left(\frac{x}{p(\log x)^2}\right) + O\left(\frac{p}{\log p}\right)\right).$$

The sum over p of the first error term is $\ll x/(\log x)^2$, and the sum over p of the second is $\ll x^{2/U}/(\log x)^2$, which is acceptable since $U \ge 2$. To estimate the contribution of the main term in the sum we write the Prime Number Theorem in the form $\pi(t) = \mathrm{li}(t) + R(t)$, apply Riemann–Stieltjes integration, and integrate the term involving $R(t)$ by parts, to see that the sum of the main term is

$$\int_{x^{1/u}}^{x^{1/U}} \frac{x w\left(\frac{\log x}{\log t} - 1\right)}{t(\log t)^2} \, dt + \left[f(t)R(t)\right]_{x^{1/u}-}^{x^{1/U}-} - \int_{x^{1/u}}^{x^{1/U}} R(t) \, df(t) \quad (7.45)$$

where

$$f(t) = \frac{x w\left(\frac{\log x}{\log t} - 1\right)}{t \log t}.$$

Since $f'(t) \ll x/(t^2 \log t)$ and $R(t) \ll t/(\log t)^A$, the terms involving $R(t)$ contribute an amount $\ll_U x/(\log x)^A$. By the change of variables $v = (\log x)/\log t - 1$ we see that the first integral in (7.45) is

$$\frac{x}{\log x} \int_{U-1}^{u-1} w(v) \, dv,$$

which by (7.39) is

$$= \frac{x}{\log x} (u w(u) - U w(U)).$$

On combining our estimates we obtain (7.44), so the inductive step is complete.   □

We now derive formulæ for $w(u)$ similar to those in Theorem 7.10 involving $\rho(u)$.

Theorem 7.12 *If $\Re s > 0$, then*

$$s + s \int_1^\infty w(u) e^{-us} \, du = \exp\left(-C_0 + \int_0^s \frac{1 - e^{-z}}{z} \, dz\right) \quad (7.46)$$

where C_0 is Euler's constant. If $u > 1$ and $\sigma_0 > 0$, then

$$w(u) = \frac{1}{2\pi i} \int_{\sigma_0 - i\infty}^{\sigma_0 + i\infty} \left(\exp\left(\int_s^\infty \frac{e^{-z}}{z} \, dz \right) - 1 \right) e^{us} \, ds. \qquad (7.47)$$

Since the right-hand side of (7.46) is an entire function, we see that the Laplace transform of $w(u)$ is entire apart from a simple pole at $s = 0$ with residue e^{-C_0}.

Proof   Let $G(s)$ denote the left-hand side of (7.46). Then

$$\left(\frac{G(s)}{s} \right)' = -\int_1^\infty w(u) u e^{-us} \, du.$$

By integrating by parts we see that this is

$$\left[\frac{w(u) u e^{-us}}{s} \right]_1^\infty - \frac{1}{s} \int_2^\infty w(u-1) e^{-us} \, du = \frac{-e^{-s} G(s)}{s^2}$$

by (7.37) and (7.38). That is,

$$G'(s) = G(s) \frac{1 - e^{-s}}{s},$$

which by the method of separation of variables implies that

$$G(s) = A \exp\left(\int_0^s \frac{1 - e^{-z}}{z} \, dz \right)$$

where A is a positive constant. To determine the value of A we note that

$$1 = \lim_{s \to \infty} \frac{G(s)}{s} = A \exp\left(\int_0^1 \frac{1 - e^{-z}}{z} \, dz - \int_1^\infty \frac{e^{-z}}{z} \, dz \right).$$

From (7.31) we deduce that $A = e^{-C_0}$, and hence we have (7.46). To obtain (7.47) it suffices to take the inverse Laplace transform, since

$$\int_0^s \frac{1 - e^{-z}}{z} \, dz = \int_s^\infty \frac{e^{-z}}{z} \, dz + \log s + C_0.$$

$\square$

7.2.1   Exercises

1. By using (7.31), or otherwise, show that

$$\int_0^s \frac{1 - e^{-z}}{z} \, dz = C_0 + \log s + \int_s^\infty \frac{e^{-z}}{z} \, dz$$

     when $\Re s > 0$.

2. (a) Show that

$$w(u) = \frac{1 + \log(u-1)}{u}$$

     for $2 \leq u \leq 3$.

(b) Show that

$$w(u) = \frac{1}{u}\left(1 + \log(u-1) + \int_3^u \frac{\log(v-2)}{v-1}\,dv\right)$$

for $3 \le u \le 4$.

(c) Show that

$$w(u) = \frac{1}{u}\left(1 + \log(u-1) + \int_3^u \frac{\log(v-2)}{v-1}\,dv\right.$$
$$\left. + \int_4^u \frac{\log\frac{u-1}{v-1}\log(v-3)}{v-2}\,dv\right)$$

for $4 \le u \le 5$.

3. (Friedlander 1972) Let $\mathcal{S}$ be a set of positive integers not exceeding X, and suppose that $(a, b) \le Y$ whenever $a \in \mathcal{S}, b \in \mathcal{S}, a \ne b$. Let $M(X, Y)$ denote the maximum cardinality of all such sets $\mathcal{S}$.

   (a) Let $\mathcal{S}_0$ be the set of those positive integers $n \le X$ such that if $d|n, d < n$, then $d \le Y$. Show that card $\mathcal{S}_0 = M(X, Y)$.

   (b) Show that if $Y \le X^{1/2}$, then

   $$M(X, Y) = 1 + \pi(X) - \pi(Y) + \sum_{p \le Y} \Phi(Y, p).$$

   (c) Show that if $X^{1/2} < Y \le X$, then

   $$M(X, Y) = 1 + \pi(X) - \pi(Y) + \sum_{p < X/Y} \Phi(Y, p) + \sum_{X/Y \le p \le Y} \Phi(X/p, p).$$

7.3   Primes in short intervals

Let Jacobsthal's function $g(q)$ be the length of the longest gap between consecutive reduced residues modulo q. We show that there are long gaps between primes by showing that there exist integers q for which $g(q)$ is large. Since the average gap between consecutive reduced residues (mod q) is $q/\varphi(q)$, it is obvious that

$$g(q) \ge \frac{q}{\varphi(q)}.$$

If $p_1 < p_2 < \cdots < p_k$ are the distinct primes dividing q, then by the Chinese Remainder Theorem there is an x such that $x \equiv -i \pmod{p_i}$ for $1 \le i \le k$. Then $(x + i, q) > 1$ for $1 \le i \le k$, and hence

$$g(q) \ge \omega(q) + 1.$$

These observations can be combined: It can be shown that

$$g(q) \gg \frac{q\omega(q)}{\varphi(q)}. \tag{7.48}$$

This is not quite enough to produce long gaps between primes, but for certain q we improve on the above to establish

Lemma 7.13 *Let* $P = P(z) = \prod_{p \le z} p$. *Then*

$$\lim_{z \to \infty} \frac{g\big(P(z)\big)}{z} = \infty.$$

This immediately yields

Theorem 7.14 (Westzynthius) *Let* p_n *denote the* n^{th} *prime number. Then*

$$\limsup_{n \to \infty} \frac{p_{n+1} - p_n}{\log p_n} = \infty.$$

Proof of Theorem 7.14 Suppose that $N = g(P) - 1$ and that M is chosen, $P \le M < 2P$, so that $(M + m, P) > 1$ for $1 \le m \le N$. But $M + m > P \ge (M + m, P)$, and hence $M + m$ is composite because it has the proper divisor $(M + m, P)$. If n is chosen so that p_n is the largest prime not exceeding M, then $p_{n+1} - p_n \ge g(P)$ and $p_n < 2P$, which is $< e^{2z}$ when z is large. Hence

$$\frac{p_{n+1} - p_n}{\log p_n} \ge \frac{g(P)}{2z}$$

which tends to infinity as $z \to \infty$. $\qquad\square$

Proof of Lemma 7.13 Let L be large and fixed, and put $N = [zL/3]$. We show that if $z > z_0(L)$, then there exists an integer M such that $(M + n, P(z)) > 1$ for $1 \le n \le N$. Put

$$P_1 = \prod_{p \le L} p, \quad P_2 = \prod_{L < p \le L^L} p, \quad P_3 = \prod_{L^L < p \le z/3} p, \quad P_4 = \prod_{z/3 < p \le z} p,$$

and let $\mathcal{N}$ be the set of those integers n, $1 \le n \le N$, such that $(n, P_1 P_3) = 1$. The members of $\mathcal{N}$ are (i) 1; (ii) integers n composed entirely of prime factors of P_2; (iii) primes p, $z/3 < p \le N$. Thus

$$\operatorname{card} \mathcal{N} \le 1 + \psi(N, L^L) + \pi(N) - \pi(z/3).$$

If z is sufficiently large, then $L^L < \log N$, so that $\psi(N, L^L) < N^\varepsilon$ by Corollary 7.9. Hence

$$\operatorname{card} \mathcal{N} < \pi(N).$$

We choose $M \equiv 0 \pmod{P_1 P_3}$, so that $(M + n, P_1 P_3) > 1$ if $1 \leq n \leq N, n \notin \mathcal{N}$. To bound the number of $n \in \mathcal{N}$ such that $(M + n, P_2) = 1$ we average as in the proof of Lemma 3.5. Clearly

$$\sum_{m=1}^{q} \sum_{\substack{n \in \mathcal{N} \\ (m+n,q)=1}} 1 = \sum_{n \in \mathcal{N}} \sum_{\substack{m=1 \\ (m+n,q)=1}}^{q} 1 = \sum_{n \in \mathcal{N}} \varphi(q) = \varphi(q) \, \text{card} \, \mathcal{N}$$

for any integer q. Hence

$$\min_{m} \sum_{\substack{n \in \mathcal{N} \\ (m+n,q)=1}} 1 \leq (\text{card} \, \mathcal{N}) \prod_{p|q} \left(1 - \frac{1}{p} \right).$$

By taking $q = P_2$ we see that there is an $M \pmod{P_2}$ such that

$$\text{card}\{n \in \mathcal{N} : (M + n, P_2) = 1\} \leq (\text{card} \, \mathcal{N}) \prod_{p|P_2} \left(1 - \frac{1}{p} \right).$$

For such an M,

$$\text{card}\{1 \leq n \leq N : (M + n, P_1 P_2 P_3) = 1\} \leq \pi(N) \prod_{p|P_2} \left(1 - \frac{1}{p} \right).$$

By Mertens' theorem (Theorem 2.7(e)), the product on the right is $\sim 1/L$ as $L \to \infty$. Suppose that L is chosen sufficiently large to ensure that this product is $\leq 3/(2L)$. Then the right-hand side above is

$$\lesssim \frac{3N}{2L \log N} \sim \frac{z}{2 \log z}.$$

The number of primes dividing P_4 is $\pi(z) - \pi(z/3) \sim 2z/(3 \log z)$ as $z \to \infty$. Thus if z is large, then there are more such primes than there are integers n, $1 \leq n \leq N$, for which $(M + n, P_1 P_2 P_3) = 1$. Hence for each such n we may associate a prime p_n, $p_n|P_4$, in a one-to-one manner, and take $M \equiv -n \pmod{p_n}$. Then $(M + n, P_4) > 1$ and we are done.   □

The success of the argument just completed can be attributed to the fact that the number of n, $1 \leq n \leq N$, for which $(n, P_1 P_3) = 1$ is considerably smaller than $N \prod_{p|P_1 P_3} (1 - 1/p)$. By considering how L may be chosen as a function of z we obtain a quantitative improvement of Lemma 7.13 and hence also of Theorem 7.14.

Theorem 7.15 (Rankin) *Let p_n denote the nth prime number in increasing order. There is a constant $c > 0$ such that*

$$\limsup_{n \to \infty} \frac{p_{n+1} - p_n}{\left(\dfrac{(\log p_n)(\log \log p_n)(\log \log \log \log p_n)}{(\log \log \log p_n)^2} \right)} \geq c.$$

Proof We repeat the argument in the proof of Lemma 7.13, with the sole change that L is allowed to depend on z. If L is chosen so that

$$\psi(N, L^L) < \frac{N}{(\log N)^2}, \tag{7.49}$$

then $L = o(\log N)$, and hence

$$\psi(N, L^L) = o\left(\frac{z}{\log N}\right).$$

Since $z/\log N \leq z/\log z \ll \pi(z/3)$, it follows that

$$\psi(N, L^L) = o(\pi(z/3)),$$

and the proof proceeds as before.

By Theorem 7.6 we see that

$$\psi\left(N, N^{1/u}\right) < \frac{N}{(\log N)^2}$$

if $u \log u \geq 3 \log \log N$, which is the case if $u \geq 4(\log \log N)/\log \log \log N$. Taking $u = (\log N)/\log L^L$, we deduce that (7.49) holds if

$$L \log L < \frac{(\log N)(\log \log \log N)}{4 \log \log N}.$$

This is satisfied if

$$L < \frac{(\log N)(\log \log \log N)}{4(\log \log N)^2},$$

since then $\log L < \log \log N$. Since $N > z$ when $L \geq 3$, we conclude that we may take

$$L = \frac{(\log z)(\log \log \log z)}{4(\log \log z)^2}.$$

Hence

$$g(P(z)) > \frac{z(\log z)(\log \log \log z)}{13(\log \log z)^2}$$

for all $z > z_0$, and this gives the stated result. $\qquad\square$

Concerning the maximum number of primes in a short interval, by the Brun–Titchmarsh inequality (Theorem 3.9) and the Prime Number Theorem we see that

$$\pi(x + y) - \pi(x) < (2 + \varepsilon)\pi(y)$$

for $y > y_0(\varepsilon)$. Let

$$\rho(y) = \limsup_{x \to \infty}(\pi(x + y) - \pi(x)). \tag{7.50}$$

Thus $\rho(y) < (2 + \varepsilon)\pi(y)$. Very little is known about $\rho(y)$. It was once conjectured that

$$\pi(M + N) \leq \pi(M) + \pi(N) \tag{7.51}$$

for $M > 1, N > 1$, but there is now serious doubt as to the validity of this inequality. Indeed, it seems likely that $\rho(y) > \pi(y)$ for all large y. To see why, let

$$\overline{\rho}(N) = \max_{M} \sum_{\substack{n=M+1 \\ p|n \Rightarrow p > N}}^{M+N} 1. \tag{7.52}$$

Clearly $\rho(N) \leq \overline{\rho}(N)$. We expect that

$$\rho(N) = \overline{\rho}(N) \tag{7.53}$$

for all N, since this would follow from the

Prime k-tuple conjecture. *Let $a_1, a_2, \ldots, a_k$, be given integers. Then there exist infinitely many positive integers n such that $n + a_1, n + a_2, \ldots, n + a_k$ are all prime, provided that for every prime number p there is an integer n such that $(n + a_i, p) = 1$ for $i = 1, 2, \ldots, k$.*

We now show that $\overline{\rho}(N) > \pi(N)$ for all large N, so that (7.51) and (7.53) are inconsistent.

Theorem 7.16 *There is an absolute constant N_0 such that if $N > N_0$ then $\overline{\rho}(N) - \pi(N) \gg N(\log N)^{-2}$.*

Proof Suppose that N is even and that $N > 2$. Then for every M,

$$\sum_{\substack{n=M+1 \\ p|n \Rightarrow p > N}}^{M+N} 1 = \sum_{\substack{n=M+1 \\ p|n \Rightarrow p \geq N}}^{M+N} 1 \geq \sum_{\substack{n=M+1 \\ p|n \Rightarrow p > N-1}}^{M+N-1} 1.$$

Hence $\overline{\rho}(N) \geq \overline{\rho}(N - 1)$ when N is even, $N > 2$, so it suffices to treat the case when N is odd, say $N = 2K + 1$. Let $\mathcal{P}(K)$ denote the set of integers n with $K/(2 \log K) < |n| \leq K$ and $|n|$ prime. Then

$$\text{card } \mathcal{P}(K) = 2(\pi(K) - \pi(K/(2 \log K))),$$

so by Theorem 6.9,

$$\text{card } \mathcal{P}(K) = \pi(2K + 1) + (c + o(1))\frac{K}{(\log K)^2}$$

where $c = 2 \log 2 - 1 > 0$. We now show that $\mathcal{P}(K)$ can be translated to form a set of integers $\{M + n : n \in \mathcal{P}(K)\}$ with each member coprime to $\prod_{p \leq N} p$. By the Chinese Remainder Theorem it suffices to show that for every prime number $p \le N$ there is a residue class r_p (mod p) that contains no element of $\mathcal{P}(K)$.

Obviously each element of $\mathcal{P}(K)$ is coprime to each prime $p \le K/(2 \log K)$, so we may take $r_p = 0$ for such primes. It remains to treat the primes p for which $K/(2 \log K) < p \le 2K + 1$. This is accomplished by means of a clever application of Lemma 7.13. Suppose that $K/(2 \log K) < p \le 2K + 1$. We show that there is an r_p such that if $|hp + r_p| \le K$, then $hp + r_p \notin \mathcal{P}(K)$. By Lemma 7.13 there is an interval $\mathcal{J} = [M_1 - 3 \log K, M_1 + 3 \log K]$ in which every integer j is divisible by a prime p_j with $p_j \le \frac{1}{3} \log K$. By the Chinese Remainder Theorem, we can choose r_p so that $r_p \equiv M_1 p$ (mod p_j) for each $j \in \mathcal{J}$. This can be done with $0 < r_p \le \exp\left(\vartheta(\frac{1}{3} \log K)\right) < K^{1/2}$. If $|h| \le 3 \log K$ then $h = j - M_1$ for some $j \in \mathcal{J}$ and so $h \equiv -M_1$ (mod p_j). Hence $hp + r_p \equiv -M_1 p + r_p \equiv 0$ (mod p_j), which implies that $hp + r_p \notin \mathcal{P}(K)$. On the other hand, if $|h| > 3 \log K$, then $|hp + r_p| \ge \left(\frac{3}{2} - o(1)\right)K > K$, so that $hp + r_p \notin \mathcal{P}(K)$ in this case also. Since the arithmetic progression $hp + r_p$ has no element in common with $\mathcal{P}(K)$ the proof is complete. $\qquad\square$

7.3.1 Exercises

1. Show that the function $\overline{p}(N)$ is weakly increasing.
2. (a) Show that in the prime k-tuple conjecture, the hypothesis that for every prime p the numbers a_j do not cover all residue classes (mod p) is satisfied for all $p > k$, so that it is enough to verify the hypothesis for $p \le k$ (a finite calculation for any given set of a_j).
   (b) Prove the converse of the prime k-tuple conjecture: If there exist infinitely many integers n for which $n + a_j$ is prime for all j, $1 \le j \le k$, then for every prime p there is a residue class x (mod p) such that $x + a_j \not\equiv 0$ (mod p)($1 \le j \le k$).
3. Show that $g(q) \gg q\omega(q)/\varphi(q)$.
4. (cf. Erdős 1951) Show that if $0 < c < 1/2$ then there exist arbitrarily large numbers x such that the interval $(x, x + c(\log x)/\log\log x)$ contains no square-free number.
5. (cf. Erdős 1946, Montgomery 1987) Suppose that $2 \le h \le x$. Let $\mathcal{P}$ denote the set of all primes $p \le h$, let $\mathcal{D}$ denote the set of positive integers composed entirely of primes in $\mathcal{P}$, and let $f(n) = \prod_{p|n, p \in \mathcal{P}}(1 - 1/p)$.
   (a) Show that $f(n) = \sum_{d|n, d \in \mathcal{D}} \mu(d)/d$.
   (b) Show that

$$\sum_{x < n \le x+h} f(n) = \frac{6}{\pi^2}h + O(\log h)$$

uniformly in x.

(c) Show that

$$\frac{\varphi(n)}{n} \geq f(n) - \sum_{\substack{p|n \\ p>h}} \frac{1}{p}.$$

(d) Among those primes $p > h$ that divide an integer in the interval $(x, x + h]$, let $\mathcal{Q}$ be those for which $p \leq h \log x$, and $\mathcal{R}$ those for which $p > h \log x$. Show that

$$\sum_{p \in \mathcal{Q}} \frac{1}{p} \ll \log \log \log x.$$

(e) Explain why

$$\prod_{\substack{p \in \mathcal{R} \\ U < p \leq 2U}} p \;\Bigg|\; \prod_{x < n \leq x+h} n,$$

and deduce that

$$\mathrm{card}\{p \in \mathcal{R} : U < p \leq 2U\} \ll \frac{h \log x}{\log U}.$$

(f) By summing over $U = 2^k h \log x$, show that

$$\sum_{p \in \mathcal{R}} \frac{1}{p} \ll \frac{1}{\log(h \log x)}.$$

(g) Show that

$$\frac{6}{\pi^2} h + O(\log h) + O(\log \log \log x) \leq \sum_{x < n \leq x+h} \frac{\varphi(n)}{n} \leq \frac{6}{\pi^2} h + O(\log h).$$

6. (cf. Pillai & Chowla 1930) Show that there is an absolute constant $c > 0$ such that there exist arbitrarily large x for which $\varphi(n)/n < 1/4$ when $x < n \leq x + c \log \log \log x$. Deduce that

$$\sum_{n \leq x} \frac{\varphi(n)}{n} - \frac{6}{\pi^2} x = \Omega(\log \log \log x).$$

7. (Hausman & Shapiro 1973; cf. Montgomery & Vaughan 1986)
(a) Show that

$$\sum_{n=1}^{q} \left(\sum_{\substack{m=1 \\ (m+n,q)=1}}^{h} 1 - \frac{\varphi(q)}{q} h \right)^2$$

$$= \frac{\varphi(q)^2}{q} \sum_{\substack{r|q \\ r>1}} \mu(r)^2 \frac{r^2}{\varphi(r)^2} \{h/r\}(1 - \{h/r\}) \prod_{\substack{p|q \\ p \nmid r}} \frac{p(p-2)}{(p-1)^2}.$$

(b) Use the inequality $\{\alpha\}(1 - \{\alpha\}) \le \alpha$ to show that

$$\sum_{n=1}^{q} \left(\sum_{\substack{m=1 \\ (m+n,q)=1}}^{h} 1 - \frac{\varphi(q)}{q}h \right)^2 \le h\varphi(q).$$

8. (Erdős 1951) (a) For a positive integer q, let $S(q)$ denote the set of those residue classes s modulo q^2 such that (s, q) is a perfect square. Show that if q is square-free, then $S(q)$ contains exactly $\prod_{p|q}(p^2 - p + 1)$ elements.

(b) Show that if q is square-free and $1 \le h \le q^2$, then there is an integer a such that the number of members of $S(q)$ in the interval $(a, a + h]$ is at most

$$h \prod_{p|q} \left(1 - \frac{1}{p} + \frac{1}{p^2} \right).$$

(c) From now on, suppose that q is the product of those primes $p \le y$ such that $p \equiv 3 \pmod 4$. By recalling Corollary 4.12, or otherwise, show that the expression above is $\asymp h/\sqrt{\log y}$.

(d) Show that if an integer n can be expressed as a sum of two squares, then $n \in S(q)$.

(e) Let $\mathcal{R}$ be the set of those primes p, $y < p \le Cy$, such that $p \equiv 3 \pmod 4$. Here C is an absolute constant, taken to be sufficiently large to ensure that $\mathcal{R}$ has at least $y/\log y$ elements. Note that such a constant exists, in view of Exercise 4.3.5(e). Let r denote the product of all members of $\mathcal{R}$. Suppose that the number of members of $S(q)$ lying in the interval $(a, a + h]$ is $< y/\log y$. For each $s \in S(q)$ satisfying $a < s \le a + h$, associate a prime $p \in \mathcal{R}$. Suppose that the integer b is chosen modulo p^2 so that $s + bq^2 \equiv p \pmod{p^2}$. Show that the interval $(a + bq^2, a + bq^2 + h]$ does not contain a sum of two squares.

(f) Show that a and b can be chosen so that $0 < a + bq^2 < (qr)^2$.

(g) Show that $\log qr \ll y$.

(h) Show that this construction succeeds with $h \asymp y/\sqrt{\log y} \gg (\log qr)/(\log\log qr)^{1/2}$.

(i) Conclude that there exist arbitrarily large x such that there is no sum of two squares between x and $x + c(\log x)/(\log\log x)^{1/2}$. Here c is a suitably small positive constant. (Note that a stronger result is established in the next exercise.)

9. (Richards 1982) For every prime $p \leq y$, let $\beta(p)$ denote the greatest positive integer such that $p^\beta \leq y$, and put

$$q = \prod_{\substack{p \leq y \\ p \equiv 3\,(4)}} p^{2\beta(p)}.$$

(a) Show that $q = \exp(2\psi(y; 4, 3))$.
(b) Show that $\log q \ll y$.
(c) Suppose that $1 \leq n \leq y$. Show that if $n \equiv 3 \pmod 4$, then there is a prime $p|q$ such that p divides n to an odd power.
(d) Let $x = (q-1)/4$. Show that x is an integer, and that $4x \equiv -1 \pmod q$.
(e) Show that if $1 \leq i \leq y/4$ and $p|q$, then the power of p that exactly divides $x + i$ is the same as the power of p that exactly divides $4i - 1$.
(f) Deduce that no integer in the interval $(x, x + y/4]$ can be expressed as a sum of two squares.
(g) Conclude that there exist arbitrarily large numbers x such that no number between x and $x + c \log x$ is a sum of two squares. Here c is a suitably small positive constant.

7.4   Numbers composed of a prescribed number of primes

Let $\sigma_k(x)$ denote the number of integers n with $1 \leq n \leq x$ and $\Omega(n) = k$. Then $\sigma_1(x) = \pi(x) \sim x/\log x$. Consider $\sigma_2(x)$. Clearly

$$\sigma_2(x) = \sum_{\substack{p_1, p_2 \\ p_1 \leq p_2 \\ p_1 p_2 \leq x}} 1 = \sum_{p \leq \sqrt{x}} (\pi(x/p) - \pi(p) + O(1)).$$

By the Prime Number Theorem this is

$$= \sum_{p \leq \sqrt{x}} (1 + o(1)) \frac{x}{p(\log x/p)} + O\left(\frac{x}{\log x}\right).$$

Thus, by partial summation and a further application of the Prime Number Theorem we find that

$$\sigma_2(x) \sim \frac{x \log \log x}{\log x}. \tag{7.54}$$

By inducting on k in this manner it can be shown that

$$\sigma_k(x) \sim \frac{x(\log \log x)^{k-1}}{(k-1)! \log x}. \tag{7.55}$$

for any fixed k. Since the sum over all $k \geq 1$ of the right-hand side is exactly x, it is tempting to think that the above holds quite uniformly in k. However this is not the case, as we shall presently discover. To obtain precise estimates that are uniform in k we apply analytic methods. In Section 2.4 we determined the asymptotic distribution of the additive function $\Omega(n) - \omega(n)$ by establishing the mean value of the multiplicative function $z^{\Omega(n)-\omega(n)}$. In the same spirit we shall derive information concerning the distribution of $\Omega(n)$ from mean value estimates of $z^{\Omega(n)}$. Since the Euler product of this latter function behaves badly when $|z|$ is large, we start not with $z^{\Omega(n)}$ but with $d_z(n)$ defined by the identities

$$\zeta(s)^z = \prod_p \left(1 - p^{-s}\right)^{-z} = \sum_{n=1}^{\infty} d_z(n) n^{-s} \qquad (\sigma > 1). \qquad (7.56)$$

Since $d_z(p) = z = z^{\Omega(p)}$, the functions $d_z(n)$ and $z^{\Omega(n)}$ are 'nearby', and hence the mean value of $z^{\Omega(n)}$ can be derived from that for $d_z(n)$ by elementary reasoning.

Theorem 7.17 *Let $D_z(x) = \sum_{n \leq x} d_z(n)$, and let R be any positive real number. If $x \geq 2$, then*

$$D_z(x) = \frac{x(\log x)^{z-1}}{\Gamma(z)} + O(x(\log x)^{\Re z - 2})$$

uniformly for $|z| \leq R$.

Proof Let $a = 1 + 1/\log x$. Then by Corollary 5.3,

$$D_z(x) - \frac{1}{2\pi i} \int_{a-iT}^{a+iT} \zeta(s)^z \frac{x^s}{s} \, ds \ll \sum_{\frac{1}{2}x < n < 2x} |d_z(n)| \min\left(1, \frac{x}{T|x-n|}\right)$$
$$+ \frac{x^a}{T} \sum_n |d_z(n)| n^{-a}. \qquad (7.57)$$

Since $|d_z(n)|$ is erratic, we must exercise some care in estimating the error terms above. Let $\mathcal{A} = \{n : |n - x| \leq x/(\log x)^{2R+1}\}$. Without loss of generality we may suppose that R is an integer. We note that $|d_z(n)| \leq d_{|z|}(n) \leq d_R(n)$. By the method of the hyperbola we see by induction on R that

$$D_R(x) = x P_R(\log x) + O_R\left(x^{1-1/R}\right)$$

where P_R is a polynomial of degree $R - 1$. Hence the contribution to the first sum in the error term in (7.57) of the $n \in \mathcal{A}$ is

$$\ll \sum_{n \in \mathcal{A}} |d_z(n)| \ll x(\log x)^{-R-2}$$

The contribution of the $n \notin \mathcal{A}$ is

$$\ll T^{-1}(\log x)^{2R+1}x(\log x)^{R-1}.$$

We take $T = \exp\left(\sqrt{\log x}\right)$ to see that this is also $\ll x(\log x)^{-R-2}$. The second sum in the error term in (7.57) is $\ll \zeta(a)^R \ll (\log x)^R$. Thus the total error term is $\ll x(\log x)^{-R-2}$.

If z is a positive integer, then $\zeta(s)^z$ has a pole at $s = 1$, and we can extract a main term by the calculus of residues, as in our proof of the Prime Number Theorem (Theorem 6.9). On the other hand, if z is not an integer, then $\zeta(s)^z$ has a branch point at $s = 1$, so greater care must be exercised in moving the path of integration. Put $b = 1 - c/\log T$ where c is a small positive constant, and replace the contour from $a - iT$ to $a + iT$ by a path consisting of $\mathcal{C}_1$, $\mathcal{C}_2$, $\mathcal{C}_3$ where $\mathcal{C}_1$ is a polygonal with vertices $a - iT$, $b - iT$, $b - i/\log x$, $\mathcal{C}_2$ begins with a line segment from $b - i/\log x$ to $1 - i/\log x$, continues with the semicircle $\{1 + e^{i\theta}/\log x : -\pi/2 \le \theta \le \pi/2\}$, and concludes with the line segment from $1 + i/\log x$ to $b + i/\log x$, and finally $\mathcal{C}_3$ is polygonal with vertices $b + i/\log x$, $b + iT$, $a + iT$. By Theorem 6.7, $\zeta(s)^z \ll (\log x)^R$ on the new path, so the integrals over $\mathcal{C}_1$ and $\mathcal{C}_3$ contribute an amount $\ll x(\log x)^{-R-2}$. On $\mathcal{C}_2$ we have $\zeta(s)^z/s = (s-1)^{-z}(1 + O(|s-1|))$. Hence

$$\frac{1}{2\pi i}\int_{\mathcal{C}_2}\zeta(s)^z\frac{x^s}{s}\,ds = \frac{1}{2\pi i}\int_{\mathcal{C}_2}(s-1)^{-z}x^s\,ds + O\left(\int_{\mathcal{C}_2}|s-1|^{1-\Re z}x^\sigma\,|ds|\right).$$

$$(7.58)$$

By the change of variables $s = 1 + w/\log x$ we see that the main term above is

$$x(\log x)^{z-1}\frac{1}{2\pi i}\int_{\mathcal{H}_2}w^{-z}e^w\,dw$$

where $\mathcal{H}_2$ starts at $-\beta - i$, loops around 0, and ends at $-\beta + i$ where $\beta = c(\log x)/\log T$. Let $\mathcal{H}_1$ be the contour $\mathcal{H}_1 = \{w = u - i : -\infty < u \le -\beta\}$, and similarly let $\mathcal{H}_3 = \{w = u + i : -\infty < u \le -\beta\}$. If we integrate over the union of the $\mathcal{H}_i$, then we obtain Hankel's formula (see Theorem C.3) for $1/\Gamma(z)$. The integral over $\mathcal{H}_1$ is $\ll_R \int_\beta^\infty e^{-u/2}\,du \ll_R e^{-\beta/2}$, which is small since $T = \exp(\sqrt{\log x})$. Thus we see that the main term in (7.58) is $x(\log x)^{z-1}/\Gamma(z) + O_R(x\exp(-c\sqrt{\log x}))$ for some constant c. On the semicircular part of $\mathcal{C}_2$ the integrand in the error term in (7.58) is $\ll x(\log x)^{\Re z-1}$, so the contribution is $\ll x(\log x)^{\Re z-2}$. By the change of variables $s = 1 + w/\log x$ we see that the linear portions of $\mathcal{C}_2$ contribute an amount

$$\ll x(\log x)^{\Re z-2}\int_0^\infty (u^2+1)^{(R-1)/2}e^{-u}\,du \ll_R x(\log x)^{\Re z-2}.$$

Thus we have the stated estimate, and the proof is complete. $\qquad\square$

We now establish a procedure by which we can pass from $d_z(n)$ to other nearby functions.

Theorem 7.18 *Suppose that $\sum_{m=1}^{\infty} |b_z(m)|(\log m)^{2R+1}/m$ is uniformly bounded for $|z| \leq R$, and for $\sigma \geq 1$ let*

$$F(s, z) = \sum_{m=1}^{\infty} b_z(m)m^{-s}.$$

Let $a_z(n)$ be defined by the relation

$$\zeta(s)^z F(s, z) = \sum_{n=1}^{\infty} a_z(n)n^{-s} \qquad (\sigma > 1)$$

and let $A_z(x) = \sum_{n \leq x} a_z(n)$. Then for $x \geq 2$,

$$A_z(x) = \frac{F(1, z)}{\Gamma(z)} x(\log x)^{z-1} + O\big(x(\log x)^{\Re z - 2}\big).$$

Proof Since $a_z(n) = \sum_{m|n} b_z(m)d_z(n/m)$, we see by Theorem 7.17 that

$$A_z(x) = \sum_{m \leq x/2} b_z(m)D_z(x/m) + \sum_{x/2 < m \leq x} b_z(m)$$

$$= \frac{x}{\Gamma(z)} \sum_{m \leq x/2} \frac{b_z(m)}{m}(\log x/m)^{z-1} + O\left(x \sum_{m \leq x} \frac{|b_z(m)|}{m}(\log 2x/m)^{\Re z - 2}\right).$$

$$(7.59)$$

The error term here is

$$\ll x(\log x)^{\Re z - 2} \sum_{m \leq \sqrt{x}} \frac{|b_z(m)|}{m} + x(\log x)^{-R-2} \sum_{m > \sqrt{x}} \frac{|b_z(m)|}{m}(\log m)^{2R}$$

$$\ll x(\log x)^{\Re z - 2}.$$

In the main term, when $m \leq x^{1/2}$ we write

$$(\log x/m)^{z-1} = (\log x)^{z-1} + O\big((\log m)(\log x)^{\Re z - 2}\big).$$

Thus the first sum on the right-hand side of (7.59) is

$$= (\log x)^{z-1} \sum_{m \leq x/2} \frac{b_z(m)}{m}$$

$$+ O\left((\log x)^{\Re z - 2} \sum_{m \leq \sqrt{x}} \frac{|b_z(m)|}{m} \log m + (\log x)^{R-1} \sum_{m > \sqrt{x}} \frac{|b_z(m)|}{m}\right)$$

$$= (\log x)^{z-1} F(1, z) + O\left((\log x)^{\Re z - 2} \sum_{m} \frac{|b_z(m)|}{m}(\log m)^{2R+1}\right),$$

which gives the result.   □

Suppose that $R < 2$, and let

$$F(s, z) = \prod_p \left(1 - \frac{z}{p^s}\right)^{-1} \left(1 - \frac{1}{p^s}\right)^z \qquad (7.60)$$

for $\sigma > 1, |z| \leq R$. Then $a_z(n) = z^{\Omega(n)}$ in the notation of Theorem 7.18. Hence, with $\sigma_k(x)$ defined as at the beginning of this section we find that

$$A_z(x) = \sum_{n \leq x} z^{\Omega(n)} = \sum_{k=0}^{\infty} \sigma_k(x) z^k.$$

Here the power series on the right is actually a polynomial, since $\sigma_k(x) = 0$ for sufficiently large k, when x is fixed. Our asymptotic estimate for $A_z(x)$ enables us to recover an estimate for the power series coefficients $\sigma_k(x)$, since Cauchy's formula asserts that

$$\sigma_k(x) = \frac{1}{2\pi i} \int_{|z|=r} \frac{A_z(x)}{z^{k+1}} \, dz \qquad (7.61)$$

for $r < 2$.

Theorem 7.19 *Suppose that $R < 2$, that $F(s, z)$ is given by (7.60), and that $G(z) = F(1, z)/\Gamma(z + 1)$. Then*

$$\sigma_k(x) = G\left(\frac{k-1}{\log \log x}\right) \frac{x(\log \log x)^{k-1}}{(k-1)! \log x} \left(1 + O_R\left(\frac{k}{(\log \log x)^2}\right)\right) \qquad (7.62)$$

uniformly for $1 \leq k \leq R \log \log x$.

Since $G(0) = G(1) = 1$, we see that (7.55) holds when $k = o(\log \log x)$, and also when $k = (1 + o(1)) \log \log x$, but that (7.55) does not hold in general. The restriction to $R < 2$ is necessary because of the contribution of the prime $p = 2$ in the Euler product (7.60) for $F(s, z)$. If $z \geq 2$, then the behaviour is different; see Exercises 7.4.5 and 7.4.6, below.

Proof Our quantitative form of the Prime Number Theorem (Theorem 6.9) gives the case $k = 1$, so we may assume that $k > 1$. We substitute the estimate of Theorem 7.18 in (7.61) with $r = (k-1)/\log \log x$. The error term contributes an amount

$$\ll x(\log x)^{r-2} r^{-k} = \frac{x}{(\log x)^2} e^{k-1} \frac{(\log \log x)^k}{(k-1)^k}$$

$$\ll \frac{x(\log \log x)^k}{(k-1)!(\log x)^2} \ll \frac{x(\log \log x)^{k-3}}{(k-1)! \log x}.$$

This is majorized by the error term in (7.62) since $G((k-1)/\log\log x) \gg 1$. The main term we obtain from (7.61) is $xI/\log x$ where

$$I = \frac{1}{2\pi i} \int_{|z|=r} G(z)(\log x)^z z^{-k}\,dz$$

$$= \frac{G(r)}{2\pi i} \int_{|z|=r} (\log x)^z z^{-k}\,dz + \frac{1}{2\pi i} \int_{|z|=r} (G(z) - G(r))(\log x)^z z^{-k}\,dz.$$

By integration by parts we find that

$$\frac{r}{2\pi i} \int_{|z|=r} (\log x)^z z^{-k}\,dz = \frac{1}{2\pi i} \int_{|z|=r} (\log x)^z z^{1-k}\,dz.$$

We multiply both sides by $G'(r)$ and combine with the former identity to see that

$$I = \frac{G(r)}{2\pi i} \int_{|z|=r} (\log x)^z z^{-k}\,dx$$

$$+ \frac{1}{2\pi i} \int_{|z|=r} (G(z) - G(r) - G'(r)(z - r))(\log x)^z z^{-k}\,dz. \quad (7.63)$$

Here the first integral is $(\log\log x)^{k-1}/(k-1)!$ by Cauchy's theorem, which gives the desired main term. On the other hand,

$$G(z) - G(r) - G'(r)(z - r) = \int_r^z (z - w)G''(w)\,dw \ll |z - r|^2,$$

so that if we write $z = re^{2\pi i\theta}$, then the second integral in (7.63) is

$$\ll r^{3-k} \int_{-1/2}^{1/2} (\sin\pi\theta)^2 e^{(k-1)\cos 2\pi\theta}\,d\theta.$$

But $|\sin x| \le |x|$ and $\cos 2\pi\theta \le 1 - 8\theta^2$ for $-1/2 \le \theta \le 1/2$, so the above is

$$\ll r^{3-k} e^{k-1} \int_0^\infty \theta^2 e^{-8(k-1)\theta^2}\,d\theta \ll r^{3-k} e^{k-1}(k-1)^{-3/2} = \frac{(\log\log x)^{k-3} e^{k-1}}{(k-1)^{k-3/2}}$$

$$\ll k(\log\log x)^{k-3}/(k-1)!.$$

This completes the proof of the theorem. $\qquad\square$

The decomposition in (7.63) is motivated by the observation that $|(\log x)^z|$ is largest, for $|z| = r$, when $z = r$. We take the Taylor expansion to the second term because

$$\left| \int (z - r)^2 (\log x)^z z^{-k}\,dz \right| \asymp \int |z - r|^2 |(\log x)^z z^{-k}|\,|dz|,$$

whereas

$$\left| \int (z - r)(\log x)^z z^{-k}\,dz \right| = o\left(\int |z - r||(\log x)^z z^{-k}|\,|dz| \right).$$

By the calculus of residues we may write

$$I = \frac{1}{(k-1)!} \frac{d^{k-1}}{dz^{k-1}} \left(G(z)(\log x)^z \right) \Big|_{z=0}$$
$$= \sum_{v=0}^{k-1} \frac{G^{(v)}(0)}{v!} \frac{(\log \log x)^{k-1-v}}{(k-1-v)!}.$$

This gives a more accurate, but more complicated, main term.

In Section 2.3 we saw that $\Omega(n)$ rarely differs very much from $\log \log n$. In particular, from Theorem 2.12 we see that if $r < 1$, then the number of $n \le x$ for which $\Omega(n) < r \log \log n$ is $\ll_r x/\log \log x$. We now give a much sharper upper bound for the number of occurrences of such large deviations.

Theorem 7.20   *Let $A(x, r)$ denote the number of $n \le x$ such that $\Omega(n) \le r \log \log x$, and let $B(x, r)$ denote the number of $n \le x$ for which $\Omega(n) \ge r \log \log x$. If $0 < r \le 1$ and $x \ge 2$, then*

$$A(x, r) \ll x(\log x)^{r-1-r \log r}.$$

If $1 \le r \le R < 2$ and $x \ge 2$, then

$$B(x, r) \ll_R x(\log x)^{r-1-r \log r}.$$

Proof   We argue directly from Theorem 7.18, using a modified form of Rankin's method. If $0 \le r \le 1$ and $\Omega(n) \le r \log \log x$, then $r^{r \log \log x} \le r^{\Omega(n)}$. Hence

$$A(x, r) \le (\log x)^{-r \log r} \sum_{n \le x} r^{\Omega(n)}.$$

By Theorem 7.18 this is

$$\sim \frac{F(1, r)}{\Gamma(r)} x(\log x)^{r-1-r \log r}$$

where $F(s, z)$ is taken as in (7.60). This gives the result since $F(1, r) \ll 1$ and $\Gamma(r) \gg 1$ uniformly for $0 < r \le 1$.

Now suppose that $1 \le r \le R < 2$ and that $\Omega(n) \ge r \log \log x$. Then $r^{\Omega(n)} \ge r^{r \log \log x}$, and hence

$$B(x, r) \le (\log x)^{-r \log r} \sum_{n \le x} r^{\Omega(n)}.$$

Thus we have only to proceed as before to obtain the result.   □

In discussing Theorem 2.12 we proposed a probabilistic model, which in conjunction with the Central Limit Theorem would predict that the quantity

$$\alpha_n = \frac{\Omega(n) - \log\log n}{\sqrt{\log\log n}} \tag{7.64}$$

is asymptotically normally distributed. We now confirm this.

Theorem 7.21 *Let α_n be given by (7.64) and suppose that $Y > 0$. Then the number of n, $3 \le n \le x$, such that $\alpha_n \le y$ is*

$$\Phi(y)x + O_Y\left(\frac{x}{\sqrt{\log\log x}}\right)$$

uniformly for $-Y \le y \le Y$ where

$$\Phi(y) = \frac{1}{\sqrt{2\pi}} \int_{-\infty}^{y} e^{-t^2/2}\, dt.$$

Proof Let

$$\beta_n = \frac{\Omega(n) - \log\log x}{\sqrt{\log\log x}}.$$

Since $\Phi'(y) \ll 1$ and $\alpha_n - \beta_n \ll 1/\sqrt{\log\log x}$ when $x^{1/2} \le n \le x$ and $\Omega(n) \le 2\log\log x$, it suffices to consider β_n in place of α_n. We may of course also suppose that x is large.

Let k be a natural number and let u be defined by writing $k = u + \log\log x$. If $|u| \le \frac{1}{2}\log\log x$, then by Stirling's formula (see (B.26) or the more general Theorem C.1) we see that

$$\frac{(\log\log x)^{k-1}}{(k-1)!}$$

$$= \frac{e^u \log x}{\sqrt{2\pi \log\log x}}\left(1 + \frac{u}{\log\log x}\right)^{\frac{1}{2} - \log\log x - u}\left(1 + O\left(\frac{1}{\log\log x}\right)\right).$$

The estimate $\log(1 + \delta) = \delta - \delta^2/2 + O(|\delta|^3)$ holds uniformly for $|\delta| \le 1/2$. By taking $\delta = u/\log\log x$ we find that

$$\left(1 + \frac{u}{\log\log x}\right)^{\frac{1}{2} - \log\log x - u}$$

$$= \exp\left(-u + \frac{u - u^2}{2\log\log x} - \frac{u^2}{4(\log\log x)^2} + O\left(\frac{|u|^3}{(\log\log x)^2}\right)\right).$$

Suppose now that $|u| \leq (\log\log x)^{2/3}$. By considering separately $|u| \leq (\log\log x)^{1/2}$ and $(\log\log x)^{1/2} < |u| \leq (\log\log x)^{2/3}$ we see that

$$\frac{u}{\log\log x} \ll \frac{1}{\sqrt{\log\log x}} + \frac{|u|^3}{(\log\log x)^2}.$$

Similarly, by considering $|u| \leq 1$ and $|u| > 1$ we see that

$$\frac{u^2}{(\log\log x)^2} \ll \frac{1}{\sqrt{\log\log x}} + \frac{|u|^3}{(\log\log x)^2}.$$

On combining these estimates we deduce that

$$\frac{(\log\log x)^{k-1}}{(k-1)!} = \frac{\log x}{\sqrt{2\pi \log\log x}} \exp\left(\frac{-u^2}{2\log\log x}\right)$$
$$\times \left(1 + O\left(\frac{1}{\sqrt{\log\log x}}\right) + O\left(\frac{|u|^3}{(\log\log x)^2}\right)\right)$$

uniformly for $|u| \leq (\log\log x)^{2/3}$. In Theorem 7.19 we have $G(1) = 1$ and

$$G\left(\frac{k-1}{\log\log x}\right) = G(1) + O\left(\frac{1+|u|}{\log\log x}\right).$$

Hence by Theorem 7.19,

$$\sigma_k(x) = \frac{x \exp\left(\frac{-(k-\log\log x)^2}{2\log\log x}\right)}{\sqrt{2\pi \log\log x}}$$
$$\times \left(1 + O\left(\frac{1}{\sqrt{\log\log x}}\right) + O\left(\frac{|k - \log\log x|^3}{(\log\log x)^2}\right)\right).$$

By Theorem 7.20 we know that the contribution of $k \leq \log\log x - (\log\log x)^{2/3}$ is negligible. We sum over the range

$$\log\log x - (\log\log x)^{2/3} \leq k \leq \log\log x + y(\log\log x)^{1/2}.$$

This gives rise to three sums, one for the main term and two for error terms. Each of these sums can be considered to be a Riemann sum for an associated integral, and the stated result follows. □

7.4.1 Exercises

1. Let $p_1, p_2, \ldots, p_K$ be distinct primes. Show that the number of $n \leq x$ composed entirely of the p_k is

$$\frac{(\log x)^K}{K! \prod_{k=1}^{K} \log p_k} + O\left((\log x)^{K-1}\right).$$

2. (a) Let $d_z(n)$ be defined as in (7.56), and suppose that $|z| \le R$. Show that $|d_z(n)| \le d_{|z|}(n) \le d_R(n)$.

   (b) Let $F(s, z)$ be defined as in (7.60). Show that if $0 < r < 1$ and $\sigma > 1/2$, then $0 < F(\sigma, r) < 1$.

   (c) Let $F(s, z)$ be defined as in (7.60). Show that if $1 < r < 2$, then the Dirichlet series coefficients of $F(s, r)$ are all non-negative.

3. (a) Show that if

$$F(s, z) = \prod_p \left(1 + \frac{z}{p^s - 1}\right)\left(1 - \frac{1}{p^s}\right)^z,$$

   then $F(s, z)$ converges for $\sigma > 1/2$, uniformly for $|z| \le R$.

   (b) Show that if $F(s, z)$ is taken as above, and if $a_z(n)$ is defined as in Theorem 7.18, then $a_z(n) = z^{\omega(n)}$.

   (c) Let $\rho_k(x)$ denote the number of $n \le x$ for which $\omega(n) = k$. Show that if $x \ge 2$, then

$$\rho_k(x) = G\left(\frac{k-1}{\log\log x}\right) \frac{x(\log\log x)^{k-1}}{(k-1)! \log x}\left(1 + O_R\left(\frac{k}{(\log\log x)^2}\right)\right)$$

   uniformly for $1 \le k \le R \log\log x$ where $G(z) = F(1, z)/\Gamma(z+1)$.

   (d) Show that $G(0) = G(1) = 1$.

   (e) Let $A(x, r)$ denote the number of $n \le x$ for which $\omega(n) \le r \log\log x$. Show that

$$A(x, r) \ll x(\log x)^{r-1-r\log r}$$

   uniformly for $0 < r \le 1$.

   (f) Let $B(x, r)$ denote the number of $n \le x$ for which $\omega(n) \ge r \log\log x$. Show that

$$B(x, r) \ll x(\log x)^{r-1-r\log r}$$

   uniformly for $1 \le r \le R$.

4. (a) Show that if

$$F(s, z) = \prod_p \left(1 + \frac{z}{p^s}\right)\left(1 - \frac{1}{p^s}\right)^z,$$

   then $F(s, z)$ converges for $\sigma > 1/2$, uniformly for $|z| \le R$.

   (b) Show that if $F(s, z)$ is taken as above, and if $a_z(n)$ is defined as in Theorem 7.18, then $a_z(n) = \mu(n)^2 z^{\omega(n)}$.

   (c) Let $\pi_k(x)$ denote the number of square-free $n \le x$ for which $\omega(n) = k$. Show that if $x \ge 2$, then

$$\pi_k(x) = G\left(\frac{k-1}{\log\log x}\right) \frac{x(\log\log x)^{k-1}}{(k-1)! \log x}\left(1 + O_R\left(\frac{k}{(\log\log x)^2}\right)\right)$$

   uniformly for $1 \le k \le R \log\log x$ where $G(z) = F(1, z)/\Gamma(z+1)$.

(d) Show that $G(0) = G(1) = 1$.

5. (a) Show that if $x \geq 2$, then

$$\sum_{n \leq x} 2^{\Omega(n)} = cx(\log x)^2 + O(x \log x)$$

where c is a positive constant.

(b) Show that if $x \geq 2$, then

$$\sum_{n \leq x} 2^{\omega(n)} = cx \log x + O(x)$$

where c is a positive constant.

6. Show that if $(2 + \varepsilon) \log \log x \leq k \leq R \log \log x$, then $\sigma_k(x) \sim c2^{-k} x \log x$.

7. Show that if $\delta \leq r \leq 1 - \delta$ (or $1 + \delta \leq r \leq 2 - \delta$), then $A(x, r)$ (or $B(x, r)$, respectively) is $\asymp x(\log x)^{r-1-r \log r} / \sqrt{\log \log r}$.

8. Show that if x is large, then there is a k such that

$$\sigma_k(x) \geq \frac{x}{3\sqrt{\log \log x}}.$$

9. Show that the mean value $\sum_{n \leq x} d(n) \sim x \log x$ is due to the numbers $n \leq x$ for which $|\omega(n) - 2 \log \log x| \ll \sqrt{\log \log x}$.

10. Suppose that $1/2 \leq r \leq R$. Show that the number of square-free $n \leq x$ that can be written as a sum of two squares and for which $\omega(n) \geq r \log \log x$ is $\ll_R x(\log x)^{r-1-r \log 2r}$.

11. (Addison 1957) Let $M_{q,k}(x)$ denote the number of $n \leq x$ such that $\Omega(n) \equiv k \pmod{q}$.

(a) Show that if q is fixed, then $M_{q,k}(x) \sim x/q$ as $x \to \infty$.

(b) Show that if q is fixed, $q > 2$, then

$$M_{q,k}(x) - \frac{x}{q} = \Omega_{\pm}\left(\frac{x}{(\log x)^{\kappa}}\right)$$

where $\kappa = 1 - \cos 2\pi/q$.

12. Show that

$$\sum_{1 < n \leq x} \frac{1}{\omega(n)} \sim \frac{x}{\log \log x}$$

as $x \to \infty$.

13. Show that if $x \geq 2$, then

$$\sum_{1 < n \leq x} \frac{\Omega(n)}{\omega(n)} = x + O\left(\frac{x}{\log \log x}\right).$$

14. Suppose that $0 \leq \alpha \leq 1$. Show that

$$\sum_{n \leq x} \frac{\text{card}\{m : m|n, m \leq n^\alpha\}}{d(n)} = \frac{2}{\pi} x \arcsin \sqrt{\alpha} + O\left(\frac{x}{\sqrt{\log x}}\right).$$

15. Show that if $x \geq 16$, then

$$\sum_{\substack{n \leq x \\ (n,\Omega(n))=1}} 1 = \frac{6}{\pi^2} x + O\left(\frac{x}{\log \log \log x}\right).$$

7.5 Notes

Section 7.1. Theorem 7.2 was first proved by Dickman (1930), and was redis-covered by Chowla & Vijayaraghavan (1947), Ramaswami (1949), and Buchstab (1949). de Bruijn (1951a) gave a more precise estimate for $\psi(x, y)$, over a longer range of y. There is a considerable range of applications of $\psi(x, y)$, such as those to the distribution of kth power residues, Waring's problem, and the complexity of arithmetical algorithms in computer science. As a reflection of this there have been two significant survey articles, by Norton (1971) and by Hildebrand & Tenenbaum (1993).

Our treatment of $\psi(x, y)$ is fairly elementary, but it would be natural to take a more analytic approach, and use Perron's formula to write

$$\psi(x, y) = \frac{1}{2\pi i} \int_{c-i\infty}^{c+i\infty} \prod_{p \leq y} (1 - p^{-s})^{-1} \frac{x^s}{s} \, ds$$

$$= \frac{1}{2\pi i} \int_{c-i\infty}^{c+i\infty} \zeta(s) \prod_{p > y} (1 - p^{-s}) \frac{x^s}{s} \, ds.$$

For s not too large, an approximation to the product over $p > y$ is provided by the Prime Number Theorem, and this suggests the main term

$$\Lambda(x, y) = \frac{1}{2\pi i} \int_{c-i\infty}^{c+i\infty} \zeta(s) \exp\left(-\int_y^\infty v^{-s} (\log v)^{-1} \, dv\right) \frac{x^s}{s} \, ds.$$

It can be shown that this is indeed a good approximation to $\psi(x, y)$ over a very long range, but the technical details are rather heavy. By Theorem 7.10 it is not hard to show that

$$\Lambda(x, y) = x \int_{0-}^\infty \rho(u - v) d([y^v] y^{-v})$$

where we use (7.30) to extend the definition of $\rho(u)$ to $u \leq 0$. It follows that

$$\Lambda(x, y) \sim \rho(u)x$$

for a large range of u. For the further development of the theory, especially on the analytic side, see Hildebrand & Tenenbaum (1993).

Section 7.2. Theorem 7.11 is due to Buchstab (1937). The finer details of the behaviour of $\Phi(x, y)$ when u is large are intimately connected with sieve theory, especially that of the linear sieve, i.e., the sieve in which on average one residue class (mod p) is removed. The standard references are Greaves (2001), Halberstam & Richert (1974), Selberg (1991).

Section 7.3. Theorem 7.14 was first proved by Westzynthius (1931). Erdős (1935a) showed that

$$\limsup_{n\to\infty} \frac{p_{n+1} - p_n}{(\log p_n)(\log\log p_n)/(\log\log\log p_n)^2} > 0,$$

and then Rankin (1938) obtained Theorem 7.15 with $c = 1/3$. The value of c has been successively improved by Schönhage (1963), Rankin (1963), Maier & Pomerance (1990), culminating in the value $c = 2e^{C_0}$ of Pintz (1997). Erdős offered a \$10,000 prize for the first proof that the limsup in Theorem 7.15 is $+\infty$.

Early studies of $g(P(z))$ were conducted by Backlund (1929), Brauer & Zeitz (1930), Ricci (1934), and Chang (1938). The size of $g(P(z))$ is not known; possibly it is $\asymp z \log z$. However, it is conceivable that infinitely often $p_{n+1} - p_n$ is as large as $(\log p_n)^{\theta}$ where $\theta > 1$. In particular, Cramér (1936) conjectured that

$$\limsup_{n\to\infty} \frac{p_{n+1} - p_n}{(\log p_n)^2} = 1.$$

Theorem 7.16 is due to Hensley & Richards (1973).

Section 7.4. The analysis of $\sigma_k(x)$ is based on Selberg's exposition (1954) of Sathe (1953a,b, 1954a,b). Sathe (1954b) also shows that the bound $R \log\log x$ cannot be replaced by $2 \log\log x + 1$. Arguments giving rise to versions of Theorem 7.20 occur in Erdős (1935b). A qualitative version of Theorem 7.21 is a special case of Erdős & Kac (1940). Quantitative versions with various weaker error terms were obtained by LeVeque (1949) and Kubilius (1956). Theorem 7.21 had been conjectured by LeVeque and was established by Rényi & Turán (1958). They also showed that the error term is both uniform in x and best-possible.

7.6 References

Addison, A. W. (1957). A note on the compositeness of numbers, *Proc. Amer. Math. Soc.* **8**, 151–154.

Alladi, K. & Erdős, P. (1977). On an additive arithmetic function, *Pacific J. Math.* **71**, 275–294.

Backlund, R. J. (1929). Über die Differenzen zwischen den Zahlen, die zu den *n* ersten Primzahlen teilerfremd sind, *Annales Acad. sci. Fennicae* **32** (Lindelöf-Festschrift), Nr. 2, 9 pp.

Brauer, A. & Zeitz, H. (1930). Über eine zahlentheoretische Behauptung von Legendre, *Sitzungsb. Math. Ges. Berlin* **29**, 116–125.

de Bruijn, N. G. (1949). The asymptotically periodic behavior of the solutions of some linear functional equations, *Amer. J. Math.* **71**, 313–330.

(1950a). On the number of uncancelled elements in the sieve of Eratosthenes, *Nederl. Akad. Wetensch. Proc.* **52**, 803–812. (= *Indag. Math.* **12**, 247–256)

(1950b). On some linear functional equations, *Publ. Math.* **1**, 129–134.

(1951a). The asymptotic behaviour of a function occurring in the theory of primes, *J. Indian Math. Soc.* **15** (A), 25–32.

(1951b). On the number of positive integers $\leq x$ and free of prime factors $> y$, *Proc. Nederl. Akad. Wetensch.* **54**, 50–60.

(1966). On the number of positive integers $\leq x$ and free of prime factors $> y$, II, *Proc. Koninkl. Nederl. Akad. Wetensch.* A **69**, 239–247. (= *Indag. Math.* **28**)

Buchstab, A. A. (1937). Asymptotic estimates of a general number-theoretic function, *Mat. Sb.* (2) **44**, 1239–1246.

(1949). On those numbers in an arithmetic progression all prime factors of which are small in magnitude, *Dokl. Akad. Nauk SSSR* (N. S.) **67**, 5–8.

Chang, T.-H. (1938). Über aufeinanderfolgende Zahlen, von denen jede mindestens einer von *n* linearen Kongruenzen genügt, deren Moduln die ersten *n* Primzahlen sind, *Schr. Math. Sem. Inst. Angew. Math. Univ. Berlin* **4**, 35–55.

Chowla, S. D. & Vijayaraghavan, T. (1947). On the largest prime divisors of numbers, *J. Indian Math. Soc.* (2) **12**, 31–37.

Cramér, H. (1936). On the order of magnitude of the difference between consecutive prime numbers, *Acta Arith.* **2**, 23–46.

DeKoninck, J.-M. (1972). On a class of arithmetical functions, *Duke Math. J.* **39**, 807–818.

Dickman, K. (1930). On the frequency of numbers containing prime factors of a certain relative magnitude, *Ark. Mat. Astr. fys.* **22**, 1–14.

Duncan, R. L. (1970). On the factorization of integers, *Proc. Amer. Math. Soc.* **25**, 191–192.

Erdős, P. (1935a). On the difference of consecutive primes, *Quart. J. Math., Oxford ser.* **6**, 124–128.

(1935b). On the normal number of prime factors of $p - 1$ and some related problems concerning Euler's ϕ- function. *Quart. J. Math., Oxford ser.* **6**, 205–213.

(1946). Some remarks about additive and multiplicative functions, *Bull. Amer. Math. Soc.* **52**, 527–537.

(1951). Some problems and results in elementary number theory, *Publ. Math. Debrecen* **2**, 103–109.

(1962). On the integers relatively prime to *n* and on a number-theoretic function considered by Jacobsthal, *Math. Scand.* **10**, 163–170.

(1963). Problem and Solution Nr. 136, *Wiskundige opgaven met de Oplossingen* **21**.

Erdős, P. & Kac, M. (1940). The Gaussian law of errors in the theory of additive number theoretic functions, *Amer. J. Math.* **62**, 738–742.

Erdős, P. & Nicolas, J.-L. (1981). Sur la fonction: nombre de facteurs premiers de n, *Enseignoment Math.* (2) **27**, 3–27.

Friedlander, J. B. (1972). Maximal sets of integers with small common divisors, *Math. Ann.* **195**, 107–113.

Greaves, G. (2001). *Sieves in Number Theory*, Ergeb. Math. (3) 43. Berlin: Springer-Verlag.

Halberstam, H. (1970). On integers all of whose prime factors are small, *Proc. London Math. Soc.* (3) **21**, 102–107.

Halberstam, H. & Richert, H.-E. (1974). *Sieve Methods*, London Mathematical Society Monographs No. 4. London: Academic Press, 1974.

Hardy, G. H. & Littlewood, J. E. (1923). Some problems of "Partitio Numerorum": III On the expression of a number as a sum of primes, *Acta Math.* **44**, 1–70.

Hausman, M. & Shapiro, H. N. (1973). On the mean square distribution of primitive roots of unity, *Comm. Pure Appl. Math.* **26**, 539–547.

Hensley, D. & Richards, I. (1973). Two conjectures concerning primes, Analytic Number Theory, Proc. Sympos. Pure Math. 24. Providence: Amer. Math. Soc., 123–128.

(1973/4). Primes in intervals, *Acta Arith.* 25, 375–391.

Hildebrand, A. (1984). Integers free of large prime factors and the Riemann Hypothesis, *Mathematika* **31**, 258–271.

(1985). Integers free of large prime divisors in short intervals, *Oxford Quart. J.* **36**, 57–69.

(1986a). On the number of positive integers $\leq x$ and free of prime factors $> y$, *J. Number Theory* **22**, 289–307.

(1986b). On the local behavior of $\psi(x, y)$, *Trans. Amer. Math. Soc.* **297**, 729–751.

(1987). On the number of prime factors of integers without large prime divisors, *J. Number Theory* **25**, 81–106.

Hildebrand, A. & Tenenbaum, G. (1986). On integers free of large prime factors, *Trans. Amer. Math. Soc.* **296**, 265–290.

(1993). Integers without large prime factors, *J. Théor. Nombres Bordeaux.* **5**, 411–484.

Kubilius, I. P. (1956). Probabilistic methods in the theory of numbers, *Uspehi Mat. Nauk* (N.S.) 11 **68**, 31–66.

Legendre, A. M. (1798). *Théorie des Nombres*, First edition, Vol. 2, pp. 71–79.

LeVeque, W. J. (1949). On the size of certain number-theoretic functions, *Trans. Amer. Math. Soc.* **66**, 440–463.

Maier, H. & Pomerance, C. (1990). Unusually large gaps between consecutive primes, *Trans. Amer. Math. Soc.* **322**, 201–237.

Montgomery, H. L. (1987). Fluctuations in the mean of Euler's phi function, *Proc. Indian Acad. Sci. (Math. Sci.)* **97**, 239–245.

Montgomery, H. L. & Vaughan, R. C. (1986). On the distribution of reduced residues, *Ann. of Math.* (2) **123** (1986), 311–333.

Norton, K. K. (1971). *Numbers with Small Factors and the Least k'th Power Non-Residues*, Memoir 106, Providence: Amer. Math. Soc.

Pillai, S. S. & Chowla, S. D. (1930). On the error terms in some asymptotic formulæ in the theory of numbers, I, *J. London Math Soc.* **5**, 95–101.

Pintz, J. (1997). Very large gaps between consecutive primes, *J. Number Theory* **63**, 286–301.

Ramaswami, V. (1949). The number of positive integers $\leq x$ and free of prime divisors $> y$, and a problem of S. S. Pillai, *Duke Math. J.* **16**, 99–109.

Rankin, R. A. (1938). The difference between consecutive primes, *J. London Math. Soc.* **13**, 242–247.

(1963). The difference between consecutive primes, V, *Proc. Edinburgh Math. Soc.* (2)**13**, 331–332.

Rényi, A. & Turán, P. (1958), On a theorem of Erdős–Kac, *Acta Arith.* **4**, 71–84.

Ricci, G. (1934). *Ricerche aritmetiche sui polinomi*, II, *Rend. Palermo* **58**, 190–208.

Richards, I. (1982). On the gaps between numbers which are sums of two squares, *Adv. in Math.* **46**, 1–2.

Sathe, L. G. (1953a,b,1954a,b). On a problem of Hardy on the distribution of integers with a given number of prime factors I, II, III, IV, *J. Indian Math. Soc.* (N.S.) **17**, 63–82 & 83–141, **18**, 27–42 & 43–81.

Schinzel, A. (1961). Remarks on the paper "Sur certaines hypothèses concernant les nombres premiers", *Acta Arith.* **7**, 1–8.

Schönhage, A. (1963). Eine Bemerkung zur Konstruktion grosser Primzahllücken, *Arch. Math.* **14**, 29–30.

Selberg, A. (1954). Note on a paper of L. G. Sathe, *J. Indian Math. Soc.* **18**, 83–87.

(1991). *Collected papers*, Vol. II. Berlin: Springer-Verlag.

Westzynthius, E. (1931). Über die Verteilung der Zahlen, die zu den n ersten Primzahlen teilerfremd sind, *Comment. Phys.–Math. Soc. Sci. Fennica* **5**, Nr. 25, 37 pp.

8

Further discussion of the Prime Number Theorem

8.1 Relations equivalent to the Prime Number Theorem

The Prime Number Theorem asserts that

$$\pi(x) \sim \frac{x}{\log x} \tag{8.1}$$

as $x \to \infty$. In this section we consider a number of asymptotic relations that are equivalent to the Prime Number Theorem in the sense that they can be derived from, and also imply the Prime Number Theorem, by means of simple elementary arguments. These relations can also be proved by using the same analytic machinery that we used to prove the Prime Number Theorem, but the elementary techniques that we use to derive one relationship from another have permanent utility.

In Corollary 2.5 we saw that $\pi(x) = \psi(x)/\log x + O(x/(\log x)^2)$ and that $\psi(x) = \vartheta(x) + O(x^{1/2})$. Hence (8.1) is equivalent to

$$\psi(x) = x + o(x), \tag{8.2}$$

and also to

$$\vartheta(x) = x + o(x). \tag{8.3}$$

These equivalences are fairly trivial, since the arithmetic functions involved are nearly the same. At a somewhat deeper level, we consider $M(x) = \sum_{n \le x} \mu(n)$, and show that the estimate

$$M(x) = o(x) \tag{8.4}$$

is equivalent to the Prime Number Theorem. As was remarked in Chapter 6, the relation (8.4) can be proved analytically, by applying the truncated Perron formula to the Dirichlet series $1/\zeta(s)$ and using the zero-free region of the zeta function, as in the proof of the Prime Number Theorem. To derive (8.4) from

(8.2) it would be natural to express $\mu(n)$ as the Dirichlet convolution of $\Lambda(n)$ with some other function. As an aid to discovering such a function we would write

$$\frac{1}{\zeta(s)} = \frac{\zeta'(s)}{\zeta(s)} \cdot \frac{1}{\zeta'(s)}.$$

Unfortunately, $1/\zeta'(s) = -1/\sum(\log n)n^{-s}$ cannot be expanded as a Dirichlet series (because $\log 1 = 0$), so we reach an impasse. To circumvent this difficulty we introduce a valuable trick. Instead of treating $M(x)$ directly, we first consider $N(x) := \sum_{n \leq x} \mu(n) \log n$. Since

$$M(x) \log x - N(x) = \sum_{n \leq x} \mu(n) \log(x/n) \ll \sum_{n \leq x} \log(x/n) \ll x,$$

it is clear that (8.4) is equivalent to the estimate

$$N(x) = o(x \log x). \tag{8.5}$$

To derive (8.5) from (8.2) we observe that the Dirichlet series generating function of $\mu(n) \log n$ is $-(1/\zeta(s))' = \zeta'(s)/\zeta(s)^2$. Alternatively, in elementary language, we recall (1.22), which asserts that

$$\sum_{d|n} \Lambda(d) = \log n \qquad \left(-\frac{\zeta'}{\zeta}(s) \cdot \zeta(s) = -\zeta'(s)\right).$$

By the Möbius inversion formula, this gives

$$\Lambda(n) = \sum_{d|n} \mu(d) \log n/d \qquad \left(-\frac{\zeta'}{\zeta}(s) = -\zeta'(s) \cdot 1/\zeta(s)\right), \tag{8.6}$$

as was already noted in the proof of Theorem 2.4. But

$$0 = (\log n) \sum_{d|n} \mu(d) \qquad \left(0 = \frac{d}{ds}(\zeta(s) \cdot 1/\zeta(s))\right)$$

for all n, and so

$$\Lambda(n) = -\sum_{d|n} \mu(d) \log d \qquad \left(-\frac{\zeta'}{\zeta}(s) = -\zeta(s) \cdot (\zeta'(s)/\zeta(s)^2)\right).$$

By Möbius inversion a second time, we deduce that

$$\mu(n) \log n = -\sum_{d|n} \mu(d)\Lambda(n/d) \qquad \left(\zeta'(s)/\zeta(s)^2 = (1/\zeta(s)) \cdot \frac{\zeta'}{\zeta}(s)\right).$$

Since $\Lambda(n/d)$ is 1 on average, we adjust by this amount:

$$\sum_{d|n} \mu(d)(1 - \Lambda(n/d)) = \begin{cases} \mu(n) \log n & (n > 1), \\ 1 & (n = 1). \end{cases}$$

We sum this over $n \leq x$ (which is to say we apply (2.7)) to see that

$$\sum_{d \leq x} \mu(d)([x/d] - \psi(x/d)) = N(x) + 1.$$

From (8.2) we know that for any $\varepsilon > 0$ there is a large number $C = C(\varepsilon)$ such that $|\psi(y) - [y]| < \varepsilon y$ provided that $y \geq C$. That is, $|\psi(x/d) - [x/d]| \leq \varepsilon x/d$ for $d \leq x/C$. Thus

$$\left| \sum_{d \leq x/C} \mu(d) (\psi(x/d) - [x/d]) \right| \leq \sum_{d \leq x/C} \frac{\varepsilon x}{d} \ll \varepsilon x \log x.$$

The remaining range we treat trivially:

$$\sum_{x/C < d \leq x} \mu(d)(\psi(x/d) - [x/d]) \ll \sum_{x/C < d \leq x} \frac{x}{d} \ll x \log 2C.$$

Since ε can be taken arbitrarily small, we see that (8.5), and hence (8.4), follows from (8.2).

It is worth pausing here to note that the choice of the main term above is extremely delicate. If we had subtracted x/d instead of $[x/d]$, then we would have had to consider the question of the size of the sum $\sum_{d \leq x} \mu(d)/d$, which will be considered later. Since $\sum_{d \leq x} \mu(d)[x/d] = 1$ for all $x \geq 1$, we avoid the problem by this judicious choice of the main term.

To complete our proof that (8.4) is equivalent to (8.2) we now assume (8.4), and derive (8.2). By summing (8.6) over n, which is to say by applying (2.7), we see that

$$\psi(x) = \sum_{d \leq x} \mu(d) T(x/d)$$

where $T(x) = \sum_{m \leq x} \log m$ as in Section 2.2. We recall that $T(x) = x \log x - x + O(\log x)$ by the integral test. The main term here is approximately the same as applies to the summatory function of the divisor function, since Theorem 2.2 asserts that $D(x) = \sum_{m \leq x} d(m) = x \log x + (2C_0 - 1)x + O(x^{1/2})$. Indeed, the arithmetic function $d(m) - 2C_0$, when summed over m, produces exactly the same main terms as $\log m$. That is, if $f(m) = \log m - d(m) + 2C_0$ and $F(x) = \sum_{m \leq x} f(m)$ then $F(x) \ll x^{1/2}$. On the other hand, $\sum_{r|n} \mu(r) d(n/r) = 1$ for all n and $\sum_{d|n} \mu(d) = 0$ for all $n > 1$, so that

$$\sum_{d|n} \mu(d) f(n/d) = \begin{cases} \Lambda(n) - 1 & (n > 1), \\ 2C_0 - 1 & (n = 1). \end{cases}$$

On summing this over $n \leq x$ we find that

$$\sum_{d \leq x} \mu(d) F(x/d) = \psi(x) - [x] + 2C_0. \tag{8.7}$$

We now use (8.4) to show that the left-hand side above is $o(x)$, which thus gives (8.2). The reasoning employed at this point is useful for other purposes, so we axiomatize the argument, as follows.

Theorem 8.1 (Axer's theorem) *Suppose that a_d is a sequence such that (i) $\sum_{d \leq x} a_d = o(x)$ and that (ii) $\sum_{d \leq x} |a_d| \ll x$. Suppose also that $F(x)$ is a function defined on $[1, \infty)$ such that (iii) $F(x)$ has bounded variation in the interval $[1, C]$ for any finite $C \geq 1$, and that (iv) $F(x) \ll x/(\log x)^c$ for some constant $c > 1$. Then*

$$\sum_{d \leq x} a_d F(x/d) = o(x).$$

By taking $a_d = \mu(d)$ and $F(x)$ as in (8.7), we see that (8.4) implies (8.2).

Proof Suppose that $1 \leq U \leq x/2$. From (ii) and (iv) we see that

$$\sum_{x/(2U) < d \leq x/U} a_d F(x/d) \ll \frac{U}{(\log U)^c} \sum_{x/(2U) < d \leq x/U} |a_d| \ll \frac{x}{(\log U)^c}.$$

On taking $U = 2^j$ and summing over $j \geq J$ we find that

$$\sum_{d \leq x/2^J} a_d F(x/d) \ll x \sum_{j=J}^{\infty} \frac{1}{j^c} \ll_c \frac{x}{J^{c-1}}.$$

This is small compared with x if J is large. Let $A(x) = \sum_{d \leq x} a_d$. To treat the remaining range, $x/2^J < d \leq x$, we sum by parts. We do not use the Riemann–Stieltjes integral here because $A(y)$ and $F(x/y)$ may have common discontinuities. Let $n_0 = [x/2^J]$ and $n_1 = [x]$. Then

$$\sum_{n_0 < d \leq n_1} a_d F(x/d) = \sum_{n_0 < d \leq n_1} (A(d) - A(d-1)) F(x/d)$$

$$= \sum_{n_0 < d \leq n_1} A(d) F(x/d) - \sum_{n_0 - 1 < d \leq n_1 - 1} A(d) F(x/(d+1))$$

$$= A(n_1) F(x/n_1) - A(n_0) F(x/(n_0 + 1))$$

$$+ \sum_{n_0 < d < n_1} A(d) (F(x/d) - F(x/(d+1))).$$

Since $A(n_i) = o(x)$ and $F(x/n_i) \ll_J 1$, the first two terms are harmless. As the points x/d are monotonically arranged in the interval $[1, 2^J]$, the sum above has absolute value not exceeding

$$\left(\max_{d \leq x} |A(d)| \right) \sum_{n_0 < d < n_1} |F(x/d) - F(x/(d+1))| \leq \left(\max_{d \leq x} |A(d)| \right) \text{var}_{[1,2^J]} F.$$

By (i) and (iii) this is $o(x)$ for any given J. Thus the proof is complete. $\qquad \square$

By means of a further application of Axer's theorem, we now show that

$$\sum_{d=1}^{\infty} \frac{\mu(d)}{d} = 0 \tag{8.8}$$

is also equivalent to the Prime Number Theorem. We take $a_d = \mu(d)$ and $F(x) = \{x\} = x - [x]$ in Axer's theorem. Thus from (8.4) we deduce that

$$\sum_{d \leq x} \mu(d)\{x/d\} = o(x).$$

But $\sum_{d \leq x} \mu(d)[x/d] = 1$ when $x \geq 1$, so the left-hand side above is

$$-1 + x \sum_{d \leq x} \frac{\mu(d)}{d}.$$

Since this is $o(x)$, we obtain (8.8). To derive (8.4) from (8.8) is easier, in view of the following useful principle:

Lemma 8.2   *If $\sum_{d=1}^{\infty} a_d/d$ converges, then $\sum_{d \leq x} a_d = o(x)$.*

Proof   Let x be given, set $r(u) = \sum_{u < d \leq x} a_d/d$, and note that

$$\sum_{d \leq x} a_d = \int_0^x r(u)\,du.$$

But $r(u)$ is bounded (independently of x), and $|r(u)| < \varepsilon$ for $u > U_0$, so the integral is $\ll U_0 + \varepsilon x$. That is, the sum is $o(x)$, as desired.   □

8.1.1   Exercises

1. As in Section 2.2, let $T(x) = \sum_{n \leq x} \log n$, and recall that $T(x) = x \log x - x + O(\log x)$.
   (a) Show that $T(x) = \sum_{d \leq x} \Lambda(d)[x/d]$.
   (b) Show that

   $$x \sum_{d \leq x} \frac{\Lambda(d)}{d} = T(x) - \sum_{d \leq x}\{x/d\} - \sum_{d \leq x}(\Lambda(d) - 1)\{x/d\}.$$

   (c) Use (8.2) and Axer's theorem to show that the last sum above is $o(x)$.
   (d) Recall Exercise 2.1.1.
   (e) Show that (8.2) implies that

   $$\sum_{d \leq x} \frac{\Lambda(d)}{d} = \log x - C_0 + o(1), \tag{8.9}$$

   and note how this compares with Theorem 2.7(a).

(f) Apply Lemma 8.2 with $a_d = \Lambda(d) - 1$ to show that (8.9) implies (8.2). Hence (8.2) and (8.9) are equivalent.

(g) Show that

$$\sum_{n \leq x} \Lambda(n)\{x/n\} = (1 - C_0)x + o(x).$$

2. (a) By recalling the proof of Theorem 2.2(c), or otherwise, show that (8.2) implies that

$$\int_1^x \frac{\psi(u)}{u^2} \, du = \log x - 1 - C_0 + o(1). \tag{8.10}$$

(b) Show that (8.10) implies (8.2).

3. Let b be defined as in Theorem 2.7. (a) Imitate the proof of Theorem 2.7(d) to show that (8.2) implies that

$$\sum_{p \leq x} \frac{1}{p} = \log \log x + b + o(1/\log x). \tag{8.11}$$

(b) Show that (8.11) implies (8.1).

4. (a) Use (8.10) and Exercise 5.2.12 to show that

$$\sum_{d \leq x} \frac{\mu(d)}{d} \log(x/d) = o(\log x). \tag{8.12}$$

(b) Show that (8.10) implies that

$$\sum_{d \leq x} \frac{\mu(d)}{d} \log d = o(\log x). \tag{8.13}$$

(c) By partial summation, derive (8.4) from (8.13), and thus show that (8.2), (8.12) and (8.13) are all equivalent. (Note that a deeper assertion concerning the sum in (8.13) was already proved in Exercise 6.2.15.)

5. Let $F(n) = \sum_{d|n} f(d)$ for all n. The opening remarks in Chapter 2 raise the possibility of a connection between the two relations
   (i) $S(x) = \sum_{n \leq x} F(n) = cx + o(x)$;
   (ii) $\sum_{d=1}^{\infty} f(d)/d = c$.
   In Exercise 6.2.19 we have seen that (i) and the hypothesis $f(n) \ll 1$ imply (ii). Apply Axer's theorem with $a_d = f(d)$, $F(x) = \{x\}$ to show that (ii) and the hypothesis $\sum_{n \leq x} |f(n)| \ll x$ imply (i).

6. Let $d_k(n)$ be the kth divisor function, as defined in Exercise 2.1.18. Put $D_0(x) = 1$, and for positive integral k let $D_k(x) = \sum_{n \leq x} d_k(n)$.
   (a) Show that if k is a positive integer, then $\sum_{d \leq x} \mu(d) D_k(x/d) = D_{k-1}(x)$.

(b) Let $g(n)$ be an arithmetic function, put $G(x) = \sum_{n \leq x} g(n)$, and suppose that

$$G(x) = x P(\log x) + O(x/(\log x)^c)$$

where $c > 1$ and P is a polynomial of degree K. Let P_k be the polynomial defined in Exercise 2.1.18, and explain why there exist constants a_k so that $P(z) = \sum_{k=1}^{K+1} a_k P_k(z)$. By applying Axer's theorem with $F(x) = G(x) - \sum_{k=1}^{K+1} a_k D_k(x)$, show that

$$\sum_{d \leq x} \mu(d) G(x/d) = x Q(\log x) + o(x)$$

where Q is a polynomial of degree $K - 1$ with leading coefficient equal to K times the leading coefficient of P.

7. Show that Axer's theorem holds with hypothesis (iv) replaced by the weaker condition that $|F(x)| \leq \omega(x)x$ for some non-negative function $\omega(x)$ satisfying $\omega(x) \searrow$ and $\int_1^\infty \omega(x)/x \, dx < \infty$.

8.2 An elementary proof of the Prime Number Theorem

As we saw in Exercise 2.1.5, a version of Möbius inversion asserts that the two relationships

$$B(x) = \sum_{n \leq x} A(x/n), \qquad A(x) = \sum_{n \leq x} \mu(n) B(x/n) \qquad (8.14)$$

are equivalent. Some familiar – and useful – examples of this pairing are displayed in Table 8.1. In many instances of (8.14), the functions $A(x)$ and $B(x)$ are summatory functions of arithmetic functions $a(n)$ and $b(n)$, respectively, in which case $a(n)$ and $b(n)$ are linked by the more common Möbius inversion

$$b(n) = \sum_{d \mid n} a(d), \qquad a(n) = \sum_{d \mid n} \mu(d) b(n/d). \qquad (8.15)$$

The linear operator that takes $A(x)$ to $B(x)$ is continuous, but the transformation is nevertheless quite unstable. For example, the choice of the functions $A(x)$ in the second and third lines of Table 8.1 are very close, and yet the corresponding functions $B(x)$ differ quite substantially.

When the asymptotic rate of growth of $A(x)$ is known, it is easy to deduce that of $B(x)$, as a form of Abelian theorem. For example, if $A(x) \sim x$ as $x \to \infty$, then $B(x) \sim x \log x$. However, from the fourth line of Table 8.1 we see that

Table 8.1

$A(x)$	$B(x)$
1	$[x]$
x	$x \sum_{n \leq x} \frac{1}{n} = x \log x + C_0 x + O(1)$
$[x]$	$\sum_{n \leq x} d(n) = x \log x + (2C_0 - 1)x + O\left(x^{1/2}\right)$
$\psi(x)$	$\sum_{n \leq x} \log n = x \log x - x + O(\log x)$
$x \log x$	$x \sum_{n \leq x} \frac{\log x/n}{n} = \frac{1}{2} x (\log x)^2 + C_1 x \log x + C_2 x + O(1)$

some sort of Tauberian converse would be useful, for the purpose of proving the Prime Number Theorem. Unfortunately, it is difficult to establish anything stronger than the trivial estimate

$$A(x) \ll \sum_{n \leq x} |B(x/n)|. \tag{8.16}$$

From this we see that if $B(x) \ll 1$, then $A(x) \ll x$. This is rather weak, since the same upper bound for $A(x)$ can be deduced from a weaker upper bound for $B(x)$: From (8.16) we see that

$$B(x) \ll x^\alpha, \ 0 \leq \alpha < 1 \quad \Longrightarrow \quad A(x) \ll_\alpha x. \tag{8.17}$$

As a first application of this, we take $A(x) = \psi(x) - x + 1 + C_0$. Then from lines 1, 2, and 4 of Table 8.1 we see that $B(x) \ll \log x$, and by (8.17) it follows that $A(x) \ll x$. That is, $\psi(x) \ll x$, which is the upper bound portion of Chebyshev's estimate. To achieve greater success we construct a prime number sum in which the main term is larger than $O(x)$.

Theorem 8.3   (Selberg) *Let*

$$\Lambda_2(n) = \Lambda(n) \log n + \sum_{bc=n} \Lambda(b) \Lambda(c).$$

Then for $x \geq 1$,

$$\sum_{n \leq x} \Lambda_2(n) = 2x \log x + O(x).$$

Clearly $\Lambda_2(n) > 0$ only when $\omega(n) \leq 2$. Thus the sum on the left above is analogous to $\psi(x)$ but with prime powers replaced by products of two prime powers, counted with suitable weights.

Proof   We begin by noting that

$$\sum_{d|n} \Lambda_2(d) = \sum_{d|n} \Lambda(d) \log d + \sum_{d|n} \sum_{bc=d} \Lambda(b)\Lambda(c)$$

$$= \sum_{d|n} \Lambda(d) \log d + \sum_{b|n} \Lambda(b) \sum_{c|n/b} \Lambda(c).$$

Here the sum over c is $\log n/b$, so the above is

$$= \log n \sum_{d|n} \Lambda(d)$$

$$= (\log n)^2. \tag{8.18}$$

Hence by Möbius inversion it follows that

$$\Lambda_2(n) = \sum_{d|n} \mu(d)(\log n/d)^2. \tag{8.19}$$

Take now

$$A(x) = \sum_{n\leq x} \Lambda_2(n) - 2x \log x + c_1 x + c_2 \tag{8.20}$$

where c_1 and c_2 are constants to be chosen later. Then by (8.18) and lines 1, 2, and 5 of Table 8.1 we see that the corresponding $B(x)$ given by (8.14) is

$$B(x) = \sum_{n\leq x}(\log n)^2 - 2x \sum_{n\leq x} \frac{\log x/n}{n} + c_1 x \sum_{n\leq x} \frac{1}{n} + c_2[x].$$

By the integral test the first sum is $\int_1^x (\log u)^2 \, du + O((\log x)^2) = x(\log x)^2 - 2x \log x + 2x + O((\log x)^2)$. Hence the above is

$$= -2x \log x + 2x - 2C_1 x \log x - 2C_2 x$$

$$+ c_1 x \log x + c_1 C_0 x + c_2 x + O((\log x)^2).$$

We now choose c_1 and c_2 so that the leading terms cancel. That is, we take $c_1 = 2 + 2C_1$ and $c_2 = -2 + 2C_2 - c_1 C_0$. Then $B(x) \ll (\log x)^2$, and hence by (8.17) it follows that $A(x) \ll x$. The desired estimate then follows from (8.20). □

Selberg's identity may be modified in a variety of ways. For example, we note that

$$\sum_{n\leq x} \Lambda(n) \log n = \int_1^x \log u \, d\psi(u) = \psi(x) \log x - \int_1^x \frac{\psi(u)}{u} \, du.$$

By Chebyshev's estimate this last integral is $\ll x$, and hence the above is

$$= \psi(x)\log x + O(x). \tag{8.21}$$

On inserting this in Selberg's identity, we find that

$$\psi(x)\log x + \sum_{n \le x} \psi(x/n)\Lambda(n) = 2x\log x + O(x). \tag{8.22}$$

Our object is to show that each term on the left above is $\sim x\log x$ as $x \to \infty$. Suppose, to the contrary, that $\psi(x)$ is somewhat larger than anticipated, say $\psi(x) = ax$ with $a > 1$. By combining Mertens' estimate $\sum_{n \le x} \Lambda(n)/n = \log x + O(1)$ with (8.22), we see that $\psi(y)/y$ is on average approximately $2 - a$ as y runs over the points x/p^k, counted with the appropriate weights. Note that $2 - a < 1$. That is, if x is chosen so that $\psi(x)$ is unusually large, then $\psi(x/p^k)$ must be unusually small for many prime powers p^k. Such an argument may be repeated, so that one finds that $\psi(x/(p^k q^\ell))$ is unusually large for many prime powers q^ℓ. The points x/p^k and $x/(p^k q^\ell)$ are highly interlacing, so that $\psi(y)$ would have to switch rapidly back and forth between large and small values. However, $\psi(x)$ is a (weakly) increasing function, which implies that if it is unusually large at one point, then it continues to be unusually large for some time after. More precisely, if $\psi(x) \ge ax$ with $a > 1$, then $\psi(y) \ge \sqrt{a}\,y$ uniformly for $x \le y \le \sqrt{a}\,x$. Similarly, if $\psi(x) \le bx$ with $b < 1$ then $\psi(y) \le \sqrt{b}\,y$ uniformly for $\sqrt{b}\,x \le y \le x$. Of course an interval on which $\psi(y)$ is large cannot overlap with one on which $\psi(y)$ is small. One expects to reach a contradiction by showing that these intervals are too numerous and too long to all fit in the interval $[1, x]$. Our remaining task is to convert this intuitive line of reasoning into a rigorous proof.

Let $R(x)$ be defined by the relation $\psi(x) = x + R(x)$. By combining the estimate of Mertens cited above with (8.22) we see that

$$R(x)\log x + \sum_{n \le x} R(x/n)\Lambda(n) \ll x. \tag{8.23}$$

Here the sum is a weighted average of values of R, but the total amount of weight, $\sum_{n \le x} \Lambda(n) = \psi(x)$, remains in doubt. To overcome this difficulty, we iterate the identity (8.23) as follows: By replacing x in (8.23) by x/m we find that

$$R(x/m)\log x/m + \sum_{n \le x/m} R(x/(mn))\Lambda(n) \ll x/m.$$

We multiply this by $\Lambda(m)$ and sum over all $m \le x$, and thus find that

$$\sum_{m \le x} R(x/m)\Lambda(m)\log x/m + \sum_{mn \le x} R(x/(mn))\Lambda(m)\Lambda(n) \ll x\log x.$$

We multiply both sides of (8.23) by $\log x$ and subtract the above to see that

$$R(x)(\log x)^2 = -\sum_{n \leq x} R(x/n)\Lambda(n)\log n$$

$$+ \sum_{mn \leq x} R(x/(mn))\Lambda(m)\Lambda(n) + O(x\log x). \quad (8.24)$$

This has the advantage over (8.23) that we know how much weight resides in the coefficients on the right-hand side, by virtue of Theorem 8.3. We now formulate a Tauberian principle that is appropriate to estimate the above expression.

Lemma 8.4 *Suppose that $a_n \geq 0$ and $b_n \geq 0$ for all n, and that*

$$\frac{1}{2}x\log x \leq \sum_{n \leq x} a_n \leq \frac{3}{2}x\log x, \quad (8.25)$$

$$\frac{1}{2}x\log x \leq \sum_{n \leq x} b_n \leq \frac{3}{2}x\log x \quad (8.26)$$

for all large x. Suppose also that

$$\sum_{n \leq x} a_n + b_n \sim 2x\log x \quad (8.27)$$

as $x \to \infty$. Finally, suppose that $r(u)$ is a function such that

$$|r(u)| \leq \beta u \quad (8.28)$$

for all large u where $0 < \beta \leq 1$, and that

$$r(v) - r(u) \geq -(v - u) \quad (8.29)$$

when $v \geq u$. Then

$$\left| \sum_{n \leq x}(a_n - b_n)r(x/n) \right| \leq \left(\beta - \frac{\beta^2}{100} + o(1) \right)x(\log x)^2.$$

Proof Without loss of generality the hypotheses hold for all $x \geq 1$, $u \geq 1$, since changes in the definitions of a_n, b_n for small n, and $r(u)$ for small u entail additional error terms of magnitude $O(x\log x)$. It suffices to show that

$$\sum_{n \leq x}(a_n - b_n)r(x/n) \leq \left(\beta - \frac{\beta^2}{100} + o(1) \right)x(\log x)^2, \quad (8.30)$$

since the reverse inequality can then be derived by exchanging the roles of a_n and b_n. By applying first (8.28) and then (8.27) we see that the left-hand side above is trivially

$$\leq \beta x \sum_{n \leq x} \frac{a_n + b_n}{n} \sim \beta x(\log x)^2. \quad (8.31)$$

We write the left-hand side of (8.30) in the form

$$\beta x \sum_{n \leq x} \frac{a_n + b_n}{n} - \sum_{n \leq x} a_n \left(\frac{\beta x}{n} - r(x/n) \right) - \sum_{n \leq x} b_n \left(\frac{\beta x}{n} + r(x/n) \right).$$

By (8.31), this is

$$\sim \beta x (\log x)^2 - S_A - S_B,$$

say. Note that both factors of the summands in S_A are non-negative, so that $S_A \geq 0$. Similarly, $S_B \geq 0$. We need to show that

$$S_A + S_B \geq \left(\frac{\beta^2}{100} + o(1) \right) x (\log x)^2. \tag{8.32}$$

To this end we show that

$$\sum_{y < n \leq 16y} a_n \left(\frac{\beta x}{n} - r(x/n) \right) + b_n \left(\frac{\beta x}{n} + r(x/n) \right) \geq \frac{1}{16} \beta^2 x \log y \tag{8.33}$$

for all large y. Then (8.32) follows on summing this over $y = x16^{-k}$, $1 \leq k \leq [(\log x)/\log 16]$. In proving (8.33) we consider three cases.

CASE 1. $r(u) \leq \frac{1}{2}\beta u$ for all $u \in [\frac{x}{16y}, \frac{x}{4y}]$. Then $r(x/n) \leq \frac{1}{2}\beta x/n$ for all $n \in [4y, 16y]$, and hence

$$\sum_{y < n \leq 16y} a_n \left(\frac{\beta x}{n} - r(x/n) \right) \geq \frac{1}{2}\beta x \sum_{4y < n \leq 16y} \frac{a_n}{n}.$$

Since the denominator does not exceed $16y$, the above is

$$\geq \frac{\beta x}{32y} \sum_{4y < n \leq 16y} a_n.$$

Here the sum is $\sum_{n \leq 16y} a_n - \sum_{n \leq 4y} a_n$, which by (8.25) is $\geq 8y \log 16y - 6y \log 4y > 2y \log y$. Thus the above is

$$\geq \frac{\beta x}{16} \log y.$$

Since $\beta \leq 1$, this gives (8.33) in this case.

CASE 2. $r(u) \geq -\frac{1}{2}\beta u$ for all $u \in [\frac{x}{4y}, \frac{x}{y}]$. Then $r(x/n) \geq -\frac{1}{2}\beta x/n$ for $n \in [y, 4y]$. Arguing as in the preceding case, but using (8.26) instead of (8.25), we find that

$$\sum_{y < n \leq 4y} b_n \left(\frac{\beta x}{n} + r(x/n) \right) \geq \frac{1}{2}\beta x \sum_{y < n \leq 4y} \frac{b_n}{n} \geq \frac{\beta x}{8y} \sum_{y < n \leq 4y} b_n \geq \frac{\beta x \log y}{16}.$$

This gives (8.33) in this case.

If neither Case 1 nor Case 2 applies, then we have

CASE 3. *There is a $u_1 \in [\frac{x}{16y}, \frac{x}{4y}]$ such that $r(u_1) \geq \frac{1}{2}\beta u_1$, and a $u_2 \in [\frac{x}{4y}, \frac{x}{y}]$ such that $r(u_2) \leq -\frac{1}{2}\beta u_2$.* Let u_4 be the inf of those $u \geq u_1$ such that $r(u) \leq -\frac{1}{2}\beta u$. We show that $r(u_4) = -\frac{1}{2}\beta u_4$. Suppose that $r(u_4) > -\frac{1}{2}\beta u_4$, say $r(u_4) + \frac{1}{2}\beta u_4 = \delta > 0$. Suppose that

$$u_4 \leq v < u_4 + \frac{\delta}{1 - \frac{1}{2}\beta}. \tag{8.34}$$

Then by (8.29) we see that

$$r(v) \geq r(u_4) - (v - u_4) = -\frac{1}{2}\beta u_4 + \delta - (v - u_4).$$

From the upper bound in (8.34) we deduce that the above expression is $> -\frac{1}{2}\beta v$. That is, the inequality $r(u) \leq -\frac{1}{2}\beta u$ holds at no point of the interval (8.34). Since this contradicts the definition of u_4, it follows that $r(u_4) \leq -\frac{1}{2}\beta u_4$. Now suppose that $r(u_4) < -\frac{1}{2}\beta u_4$, say $-r(u_4) - \frac{1}{2}\beta u_4 = \delta > 0$. Suppose also that

$$u_4 - \frac{\delta}{1 - \frac{1}{2}\beta} \leq u \leq u_4. \tag{8.35}$$

Then by (8.29) we see that

$$r(u) \leq r(u_4) + (u_4 - u) = -\frac{1}{2}\beta u_4 - \delta + (u_4 - u).$$

From the lower bound in (8.35) we deduce that this expression is $\leq -\frac{1}{2}\beta u$. That is, the inequality $r(u) \leq -\frac{1}{2}\beta u$ holds throughout the interval (8.35). Since this contradicts the definition of u_4, we conclude that $r(u_4) = -\frac{1}{2}\beta u_4$. Put

$$u_3 = \frac{1 - \frac{1}{2}\beta}{1 + \frac{1}{2}\beta} u_4,$$

and suppose that

$$u_3 < u \leq u_4. \tag{8.36}$$

Then by (8.29) we see that

$$r(u) \leq r(u_4) + (u_4 - u) = -\frac{1}{2}\beta u_4 + (u_4 - u).$$

From the lower bound in (8.36) we deduce that this expression is $< \frac{1}{2}\beta u$. That is, the inequality $r(u) \geq \frac{1}{2}\beta u$ holds at no point of the interval (8.36), and hence $u_1 \leq u_3$.

To summarize, we have $\frac{x}{16y} \le u_1 \le u_3 \le u_4 \le \frac{x}{y}$ and $|r(u)| \le \frac{1}{2}\beta u$ for $u_3 < u \le u_4$. Hence

$$\sum_{x/u_4 \le n < x/u_3} a_n \left(\frac{\beta x}{n} - r(x/n) \right) + b_n \left(\frac{\beta x}{n} + r(x/n) \right)$$

$$\ge \frac{1}{2}\beta x \sum_{x/u_4 \le n < x/u_3} \frac{a_n + b_n}{n}$$

$$= \left(\frac{1}{2}\beta + o(1) \right) x \left((\log x/u_3)^2 - (\log x/u_4)^2 \right). \tag{8.37}$$

To estimate the last factor above we note that

$$\log \frac{x}{u_3} - \log \frac{x}{u_4} = \log \frac{1 + \frac{1}{2}\beta}{1 - \frac{1}{2}\beta} = \sum_{r=0}^{\infty} \frac{\beta^{2r+1}}{(2r+1)2^{2r}} > \beta.$$

Also, since u_3 and u_4 do not exceed x/y, it follows that

$$\log \frac{x}{u_3} + \log \frac{x}{u_4} \ge 2 \log y.$$

Hence the expression (8.37) is

$$\ge \left(\beta^2 + o(1) \right) x \log y.$$

Thus we have (8.33) in this case also, and the proof of Lemma 8.4 is complete.

$\square$

To complete the proof of the Prime Number Theorem we apply Lemma 8.4 with

$$a_n = \Lambda(n) \log n, \qquad b_n = \sum_{bc=n} \Lambda(b)\Lambda(c).$$

We combine Chebyshev's estimates in the form

$$(\log 2 + o(1))x \le \psi(x) \le (2 \log 2 + o(1))x$$

with (8.21) to see that

$$(\log 2 + o(1))x \log x \le \sum_{n \le x} a_n \le (2 \log 2 + o(1))x \log x. \tag{8.38}$$

This gives (8.25), and (8.27) is Selberg's identity as expressed in Theorem 8.3. To obtain (8.26) it suffices to subtract (8.38) from (8.27). We apply the lemma with $r(u) = R(u) = \psi(u) - u$. Then

$$r(v) - r(u) = \sum_{u < n \le v} \Lambda(n) \quad - (v - u) \ge -(v - u),$$

so we have (8.28). Let $\alpha = \limsup |r(u)|/u$. Our object is to show that $\alpha = 0$. We know that $\alpha \le 1/2$, by Chebyshev's estimates. Suppose that $\alpha > 0$, and choose β, $0 < \beta \leq 1$ so that

$$\beta - \frac{\beta^2}{100} < \alpha < \beta.$$

By combining the conclusion of Lemma 8.4 with (8.24) we deduce that $\alpha \leq \beta - \beta^2/100$, a contraction. Thus $\alpha = 0$, and the proof of the Prime Number Theorem is complete.

8.2.1 Exercises

1. For which entries in Table 8.1 are $A(x)$ and $B(x)$ summatory functions of arithemtic functions $a(n)$ and $b(n)$ related as in (8.15) ?
2. If $A(x) = M(x) := \sum_{n \leq x} \mu(n)$ in (8.14), then what is the function $B(x)$?
3. (a) Verify the Dirichlet series identity

$$\left(\frac{\zeta'}{\zeta}(s)\right)' + \left(\frac{\zeta'}{\zeta}(s)\right)^2 = \frac{\zeta''}{\zeta}(s).$$

   (b) Compute the Dirichlet series coefficients of the three functions in the above identity, and thus give a proof of (8.18) by means of formal Dirichlet series.
   (c) Compute the leading term of the Laurent expansions of the three functions above, at the point $s = 1$.
   (d) Suppose that ρ is a zero of $\zeta(s)$ of multiplicity $m > 0$. Compute the singular portion of the Laurent expansions of the three functions above, at $s = \rho$. Note that the pole of ζ''/ζ at $s = \rho$ is simple if and only if ρ is a simple zero of $\zeta(s)$.
4. Let $a = \limsup_{x \to \infty} \psi(x)/x$ and $b = \liminf_{x \to \infty} \psi(x)/x$. Suppose that a sequence x_ν tending to infinity is chosen so that $\lim_{\nu \to \infty} \psi(x_\nu)/x_\nu = a$. Use (8.22) to show that for each ν a prime p_ν can be selected so that $x_\nu/p_\nu \to \infty$ and $\liminf_{\nu \to \infty} \psi(x_\nu/p_\nu)/(x_\nu/p_\nu) \leq 2 - a$. Thus show that $a + b \leq 2$. By a similar argument, show that $a + b \geq 2$. Hence demonstrate that the relation $a + b = 2$ is a consequence of (8.22).
5. (a) Show that

$$\log x \sum_{\substack{p^k \leq x \\ k \geq 2}} \log p + \sum_{\substack{p^k q^\ell \leq x \\ k+\ell \geq 3}} (\log p) \log q \ll x.$$

   Here p and q denote prime numbers.
   (b) As usual, let $\vartheta(x) = \sum_{p \leq x} \log p$, and use Selberg's identity to show that

$$\vartheta(x) \log x + \sum_{p \leq x} \vartheta(x/p) \log p = 2x \log x + O(x).$$

6. Show that $\sum_{d|n} \mu(d)(\log n/d)^2 = \Lambda(n)\log n + \sum_{d|n} \Lambda(d)\Lambda(n/d)$.
7. Let k be a positive integer, and put

$$\Lambda_k(n) = \sum_{d|n} \mu(d)(\log n/d)^k.$$

(a) Show that

$$\Lambda_{k+1}(n) = \Lambda_k(n)\log n + \sum_{d|n} \Lambda_k(d)\Lambda(n/d).$$

(b) Show that $\Lambda_k(n) \geq 0$ for all n, and that if $\Lambda_k(n) > 0$, then $\omega(n) \leq k$.
8. Let c and M be positive constants, and suppose that $f(x)$ is a function defined on $[1, \infty)$ such that (i) $|\int_1^x f(u)u^{-2} du| \leq M$ for all $x \geq 1$, and also (ii) $|f(u) - f(v)| \leq c|u - v|$ whenever $u \geq 1$ and $v \geq 1$. Put

$$\alpha = \limsup_{x \to \infty} \frac{|f(x)|}{x}, \qquad \beta = \limsup_{x \to \infty} \frac{1}{\log x} \int_1^x \frac{|f(u)|}{u^2} du.$$

Show that $\beta \leq \alpha(1 - \alpha^2/(32cM))$.

8.3   The Wiener–Ikehara Tauberian theorem

In Chapter 6 we developed some understanding of the analytic behaviour of the zeta function, which allowed us to show that $\zeta(s) \neq 0$ for $\sigma \geq 1 - c/\log \tau$, which in turn permitted us to establish the Prime Number Theorem with an error term $\ll x \exp(-c\sqrt{\log x})$. On the other hand, it is reasonable to ask what is the least information concerning the zeta function that would suffice to establish the Prime Number Theorem in the weak form (8.1). In this section we establish a general Tauberian theorem, from which the Prime Number Theorem follows from the information that the functions

$$\zeta(s) - \frac{1}{s-1}, \qquad \zeta'(s) + \frac{1}{(s-1)^2}$$

are continuous in the closed half-plane $\sigma \geq 1$, and that

$$\zeta(1+it) \neq 0 \tag{8.39}$$

for all real t. Conversely from (8.2) we see that

$$-\frac{\zeta'}{\zeta}(s) = \frac{s}{s-1} + s \int_1^\infty \frac{\psi(x) - x}{x^{s+1}} dx = o\left(\frac{1}{\sigma - 1}\right)$$

as $\sigma \to 1^+$ with t fixed, $t \neq 0$. But if $\zeta(s)$ had a zero of multiplicity m at $1 + it$, then

$$\frac{\zeta'}{\zeta}(s) \sim \frac{m}{s-1}$$

when s is near $1 + it$. Since this is possible only when $m = 0$, we have (8.39). The above observations can be paraphrased as 'the Prime Number Theorem is equivalent to the assertion (8.39)', although one needs to bear in mind the continuity conditions also.

Suppose that $\alpha(s) = \sum_{n=1}^{\infty} a_n n^{-s}$. In Section 5.2 we derived information concerning partial sums of this series at $s = 1$ from the behaviour of $\alpha(\sigma)$ as $\sigma \to 1^+$. We now take much stronger hypotheses that concern $\alpha(s)$ throughout the closed half-plane $\sigma \geq 1$, but we obtain from them much stronger conclusions, concerning partial sums of the series at $s = 0$. Our proof of the Hardy–Littlewood Tauberian theorem (Theorem 5.7) depended on a simple lemma concerning one-sided polynomial approximation (Lemma 5.8). Our new approach depends similarly on a corresponding lemma concerning one-sided trigonometric approximation, as follows.

Lemma 8.5   *Let $E(x) = e^x$ for $x \leq 0$, and $E(x) = 0$ for $x > 0$. For any given $\varepsilon > 0$ there is a T and continuous functions $f_+(x)$, $f_-(x)$ with $f_{\pm} \in L^1(\mathbb{R})$ such that*
   (i) $f_-(x) \leq E(x) \leq f_+(x)$ *for all real x;*
   (ii) $\widehat{f}_{\pm}(t) = 0$ *for $|t| \geq T$;*
   (iii) $\int_{-\infty}^{\infty} f_+(x)\, dx < 1 + \varepsilon$, $\quad \int_{-\infty}^{\infty} f_-(x)\, dx > 1 - \varepsilon$.

Before proving the above, we first explore its consequences.

Since the $f_{\pm} \in L^1(\mathbb{R})$, it follows that the Fourier transforms $\widehat{f}_{\pm}(t)$ are uniformly continuous. Thus from (ii) above it follows that $\widehat{f}_{\pm}(\pm T) = 0$, so that $\widehat{f}_{\pm}(t) = 0$ for all t with $|t| \geq T$. Since the $f_{\pm}$ are also continuous, it follows by the Fourier integral theorem that

$$\lim_{\tau \to \infty} \int_{-\tau}^{\tau} (1 - |t|/\tau)\widehat{f}_{\pm}(t)e(tx)\, dt = f_{\pm}(x)$$

for all x. But the functions $\widehat{f}_{\pm}$ are supported on the fixed interval $[-T, T]$, so the limit on the left above is simply $\int_{-T}^{T} \widehat{f}_{\pm}(t)e(tx)\, dt$. That is,

$$f_{\pm}(x) = \int_{-T}^{T} \widehat{f}_{\pm}(t)e(tx)\, dt \tag{8.40}$$

for all x. It may be further noted that $\int_{-T}^{T} \widehat{f}_{\pm}(t)e^{2\pi itz}\, dt$ is an entire function of z. Thus $f_{\pm}(x)$ is the restriction to the real axis of an entire function.

Theorem 8.6   (Wiener–Ikehara) *Suppose that the function $a(u)$ is nonnegative and increasing on $[0, \infty)$, that*

$$\alpha(s) = \int_{0}^{\infty} e^{-us}\, da(u)$$

converges for all s with σ > 1, and that r(s) := α(s) − c/(s − 1) extends to a continuous function in the closed half-plane σ ≥ 1. Then

$$\int_0^x 1 \, da(u) = ce^x + o(e^x)$$

as x → ∞.

By making the change of variable $a(u) = A(e^u)$, we obtain the following equivalent formulation.

Corollary 8.7 (Wiener–Ikehara) *Suppose that A(v) is non-negative and increasing on [1, ∞), that*

$$\alpha(s) = \int_1^\infty v^{-s} \, dA(v)$$

converges for all s with σ > 1, and that r(s) := α(s) − c/(s − 1) extends to a continuous function in the closed half-plane σ ≥ 1. Then

$$\int_1^x 1 \, dA(v) = cx + o(x)$$

as x → ∞.

By setting $A(v) = \sum_{n<v} a_n$ we obtain a useful Tauberian theorem for Dirichlet series.

Corollary 8.8 (Wiener–Ikehara) *Suppose that $a_n \geq 0$ for all n, that*

$$\alpha(s) = \sum_{n=1}^\infty a_n n^{-s}$$

converges for all s with σ > 1, and that r(s) := α(s) − c/(s − 1) extends to a continuous function in the closed half-plane σ ≥ 1. Then

$$\sum_{n \leq x} a_n = cx + o(x)$$

as x → ∞.

By taking $a_n = \Lambda(n)$, we see that (8.39) gives the hypotheses with $c = 1$, and hence we obtain the Prime Number Theorem in the form (8.2).

Proof of Theorem 8.6   Take δ > 0, and let $E(u)$ be as in Lemma 8.5. Then

$$\int_0^x e^{-\delta u} \, da(u) = e^x \int_0^\infty E(u - x)e^{-(1+\delta)u} \, da(u),$$

which by Lemma 8.5(i) is

$$\leq e^x \int_0^\infty f_+(u - x)e^{-(1+\delta)u} \, da(u).$$

By (8.40) this is

$$= e^x \int_0^\infty \int_{-T}^T \widehat{f}_+(t)e(tu - tx)\,dt\, e^{-(1+\delta)u}\,da(u).$$

By Fubini's theorem we may interchange the order of integration. Thus the above is

$$= e^x \int_{-T}^T \widehat{f}_+(t)e(-tx)\int_0^\infty e^{-(1+\delta-2\pi it)u}\,da(u)\,dt$$

$$= e^x \int_{-T}^T \widehat{f}_+(t)e(-tx)\alpha(1 + \delta - 2\pi it)\,dt. \qquad (8.41)$$

If $a(u) = e^u$, then $\alpha(s) = 1/(s - 1)$, and thus from the above calculation we see in particular that

$$\int_0^\infty f_+(u - x)e^{-\delta u}\,du = \int_{-T}^T \widehat{f}_+(t)e(-tx)\frac{1}{\delta - 2\pi it}\,dt.$$

On multiplying both sides by ce^x and combining this with (8.41), we deduce that

$$\int_0^x e^{-\delta u}\,da(u) \le e^x \int_{-T}^T \widehat{f}_+(t)e(-tx)r(1 + \delta - 2\pi it)\,dt$$

$$+ ce^x \int_0^\infty f_+(u - x)e^{-\delta u}\,du.$$

Since $r(s)$ is uniformly continuous in the closed rectangle $1 \le \sigma \le 1+\delta$, $|t| \le 2\pi T$, each of the above three terms tends to a limit as $\delta \to 0^+$. Thus

$$\int_0^x 1\,da(u) \le e^x \int_{-T}^T \widehat{f}_+(t)e(-tx)\,r(1 - 2\pi it)\,dt + ce^x \int_0^\infty f_+(u - x)\,du.$$

We divide through by e^x and let x tend to infinity. The first integral on the right tends to 0 by the Riemann–Lebesgue lemma, and the second integral on the right tends to $\int_{-\infty}^\infty f_+(u)\,du$. Thus we see that

$$\limsup_{x \to \infty} e^{-x} \int_0^x 1\,da(u) \le c\int_{-\infty}^\infty f_+(u)\,du \le c(1 + \varepsilon)$$

by Lemma 8.5(iii). By using f_- similarly we may also show that

$$\liminf_{x \to \infty} e^{-x} \int_0^x 1\,da(u) \ge c(1 - \varepsilon).$$

Since ε may be taken arbitrarily small, we obtain the stated result, apart from the need to prove Lemma 8.5.   □

Proof of Lemma 8.5   We assume, as we may, that $T \geq 1$. Let

$$\Delta_T(x) = T \left(\frac{\sin \pi T x}{\pi T x} \right)^2, \qquad J_T(x) = \frac{3T}{4} \left(\frac{\sin \pi T x/2}{\pi T x/2} \right)^4$$

be the Fejér and Jackson kernels, respectively. These functions have a peak of height $\asymp T$ and width $\asymp 1/T$ at 0, and have total mass 1. Set

$$f(x) = (E \star J_T)(x) = \int_{-\infty}^{\infty} E(u) J_T(x - u) \, du.$$

This is a weighted average of the values of $E(u)$ with special emphasis on those u near x. We show that

$$f(x) = E(x) + O(\min(1, 1/(Tx)^2)). \tag{8.42}$$

To establish this we consider several cases. If $|x| \leq 1/T$ we simply observe that $0 \leq f(x) \leq \int_{-\infty}^{\infty} J_T(u) \, du = 1$. If $x \geq 1/T$ we observe that $0 \leq f(x) \ll T^{-3} \int_{-\infty}^{0} (x - u)^{-4} \, du \ll 1/(Tx)^3$. By the calculus of residues it is easy to show that $\int_{-\infty}^{\infty} J_T(u) \, du = 1$. Hence

$$f(x) - E(x) = \int_{-\infty}^{\infty} (E(u) - E(x)) J_T(x - u) \, du.$$

Next, suppose that $-1 \leq x \leq -1/T$. If $2x \leq u \leq 0$, then $E(u) - E(x) = e^x(e^{u-x} - 1) = e^x(u - x + O((u - x)^2))$. Thus

$$\int_{2x}^{0} (E(u) - E(x)) J_T(x - u) \, du = -e^x \int_{x}^{-x} u J_T(u) \, du$$

$$+ O\left(\int_{x}^{-x} u^2 J_T(u) \, du \right).$$

Here the first integral on the right vanishes because the integrand is an odd function, and the second integral is $\ll 1/T^2$. On the other hand,

$$\int_{0}^{\infty} (E(u) - E(x)) J_T(x - u) \, du \ll T^{-3} \int_{-x}^{\infty} u^{-4} \, du \ll 1/|Tx|^3,$$

and similarly $\int_{-\infty}^{2x} \ll 1/|Tx|^3$, so we have (8.42) in this case also. Finally, suppose that $x \leq -1$. Then $E(u) - E(x) = e^x(u - x + O((u - x)^2))$ for $x - 1 \leq u \leq x + 1$, so that

$$\int_{x-1}^{x+1} (E(u) - E(x)) J_T(x - u) \, du = -e^x \int_{-1}^{1} u J_T(u) \, du$$

$$+ O\left(e^x \int_{-1}^{1} u^2 J_T(u) \, du \right) \ll e^x T^{-2},$$

which is $\ll 1/(Tx)^2$. On the other hand,

$$\int_{-\infty}^{x-1} (E(u) - E(x))J_T(x - u)\,du \ll e^x T^{-3} \int_1^\infty u^{-4}\,du \ll (Tx)^{-2},$$

and

$$\int_{x+1}^\infty (E(u) - E(x))J_T(x - u)\,du \ll T^{-3}x^{-4},$$

so we again have (8.42).

Clearly $\Delta_T(x) \ll T\min(1, 1/(Tx)^2)$, but there is no inequality in the reverse direction because $\Delta_T(x)$ vanishes at integral multiples of $1/T$. To overcome this difficulty we consider also a translate of the Fejér kernel. Since

$$\Delta_T(x) + \Delta_T(x + 1/(2T)) \gg T\min(1, 1/(Tx)^2),$$

we take

$$f_\pm(x) = f(x) \pm \frac{C}{T}\left(\Delta_T(x) + \Delta_T(x + 1/(2T))\right).$$

By (8.42) we see that if C is taken large enough, then $f_-(x) \le E(x) \le f_+(x)$ for all x.

By Fubini's theorem it is easy to see that if $f_1, f_2 \in L^1(\mathbb{R})$ then the convolution $f_1 \star f_2$ is also in $L^1(\mathbb{R})$, and also that $\widehat{f_1 \star f_2}(t) = \widehat{f_1}(t)\widehat{f_2}(t)$. Hence in particular, $f \in L^1(\mathbb{R})$ and $\widehat{f}(t) = \widehat{E}(t)\widehat{J}_T(t)$. But $\widehat{J}_T(t) = 0$ for $|t| \ge T$, so $\widehat{f}(t) = 0$ for $|t| \ge T$. Also, $\widehat{\Delta}_T(t) = 0$ for $|t| \ge T$, and we see that the functions $f_\pm$ have the property (ii).

Finally, we note by Fubini's theorem that

$$\int_{-\infty}^\infty f(x)\,dx = \left(\int_{-\infty}^\infty E(x)\,dx\right)\left(\int_{-\infty}^\infty J_T(u)\,du\right) = 1 \cdot 1 = 1,$$

and hence $\int_{-\infty}^\infty f_\pm(x)\,dx = 1 \pm 2C/T$. Thus we have (iii) if $T \ge C/\varepsilon$, so the proof is complete. $\qquad\square$

8.3.1 Exercises

1. Use the Wiener–Ikehara theorem (Theorem 8.6) to show that $M(x) = o(x)$.
2. (Dressler 1970; cf. Bateman 1972) Let $f(n)$ denote the number of positive integers k such that $\varphi(k) = n$.
   (a) Show that if $\sigma > 1$, then

   $$\sum_{n=1}^\infty \frac{f(n)}{n^s} = \sum_{k=1}^\infty \frac{1}{\varphi(k)^s} = \prod_p \left(1 + \frac{1}{\varphi(p)^s} + \frac{1}{\varphi(p^2)^s} + \cdots\right),$$

   and explain why this is not an Euler product in the usual sense.

(b) Let the above Dirichlet series be $F(s)$. Show that $F(s) = \zeta(s)G(s)$ for $\sigma > 1$, where

$$G(s) = \prod_p \left(1 - \frac{1}{p^s} + \frac{1}{(p-1)^s}\right).$$

(c) By writing

$$\frac{1}{(p-1)^s} - \frac{1}{p^s} = s \int_{p-1}^{p} u^{-s-1}\, du,$$

show that the above is $\ll p^{-\sigma-1}$ for any fixed s.

(d) Let $\mathcal{K}$ be a compact set in the complex plane, and let $\sigma_0 = \min_{s \in \mathcal{K}} \sigma$. Show that $(p-1)^{-s} - p^{-s} \ll p^{-\sigma_0-1}$ uniformly for $s \in \mathcal{K}$.

(e) Show that the product $G(s)$ converges locally uniformly in the half-plane $\sigma > 0$, and hence represents an analytic function in this region.

(f) Show that $G(1) = \zeta(2)\zeta(3)/\zeta(6)$.

(g) Use the Wiener–Ikehara theorem (Theorem 8.6) to show that the number of integers k such that $\varphi(k) \le x$ is asymptotic to $G(1)x$ as $x \to \infty$.

3. Show that Corollary 8.8 still holds if the hypothesis $a_n \ge 0$ is replaced by the weaker hypothesis that there is a constant C such that $a_n \ge C$ for all n.

4. Let $\sigma_s(n) = \sum_{d \mid n} d^s$, and let $c_q(n)$ be Ramanujan's sum, as discussed in Section 4.1.

(a) Show that if n is a positive integer, then

$$\sum_{q=1}^{\infty} \frac{c_q(n)}{q^s} = \frac{\sigma_{1-s}(n)}{\zeta(s)} \qquad (\sigma > 1).$$

(b) Show that if n is a fixed positive integer, then $\sum_{q \le x} c_q(n) = o(x)$ as $x \to \infty$.

(c) Show that if n is a positive integer, then

$$\sum_{q \le x} c_q(n)\left[\frac{x}{q}\right] = \sum_{\substack{d \mid n \\ d \le x}} d.$$

(d) By Axer's theorem, or otherwise, show that if n is a positive integer, then

$$\sum_{q=1}^{\infty} \frac{c_q(n)}{q} = 0.$$

(See also Exercise 4.1.8.)

5. (Graham & Vaaler 1981) Let $f_+(x)$ and $f_-(x)$ be as in Lemma 8.5.

(a) Use the Poisson summation formula to show that

$$\sum_{n=-\infty}^{\infty} f_+(n/T) = T \sum_{k=-\infty}^{\infty} \widehat{f}_+(kT).$$

(b) Explain why the right-hand side above is $= T\widehat{f}_+(0) = T\int_{\mathbb{R}} f_+(x)\,dx$.
(c) Explain why the left-hand side above is $\geq (1 - e^{-1/T})^{-1}$.
(d) Deduce that

$$\int_{\mathbb{R}} f_+(x)\,dx \geq \frac{1}{T(1 - e^{-1/T})}\,.$$

(e) Suppose that $T \geq 2$. Show that the right-hand side above is $= 1 + 1/(2T) + O(1/T^2)$.
(f) Show similarly that

$$\int_{\mathbb{R}} f_-(x)\,dx \leq \frac{1}{T(e^{1/T} - 1)},$$

and that the right-hand side is $= 1 - 1/(2T) + O(1/T^2)$ when $T \geq 2$.

8.4 Beurling's generalized prime numbers

One of the most valuable generalizations of the Prime Number Theorem is to algebraic number fields. Suppose that K is an algebraic number field of degree d over the rationals, and let $\mathcal{O}_K$ denote the ring of algebraic integers in K. For some fields K the members of $\mathcal{O}_K$ factor uniquely into primes, but in general this is not the case. However, it is always true that ideals in $\mathcal{O}_K$ factor uniquely into prime ideals. For an ideal $\mathfrak{a}$ of $\mathcal{O}_K$, let $N(\mathfrak{a})$ denote its norm, which is to say the size of the quotient ring $\mathcal{O}_K/\mathfrak{a}$. For $\sigma > 1$ we can define the *Dedekind zeta function* of K by the absolutely convergent series

$$\zeta_K(s) = \sum_{\mathfrak{a}} N(\mathfrak{a})^{-s}.$$

This is an ordinary Dirichlet series, since the $N(\mathfrak{a})$ are positive integers, and thus the above can be written in the form $\sum a_n n^{-s}$ where a_n is the number of ideals with norm n.

Counting ideals $\mathfrak{a}$ with $N(\mathfrak{a}) \leq x$ is rather like counting rational integers. The ideals can be parametrized by the points of a lattice in $\mathbb{R}^d$, so one is counting lattice points in a certain region, which is approximately the volume of that region, and thus it can be shown that the number $I(x)$ of ideals $\mathfrak{a}$ with $N(\mathfrak{a}) \leq x$ is

$$I(x) = cx + O\left(x^{1-1/d}\right) \tag{8.43}$$

where $c = c(K)$ is a certain positive constant, called the *ideal density*. Here the implicit constant may also depend on K, which we assume is fixed. By Theorem 1.3 it follows that

$$\zeta_K(s) = s\int_1^\infty I(x)x^{-s-1}\,dx = \frac{cs}{s-1} + s\int_1^\infty (I(x) - cx)x^{-s-1}\,dx.$$

Since this latter integral is uniformly convergent for $\sigma > 1 - 1/d + \delta$, we deduce that $\zeta_K(s)$ is analytic in the half-plane $\sigma > 1 - 1/d$ apart from a simple pole at $s = 1$ with residue c. Moreover, we see that if δ is fixed, $\delta > 0$, then $\zeta_K(s) \ll |t|$ uniformly for $\sigma \geq 1 - 1/d + \delta, |t| \geq 1$.

If $\mathfrak{a}$ and $\mathfrak{b}$ are two ideals in $\mathcal{O}_K$, then

$$N(\mathfrak{a}\mathfrak{b}) = N(\mathfrak{a})N(\mathfrak{b}). \tag{8.44}$$

Hence $\zeta_K(s)$ has an Euler product formula

$$\zeta_K(s) = \prod_{\mathfrak{p}} (1 - N(\mathfrak{p})^{-s})^{-1}$$

for $\sigma > 1$. On taking logarithmic derivatives we also see that

$$-\frac{\zeta_K'}{\zeta_K}(s) = \sum_{\mathfrak{a}} \Lambda(\mathfrak{a})N(\mathfrak{a})^{-s}$$

where $\Lambda(\mathfrak{a}) = \log N(\mathfrak{p})$ if $\mathfrak{a} = \mathfrak{p}^k$, $\Lambda(\mathfrak{a}) = 0$ otherwise. Thus, as in Lemma 6.5,

$$\Re\left(-3\frac{\zeta_K'}{\zeta_K}(\sigma) - 4\frac{\zeta_K'}{\zeta_K}(\sigma + it) - \frac{\zeta_K'}{\zeta_K}(\sigma + 2it)\right) \geq 0$$

for $\sigma > 1$ and any real t. Also as in Chapter 6 we may derive a zero-free region for $\zeta_K(s)$, namely that $\zeta_K(s) \neq 0$ provided that $\sigma > 1 - c/\log \tau$. Here, as before, $\tau = |t| + 4$, and c is a constant depending on K. Continuing as in Chapter 6, we can derive estimates analogous to those in Theorem 6.7, but with constants depending on K, and we may use our quantitative version of Perron's formula (Theorem 5.2) to establish a quantitative version of the *Prime Ideal Theorem*:

Theorem 8.9 (Landau) *Let K be an algebraic number field of finite degree over $\mathbb{Q}$, and let $\mathcal{O}_K$ denote the ring of algebraic integers in K. Then for $x \geq 2$ the number of prime ideals $\mathfrak{p}$ in $\mathcal{O}_K$ such that $N(\mathfrak{p}) \leq x$ is* $\mathrm{li}(x) + O_K(x \exp(-c\sqrt{\log x}))$ *where c depends on K.*

It is notable that the chain of reasoning we have just described depends only on the estimate (8.43) and the identity (8.44). Thus the entire situation could be abstracted as follows. Suppose we have a sequence $\mathcal{P}$ of real numbers p_i such that $1 < p_1 \leq p_2 \leq \cdots$ and $p_i \to \infty$. We call these numbers 'generalized primes'. We form products of powers of these numbers, $p_1^{a_1} p_2^{a_2} \cdots p_k^{a_k}$, and call such products 'generalized integers'. Let $N(x)$ denote the number of such products whose value does not exceed x. If

$$N(x) = cx + O(x^\theta) \tag{8.45}$$

for some $c > 0$ and $\theta < 1$, then by the reasoning we have outlined it follows that the number $P(x)$ of generalized primes p_i such that $p_i \leq x$ is $\mathrm{li}(x) + O(x \exp(-c\sqrt{\log x}))$.

The integers $\mathbb{Z}$ form an additive group, a cyclic group generated by the number 1. Moreover, the positive integers form a multiplicative semigroup with the primes as generators. From the additive property of the integers we know that $[x] = x + O(1)$, which is a strong form of (8.45). However, it is now quite clear that our proof of the Prime Number Theorem requires no further knowledge of the additive nature of the integers beyond this estimate.

We have seen that the estimate (8.45) gives a generalization of the Prime Number Theorem with the classical error term. We now consider the issue of how much this hypothesis can be weakened, if the goal is only to obtain a generalization of (8.1), namely that $P(x) \sim x/\log x$ as $x \to \infty$.

Theorem 8.10   (Beurling) *Let* $\mathcal{P} = \{p_i\}$ *where* $1 < p_1 \leq p_2 \leq \cdots$ *and* $p_i \to \infty$, *and let* $N(x)$ *denote the number of products* $p_1^{a_1} p_2^{a_2} \cdots p_k^{a_k} \leq x$ *where the* a_i *are non-negative integers. Suppose that there is a positive constant c such that*

$$N(x) = cx + O\left(\frac{x}{(\log x)^\gamma}\right) \tag{8.46}$$

for $x \geq 2$. *Let* $P(x)$ *denote the number of members of* $\mathcal{P}$ *not exceeding* x. *If* $\gamma > 3/2$, *then*

$$P(x) \sim \frac{x}{\log x} \tag{8.47}$$

as $x \to \infty$.

Proof Let $\mathcal{N} = \{n_j\}$ where $1 = n_1 < n_2 \leq n_3 \leq \cdots$ are the generalized integers, and for $\sigma > 1$ let

$$\zeta_{\mathcal{P}}(s) = \sum_{n \in \mathcal{N}} n^{-s}.$$

Since the $n \in \mathcal{N}$ are not necessarily rational integers, the above is not necessarily an ordinary Dirichlet series, but it is an example of a 'generalized Dirichlet series'. In any case it is an absolutely convergent series and by integration by parts as in the proof of Theorem 1.3 we see that

$$\zeta_{\mathcal{P}}(s) = \int_{1^-}^{\infty} u^{-s} \, dN(u) = s \int_1^{\infty} N(u)u^{-s-1} \, du.$$

We subtract cu from $N(u)$ to see that

$$\zeta_{\mathcal{P}}(s) = \frac{cs}{s-1} + s \int_1^{\infty} (N(u) - cu)u^{-s-1} \, du.$$

From (8.46) we know that $\int_1^\infty |N(u) - cu|u^{-2}\,du < \infty$. Hence the integral above is uniformly convergent for $\sigma \geq 1$, and consequently it is continuous in this closed half-plane. Thus we can extend the definition of $\zeta_{\mathcal{P}}(s)$ so that $\zeta_{\mathcal{P}}(s) = c/(s-1) + r_0(s)$ and $r_0(s)$ is continuous for $\sigma \geq 1$. To bound the modulus of continuity of $r_0(s)$ we differentiate. Thus $\zeta_{\mathcal{P}}'(s) = -c/(s-1)^2 + r_1(s)$ for $\sigma > 1$ where

$$r_1(s) = r_0'(s) = \int_1^\infty (N(u) - cu)u^{-s-1}\,du - s\int_1^\infty (N(u) - cu)(\log u)u^{-s-1}\,du.$$

If (8.46) holds with $\gamma > 2$, then $\int_1^\infty |N(u) - cu|(\log u)u^{-2}\,du < \infty$ and then $r_1(s)$ is continuous in the closed half-plane $\sigma \geq 1$. When γ is smaller, however, the situation is more delicate. From now on we assume, as we may, that $3/2 < \gamma \leq 2$. Since

$$\int_2^\infty (\log u)^{1-\gamma} u^{-\sigma}\,du = \int_{\log 2}^\infty v^{1-\gamma} e^{-(\sigma-1)v}\,dv$$

$$= (\sigma - 1)^{\gamma-2} \int_{(\sigma-1)\log 2}^\infty u^{1-\gamma} e^{-u}\,du$$

$$\ll (\sigma - 1)^{-\frac{1}{2}+\eta},$$

where $\eta = \eta(\gamma) > 0$, from (8.46) we deduce that $r_1(s) \ll (\sigma - 1)^{-\frac{1}{2}+\eta}$ uniformly for $\sigma > 1$. Consequently, if t is fixed, $t \neq 0$, then

$$\zeta_{\mathcal{P}}(\sigma + it) - \zeta_{\mathcal{P}}(1 + it) = \int_1^\sigma \zeta_{\mathcal{P}}'(\alpha + it)\,d\alpha \ll (\sigma - 1)^{\frac{1}{2}+\eta} \quad (8.48)$$

for $\sigma > 1$, σ near 1.

Next we use the above estimate to show that

$$\zeta_{\mathcal{P}}(1 + it) \neq 0 \quad (8.49)$$

when t is real, $t \neq 0$. By mimicking the proof of the usual Euler product formula for $\zeta(s)$, we see that

$$\zeta_{\mathcal{P}}(s) = \prod_{p \in \mathcal{P}} (1 - p^{-s})^{-1}$$

for $\sigma > 1$. This product is absolutely convergent, and each factor is non-zero, so $\zeta_{\mathcal{P}}(s) \neq 0$ for $\sigma > 1$, and indeed we may write

$$\log \zeta_{\mathcal{P}}(s) = \sum_{p \in \mathcal{P}} \sum_{r=1}^\infty \frac{1}{r} p^{-rs}. \quad (8.50)$$

Instead of the cosine polynomial $3 + 4\cos\theta + \cos 2\theta$ used in Chapter 6, we must now employ a non-negative cosine polynomial $a_0 + \sum_{k=1}^K a_k \cos k\theta$ for which the ratio a_1/a_0 is larger. As we observed in Section 6.1, it is always the case that $a_1 < 2a_0$, but we can make a_1 as close to $2a_0$ as we wish by using the Fejér kernel $\Delta_K(\theta)$ with K large, since

$$\Delta_K(\theta) = 1 + 2\sum_{k=1}^{K}\left(1 - \frac{k}{K}\right)\cos 2\pi k\theta = \frac{1}{K}\left(\frac{\sin \pi K\theta}{\sin \pi\theta}\right)^2 \geq 0.$$

Hence if $\sigma > 1$, then

$$\prod_{k=-K}^{K} \zeta_{\mathcal{P}}(\sigma + ikt)^{(1-|k|/K)} = \exp\left(\sum_{p\in\mathcal{P}}\sum_{r=1}^{\infty}\frac{1}{rp^{r\sigma}}\sum_{k=-K}^{K}(1 - |k|/K)p^{-irkt}\right)$$

$$= \exp\left(\sum_{p\in\mathcal{P}}\sum_{r=1}^{\infty}\frac{1}{rp^{r\sigma}}\Delta_K\left(rt(\log p)/(2\pi)\right)\right).$$

Now $\zeta_{\mathcal{P}}(\sigma - it) = \overline{\zeta_{\mathcal{P}}(\sigma + it)}$, so that $|\zeta_{\mathcal{P}}(\sigma - it)| = |\zeta_{\mathcal{P}}(\sigma + it)|$. Also, $\Delta_K(\theta) \geq 0$ for all θ. Hence from the above we see that

$$\zeta_{\mathcal{P}}(\sigma)\prod_{k=1}^{K} |\zeta_{\mathcal{P}}(\sigma + ikt)|^{2(1-k/K)} \geq 1.$$

Suppose that t is a fixed, non-zero real number. As σ tends to 1 from above, the numbers $|\zeta_{\mathcal{P}}(\sigma + ikt)|$ tend to finite limits, and $\zeta_{\mathcal{P}}(\sigma) \asymp 1/(\sigma - 1)$. Thus

$$|\zeta_{\mathcal{P}}(\sigma + it)| \gg (\sigma - 1)^{\frac{K}{2(K-1)}}$$

as $\sigma \to 1^+$. Here the implicit constant may depend not only on $\mathcal{P}$ but also on t. Suppose now that $\zeta_{\mathcal{P}}(1 + it) = 0$. Then from (8.48) we have $\zeta_{\mathcal{P}}(\sigma + it) \ll (\sigma - 1)^{\frac{1}{2}+\eta}$ as $\sigma \to 1^+$. This contradicts the lower bound above if K is large enough, say $K > 1 + \frac{1}{2\eta}$. Hence $\zeta(1 + it) \neq 0$, as desired.

For $n \in \mathcal{N}$ let $\Lambda(n) = \log p$ if $n = p^r$ and $p \in \mathcal{P}$, $\Lambda(n) = 0$ otherwise. On differentiating (8.50) we see that

$$-\frac{\zeta_{\mathcal{P}}'}{\zeta_{\mathcal{P}}}(s) = \sum_{n\in\mathcal{N}}\Lambda(n)n^{-s}$$

for $\sigma > 1$. Set

$$S(x) = \sum_{\substack{n\in\mathcal{N}\\n\leq x}}\Lambda(n).$$

Suppose for the moment that $\gamma > 2$. Then $r_0(s)$ and $r_1(s)$ are both continuous in the closed half-plane $\sigma \geq 1$, and then

$$-\frac{\zeta_{\mathcal{P}}'}{\zeta_{\mathcal{P}}}(s) = \frac{1}{s-1} + r(s)$$

where

$$r(s) = -\frac{r_0(s) + (s-1)r_1(s)}{(s-1)\zeta_{\mathcal{P}}(s)}$$

is continuous in the closed half-plane $\sigma \geq 1$. Then by the Wiener–Ikehara theorem it follows that $S(x) \sim x$ as $x \to \infty$. Under the weaker hypothesis that $3/2 < \gamma \leq 2$ we are no longer able to guarantee that $r_1(s)$ is continuous, but by Plancherel's identity it is bounded in mean-square. Thus, below, we follow the lines of the proof of the Wiener–Ikehara theorem, but with an appeal to Plancherel's identity where continuity had sufficed before.

Suppose that $\delta > 0$, that T is a large positive number, and that $E(u)$ is defined as in Lemma 8.5. Then

$$\sum_{\substack{n \in \mathcal{N} \\ n \leq x}} \Lambda(n) n^{-\delta} = x \sum_{n \in \mathcal{N}} \Lambda(n) n^{-1-\delta} E(\log n - \log x)$$

which by Lemma 8.5 is

$$\leq x \sum_{n \in \mathcal{N}} \Lambda(n) n^{-1-\delta} f_+(\log n - \log x)$$

$$\leq x \sum_{n \in \mathcal{N}} \Lambda(n) n^{-1-\delta} \int_{-T}^{T} \widehat{f}_+(t) \left(\frac{x}{n}\right)^{-2\pi i t} dt$$

$$= -x \int_{-T}^{T} \widehat{f}_+(t) x^{-2\pi i t} \frac{\zeta'_P}{\zeta_P} (1 + \delta - 2\pi i t) \, dt. \qquad (8.51)$$

As for the main term, we note that similarly

$$\int_1^{\infty} u^{-1-\delta} f_+(\log u - \log x) \, du = \int_1^{\infty} u^{-1-\delta} \int_{-T}^{T} \widehat{f}_+(t) \left(\frac{x}{u}\right)^{-2\pi i t} du \, dt$$

$$= \int_{-T}^{T} \widehat{f}_+(t) x^{-2\pi i t} \int_1^{\infty} u^{-1-\delta+2\pi i t} \, du \, dt$$

$$= \int_{-T}^{T} \widehat{f}_+(t) x^{-2\pi i t} \frac{1}{\delta - 2\pi i t} \, dt.$$

We multiply both sides of this by x and combine with (8.51) to see that

$$\sum_{\substack{n \in \mathcal{N} \\ n \leq x}} \Lambda(n) n^{-\delta} \leq x \int_1^{\infty} u^{-1-\delta} f_+(\log u - \log x) \, du$$

$$\qquad\qquad\qquad\qquad\qquad\qquad\qquad\qquad\qquad (8.52)$$

$$+ x \int_{-T}^{T} \widehat{f}_+(t) x^{-2\pi i t} \left(-\frac{\zeta'_P}{\zeta_P} (1 + \delta - 2\pi i t) - \frac{1}{\delta - 2\pi i t} \right) dt.$$

By using our formulæ for $r_i(s)$ in terms of integrals we see that we may write

$$r_1(s) = r'_0(s) = -s J(s) + \frac{r_0(s) - c}{s}$$

where

$$J(s) = \int_1^{\infty} (N(u) - cu) (\log u) u^{-s-1} \, du,$$

and

$$-\zeta_P'(s) = \frac{c}{(s-1)^2} - \frac{r_0(s) - c}{s} + s J(s).$$

Thus

$$-\frac{\zeta_P'}{\zeta_P}(s) - \frac{1}{s-1} = \frac{c(s-1) + (1-2s)r_0(s)}{s(s-1)\zeta_P(s)} + \frac{s}{\zeta_P(s)} J(s)$$

and by splitting the integral at X, where X is a large parameter we have

$$-\frac{\zeta_P'}{\zeta_P}(s) - \frac{1}{s-1} = C(s) + R(s)$$

where

$$R(s) = \int_X^\infty (N(u) - cu)(\log u) u^{-s-1} \, du$$

and $C(s)$ is continuous for $\sigma \geq 1$. We consider first the contribution of the remainder $R(s)$ to (8.52). By the Cauchy–Schwartz inequality we see that

$$\left| \int_{-T}^T \widehat{f}_+(t) x^{-2\pi i t} R(1 + \delta - 2\pi i t) \, dt \right|^2$$

(8.53)

$$\leq \int_{-T}^T \left| \widehat{f}_+(t) \frac{1 + \delta - 2\pi i t}{\zeta_P(1 + \delta - 2\pi i t)} \right|^2 dt \int_{-T}^T \left| \int_X^\infty \frac{(N(u) - cu)(\log u)}{u^{2+\delta-2\pi i t}} \, du \right|^2 dt.$$

In Theorem 5.4 we take $\sigma = 1 + \delta$ and $w(u) = (N(u) - cu)\log u$ for $u \geq X$, $w(u) = 0$ otherwise. Thus we see that

$$\int_{-\infty}^\infty \left| \int_X^\infty (N(u) - cu)(\log u) u^{-2-\delta+2\pi i t} \, du \right|^2 dt$$
$$= \int_X^\infty (N(u) - cu)^2 (\log u)^2 u^{-3-2\delta} \, du,$$

which by (8.46) is

$$\ll \int_X^\infty u^{-1} (\log u)^{2-2\gamma} \, du \ll_\gamma (\log X)^{3-2\gamma}$$

uniformly for $\delta > 0$. The first integral on the right-hand side of (8.53) is also uniformly bounded as δ tends to 0, since $\zeta_P(1 + it) \neq 0$. Thus the contribution of $R(s)$ to (8.52) is $\ll_\gamma (\log X)^{3/2-\gamma}$, uniformly for $\delta > 0$. Hence if we let δ tend to 0 from above in (8.52), and divide through by x, we find that

$$\frac{S(x)}{x} \leq \int_1^\infty u^{-1} f_+(\log u - \log x) \, du + \int_{-T}^T \widehat{f}_+(t) x^{-2\pi i t} C(1 - 2\pi i t) \, dt$$
$$+ O_\gamma\left((\log X)^{3/2-\gamma}\right).$$

As x tends to infinity, the first integral on the right tends to $\int_{-\infty}^{\infty} f_+(v)\,dv$. Since $\widehat{f}_+(t)C(1 - 2\pi i t)$ is a continuous function of t, by the Riemann–Lebesgue lemma the second integral on the right tends to 0 as x tends to infinity. Hence

$$\limsup_{x\to\infty} \frac{S(x)}{x} \le \int_{-\infty}^{\infty} f_+(v)\,dv + O_\gamma\big((\log X)^{3/2-\gamma}\big).$$

By Lemma 8.5 we know that the integral on the right is $< 1 + \varepsilon$ if T is sufficiently large. Since X may also be taken arbitrarily large, we conclude that the limsup above is ≤ 1. By a similar argument with f_+ replaced by f_-, we find that the corresponding liminf is ≥ 1, so we have the generalized Prime Number Theorem in the form $S(x) \sim x$. By integrating by parts we obtain the desired relation (8.47). $\square$

We now show that the exponent $3/2$ is critical in Beurling's theorem.

Theorem 8.11 *The primes $\mathcal{P}$ can be chosen in such a way that* (8.46) *holds with* $\gamma = 3/2$ *but* (8.47) *fails.*

The general idea is that if $\zeta_\mathcal{P}(s)$ has a simple pole at $s = 1$ and zeros of multiplicity $1/2$ at $1 \pm ia$, say

$$\zeta_\mathcal{P}(s) = \frac{(s - 1 - ia)^{1/2}(s - 1 + ia)^{1/2}}{s - 1} H(s) \tag{8.54}$$

where $H(s)$ is analytic for $\sigma > \theta, \theta < 1$, then we can express $N(x)$ by Perron's formula applied to $\zeta_\mathcal{P}(s)$. After moving the contour to the left, we would find that the residue at $s = 1$ gives rise to the main term cx, and the loop of contour around the branch points at $1 \pm ia$ give oscillatory terms of size $x/(\log x)^{3/2}$. On the other hand,

$$-\frac{\zeta_\mathcal{P}'}{\zeta_\mathcal{P}}(s) = \frac{1}{s - 1} - \frac{1}{2(s - 1 - ia)} - \frac{1}{2(s - 1 + ia)} - \frac{H'}{H}(s),$$

which suggests that S is approximately

$$x - \frac{x^{1+ia}}{2(1 + ia)} - \frac{x^{1-ia}}{2(1 - ia)}.$$

This is of the order of magnitude x but not asymptotic to x. It is of course essential that the above main term should be increasing; we note that its derivative is $1 - \cos(a \log x) \ge 0$. For a rigorous construction we begin by defining primes so that $S(x)$ approximates this main term, and then we show that the resulting $\zeta_\mathcal{P}(s)$ satisfies (8.54).

Proof   Let a be a fixed positive real number, and set

$$f(x) = \int_1^x \frac{1 - \cos(a \log u)}{\log u}\,du.$$

We note that this function is increasing and tends to infinity with x. Hence for each positive integer j there is a unique real number p_j such that $f(p_j) = j$. If $p_j \leq x < p_{j+1}$, then $P(x) = j$ and $j \leq f(x) < j + 1$; hence $P(x) = [f(x)]$. By integration by parts we see that

$$\int_2^x \frac{u^{i\alpha}}{\log u}\, du = \frac{x^{1+i\alpha}}{(1+i\alpha)\log x} + O\left(\frac{x}{(\log x)^2}\right).$$

By taking $\alpha = -a, 0, a$, and combining, we see that

$$f(x) = \left(1 - \frac{x^{ia}}{2(1+ia)} - \frac{x^{-ia}}{2(1-ia)}\right)\frac{x}{\log x} + O\left(\frac{x}{(\log x)^2}\right),$$

and consequently

$$\liminf_{x\to\infty} \frac{P(x)}{x/\log x} = 1 - \frac{1}{\sqrt{1+a^2}}, \qquad \limsup_{x\to\infty} \frac{P(x)}{x/\log x} = 1 + \frac{1}{\sqrt{1+a^2}}.$$

Clearly

$$\sum_{\substack{p\in P \\ p\leq x}} \log p = \int_1^x \log u\, d[f(u)]$$

$$= \int_1^x \log u\, df(u) - \int_1^x \log u\, d\{f(u)\}$$

$$= \int_1^x 1 - \cos(a\log u)\, du - \left[\{f(u)\}\log u\Big|_1^x + \int_1^x \frac{\{f(u)\}}{u}\, du\right]$$

$$= x - \frac{x^{1+ia}}{2(1+ia)} - \frac{x^{1-ia}}{2(1-ia)} + O(\log x),$$

and hence

$$S(x) = x - \frac{x^{1+ia}}{2(1+ia)} - \frac{x^{1-ia}}{2(1-ia)} + O\left(x^{1/2}\right).$$

Let $r(x)$ denote this last error term. Then for $\sigma > 1$,

$$-\frac{\zeta_P'}{\zeta_P}(s) = \int_1^\infty u^{-s}\, dS(u)$$

$$= \frac{1}{s-1} - \frac{1}{2(s-1-ia)} - \frac{1}{2(s-1+ia)} + g(s)$$

where $g(s)$ is analytic for $\sigma > 1/2$. Hence

$$\log \zeta_P(s) = -\log(s-1) + \frac{1}{2}\log(s-1-ia) + \frac{1}{2}\log(s-1+ia) + G(s)$$

where $G'(s) = -g(s)$, and so we have (8.54) with $H(s) = e^{G(s)}$.

To complete the proof we need not only (8.54) but also an estimate of the size of $\zeta_P(s)$ when $\sigma < 1$. To this end we mimic the approach used to estimate

$1/\zeta(s)$ in Theorem 6.7. Since $P(x) \ll x/\log x$ it follows that $\log \zeta_P(1 + \delta + it) \ll \log 1/\delta$ uniformly for $0 < \delta \le 1/2$. If $t \ge 4 + a$ and $1 - 1/\log t \le \sigma \le 1 + 1/\log t$, then

$$-\frac{\zeta_P'}{\zeta_P}(s) = \sum_{\substack{n \le t^2 \\ n \in \mathcal{N}}} \Lambda(n)n^{-s} + \int_{t^2}^{\infty} u^{-s} \, dS(u).$$

Here the sum is

$$\ll \sum_{\substack{n \le t^2 \\ n \in \mathcal{N}}} \frac{\Lambda(n)}{n} \ll \log t,$$

and the integral is

$$\frac{t^{2(1-s)}}{s-1} - \frac{t^{2(1+ia-s)}}{2(s-1-ia)} - \frac{t^{2(1-ia-s)}}{2(s-1+ia)} - \frac{r(t^2)}{t^{2s}} + s \int_{t^2}^{\infty} r(u)u^{-s-1} \, du \ll 1,$$

so that

$$\log \zeta_P(s) = -\int_{\sigma}^{1+1/\log t} \frac{\zeta_P'}{\zeta_P}(\alpha + it)d\alpha + \log \zeta_P(1 + 1/\log t + it)$$
$$\ll 1 + \log \log t$$

for $\sigma \ge 1 - 1/\log t$. Hence there is a constant A such that $\zeta_P(s) \ll (\log t)^A$ for $\sigma \ge 1 - 1/\log t, t \ge 4 + a$.

We now estimate $N(x)$ by taking an inverse Mellin transform of $\zeta_P(s)$. However, the truncated Perron formula (Corollary 5.3) is not so useful since we lack information concerning the number of generalized integers in a short interval. To avoid this difficulty we use Cesàro weights as discussed in Section 5.1, by means of which we see that if $b > 1$ and $h > 0$, then

$$\frac{1}{2\pi i h} \int_{b-i\infty}^{b+i\infty} \zeta_P(s)\frac{(x+h)^{s+1} - x^{s+1}}{s(s+1)} \, ds = \sum_{n \in \mathcal{N}} w_+(n)$$

where

$$w_+(u) = \begin{cases} 1 & (u \le x), \\ (x+h-u)/h & (x < u \le x+h), \\ 0 & (u > x+h). \end{cases}$$

We now pull the contour to the left. In view of (8.54), at $s = 1$ we encounter a simple pole with residue $c(x + h/2)$ where $c = aH(1)$. Because of the branch points at $1 \pm ia$, we slit the plane by the segments $\sigma \pm ia$ for $-\infty < \sigma \le 1$. Our contour follows the upper and lower sides of these segments; the integral along these loops is $\ll \int_{-\infty}^{1}(x+h)^\sigma (1-\sigma)^{1/2} \, d\sigma \ll x/(\log x)^{3/2}$. By taking more care, and using Theorem C.3, we could obtain oscillatory main terms of this order of magnitude. On the rest of the contour we estimate the integral as in the proof of the Prime Number Theorem, and thus we see that

$$N(x) \le \sum_{n \in \mathcal{N}} w_+(n) = cx + \frac{1}{2}ch + O\left(\frac{x}{(\log x)^{3/2}}\right)$$
$$+ O\left(\frac{x^2}{h}\exp\left(-C\sqrt{\log x}\right)\right).$$

On taking $h = x/(\log x)^2$ we obtain an upper bound of the desired type. To obtain a corresponding lower bound we argue similarly from the formula

$$\frac{1}{2\pi i h}\int_{b-i\infty}^{b+i\infty} \zeta_P(s)\frac{x^{s+1} - (x-h)^{s+1}}{s(s+1)}\,ds = \sum_{n \in \mathcal{N}} w_-(n)$$

where

$$w_-(u) = \begin{cases} 1 & (u \le x - h), \\ (x-u)/h & (x - h < u \le x), \\ 0 & (u \ge x). \end{cases}$$

$\square$

8.5 Notes

Section 8.1. Historical accounts of the development of prime number theory and of the various proofs of the Prime Number Theorem have been given by Bateman & Diamond (1996), Narkiewicz (2000), and by Schwarz (1994). Axer's theorem originates in Axer (1911). The definitive account of Axer's theorem is that of Landau (1912).

Section 8.2. In former times, an argument was considered to be 'non-elementary' if it involved Cauchy's theorem or Fourier inversion. Prior to Selberg's elementary proof of the Prime Number Theorem, a distinction was drawn between those results that could be obtained by elementary arguments, and those that could not. Selberg's elementary proof rendered the terminology nugatory.

Theorem 8.3 and a deduction of the Prime Number Theorem occur in Selberg (1949). There are a number of variants of the less than straightforward Tauberian process used in the deduction; see, for example, Erdős (1949), Wright (1952), and Levinson (1969). For a historical review of elementary proofs of the Prime Number Theorem see Goldfeld (2004).

Quantitative estimates of the form

$$\pi(x) = \mathrm{li}(x)(1 + O((\log x)^{-a}))$$

have been derived by elementary methods. van der Corput (1956) obtained $a = 1/200$, Kuhn (1955) obtained $a = 1/10$, Breusch $a = 1/6 - \varepsilon$, and

Wirsing (1962) $a = 3/4$. Then Bombieri (1962a,b) and Wirsing (1964) showed that the above is true for any fixed positive a. Subsequently, elementary techniques have been used to show that

$$\pi(x) = \text{li}(x) + O(x \exp(-c(\log x)^{-b}))$$

for various values of b. Diamond & Steinig (1970) obtained $b = 1/7 - \varepsilon$, Lavrik & Sobirov (1973) $b = 1/6 - \varepsilon$, and Srinivasan & Sampath (1988) $b = 1/6$. Although the estimates obtained by elementary methods have thus far been weaker than those derived by analytic means, we have no reason to believe that this will always be the case.

Section 8.3. The theorem of Ikehara (1931) represented a major advance, because it gave for the first time a Tauberian theorem that could be used to prove the Prime Number Theorem without imposing growth conditions on the Dirichlet series generating function. Ikehara assumed that $\alpha(s) - c/(s - 1)$ is analytic in the closed half-plane $\sigma \geq 1$. Wiener (1932) showed that mere continuity is enough, but this is of lesser significance, since still weaker hypotheses are sufficient – see Korevaar (2006).

The heart of the Wiener–Ikehara proof of the Prime Number Theorem is Lemma 8.5, which has the effect of enabling one to reduce directly to a use of the Riemann–Lebesgue lemma on a finite section of the line $\Re s = 1$. In the proof of Lemma 8.5 we see that it suffices to take $T = C/\varepsilon$, and from Exercise 8.3.5 we see that it is necessary to take $T \geq 1/(2\varepsilon) + O(1)$. Graham & Vaaler (1981) have shown that f_+ and f_- can be constructed so that equality is achieved in Exercise 8.3.5(e),(g).

Lemma 8.5, with T small and ε large, is also useful for proving interesting theorems of Fatou and Riesz. Fatou (1906) showed that if $a_n = o(1)$, then the series $f(z) = \sum a_n z^n$ converges at any point of the circle $|z| = 1$ at which f is analytic. Landau (1910, Section 10) gives Riesz's proof that if $\sum_{n \leq x} a_n = o(x)$, then the Dirichlet series $\alpha(s) = \sum a_n n^{-s}$ converges at every point of the line $\sigma = 1$ at which $\alpha(s)$ is analytic. Riesz (1916) extended this to generalized Dirichlet series.

For detailed discussion of Wiener's Tauberian theorem, the Ikehara theorem, and Tauberian theorems associated with the elementary proof of the Prime Number Theorem see Pitt (1958).

Section 8.4. The concept of generalized primes are introduced in Beurling (1937). The hypothesis of Theorem 8.10 can be weakened: Kahane (1997) has shown that if

$$\int_1^\infty (N(x) - cx)^2 x^{-3} (\log x)^2 \, dx < \infty,$$

then (8.47) still follows.

Theorem 8.11 is due to Diamond (1970b). Diamond (1973) also showed that if (8.46) holds with $\gamma > 1$, then one has an estimate $P(x) \ll x/\log x$ of the Chebyshev kind. Zhang (1993) showed that the hypothesis here can be weakened to

$$\int_1^\infty \sup_{y \leq x} \frac{|N(y) - cy|}{y} \frac{dx}{x} < \infty.$$

In the negative direction, Hall (1973) showed that if $\gamma < 1$, then the hypothesis (8.46) is not sufficient to imply a Chebyshev estimate. Also, Kahane (1998) has shown that the hypothesis

$$\int_1^\infty \frac{|N(x) - cx|}{x^2} dx < \infty$$

does not imply a Chebyshev estimate. Zhang (1987b) has shown that if (8.46) holds with $\gamma > 1$, then

$$\sum_{\substack{n \leq x \\ n \in \mathcal{N}}} \mu(n) = o(x).$$

In the classical context, the above is equivalent – by Axer's theorem – to the Prime Number Theorem. However, in the Beurling situation, if $1 < \gamma \leq 3/2$, the above holds but PNT may fail.

Nyman (1949) showed that if (8.46) holds for all γ (with the implicit constant depending on γ), then $P(x) = \mathrm{li}(x) + O_c(x/(\log x)^c)$ for all c. Malliavin (1961) showed that if $N(x) = cx + O(x \exp(-(\log x)^a))$ where $0 < a < 1$, then $\pi(x) = \mathrm{li}(x) + O(x \exp(-(\log x)^b))$ with $b = a/10$. Both these authors proved converse theorems in which an estimate for $P(x)$ is used to establish a corresponding estimate for $N(x)$, but those results have since been sharpened by Diamond (1970a). It is now known that the method of Landau, in which one starts from (8.45) to derive the indicated error term, is sharp: Diamond, Montgomery & Vorhauer (2006) have shown that if θ is given, $1/2 < \theta < 1$, then there exists a Beurling system for which (8.45) holds, but $P(x) - \mathrm{li}(x) = \Omega_\pm(x \exp(-c\sqrt{\log x}))$.

Some of the ideas and themes developed in connection with the Prime Number Theorem have had ramifications in surprisingly diverse areas. See, for example, Hejhal's expositions (1976, 1983) of Selberg's trace formula for $PSL(2, \mathbb{R})$, and the monograph of Parry & Pollicott (1990) on the periodic orbit structure of hyperbolic dynamics.

Some writers avoid the term 'Beurling', and instead discuss 'arithmetic semigroups'. The mathematics is the same in either case. For more on this topic see Bateman & Diamond (1969), and Knopfmacher (1990).

8.6 References

Axer, A. (1911). Über einige Grenzwertsätze, *Sitz. Kais. Akad. Wiss. Wien. math-natur. Klasse* **120**, 1253–1298.

Balanzario, E. P. (2000). On Chebyshev's inequalities for Beurling's generalized primes, *Math. Slovaca* **50**, No.4, 415–436.

Bateman, P. T. (1972). The distribution of values of the Euler function, *Acta Arith.* **21**, 329–345.

Bateman, P. T. & Diamond, H. G. (1969). Asymptotic distribution of Beurling's generalized prime numbers, *Studies in Number Theory*, W. J. LeVeque, Ed., MAA Studies in math. 6. Washington: Mathematical Association of America, pp. 152–210.

—— (1996). A hundred years of prime numbers, *Amer. Math. Monthly* **103**, 729–741.

Beurling, A. (1937). Analyse de la loi asymptotique de la distribution des nombres premiers généralisés, I, *Acta Math.* **68**, 255–291.

Bombieri, E. (1962a). Maggiorazione del resto nel "Primzahlsatz" col metodo di Erdős–Selberg, *Ist. Lombardo Accad. Sci. Lett. Rend.* A **96**, 343–350.

—— (1962b). Sulle formule di A. Selberg generalizzate per classi di funzioni aritmetiche e le applicazioni al problema del resto nel "Primzahlsatz", *Riv. Mat. Univ. Parma* (2) **3**, 393–440.

Borel, J.-P. (1980/81). Quelques résultats d'équirépartition liés aux nombres généralisés de Beurling, *Acta Arith.* **38**, 255–272.

—— (1984). Sur le prolongement des fonctions ζ associées à un système des nombres premiers généralisés de Beurling, *Acta Arith.* **43**, 273–282.

Breusch, R. (1960). An elementary proof of the prime number theorem with remainder term, *Pacific J. Math.* **10**, 487–497.

van der Corput, J. G. (1956). Sur le reste dans la démonstration élémentaire du theorème des nombres premiers, *Colloque sur la Théorie des Nombres* (Bruxelles, 1955). Paris: Masson & Cie, pp. 163–182.

Diamond, H. G. (1969). The prime number theorem for Beurling's generalized numbers, *J. Number Theory* **1**, 200–207.

—— (1970a). Asymptotic distribution of Beurling's generalized integers, *Illinois J. Math.* **14**, 12–28.

—— (1970b). A set of generalized numbers showing Beurling's theorem to be sharp, *Illinois J. Math.* **14**, 29–34.

—— (1973). Chebyshev estimates for Beurling generalized prime numbers, *Proc. Amer. Math. Soc.* **39**, 503–508.

—— (1977). When do Beurling generalized integers have a density?, *J. Reine Angew. Math.* **295**, 22–39.

Diamond, H. G., Montgomery, H. L., & Vorhauer, U. M. A. (2006). Beurling primes with large oscillation, *Math. Ann.*, **334**, 1–36.

Diamond, H. G. & Steinig, J. (1970). An elementary proof of the prime number theorem with a remainder term, *Invent. Math.* **11**, 199–258.

Dressler, R. E. (1970). A density which counts multiplicity, *Pacific Math. J.* **34**, 371–378.

Erdős, P. (1949). On a new method in elementary number theory which leads to an elementary proof of the prime number theorem, *Proc. Natl. Acad. Sci. USA* **35**, 374–384.

Fatou, P. (1906). Séries trigonométriques et séries de Taylor, *Acta Math.* **30**, 335–400.

Goldfeld, D. (2004). The elementary proof of the prime number theorem: an historical perspective, *Number Theory* (New York, 2003). New York: Springer-Verlag, pp. 179–192.

Graham, S. W. & Vaaler, J. D. (1981). A class of extremal functions for the Fourier transform, *Trans. Amer. Math. Soc.* **265**, 283–302.

Hall, R. S. (1972). The prime number theorem for generalized primes, *J. Number Theory* **4**, 313–320.

(1973). Beurling generalized prime number systems in which the Chebyshev inequalities fail, *Proc. Amer. Math. Soc.* **40**, 79–82.

Hejhal, D. A. (1976). *The Selberg Trace Formula for* $PSL(2, \mathbb{R})$. Vol. I, Lecture Notes Math. 548. Berlin: Springer-Verlag.

(1983). *The Selberg Trace Formula for* $PSL(2, \mathbb{R})$. Vol. 2, Lecture Notes Math. 1001. Berlin: Springer-Verlag.

Ikehara, S. (1931). An extension of Landau's theorem in the analytic theory of numbers, *J. Math. Phys.* **10**, 1–12.

Ingham, A. E. (1945). Some Tauberian theorems connected with the prime number theorem, *J. London Math. Soc.* **20**, 171–180.

Kahane, J.-P. (1995). Sur travaux de Beurling et Malliavin, *Séminaire Bourbaki Vol. 7* Exp. 225, Paris: Soc. Math. France, 27–39.

(1996). Une formula de Fourier pour les nombres premiers. Application aux nombres premiers généralisés de Beurling, *Harmonic analysis from the Pichorides viewpoint* (Anogia, 1995) Publ. Math. Orsay, 96–01, Orsay: Univ. Paris XI, 41–49.

(1997). Sur les nombres premiers généralisés de Beurling. Preuve d'une conjecture de Bateman et Diamond, *J. Théor. Nombres Bordeaux* **9**, 251–266.

(1998). Le rôle des algèbres A de Wiener, A^∞ de Beurling et H^1 de Sobolev dans la théorie des nombres premiers généralisés de Beurling, *Ann. Inst. Fourier (Grenoble)* **48**, 611–648.

(1999). Un théorème de Littlewood pour les nombres premiers de Beurling *Bull. London Math. Soc.* **31**, 424–430.

Knopfmacher, J. (1990). *Abstract Analytic Number Theory*, Second Edition. New York: Dover.

Korevaar, J. (2006). The Wiener–Ikehara theorem by complex analysis, *Proc. Amer. Math. Soc.* **134**, 1107–1116.

Kuhn, P. (1955). Eine Verbesserung des Restgliedes beim elementaren Beweis des Primzahlsatzes, *Math. Scand.* **3**, 75–89.

Landau, E. (1910). Über die Bedeutung einiger neuen Grenzwertsätze der Herren Hardy und Axer, *Prace mat.-fiz.* **21**, 97–177; *Collected Works*, Vol. 4. Essen: Thales Verlag, 1986, pp. 267–347.

(1912). Über einige neuere Grenzwertsätze, *Rend. Circ. Mat. Palermo* **34**, 121–131; *Collected Works*, Vol. 5. Essen: Thales Verlag, 1986, pp. 145–155.

Lavrik, A. F. & Sobirov, A. Š. (1973). The remainder term in the elementary proof of the Prime Number Theorem, *Dokl. Akad. Nauk SSSR* **211**, 534–536.

Levinson, N. (1969). A motivated account of an elementary proof of the Prime Number Theorem, *Amer. Math. Monthly* **76**, 225–245.

Malliavin, P. (1961). Sur le reste de la loi asymptotique de répartition des nombres premiers généralisés de Beurling, *Acta Math.* **106**, 281–298.

Narkiewicz, W. (2000). *The Development of Prime Number Theory*. Berlin: Springer-Verlag.

Nyman, B. (1949). A general Prime Number Theorem, *Acta Math.* **81**, 299–307.

Parry, W. & Pollicott, M. (1990). Zeta functions and the periodic orbit structure of hyperbolic dynamics, *Astérisque No. 268*, pp. 187–188.

Pitt, H. R. (1958). *Tauberian Theorems*. Oxford: Oxford University Press.

Riesz, M. (1916). Ein Konvergenzsatz für Dirichletsche Reihen, *Acta Math.* **40**, 349–361.

Schwarz, W. (1994). Some remarks on the history of the Prime Number Theorem from 1896 to 1960, *Development of mathematics 1900–1950* (Luxembourg, 1992). Basel: Birkhäuser, pp. 565–616.

Selberg, A. (1949). An elementary proof of the prime-number theorem, *Ann. Math.* (2) **50**, 305–313.

Srinivasan, B. R. & Sampath, A. (1988). An elementary proof of the Prime Number Theorem with a remainder term, *J. Indian, Math. Soc.*, New Ser. **53**, No.1-4, 1-50.

Widder, D. V. (1971). *An Introduction to Transform Theory*. New York: Academic Press.

Wiener, N. (1932). Tauberian theorems, *Ann. of Math.* (2) **33**, 1–100; *Collected Works*, Vol. 2. Cambridge: MIT, 1979, pp. 519–619.

Wirsing, E. (1962). Elementare Beweise des Primzahlsatzes mit Restglied, I, *J. Reine Angew. Math.* **211**, 205–214.

(1964). Elementare Beweise des Primzahlsatzes mit Restglied, II, *Reine Angew., J. Math.* **214/215**, 1–18.

Wright, E. M. (1952). The elementary proof of the Prime Number Theorem, *Proc. Roy. Soc. Edinbugh* A **63**, 257–267.

Zhang, W. B. (1987a). Chebyshev type estimates for Beurling generalized prime numbers, *Proc. Amer. Math. Soc.* **101**, 205–212.

(1987b). A generalization of Halász's theorem to Beurling's generalized integers and its application, *Illinois J. Math.* **31**, 645–664.

(1988). Density and *O*-density of Beurling generalized integers, *J. Number Theory* **30**, 120–139.

(1993). Chebyshev type estimates for Beurling generalized prime numbers, II, *Trans. Amer. Math. Soc.* **337**, 651–675.

9

Primitive characters and Gauss sums

9.1 Primitive characters

Suppose that $d \mid q$ and that χ^* is a character (mod d), and set

$$\chi(n) = \begin{cases} \chi^*(n) & (n, q) = 1; \\ 0 & \text{otherwise.} \end{cases} \tag{9.1}$$

Then $\chi(n)$ is multiplicative and has period q, so by Theorem 4.7 we deduce that $\chi(n)$ is a Dirichlet character (mod q). In this situation we say that χ^* *induces* χ. If q is composed entirely of primes dividing d, then $\chi(n) = \chi^*(n)$ for all n, but if there is a prime factor of q not found in d, then $\chi(n)$ does not have period d. Nevertheless, χ and χ^* are nearly the same in the sense that $\chi(p) = \chi^*(p)$ for all but at most finitely many primes, and hence

$$L(s, \chi) = L(s, \chi^*) \prod_{p \mid q} \left(1 - \frac{\chi^*(p)}{p^s} \right). \tag{9.2}$$

Our immediate task is to determine when one character induces another.

Lemma 9.1 *Let χ be a character* (mod q). *We say that d is a* quasiperiod *of χ if $\chi(m) = \chi(n)$ whenever $m \equiv n$ (mod d) and $(mn, q) = 1$. The least quasiperiod of χ is a divisor of q.*

Proof Let d be a quasiperiod of χ, and put $g = (d, q)$. We show that g is also a quasiperiod of χ. Suppose that $m \equiv n$ (mod g) and that $(mn, q) = 1$. Since g is a linear combination of d and q, and $m - n$ is a multiple of g, it follows that there are integers x and y such that $m - n = dx + qy$. Then $\chi(m) = \chi(m - qy) = \chi(n + dx) = \chi(n)$. Thus g is a quasiperiod of χ. □

With more effort (see Exercise 9.1.1) it can be shown that if d_1 and d_2 are quasiperiods of χ, then (d_1, d_2) is also a quasiperiod, and hence the least quasiperiod divides all other quasiperiods, and in particular it divides q (since q is a quasiperiod of χ).

The least quasiperiod d of χ is called the *conductor* of χ. Suppose that d is the conductor of χ. If $(n, d) = 1$, then $(n + kd, d) = 1$. Also, if $(r, d) = 1$ then there exist values of k (mod r) for which $(n + kd, r) = 1$. Hence there exist integers k for which $(n + kd, q) = 1$. For such a k put $\chi^*(n) = \chi(n + kd)$. Although there are many such k, there is only one value of $\chi(n + kd)$ when $(n + kd, q) = 1$. We extend the definition of χ^* by setting $\chi^*(n) = 0$ when $(n, d) > 1$. It is readily seen that χ^* is multiplicative and that χ^* has period d. Thus by Theorem 4.7, χ^* is a character modulo d. Moreover, if χ_0 is the principal character modulo q, then $\chi(n) = \chi^*(n)\chi_0(n)$. Thus χ^* induces χ. Clearly χ^* has no quasiperiod smaller than d, for otherwise χ would have a smaller quasiperiod, contradicting the minimality of d. In addition, χ^* is the only character (mod d) that induces χ, for if there were another, say χ_1, then for any n with $(n, d) = 1$ we would have $\chi^*(n) = \chi^*(n + kd) = \chi(n + kd) = \chi_1(n + kd) = \chi_1(n)$, on choosing k as above.

A character χ modulo q is said to be *primitive* when q is the least quasiperiod of χ. Such χ are not induced by any character having a smaller conductor. We summarize our discussion as follows.

Theorem 9.2 *Let χ denote a Dirichlet character modulo q and let d be the conductor of χ. Then $d \mid q$, and there is a unique primitive character χ^* modulo d that induces χ.*

We now identify the primitive characters in such a way that we can describe them in terms of the explicit construction of Section 5.2.

Lemma 9.3 *Suppose that $(q_1, q_2) = 1$ and that χ_1 and χ_2 are characters modulo q_1 and q_2, respectively. Put $\chi(n) = \chi_1(n)\chi_2(n)$. Then the character χ is primitive modulo $q_1 q_2$ if and only if both χ_1 and χ_2 are primitive.*

Proof For convenience write $q = q_1 q_2$. Suppose that χ is primitive modulo q, and for $i = 1, 2$ let d_i be the conductor of χ_i. If $(mn, q) = 1$ and $m \equiv n$ (mod $d_1 d_2$) then $\chi_i(m) = \chi_i(n)$ for $i = 1, 2$, and hence $d_1 d_2$ is a quasiperiod of χ. Since χ is primitive, this means that $d_1 d_2 = q$. But $d_i \mid q_i$, so this implies that $d_i = q_i$, which is to say that the characters χ_i are primitive.

Now suppose that χ_i is primitive modulo q_i for $i = 1, 2$, and let d be the conductor of χ. Put $d_i = (d, q_i)$. We show that d_1 is a quasiperiod of χ_1. Suppose that $m \equiv n$ (mod d_1) and that $(mn, q_1) = 1$. Choose m' so that $m' \equiv m$ (mod q_1), $m' \equiv 1$ (mod q_2). Similarly, choose n' so that $n' \equiv n$ (mod q_1) and $n' \equiv 1$ (mod q_2). Thus $m' \equiv n'$ (mod d) and $(m'n', q) = 1$, and hence $\chi(m') = \chi(n')$. But $\chi(m') = \chi_1(m)$ and $\chi(n') = \chi_1(n)$, so $\chi_1(m) = \chi_1(n)$. Thus

d_1 is a quasiperiod of χ_1. Since χ_1 is primitive, it follows that $d_1 = q_1$. Similarly $d_2 = q_2$. Thus $d = q$, which is to say that χ is primitive. $\qquad\square$

By Lemma 9.3 we see that in order to exhibit the primitive characters explicitly it suffices to determine the primitive characters (mod p^α). Suppose first that p is odd, and let g be a primitive root of p^α. Then by (4.16) we know that any character χ (mod p^α) is given by

$$\chi(n) = e\left(\frac{k \operatorname{ind}_g n}{\varphi(p^\alpha)}\right)$$

for some integer k. If $\alpha = 1$, then χ is primitive if and only if it is non-principal, which is to say that $(p-1) \nmid k$. If $\alpha > 1$, then χ is primitive if and only if $p \nmid k$. Now consider primitive characters (mod 2^α). When $\alpha = 1$ we have only the principal character, which is imprimitive. When $\alpha = 2$ we have two characters, namely the principal character, which is imprimitive, and the primitive character χ given by $\chi(4k+1) = 1$, $\chi(4k-1) = -1$. When $\alpha \geq 3$, we write an odd integer n in the form $n \equiv (-1)^\mu 5^\nu$ (mod 2^α), and then characters (mod 2^α) are of the form

$$\chi(n) = e\left(\frac{j\mu}{2} + \frac{k\nu}{2^{\alpha-2}}\right)$$

where j is determined (mod 2) and k is determined (mod $2^{\alpha-2}$). Here χ is primitive if and only if k is odd.

We now give two useful criteria for primitivity.

Theorem 9.4 *Let χ be a character modulo q. Then the following are equivalent:*

(1) *χ is primitive.*

(2) *If $d \mid q$ and $d < q$, then there is a c such that $c \equiv 1$ (mod d), $(c, q) = 1$, $\chi(c) \neq 1$.*

(3) *If $d \mid q$ and $d < q$, then for every integer a,*

$$\sum_{\substack{n=1 \\ n \equiv a \,(\mathrm{mod}\, d)}}^{q} \chi(n) = 0.$$

Proof (1) $\Rightarrow$ (2). Suppose that $d \mid q$, $d < q$. Since χ is primitive, there exist integers m and n such that $m \equiv n$ (mod d), $\chi(m) \neq \chi(n)$, $\chi(mn) \neq 0$. Choose c so that $(c, q) = 1$, $cm \equiv n$ (mod q). Thus we have (2).

(2) $\Rightarrow$ (3). Let c be as in (2). As k runs through a complete residue system (mod q/d), the numbers $n = ac + kcd$ run through all residues (mod q) for

which $n \equiv a \pmod{d}$. Thus the sum S in question is

$$S = \sum_{k=1}^{q/d} \chi(ac + kcd) = \chi(c)S.$$

Since $\chi(c) \neq 1$, it follows that $S = 0$.

(3) $\Rightarrow$ (1). Suppose that $d \mid q$, $d < q$. Take $a = 1$ in (3). Then $\chi(1) = 1$ is one term in the sum, but the sum is 0, so there must be another term $\chi(n)$ in the sum such that $\chi(n) \neq 1$, $\chi(n) \neq 0$. But $n \equiv 1 \pmod{d}$, so d is not a quasiperiod of χ, and hence χ is primitive. $\qquad\square$

9.1.1 Exercises

1. Let $f(n)$ be an arithmetic function with period q such that $f(n) = 0$ whenever $(n, q) > 1$. Call d a quasiperiod of f if $f(m) = f(n)$ whenever $m \equiv n \pmod{d}$ and $(mn, q) = 1$.
   (a) Suppose that d_1 and d_2 are quasiperiods, put $g = (d_1, d_2)$, and suppose that $m \equiv n \pmod{g}$ and $(mn, q) = 1$. Show that there exist integers a and b such that $m = n + ad_1 + bd_2$ and $(n + ad_1, q) = 1$.
   (b) Show that if d_1 and d_2 are quasiperiods of f then so also is (d_1, d_2).
   (c) Show that the least quasiperiod of f divides all quasiperiods.
2. Let $\mathcal{S}(q)$ denote the set of all Dirichlet characters $\chi \pmod{q}$, and put $\mathcal{T}(q) = \bigcup_{d|q} \mathcal{S}(d)$. Show that the members of $\mathcal{T}(q)$ form a basis of the vector space of all arithmetic functions with period q if and only if q is square-free.
3. For $d|q$ let $\mathcal{U}(d, q)$ denote the set of $\varphi(q/d)$ functions

   $$f(a) = \begin{cases} \chi(a/d) & (a, q) = d, \\ 0 & \text{otherwise} \end{cases}$$

   where χ runs over all Dirichlet characters $\pmod{q/d}$. Set $\mathcal{V}(q) = \bigcup_{d|q} \mathcal{U}(d, q)$. Show that the members of $\mathcal{V}(q)$ form a basis for the vector space of arithmetic functions with period q.
4. For $i = 1, 2$ let χ_i be a character $\pmod{q_i}$ where $(q_1, q_2) = 1$, and suppose that d_i is the conductor of χ_i. Show that $d_1 d_2$ is the conductor of $\chi_1 \chi_2$.
5. For $i = 1, 2$ suppose that χ_i is a character $\pmod{q_i}$. Show that the following two assertions are equivalent:
   (a) The characters χ_1 and χ_2 are induced by the same primitive character.
   (b) $\chi_1(p) = \chi_2(p)$ for all but at most finitely many primes p.
6. Let $\varphi_2(q)$ denote the number of primitive characters $\pmod{q}$.
   (a) Show that $\varphi_2(q)$ is a multiplicative function.
   (b) Show that $\sum_{d|q} \varphi_2(d) = \varphi(q)$.

(c) Show that

$$\varphi_2(q) = q \prod_{p \| q} \left(1 - \frac{2}{p}\right) \prod_{p^2 | q} \left(1 - \frac{1}{p}\right)^2.$$

(d) Show that $\varphi_2(q) > 0$ if and only if $q \not\equiv 2 \pmod 4$.

7. Suppose that χ is a character (mod q), and that d is the conductor of χ. Show that if $(a, q) = 1$, then

$$\left| \sum_{\substack{n=1 \\ n \equiv a \pmod d}}^{q} \chi(n) \right| = \frac{\varphi(q)}{\varphi(d)}.$$

8. (Martin 2006; Vorhauer 2006) Let $d(\chi)$ denote the conductor of χ.

(a) Use the identity $\log d = \sum_{r | d} \Lambda(r)$ to show that

$$\sum_{\chi} \log d(\chi) = \varphi(q) \log q - \sum_{r | q} \Lambda(r) \sum_{\substack{\chi \\ r \nmid d(\chi)}} 1.$$

(b) Show that if $p^a \| q$ and $1 \le b \le a$, then the number of χ modulo q such that $p^b \nmid d(\chi)$ is exactly $\varphi(q) \varphi(p^{b-1}) / \varphi(p^a)$.

(c) Conclude that

$$\sum_{\chi} \log d(\chi) = \varphi(q) \left(\log q - \sum_{p | q} \frac{\log p}{p - 1} \right).$$

9.2 Gauss sums

Given a character χ modulo q, we define the Gauss sum $\tau(\chi)$ of χ to be

$$\tau(\chi) = \sum_{a=1}^{q} \chi(a) e(a/q). \tag{9.3}$$

This may be regarded as the inner product of the multiplicative character $\chi(a)$ with the additive character $e(a/q)$. As such, it is analogous to the gamma function $\Gamma(s) = \int_0^\infty x^{s-1} e^{-x}\, dx$, which is the inner product of the multiplicative character x^s with the additive character e^{-x} with respect to the invariant measure dx/x. Gauss sums are invaluable in transferring questions concerning Dirichlet characters to questions concerning additive characters, and vice versa.

The Gauss sum is a special case of the more general sum

$$c_\chi(n) = \sum_{a=1}^{q} \chi(a) e(an/q). \tag{9.4}$$

When χ is the principal character, this is Ramanujan's sum

$$c_q(n) = \sum_{\substack{a=1 \\ (a,q)=1}}^{q} e(an/q), \tag{9.5}$$

whose properties were discussed in Section 4.1. We now show that the sum $c_\chi(n)$ is closely related to $\tau(\chi)$.

Theorem 9.5 *Suppose that χ is a character modulo q. If $(n, q) = 1$, then*

$$\chi(n)\tau(\overline{\chi}) = \sum_{a=1}^{q} \overline{\chi}(a)e(an/q), \tag{9.6}$$

and in particular

$$\overline{\tau(\chi)} = \chi(-1)\tau(\overline{\chi}). \tag{9.7}$$

Proof If $(n, q) = 1$, then the map $a \mapsto an$ permutes the residues modulo q, and hence

$$\chi(n)c_\chi(n) = \sum_{a=1}^{q} \chi(an)e(an/q) = \tau(\chi).$$

On replacing χ by $\overline{\chi}$, this gives (9.6), and (9.7) follows by taking $n = -1$. □

Theorem 9.6 *Suppose that $(q_1, q_2) = 1$, that χ_i is a character modulo q_i for $i = 1, 2$, and that $\chi = \chi_1\chi_2$. Then*

$$\tau(\chi) = \tau(\chi_1)\tau(\chi_2)\chi_1(q_2)\chi_2(q_1).$$

Proof By the Chinese Remainder Theorem, each $a \pmod{q_1 q_2}$ can be written uniquely as $a_1q_2 + a_2q_1$ with $1 \le a_i \le q_i$. Thus the general term in (9.3) is $\chi_1(a_1q_2)\chi_2(a_2q_1)e(a_1/q_1)\,e(a_2/q_2)$, so the result follows. □

For primitive characters the hypothesis that $(n, q) = 1$ in Theorem 9.5 can be removed.

Theorem 9.7 *Suppose that χ is a primitive character modulo q. Then (9.6) holds for all n, and $|\tau(\chi)| = \sqrt{q}$.*

Proof It suffices to prove (9.6) when $(n, q) > 1$. Choose m and d so that $(m, d) = 1$ and $m/d = n/q$. Then

$$\sum_{a=1}^{q} \chi(a)e(an/q) = \sum_{h=1}^{d} e(hm/d) \sum_{\substack{a=1 \\ a \equiv h \,(\mathrm{mod}\ d)}}^{q} \chi(a).$$

Since $d \mid q$ and $d < q$, the inner sum vanishes by Theorem 9.4. Thus (9.6) holds also in this case.

We replace χ in (9.6) by $\overline{\chi}$, take the square of the absolute value of both sides, and sum over n to see that

$$\varphi(q)|\tau(\chi)|^2 = \sum_{n=1}^{q}\left|\sum_{a=1}^{q}\chi(a)e(an/q)\right|^2 = \sum_{a=1}^{q}\sum_{b=1}^{q}\chi(a)\overline{\chi}(b)\sum_{n=1}^{q}e((a-b)n/q).$$

The innermost sum on the right is 0 unless $a \equiv b \pmod{q}$, in which case it is equal to q. Thus $\varphi(q)|\tau(\chi)|^2 = \varphi(q)q$, and hence $|\tau(\chi)| = \sqrt{q}$. $\square$

If χ is primitive modulo q, then not only does (9.6) hold for all n but also $\tau(\overline{\chi}) \neq 0$, and hence we have

Corollary 9.8 *Suppose that χ is a primitive character modulo q. Then for any integer n,*

$$\chi(n) = \frac{1}{\tau(\overline{\chi})}\sum_{a=1}^{q}\overline{\chi}(a)e(an/q).$$

This is very useful, since it allows us to express the multiplicative character χ as a linear combination of additive characters $e(an/q)$. As a first application, we use this formula to express $L(1,\chi)$ in closed form.

Theorem 9.9 *Suppose that χ is a primitive character modulo q with $q > 1$. If $\chi(-1) = 1$, then*

$$L(1,\chi) = \frac{-\tau(\chi)}{q}\sum_{a=1}^{q-1}\overline{\chi}(a)\log(\sin\pi a/q), \tag{9.8}$$

while if $\chi(-1) = -1$, then

$$L(1,\chi) = \frac{i\pi\tau(\chi)}{q^2}\sum_{a=1}^{q-1}a\overline{\chi}(a). \tag{9.9}$$

Proof Since $L(1,\chi) = \sum_{n=1}^{\infty}\chi(n)/n$, by Corollary 9.8,

$$L(1,\chi) = \frac{1}{\tau(\overline{\chi})}\sum_{n=1}^{\infty}\frac{1}{n}\sum_{a=1}^{q-1}\overline{\chi}(a)e(an/q) = \frac{1}{\tau(\overline{\chi})}\sum_{a=1}^{q-1}\overline{\chi}(a)\sum_{n=1}^{\infty}\frac{e(an/q)}{n}.$$

But $\log(1-z)^{-1} = \sum_{n=1}^{\infty}z^n/n$ for $|z| \leq 1$, $z \neq 1$, where the logarithm is the principal branch. We take $z = e(\theta)$ where $0 < \theta < 1$. Since $1 - e(\theta) = -2ie(\theta/2)\sin\pi\theta$, it follows that $\log(1-e(\theta)) = \log(2\sin\pi\theta) + i\pi(\theta - 1/2)$. Thus

$$L(1,\chi) = \frac{-1}{\tau(\overline{\chi})}\sum_{a=1}^{q-1}\overline{\chi}(a)(\log(2\sin\pi a/q)+i\pi(a/q-1/2)).$$

Since $\sum_{a=1}^{q-1} \overline{\chi}(a) = 0$, this is

$$\frac{-1}{\tau(\overline{\chi})}(S + iT)$$

where $S = \sum_{a=1}^{q-1} \overline{\chi}(a) \log(\sin \pi a/q)$ and $T = \pi/q \sum_{a=1}^{q-1} \overline{\chi}(a)a$. On replacing a by $q - a$ we see that $S = \chi(-1)S$ and $T = -\chi(-1)T$. Thus if $\chi(-1) = 1$, then $T = 0$ and so

$$L(1, \chi) = \frac{-1}{\tau(\overline{\chi})} \sum_{a=1}^{q-1} \overline{\chi}(a) \log(\sin \pi a/q).$$

Then by (9.7) we obtain (9.8). If $\chi(-1) = -1$ then $S = 0$ and so

$$L(1, \chi) = \frac{-i\pi}{\tau(\overline{\chi})q} \sum_{a=1}^{q-1} \overline{\chi}(a)a.$$

Then by (9.7) we obtain (9.9). $\qquad\square$

We next show that $\tau(\chi)$ can be expressed in terms of $\tau(\chi^\star)$ where $\chi^\star$ is the primitive character that induces χ.

Theorem 9.10 *Let χ be a character modulo q that is induced by the primitive character $\chi^\star$ modulo d. Then $\tau(\chi) = \mu(q/d)\chi^\star(q/d)\tau(\chi^\star)$.*

Proof If $(d, q/d) > 1$, then $\chi^\star(q/d) = 0$, so we begin by showing that $\tau(\chi) = 0$ in this case. Let p be a prime such that $p \mid d$, $p \mid q/d$, and write $a = jq/p + k$ with $0 \le j < p$, $0 \le k < q/p$. Then

$$\tau(\chi) = \sum_{a=0}^{q-1} \chi(a)e(a/q) = \sum_{k=1}^{q/p} \sum_{j=1}^{p} \chi(jq/p + k)e(j/p + k/q).$$

But $p \mid (q/p)$, so $(jq/p + k, q) = 1$ if and only if $(jq/p + k, q/p) = 1$, which in turn is equivalent to $(k, q/p) = 1$. Also, $d \mid q/p$, so the above is

$$= \sum_{\substack{k=1 \\ (k,q/p)=1}}^{q/p} \chi^\star(k)e(k/q) \sum_{j=1}^{p} e(j/p).$$

Here the inner sum vanishes, so $\tau(\chi) = 0$ when $(d, q/d) > 1$.

Now suppose that $(d, q/d) = 1$, and let χ_0 denote the principal character modulo q/d. Then by Theorem 9.6,

$$\tau(\chi) = \tau(\chi_0\chi^\star) = \tau(\chi_0)\tau(\chi^\star)\chi_0(d)\chi^\star(q/d).$$

By taking $n = 1$ in Theorem 4.1 we find that $\tau(\chi_0) = \mu(q/d)$. Thus we have the stated result. $\qquad\square$

We now turn our attention to the more general $c_\chi(n)$. To this end we begin with an auxiliary result.

Lemma 9.11 *Let χ be a character modulo q induced by the primitive character χ^* modulo d. Suppose that $r \mid q$. Then*

$$\sum_{\substack{n=1 \\ n \equiv b \,(\mathrm{mod}\, r)}}^{q} \chi(n) = \begin{cases} \chi^*(b)\varphi(q)/\varphi(r) & \text{if } (b,r) = 1 \text{ and } d \mid r, \\ 0 & \text{otherwise.} \end{cases}$$

Proof Let $S(b,r)$ denote the sum in question. If $p \mid (b,r)$ and $n \equiv b \pmod r$, then $p \mid n$, and so $(n,q) > 1$. Thus each term in $S(b,r)$ is 0. Thus we are done when $(b,r) > 1$, so we suppose that $(b,r) = 1$. Consider next the case when $d \nmid r$. Then r is not a quasiperiod of χ. Hence there exist m and n such that $(mn,q) = 1$, $m \equiv n \pmod r$, and $\chi(m) \neq \chi(n)$. Choose c so that $cn \equiv m \pmod q$. Then $c \equiv 1 \pmod r$ and $\chi(c) \neq 1$. Hence $\chi(c)S(b,r) = S(b,r)$, as in the proof of Theorem 9.4, so $S(b,r) = 0$ in this case. Finally suppose that $d \mid r$. Let χ_0 be the principal character modulo q. If $n \equiv b \pmod r$, then $\chi^*(n) = \chi^*(b)$. Thus

$$S(b,r) = \chi^*(b) \sum_{\substack{n=1 \\ n \equiv b \,(\mathrm{mod}\, r)}}^{q} \chi_0(n).$$

Write $q/r = q_1 q_2$ where q_1 is the largest divisor of q/r that is relatively prime to r. Then the sum on the right above is

$$\sum_{\substack{k=1 \\ (kr+b,q_1)=1}}^{q_1 q_2} 1 = q_2\varphi(q_1) = \varphi(q)/\varphi(r),$$

as required. $\square$

We are now in a position to deal with $c_\chi(n)$.

Theorem 9.12 *Let χ be a character modulo q induced by the primitive character χ^* modulo d. Put $r = q/(q,n)$. Then $c_\chi(n) = 0$ if $d \nmid r$, while if $d \mid r$, then*

$$c_\chi(n) = \overline{\chi}^*(n/(q,n))\chi^*(r/d)\mu(r/d)\frac{\varphi(q)}{\varphi(r)}\tau(\chi^*).$$

Proof If $(n,q) = 1$, then by Theorem 9.5 and Theorem 9.10 we see that

$$c_\chi(n) = \overline{\chi}(n)\tau(\chi) = \overline{\chi}^*(n)\mu(q/d)\chi^*(q/d)\tau(\chi^*).$$

Since $r = q$, we have $d \mid r$, so we have the correct result. Now suppose that $(n,q) > 1$. In the definition (9.4) of $c_\chi(n)$, let $a = br + k$ with $0 \le b < q/r$,

$1 \le k \le r$. Then

$$c_\chi(n) = \sum_{k=1}^{r} e(kn/q) \sum_{b=1}^{q/r} \chi(br + k).$$

By Lemma 9.11 this is 0 when $d \nmid r$. Thus we may suppose that $d \mid r$. Then, by Lemma 9.11,

$$c_\chi(n) = \sum_{\substack{k=1 \\ (k,r)=1}}^{r} e(kn/q)\chi^*(k)\varphi(q)/\varphi(r).$$

Put $m = n/(q, n)$, and let χ_1 denote the character modulo r induced by χ^*. Then the above is

$$= \frac{\varphi(q)}{\varphi(r)} \sum_{k=1}^{r} e(km/r)\chi_1(k).$$

Since $(m, r) = 1$, we see by the first case treated that the above is

$$\frac{\varphi(q)}{\varphi(r)}\overline{\chi^*}(m)\mu(r/d)\chi^*(r/d)\tau(\chi^*),$$

which suffices. □

9.2.1 Exercises

1. (a) Show that

$$\frac{1}{\varphi(q)} \sum_{\chi} \overline{\chi}(a)\tau(\chi) = \begin{cases} e(a/q) & (a, q) = 1, \\ 0 & \text{otherwise.} \end{cases}$$

(b) Show that for all integers a,

$$e(a/q) = \sum_{\substack{d\mid q \\ d\mid a}} \frac{1}{\varphi(q/d)} \sum_{\chi \,(\mathrm{mod}\, q/d)} \overline{\chi}(a/d)\tau(\chi).$$

2. Let

$$G_k(a) = \sum_{n=1}^{p} e\left(\frac{an^k}{p}\right).$$

(a) Let $N_k(h)$ denote the number of solutions of the congruence $x^k \equiv h$ (mod p). Explain why

$$G_k(a) = \sum_{h=1}^{p} N_k(h)e\left(\frac{ah}{p}\right).$$

(b) Let $l = (k, p - 1)$. Show that if k is a positive integer, then $N_k(h) = N_l(h)$ for all h, and hence that $G_k(a) = G_l(a)$.

(c) Suppose that $k \mid (p - 1)$. Explain why

$$\sum_{a=1}^{p} |G_k(a)|^2 = p \sum_{h=1}^{p} N_k(h)^2.$$

(d) Suppose that $k \mid (p - 1)$. Show that there are $(p - 1)/k$ residues h (mod p) for which $N_k(h) = k$, that $N_k(0) = 1$, and that $N_k(h) = 0$ for all other residue classes (mod p). Hence show that the right-hand side above is $p(1 + (p - 1)k)$.

(e) Let k be a divisor of $p - 1$. Suppose that $p \nmid a$, $p \nmid c$, and that $b \equiv ac^k$ (mod p). Show that $G_k(a) = G_k(b)$.

(f) Suppose that $k \mid (p - 1)$. Show that if $p \nmid a$ then $|G_k(a)| < k\sqrt{p}$.

3. Suppose that $k \mid \varphi(q)$ and that $(h, q) = 1$.

(a) Explain why

$$\frac{1}{\varphi(q)} \sum_{\chi} \chi(x^k)\overline{\chi}(h) = \begin{cases} 1 & \text{if } x^k \equiv h \text{ (mod } q), \\ 0 & \text{otherwise.} \end{cases}$$

(b) Let $N_k(h)$ be as in Exercise 2(a). Show that

$$N_k(h) = \sum_{\substack{\chi \\ \chi^k = \chi_0}} \chi(h).$$

4. Suppose that $k \mid (p - 1)$, that $N_k(h)$ is as in Exercise 2(a), and let χ be a character of order k, say $\chi(n) = e((\text{ind } n)/k)$.

(a) Show that for all h,

$$N_k(h) = 1 + \sum_{j=1}^{k-1} \chi^j(h).$$

(b) Show that if $p \nmid a$, then

$$G_k(a) = \sum_{j=1}^{k-1} \overline{\chi}^j(a)\tau(\chi^j).$$

(c) Show that if $p \nmid a$, then $|G_k(a)| \le (k - 1)\sqrt{p}$.

5. Suppose that χ_i is a character (mod q_i) for $i = 1, 2$, with $(q_1, q_2) = 1$. Show that

$$c_{\chi_1 \chi_2}(n) = \chi_1(q_2)\chi_2(q_1)c_{\chi_1}(n)c_{\chi_2}(n).$$

6. (Apostol 1970) Let χ be a character modulo q such that the identity (9.6) holds for all integers n. Show that χ is primitive (mod q).

7. Let $N(q)$ denote the number of pairs x, y of residue classes (mod q) such that $y^2 \equiv x^3 + 7 \pmod{q}$.

   (a) Show that $N(q)$ is a multiplicative function of q, that $N(2) = 2, N(3) = 3$, $N(7) = 7$, and that $N(p) = p$ when $p \equiv 2 \pmod 3$.

   (b) Suppose that $p \equiv 1 \pmod 3$. Let $\chi_1(n)$ be a cubic character modulo p, and let $\chi_2(n) = \left(\frac{n}{p}\right)$ be the quadratic character modulo p. Show that

   $$N(p) = \frac{1}{p} \sum_{a=1}^{p} e(7a/p) \left(\sum_{h=1}^{p} \left(1 + \chi_1(h) + \chi_1^2(h)\right) e(ah/p) \right)$$

   $$\times \left(\sum_{k=1}^{p} (1 + \chi_2(k)) e(-ak/p) \right)$$

   $$= p + \frac{2}{p} \Re\left(\tau(\chi_1)\tau(\chi_2)\tau\left(\chi_1^2\chi_2\right)\chi_1\chi_2(-7) \right),$$

   and deduce that $|N(p) - p| \leq 2\sqrt{p}$.

   (c) Deduce that $N(p) > 0$ for all p.

   (d) Show that $N(2^k) = 2^{k-1}$ for $k \geq 2$, that $N(3^k) = 2 \cdot 3^{k-1}$ for $k \geq 2$, that $N(7^k) = 6 \cdot 7^{k-1}$ for $k \geq 2$, and that $N(p^k) = N(p)p^{k-1}$ for all other primes.

   (e) Conclude that the congruence $y^2 \equiv x^3 + 7 \pmod{q}$ has solutions for every positive integer q.

   (f) Suppose that x and y are integers such that $y^2 = x^3 + 7$. Show that $2 \mid y$, $x \equiv 1 \pmod 4$, and that $x > 0$. Note that $y^2 + 1 = (x + 2)(x^2 - 2x + 4)$, so that $y^2 + 1$ is composed of primes $\equiv 1 \pmod 4$, and yet $x + 2 \equiv 3 \pmod 4$. Deduce that this equation has no solution in integers.

8. (Mordell 1933) Explain why the number N of solutions of the congruence $c_1 x_1^{k_1} + \cdots + c_m x_m^{k_m} \equiv c \pmod{p}$ is

   $$N = \frac{1}{p} \sum_{a=1}^{p} e(-ac/p) \prod_{j=1}^{m} G_{k_j}(ac_j)$$

   where G_k is defined as in Exercise 2.

   (b) Suppose that $c = 0$ but that p does not divide any of the numbers c_j. Show that $|N - p^{m-1}| \leq Cp^{m/2}$ where $C = \prod_{j=1}^{m}((k_j, p - 1) - 1)$.

   (c) Suppose that $c \not\equiv 0 \pmod{p}$ and that for all j, $c_j \not\equiv 0 \pmod{p}$. Show that $|N - p^{m-1}| \leq Cp^{(m-1)/2}$ where C is defined as above.

9. (Mattics 1984) Suppose that h has order $(p - 1)/k$ modulo p. Show that

   $$\left| \sum_{m=1}^{p-1} e\left(\frac{h^m}{p}\right) \right| \leq 1 + (k - 1)\sqrt{p}.$$

10. Let χ_1 and χ_2 be primitive characters (mod q).

(a) Show that if $(a, q) = 1$, then

$$\sum_{n=1}^{q} \chi_1(n)\chi_2(a - n) = \chi_1\chi_2(a)q\frac{\tau(\overline{\chi_1\chi_2})}{\tau(\overline{\chi_1})\tau(\overline{\chi_2})}.$$

(b) Show that if $\chi_1\chi_2$ is primitive, then

$$\sum_{n=1}^{q} \chi_1(n)\chi_2(a - n) = \chi_1\chi_2(a)\frac{\tau(\chi_1)\tau(\chi_2)}{\tau(\chi_1\chi_2)} \tag{9.10}$$

for all a.

When $a = 1$, the sum (9.10) is known as the *Jacobi sum* $J(\chi_1, \chi_2)$. In the same way that the Gauss sum is analogous to the gamma function, the Jacobi sum (and its evaluation in terms of Gauss sums) is analogous to the beta function

$$B(\alpha, \beta) = \int_0^1 x^{\alpha-1}(1 - x)^{\beta-1}\, dx = \frac{\Gamma(\alpha)\Gamma(\beta)}{\Gamma(\alpha + \beta)}.$$

11. Let C be the smallest field that contains the field $\mathbb{Q}$ of rational numbers and is closed under square roots. Thus C is the set of complex numbers that are constructible by ruler-and-compass. We show that if p is of the form $p = 2^k + 1$, then $\zeta = e(1/p) \in C$, which is to say that a regular p-gon can be constructed.

   (a) Let p be any prime, and χ any non-principal character modulo p. Explain why

$$\tau(\chi)^2 \sum_{n=1}^{p} \overline{\chi}(n)\overline{\chi}(1 - n) = p\tau(\chi^2).$$

   (b) From now on assume that p is of the form $p = 2^k + 1$. Explain why $\chi^{2^k} = \chi_0$ for any character modulo p, and deduce that $\chi(n) \in C$ for all χ and all integers n.

   (c) Deduce that if $\tau(\chi^2) \in C$, then $\tau(\chi) \in C$.

   (d) Suppose that χ has order 2^r. Show successively that the numbers

$$-1 = \tau(\chi^{2^r}), \ \tau(\chi^{2^{r-1}}), \ \ldots, \ \tau(\chi^2), \ \tau(\chi)$$

   lie in C.

   (e) Explain why $\sum_{\chi} \tau(\chi) = (p - 1)\zeta$.

   (f) (Gauss) If $p = 2^k + 1$, then $\zeta \in C$.

12. Let χ be a character modulo p and put $J(\chi) = \sum_{n=1}^{p} \chi(n)\chi(1 - n)$.

   (a) Show that if $\chi^2 \neq \chi_0$, then $|J(\chi)| = \sqrt{p}$.

   (b) Suppose that $p \equiv 1 \pmod 4$. Show that there is a quartic character χ modulo p.

(c) Show that if χ is a quartic character, then $J(\chi)$ is a Gaussian integer. That is, $J(\chi) = a + ib$ where a and b are rational integers.

(d) Deduce that $a^2 + b^2 = p$.

13. (a) Write

$$|\tau(\chi)|^2 = \sum_{m=1}^{q} \chi(m)e(m/q) \sum_{n=1}^{q} \overline{\chi}(n)e(-n/q),$$

and in the second sum replace n by mn where $(m, q) = 1$, to see that the above is

$$= \sum_{n=1}^{q} \overline{\chi}(n)c_q(n - 1).$$

(b) Use Theorem 4.1 to show that the above is

$$= \sum_{d|q} d\mu(q/d) \sum_{\substack{n=1 \\ n \equiv 1 \pmod{d}}}^{q} \overline{\chi}(n).$$

(c) Use Theorem 9.4 to show that if χ is primitive, then $|\tau(\chi)| = \sqrt{q}$.

9.3 Quadratic characters

A character is *quadratic* if it has order 2 in the group of characters modulo q. That is, the character takes on only the values -1, 0, and 1, with at least one -1. Similarly, a character is *real* if all its values are real. Hence a real character is either the principal character or a quadratic character. The Legendre symbol $\left(\frac{n}{p}\right)_L$ is a primitive quadratic character modulo p, and further quadratic characters arise from the Jacobi and Kronecker symbols. We now determine all quadratic characters modulo q. If χ is a character modulo q induced by the primitive character χ^* modulo d, $d \mid q$, then χ is quadratic if and only if χ^* is quadratic. Hence it suffices to determine the primitive quadratic characters.

Suppose that χ is a character modulo q, that $q = q_1q_2$, $(q_1, q_2) = 1$, $\chi = \chi_1\chi_2$ as in Lemma 9.3. By the Chinese Remainder Theorem we see that χ is a real character if and only if both χ_1 and χ_2 are real characters. Hence by Lemma 9.3, χ is a primitive quadratic character if and only if χ_1 and χ_2 are. Thus it suffices to determine the primitive quadratic characters modulo a prime power.

In Section 5.2 we saw that a character χ modulo p may be written in the form $\chi(n) = e(k \text{ ind } n/(p - 1))$. Such a character is primitive provided that it is non-principal, which is to say that $k \not\equiv 0 \pmod{p - 1}$. Similarly, χ is quadratic if and only if the least denominator of the fraction $k/(p - 1)$ is 2. If

$p = 2$ then this is impossible, but for $p > 2$ this is equivalent to the condition
$k \equiv (p-1)/2 \pmod{p-1}$. Thus there is no quadratic character modulo 2,
but for each odd prime p there is a unique quadratic character, given by the
Legendre symbol.

Now suppose that p is an odd prime and that $q = p^m$ with $m > 1$. We have
seen that a character χ modulo such a q is of the form $\chi(n) = e(k \text{ ind } n/\varphi(q))$,
and that χ is primitive if and only if $p \nmid k$. This character is quadratic only when
$k \equiv \varphi(q)/2 \pmod{\varphi(q)}$, so there is a unique quadratic character modulo q, but
it is not primitive because $p \mid k$ for this k. That is, the only quadratic character
modulo p^m is induced by the primitive quadratic character modulo p.

Finally, suppose that $q = 2^m$. For the modulus 2 there is only the principal
character, but for $q = 4$ we have a primitive quadratic character

$$\chi_1(n) = \begin{cases} (-1)^{(n-1)/2} & (n \text{ odd}), \\ 0 & (n \text{ even}). \end{cases}$$

For $m > 2$ we write $\chi((-1)^\mu 5^\nu) = e(j\mu/2 + k\nu/2^{m-2})$, and we see that this
character is real if and only if $2^{m-3} \mid k$. However, the character is primitive if and
only if k is odd, so primitive quadratic characters arise only when $m = 3$, and for
this modulus we have two different characters (corresponding to $j = 0, j = 1$).
Let $\chi_2((-1)^\mu 5^\nu) = e(\nu/2)$. That is, $\chi_2(n) = (-1)^{(n^2-1)/8}$. Then the characters
modulo 8 are χ_0, χ_1, χ_2, and $\chi_1 \chi_2$, of which the latter two are primitive.

We next show that the primitive quadratic characters arise precisely from
the Kronecker symbol $\left(\frac{d}{n}\right)_K$. We say that d is a *quadratic discriminant* if either

(a) $d \equiv 1 \pmod 4$ and d is square-free

or

(b) $4 \mid d$, $d/4 \equiv 2$ or $3 \pmod 4$, and $d/4$ is square-free.

For each quadratic discriminant d we define the Kronecker symbol $\left(\frac{d}{n}\right)_K$ by the
following relations:

(i) $\left(\frac{d}{p}\right)_K = 0$ when $p \mid d$;

(ii) $\left(\frac{d}{2}\right)_K = \begin{cases} 1 & \text{when } d \equiv 1 \pmod 8, \\ -1 & \text{when } d \equiv 5 \pmod 8; \end{cases}$

(iii) $\left(\frac{d}{p}\right)_K = \left(\frac{d}{p}\right)_L$, the Legendre symbol, when $p > 2$;

(iv) $\left(\frac{d}{-1}\right)_K = \begin{cases} 1 & \text{when } d > 0, \\ -1 & \text{when } d < 0; \end{cases}$

(v) $\left(\frac{d}{n}\right)_K$ is a totally multiplicative function of n.

It is not immediately apparent that this definition of the Kronecker symbol gives
rise to a character, but we now show that this is the case.

Theorem 9.13 *Let d be a quadratic discriminant. Then $\chi_d(n) = \left(\frac{d}{n}\right)_K$ is a primitive quadratic character modulo $|d|$, and every primitive quadratic character is given uniquely in this way.*

Proof It is easy to see that $\left(\frac{-4}{n}\right)_K$ is the primitive quadratic character modulo 4. Similarly, $\left(\frac{8}{n}\right)_K$ and $\left(\frac{-8}{n}\right)_K$ are the primitive quadratic characters modulo 8.

Suppose that p is a prime, $p \equiv 1 \pmod 4$. We show that $\left(\frac{p}{n}\right)_K = \left(\frac{n}{p}\right)_L$ for all n. To see this, note that if q is an odd prime, then by (iii) and quadratic reciprocity, $\left(\frac{p}{q}\right)_K = \left(\frac{p}{q}\right)_L = \left(\frac{q}{p}\right)_L$. Also, $\left(\frac{p}{2}\right)_K = (-1)^{(p^2-1)/8} = \left(\frac{2}{p}\right)_L$, and $\left(\frac{p}{-1}\right)_K = 1 = \left(\frac{-1}{p}\right)_L$. Since these two functions agree on all primes, and also on -1, and both are totally multiplicative, it follows that $\left(\frac{p}{n}\right)_K = \left(\frac{n}{p}\right)_L$ for all integers n.

Suppose that p is a prime, $p \equiv 3 \pmod 4$. We show that $\left(\frac{-p}{n}\right)_K = \left(\frac{n}{p}\right)_L$ for all n. To see this, note that if q is an odd prime, then by (iii) and quadratic reciprocity, $\left(\frac{-p}{q}\right)_K = \left(\frac{-p}{q}\right)_L = \left(\frac{q}{p}\right)_L$. Also, $\left(\frac{-p}{2}\right)_K = (-1)^{((-p)^2-1)/8} = (-1)^{(p^2-1)/8} = \left(\frac{2}{p}\right)_L$, and $\left(\frac{-p}{-1}\right)_K = -1 = \left(\frac{-1}{p}\right)_L$. Since these two functions agree on all primes, and also on -1, and both are totally multiplicative, it follows that $\left(\frac{-p}{n}\right)_K = \left(\frac{n}{p}\right)_L$ for all integers n.

Suppose next that d_1 and d_2 are quadratic discriminants with $(d_1, d_2) = 1$. Put $d = d_1 d_2$. Supposing that $\left(\frac{d_i}{n}\right)_K$ is a primitive quadratic character modulo $|d_i|$ for $i = 1, 2$, we shall show that $\left(\frac{d}{n}\right)_K$ is a primitive quadratic character modulo $|d|$. If q is an odd prime, then by (iii), $\left(\frac{d}{q}\right)_K = \left(\frac{d}{q}\right)_L = \left(\frac{d_1}{q}\right)_L \left(\frac{d_2}{q}\right)_L = \left(\frac{d_1}{q}\right)_K \left(\frac{d_2}{q}\right)_K$. Also, by (ii) we see that $\left(\frac{d}{2}\right)_K = \left(\frac{d_1}{2}\right)_K \left(\frac{d_2}{2}\right)_K$, and by (iv) that $\left(\frac{d}{-1}\right)_K = \left(\frac{d_1}{-1}\right)_K \left(\frac{d_2}{-1}\right)_K$. Since $\left(\frac{d}{n}\right)_K = \left(\frac{d_1}{n}\right)_K \left(\frac{d_2}{n}\right)_K$ when n is a prime or $n = -1$, and since both sides are totally multiplicative functions, it follows that this identity holds for all integers n. Hence by Lemma 9.3, $\left(\frac{d}{n}\right)_K$ is a primitive character modulo $|d|$.

This allows us to account for all primitive quadratic characters, so the proof is complete. $\square$

Since the Kronecker symbol and Legendre symbol agree whenever both are defined, we may omit the subscripts. The same remark applies to the Jacobi symbol $\left(\frac{n}{q}\right)_J$, which for odd positive $q = p_1 p_2 \cdots p_r$ is defined to be $\left(\frac{n}{q}\right)_J = \prod_{i=1}^r \left(\frac{n}{p_i}\right)_L$. Sometimes we let $\chi_d(n)$ denote the character $\left(\frac{d}{n}\right)$.

A character χ modulo q is an even function, $\chi(-n) = \chi(n)$, if $\chi(-1) = 1$; for the primitive quadratic character χ_d this occurs if $d > 0$. In the case of the Legendre symbol, if $p \equiv 1 \pmod 4$, then $\left(\frac{n}{p}\right)_L = \chi_p(n)$ is even. Similarly, χ is odd, $\chi(-n) = -\chi(n)$, if $\chi(-1) = -1$. For χ_d this occurs when $d < 0$. For the Legendre symbol, if $p \equiv 3 \pmod 4$, then $\left(\frac{n}{p}\right)_L = \chi_{-p}(n)$ is odd.

We have taken the quadratic reciprocity law for the Legendre symbol for granted, since it is treated in a variety of ways in elementary texts. In Exercise 9.3.6 below we outline a proof of quadratic reciprocity that is unusual that it applies directly to the Jacobi symbol, without first being restricted to the Legendre symbol. For future purposes it is convenient to formulate quadratic reciprocity also for the Kronecker symbol.

Theorem 9.14   *Suppose that d_1 and d_2 are relatively prime quadratic discriminants. Then*

$$\left(\frac{d_1}{d_2}\right)\left(\frac{d_2}{d_1}\right) = \varepsilon(d_1, d_2) \tag{9.11}$$

where $\varepsilon(d_1, d_2) = 1$ if $d_1 > 0$ or $d_2 > 0$, and $\varepsilon(d_1, d_2) = -1$ if $d_1 < 0$ and $d_2 < 0$.

For odd n let m^2 be the largest square dividing n. Then there is a unique choice of sign and a unique quadratic discriminant d_2 such that $n = \pm m^2 d_2$, and then if $(n, d_1) = 1$ the above can be applied to express $\left(\frac{d_1}{n}\right)$ in terms of $\left(\frac{d_2}{d_1}\right)$. If n is even, then $4n = m^2 d_2$ for unique m and quadratic discriminant d_2, so if $(n, d_1) = 1$ we can again express $\left(\frac{d_1}{n}\right)$ in terms of $\left(\frac{d_2}{d_1}\right)$.

Proof   Suppose that $d_1 = p \equiv 1 \pmod 4$. Then

$$\left(\frac{p}{d_2}\right)_K = \left(\frac{d_2}{p}\right)_L = \left(\frac{d_2}{p}\right)_K,$$

so (9.11) holds in this case. Next suppose that $d_1 = -p$ where $p \equiv 3 \pmod 4$. Then

$$\left(\frac{-p}{d_2}\right)_K = \left(\frac{d_2}{p}\right)_L = \left(\frac{d_2}{-1}\right)_K\left(\frac{d_2}{-p}\right)_K,$$

so (9.11) holds in this case also. Next consider the case $d_1 = -4$. Then d_2 is odd, and hence $d_2 \equiv 1 \pmod 4$, so that $\left(\frac{-4}{d_2}\right)_K = \left(\frac{-4}{1}\right)_K = 1$, while $\left(\frac{d_2}{-4}\right)_K = \left(\frac{d_2}{-1}\right)_K$, and (9.11) again holds. If $d_1 = 8$ then d_2 is odd and $\left(\frac{8}{d_2}\right)_K = (-1)^{(d_2^2-1)/8} = \left(\frac{d_2}{8}\right)_K$, so (9.11) holds. Similarly, if d_2 is odd, then $\left(\frac{-8}{d_2}\right)_K = \left(\frac{-4}{d_2}\right)_K\left(\frac{8}{d_2}\right)_K = \left(\frac{8}{d_2}\right)_K = \left(\frac{d_2}{8}\right)_K = \left(\frac{d_2}{-1}\right)_K\left(\frac{d_2}{-8}\right)_K$, so again (9.11) holds.

Now let d_1, d_2 and d be pairwise coprime quadratic discriminants. Then

$$\left(\frac{d_1 d_2}{d}\right)_K = \left(\frac{d_1}{d}\right)_K\left(\frac{d_2}{d}\right)_K.$$

Suppose that (9.11) holds for the pair d_1, d, and also for the pair d_2, d. Then the above is

$$= \varepsilon(d_1, d)\left(\frac{d}{d_1}\right)_K \varepsilon(d_2, d)\left(\frac{d}{d_2}\right)_K$$

$$= \varepsilon(d_1, d)\varepsilon(d_2, d)\left(\frac{d}{d_1 d_2}\right)_K.$$

But $\varepsilon(d_1, d)\varepsilon(d_2, d) = \varepsilon(d_1 d_2, d)$, so it follows that (9.11) holds also for the pair $d_1 d_2, d$. Since all quadratic discriminants can be constructed as the product of smaller quadratic discriminants, or by appealing to the special cases already considered, it follows now that (9.11) holds for all quadratic discriminants. $\square$

Let χ be a character modulo q. By means of Theorems 9.7 and 9.10 we can describe $|\tau(\chi)|$. By Theorem 9.5 we may also relate the argument of $\tau(\chi)$ to that of $\tau(\overline{\chi})$, but otherwise there is little in general that we can say about the argument of $\tau(\chi)$. However, in the special case of quadratic characters, a striking phenomenon arises, which was first noted and established by Gauss. Suppose that χ_d is a primitive quadratic character. Then $\overline{\chi}_d = \chi_d$, so by multiplying both sides of (9.7) by $\tau(\chi_d)$, and using Theorem 9.7, we see that $\tau(\chi_d)^2 = \chi_d(-1)|d| = d$. Thus $\tau(\chi_d) = \pm\sqrt{d}$ if $d > 0$ and $\tau(\chi_d) = \pm i\sqrt{-d}$ if $d < 0$. We show below that in both cases it is always the positive sign that occurs. We begin with the following fundamental result.

Theorem 9.15 *Let*

$$S(a, q) = \sum_{n=1}^{q} e\left(\frac{an^2}{2q}\right).$$

If a and q are positive integers and at least one of them is even, then

$$S(a, q) = \overline{S(q, a)}e(1/8)\sqrt{q/a}.$$

Proof We apply the Poisson summation formula, in the form of Theorem D.3, to the function $f(x) = e(ax^2/(2q))$ for $1/2 < x < q + 1/2$, with $f(x) = 0$ otherwise. Thus

$$S(a, q) = \sum_{n} f(n) = \lim_{K \to \infty} \sum_{k=-K}^{K} \widehat{f}(k)$$

where

$$\widehat{f}(k) = \int_{1/2}^{q+1/2} e(ax^2/(2q) - kx)\,dx.$$

We complete the square by writing

$$\frac{ax^2}{2q} - kx = \frac{a}{2q}(x - kq/a)^2 - \frac{k^2 q}{2a},$$

and make the change of variable $u = (x - kq/a)/q$, to see that

$$\widehat{f}(k) = qe(-k^2 q/(2a)) \int_{1/(2q)-k/a}^{1/(2q)+1-k/a} e(aqu^2/2)\,du.$$

By integrating by parts we see that

$$\widehat{f}(k) \ll_{a,q} 1/(|k| + 1).$$

Since at least one of a and q is even, if $k \equiv r \pmod{a}$ then $qk^2 \equiv qr^2 \pmod{2a}$. Thus if we write $k = am + r$, then

$$\sum_{k=-K}^{K} \widehat{f}(k) = q \left(\sum_{r=1}^{a} e\left(\frac{-qr^2}{2a} \right) \right) \left(\sum_{m=-K/a}^{K/a} \int_{1/(2q)-m-r/a}^{1/(2q)+1-m-r/a} e(aqu^2/2)\,du \right)$$
$$+ O_{q,a}(1/K).$$

Here the integrals may be combined to form one integral, which, as K tends to infinity tends to $I(aq/2)$ where $I(c) = \int_{-\infty}^{\infty} e(cu^2)\,du$. This is a conditionally convergent improper Riemann integral, but it is not necessary to evaluate this symmetrically as $\lim_{U \to \infty} \int_{-U}^{U}$, since $\int_{U}^{\infty} e(cu^2)\,du \ll 1/U$, by integration by parts. Thus we have shown that

$$S(a, q) = q\overline{S(q, a)}I(aq/2).$$

We take $a = 2$ and $q = 1$, and note that $S(2, 1) = 1$ and $S(1, 2) = 1 + i$. Hence $I(1) = 1/(1 - i) = e(1/8)/\sqrt{2}$. By a linear change of variables it is clear that if $c > 0$ then $I(c) = I(1)/\sqrt{c}$. On combining this information in the above, we obtain the stated identity. $\qquad\qquad\qquad\square$

By taking $a = 2$ we immediately obtain

Corollary 9.16 (Gauss) *For any positive integer q,*

$$\sum_{n=1}^{q} e(n^2/q) = q^{1/2}\frac{1+i^{-q}}{1+i^{-1}} = \begin{cases} q^{1/2} & \text{if } q \equiv 1 \pmod 4, \\ 0 & \text{if } q \equiv 2 \pmod 4, \\ iq^{1/2} & \text{if } q \equiv 3 \pmod 4, \\ (1+i)q^{1/2} & \text{if } q \equiv 0 \pmod 4. \end{cases}$$

This in turn enables us to reach our goal.

Theorem 9.17 *Let $\chi_d(n) = \left(\frac{d}{n}\right)$ be a primitive quadratic character. If $d > 0$, then $\tau(\chi_d) = \sqrt{d}$. If $d < 0$ then $\tau(\chi_d) = i\sqrt{-d}$.*

In the special case of the Legendre symbol, if we write $\tau_p = \sum_{n=1}^{p} \left(\frac{n}{p}\right)e(n/p)$, then this asserts that $\tau_p = \sqrt{p}$ for $p \equiv 1 \pmod 4$, while $\tau_p = i\sqrt{p}$ for $p \equiv 3 \pmod 4$.

Proof As in some of the preceding proofs, we establish the identities when the modulus is an odd prime or power of 2, and then write $d = d_1d_2$ to extend to the general primitive quadratic character.

Let

$$G(a, q) = \sum_{x=1}^{q} e\left(\frac{ax^2}{q}\right). \tag{9.12}$$

If p is an odd prime, then the number of solutions of the congruence $x^2 \equiv n \pmod{p}$ is $1 + \left(\frac{n}{p}\right)_L$, so $G(a, p) = \sum_{n=1}^{p} \left(1 + \left(\frac{n}{p}\right)\right)e(an/p)$. Thus if $p \nmid a$, then

$$G(a, p) = \sum_{n=1}^{p} \left(\frac{n}{p}\right) e(an/p). \tag{9.13}$$

Suppose that $p \equiv 1 \pmod 4$. Then from the above we see that $\tau(\chi_p) = G(1, p)$, and then by taking $q = p$ in Corollary 9.16 it follows that $G(1, p) = \sqrt{p}$ in this case.

Now suppose that $p \equiv 3 \pmod 4$. Then from the above we see that $\tau(\chi_{-p}) = G(1, p)$, and then by taking $q = p$ in Corollary 9.16 it follows that $G(1, p) = i\sqrt{p}$ in this case.

Clearly $\tau(\chi_{-4}) = e(1/4) - e(3/4) = 2i$, $\tau(\chi_8) = e(1/8) - e(3/8) - e(5/8) + e(7/8) = \sqrt{8}$, and $\tau(\chi_{-8}) = e(1/8) + e(3/8) - e(5/8) - e(7/8) = i\sqrt{8}$. Thus we have the stated result when d is a power of 2.

Next suppose that $d = d_1 d_2$ where d_1 and d_2 are quadratic discriminants and $(d_1, d_2) = 1$. Then by Theorem 9.6, $\tau(\chi_d) = \tau(\chi_{d1})\tau(\chi_{d2})\chi_{d1}(|d_2|)\chi_{d_2}(|d_1|)$. By considering the possible combinations of signs of d_1 and of d_2 we find that $\chi_{d_1}(|d_2|)\chi_{d_2}(|d_1|) = \chi_{d_1}(d_2)\chi_{d_2}(d_1)$ in all cases. This product is $\varepsilon(d_1, d_2)$ in the notation of Theorem 9.14. That is,

$$\tau(\chi_d) = \varepsilon(d_1, d_2)\tau(\chi_{d_1})\tau(\chi_{d_2}).$$

Thus if $\tau(\chi_{d_1})$ and $\tau(\chi_{d_2})$ have the asserted values, then so also does $\tau(\chi_d)$. Since every primitive quadratic character can be constructed this way, the proof is complete. $\qquad\square$

9.3.1   Exercises

1. (a) Show that if $p > 2$ and $p \nmid b$, then

$$\sum_{n=1}^{p} \left(\frac{n}{p}\right)\left(\frac{n+b}{p}\right) = -1.$$

(b) Suppose that $p > 2$ and that $p \nmid d$. Explain why

$$\sum_{x=1}^{p} \left(\frac{x^2 - d}{p}\right) = \sum_{n=1}^{p} \left(1 + \left(\frac{n}{p}\right)\right)\left(\frac{n-d}{p}\right),$$

and deduce that this sum is -1.

(c) Put $d = b^2 - 4ac$, and suppose that $p > 2$, $p \nmid d$. Show that

$$\sum_{x=1}^{p} \left(\frac{ax^2 + bx + c}{p} \right) = \left(\frac{a}{p} \right).$$

2. Let p be a prime, $p \equiv 1 \pmod 4$, and let $\mathcal{N}$ be a set of Z residue classes modulo p.
   (a) Explain why

   $$\sum_{m \in \mathcal{N}} \sum_{n \in \mathcal{N}} \left(\frac{m - n}{p} \right) = \frac{1}{\sqrt{p}} \sum_{a=1}^{p} \left(\frac{a}{p} \right) \left| \sum_{n \in \mathcal{N}} e(an/p) \right|^2.$$

   (b) Suppose that $\left(\frac{m-n}{p} \right) = 1$ whenever $m \in \mathcal{N}$, $n \in \mathcal{N}$, and $m \neq n$. Show that $Z \leq \sqrt{p}$.

3. Put $f_a(r) = r^2 + a_1 r + a_0$ where $a = (a_0, a_1)$. Show that if r_1, r_2, r_3 are distinct modulo p, then

   $$\sum_{a_0=1}^{p} \sum_{a_1=1}^{p} \left(\frac{f_a(r_1)}{p} \right) \left(\frac{f_a(r_2)}{p} \right) \left(\frac{f_a(r_3)}{p} \right) = p.$$

4. We used Corollary 9.16 to determine the sign of $\tau(\chi_{\pm p})$, and then used quadratic reciprocity to determine the sign of $\tau(\chi_d)$ for the general quadratic discriminant d. We now show that quadratic reciprocity for the Legendre symbol can be derived from Theorem 9.15 (mainly Corollary 9.16). Let $G(a, q) = \sum_{n=1}^{q} e(an^2/q)$.
   (a) Suppose that p is an odd prime. Explain why

   $$G(a, p) = \left(\frac{a}{p} \right)_L \sum_{n=1}^{p} \left(\frac{n}{p} \right) e(n/p)$$

   when $(a, p) = 1$.
   (b) Suppose that $(q_1, q_2) = 1$. By writing n modulo $q_1 q_2$ in the form $n = n_1 q_2 + n_2 q_1$, show that $G(a, q_1 q_2) = G(aq_2, q_1)G(aq_1, q_2)$.
   (c) Let p and q denote odd primes. Show that

   $$G(1, pq) = \left(\frac{p}{q} \right)_L \left(\frac{q}{p} \right)_L G(1, p)G(1, q),$$

   and use Corollary 9.16 to show that

   $$\left(\frac{p}{q} \right)_L \left(\frac{q}{p} \right)_L = (-1)^{\frac{p-1}{2} \cdot \frac{q-1}{2}}.$$

   (d) By taking $a = -1$ in (a), and using Corollary 9.16, show that $\left(\frac{-1}{p} \right) = (-1)^{(p-1)/2}$.
   (e) By taking $a = 4$ in Theorem 9.15, show that $\left(\frac{2}{p} \right)_L = (-1)^{(p^2-1)/8}$.

(f) Suppose that p is an odd prime, and k is an integer, $k \geq 2$. Show that $G(a, p^k) = pG(a, p^{k-2})$.

5. Let $\mathcal{L}_1$ denote the contour $z = u$, $-\infty < u < \infty$ in the complex plane, let $\mathcal{L}_2$ denote the contour $z = (1 + i)u$, $-\infty < u < \infty$, and let $I(c) = \int_{-\infty}^{\infty} e(cu^2)\, du$, as in the proof of Theorem 9.15.

(a) Note that $I(c) = \int_{\mathcal{L}_1} e^{2\pi i c z^2}\, dz$.

(b) Explain why $\int_{\mathcal{L}_1} e^{2\pi i c z^2}\, dz = \int_{\mathcal{L}_2} e^{2\pi i c z^2}\, dz$.

(c) Show that

$$\int_{\mathcal{L}_2} e^{2\pi i c z^2}\, dz = (1 + i) \int_{-\infty}^{\infty} e^{-4\pi c u^2}\, du = \frac{1 + i}{2\sqrt{\pi c}} \int_{-\infty}^{\infty} e^{-v^2}\, dv = \frac{1 + i}{2\sqrt{c}}.$$

(d) Thus give a proof, independent of that found in the proof of Theorem 9.15, that

$$\int_{-\infty}^{\infty} e(cu^2)\, du = \frac{1}{(1 - i)\sqrt{c}}.$$

6. Quadratic reciprocity à la Conway (1997, pp. 127–133). If $(a, n) = 1$ and n is an odd positive integer, then we define the *Zolotarev symbol* (not a standard term) $\left(\frac{a}{n}\right)_Z$ to be 1 if the map $x \mapsto ax$ is an even permutation of a complete residue system modulo n, and $\left(\frac{a}{n}\right)_Z = -1$ if it is odd.

(a) Compute the decomposition of the permutation $x \mapsto 7x \pmod{15}$ into disjoint cycles, and thus show that $\left(\frac{7}{15}\right)_Z = -1$.

(b) Suppose that p is an odd prime and that a has order h modulo p. Show that the map $x \mapsto ax \pmod{p}$ consists of one 1-cycle (0) and $(p - 1)/h$ h-cycles. Deduce that $\left(\frac{a}{p}\right)_Z = (-1)^{(p-1)/h}$.

(c) Continue in the same notation, and show that $(p - 1)/h$ is even if and only if $a^{(p-1)/2} \equiv 1 \pmod{p}$. Deduce that $\left(\frac{a}{p}\right)_Z = \left(\frac{a}{p}\right)_L$.

(d) If n is odd and positive, then the permutation $x \mapsto -x \pmod{n}$ consists of one 1-cycle and $(n - 1)/2$ 2-cycles of the form $(x - x)$. Hence deduce that $\left(\frac{-1}{n}\right)_Z = (-1)^{(n-1)/2}$.

(e) If $(ab, n) = 1$, then the map $x \mapsto abx \pmod{n}$ is the composition of the map $x \mapsto ax \pmod{n}$ and the map $x \mapsto bx \pmod{n}$. Deduce that $\left(\frac{ab}{n}\right)_Z = \left(\frac{a}{n}\right)_Z \left(\frac{b}{n}\right)_Z$.

(f) Let p be a prime, $p > 2$, and let g be a primitive root of p. By (b) with $h = p - 1$, deduce that $\left(\frac{g}{p}\right)_Z = -1$. Then by (e) deduce that $\left(\frac{g^k}{p}\right)_Z = (-1)^k$, and hence give a second proof of (c).

(g) Suppose that n is odd and positive, and that $(a, n) = 1$. Let

$$\mathcal{P} = \{1, 2, \ldots, (n - 1)/2\}, \qquad \mathcal{N} = \{-1, -2, \ldots, -(n - 1)/2\}.$$

Let K be the number of $k \in \mathcal{P}$ such that $ak \in \mathcal{N} \pmod{n}$. Put $\varepsilon_k = 1$

if k and ak lie in the same subset, otherwise put $\varepsilon_k = -1$. Note that $\varepsilon_k = \varepsilon_{-k}$. Let π^+ be the permutation that leaves $\mathcal{N}$ fixed and maps $\mathcal{P}$ to itself by the formula $k \mapsto \varepsilon_k ak \pmod{n}$. Let π^- be the map that leaves $\mathcal{P}$ fixed and maps $\mathcal{N}$ to itself by the formula $k \mapsto \varepsilon_k ak \pmod{n}$. Finally let π^* be the product of those transpositions $(ak - ak)$ for which $k \in \mathcal{P}$ and $ak \in \mathcal{N}$. Show that the map $x \mapsto ax \pmod{n}$ is the permutation $\pi^*\pi^+\pi^-$. Let σ be the 'sign change permutation' $x \mapsto -x \pmod{n}$. Show that $\pi^- = \sigma\pi^+\sigma$. That is, π^+ and π^- are conjugate permutations. They are the same apart from the fact that they operate on different sets. Thus they have the same cycle structure, and hence the same parity. Deduce that $\left(\frac{a}{n}\right)_Z = (-1)^K$.

(h) Suppose that n is odd and positive, that $(a, n) = 1$, and that $a > 0$. Show that $\left(\frac{a}{n}\right)_Z = (-1)^K$ where K is the number of integers lying in the intervals $((r - \frac{1}{2})\frac{n}{a}, \frac{rn}{a})$ for $r = 1, 2, \dots [a/2]$.

(i) Show that if $a > 0$, $(2a, n) = 1$, $m \equiv n \pmod{4a}$, then $\left(\frac{a}{m}\right)_Z = \left(\frac{a}{n}\right)_Z$.

(j) Show that if n is odd and positive, then $\left(\frac{2}{n}\right)_Z = (-1)^{(p^2-1)/8}$.

(k) Suppose that m and n are odd and positive, and that $m \equiv -n \pmod{4}$, say $m + n = 4a$. Justify the following manipulations:

$$\left(\frac{m}{n}\right)_Z = \left(\frac{4a}{n}\right)_Z = \left(\frac{a}{n}\right)_Z = \left(\frac{a}{m}\right)_Z = \left(\frac{4a}{m}\right)_Z = \left(\frac{n}{m}\right)_Z.$$

(l) Suppose that m and n are odd and positive, and that $m \equiv n \pmod{4}$, say $m > n$ and $m - n = 4a$. Justify the following manipulations:

$$\left(\frac{m}{n}\right)_Z = \left(\frac{4a}{n}\right)_Z = \left(\frac{a}{n}\right)_Z = \left(\frac{a}{m}\right)_Z = \left(\frac{4a}{m}\right)_Z$$
$$= \left(\frac{-n}{m}\right)_Z = \left(\frac{n}{m}\right)_Z (-1)^{(m-1)/2}.$$

(m) Suppose that a is odd and positive and that $(2a, mn) = 1$. Show that

$$\left(\frac{a}{mn}\right)_Z = \left(\frac{mn}{a}\right)_Z (-1)^{\frac{a-1}{2}\frac{mn-1}{2}} = \left(\frac{m}{a}\right)_Z \left(\frac{n}{a}\right)_Z (-1)^{\frac{a-1}{2}\frac{mn-1}{2}}$$
$$= \left(\frac{a}{m}\right)_Z \left(\frac{a}{n}\right)_Z (-1)^{\frac{a-1}{2}\frac{mn-1}{2}+\frac{a-1}{2}\frac{m-1}{2}+\frac{a-1}{2}\frac{n-1}{2}}.$$

Show that this last exponent is even, so that $\left(\frac{a}{mn}\right)_Z = \left(\frac{a}{m}\right)_Z\left(\frac{a}{n}\right)_Z$ in this case.

(n) Suppose that a is odd and negative and that $(a, mn) = 1$. Use (m) to show that the identity $\left(\frac{a}{mn}\right)_Z = \left(\frac{a}{m}\right)_Z\left(\frac{a}{n}\right)_Z$ holds in this case also. Thus this holds for all odd a.

(o) Suppose that a is even and that $(a, mn) = 1$. Justify the following manipulations:

$$\left(\frac{a}{mn}\right)_Z = \left(\frac{-a}{mn}\right)_Z (-1)^{\frac{mn-1}{2}} = \left(\frac{mn-a}{mn}\right)_Z (-1)^{\frac{mn-1}{2}}$$

$$= \left(\frac{mn-a}{m}\right)_Z \left(\frac{mn-a}{n}\right)_Z (-1)^{\frac{mn-1}{2}}$$

$$= \left(\frac{-a}{m}\right)_Z \left(\frac{-a}{n}\right)_Z (-1)^{\frac{mn-1}{2}} = \left(\frac{a}{m}\right)_Z \left(\frac{a}{n}\right)_Z (-1)^{\frac{mn-1}{2} + \frac{m-1}{2} \cdot 2 + \frac{n-1}{2}}.$$

Show that this last exponent is even, and thus deduce that

$$\left(\frac{a}{mn}\right)_Z = \left(\frac{a}{m}\right)_Z \left(\frac{a}{n}\right)_Z$$

holds in all cases.

(p) Suppose that $(a, m) = 1$ and that m is odd, composite, and square-free. Show that the permutation $x \mapsto ax \pmod{m}$ of reduced residues modulo m is always even. (Hence it is essential that we used complete residue systems in the above.)

7. Let p be a prime number, $p > 2$. (a) Show that

$$\prod_{k=1}^{p-1} (1 - e(k/p))^{\left(\frac{k}{p}\right)} = \exp(-\tau(\chi_p)L(1, \chi_p))$$

where $\chi_p(n) = \left(\frac{k}{p}\right)$.

Let $\mathcal{R} = \{r : 0 < r < p, \left(\frac{r}{p}\right) = 1\}$, $\mathcal{N} = \{n : 0 < n < p, \left(\frac{n}{p}\right) = -1\}$, and set

$$Q = \frac{\prod_{n \in \mathcal{N}} \sin \pi n / p}{\prod_{r \in \mathcal{R}} \sin \pi r / p}.$$

(b) Show that if $p \equiv 3 \pmod 4$, then $Q = 1$.

(c) Show that if $p \equiv 1 \pmod 4$, then $Q = \exp(\sqrt{p}\, L(1, \chi_p))$.

8. (Chowla & Mordell 1961) Continue with the notation of the preceding problem, let c be chosen, $0 < c < p$, so that $\left(\frac{c}{p}\right) = -1$, and put

$$f(z) = \prod_{r \in \mathcal{R}} \frac{1 - z^{cr}}{1 - z^r} - 1.$$

(a) Show that if $L(1, \chi_p) = 0$, then $f(e(1/p)) = 0$.

(b) Explain why f is a polynomial with integral coefficients.

(c) Show that if $L(1, \chi_p) = 0$, then there exists a polynomial $g \in \mathbb{Z}[z]$ such that $f(z) = g(z)(1 + z + \cdots + z^{p-1})$.

(d) By taking $z = 1$ in the above, show that it would follow that $c^{(p-1)/2} \equiv 1 \pmod{p}$.

(e) Explain why $c^{(p-1)/2} \equiv -1 \pmod{p}$; deduce that $L(1, \chi_p) \neq 0$.

9.4 Incomplete character sums

Let χ be a character modulo q. We call the sum $\sum_{n=M+1}^{M+N} \chi(n)$ *incomplete* if $N < q$. Such a sum trivially has absolute value at most N. We now use our knowledge of Gauss sums to show that if χ is non-principal, then this sum is $o(N)$ provided that N is not too small compared with q. Suppose first that χ is a primitive character modulo q with $q > 1$. Then by Corollary 9.8,

$$\sum_{n=M+1}^{M+N} \chi(n) = \frac{1}{\tau(\overline{\chi})} \sum_{a=1}^{q} \overline{\chi}(a) \sum_{n=M+1}^{M+N} e(an/q).$$

Here the inner sum is a geometric series. We note that

$$\sum_{n=M+1}^{M+N} e(n\alpha) = \frac{e((M+N+1)\alpha) - e((M+1)\alpha)}{e(\alpha) - 1}$$

$$= e((2M+N+1)\alpha/2) \frac{\sin \pi N\alpha}{\sin \pi \alpha} \qquad (9.14)$$

if α is not an integer. (If $\alpha \in \mathbb{Z}$, then the sum is N.) On combining this with the above, we see that

$$\sum_{n=M+1}^{M+N} \chi(n) = \frac{1}{\tau(\overline{\chi})} \sum_{a=1}^{q} \overline{\chi}(a) e\left(\frac{a(2M+N+1)}{2q}\right) \frac{\sin \pi a N/q}{\sin \pi a/q}. \qquad (9.15)$$

By Theorem 9.7 and the triangle inequality the right-hand side has absolute value

$$< \frac{1}{\sqrt{q}} \sum_{\substack{a=1 \\ (a,q)=1}}^{q-1} \frac{1}{\sin \pi a/q}.$$

Here the second half of the range of summation contributes the same amount as the first. Hence it suffices to multiply by 2 and sum over $1 \leq a \leq q/2$. However, if q is odd, then $q/2$ is not an integer and hence the sum is actually over the range $1 \leq a \leq (q-1)/2$, while if q is even, then $4 \mid q$ (since if $q \equiv 2 \pmod 4$, then there is no primitive character modulo q), and hence $(q/2, q) > 1$, and so it suffices to sum over $1 \leq a \leq q/2 - 1$ in this case. Hence in either case the expression above is

$$\leq \frac{2}{\sqrt{q}} \sum_{a=1}^{(q-1)/2} \frac{1}{\sin \pi a/q}.$$

The function $f(\alpha) = \sin \pi \alpha$ is concave downward in the interval $[0, 1/2]$, and hence it lies above the chord through the points $(0, 0), (1/2, 1)$. That is, $\sin \pi \alpha \geq 2\alpha$ for $0 \leq \alpha \leq 1/2$. Thus the above is

$$\leq \sqrt{q} \sum_{a=1}^{(q-1)/2} \frac{1}{a} < \sqrt{q} \sum_{a=1}^{(q-1)/2} \log \frac{1 + \frac{1}{2a}}{1 - \frac{1}{2a}} = \sqrt{q} \sum_{a=1}^{(q-1)/2} \log \frac{2a+1}{2a-1} = \sqrt{q} \log q.$$

That is,

$$\left| \sum_{n=M+1}^{M+N} \chi(n) \right| < \sqrt{q} \log q \qquad (9.16)$$

when χ is primitive. We now extend this to imprimitive non-principal characters. Suppose that χ is induced by χ^* modulo d. Let r be the product of those primes that divide q but not d. Then

$$\sum_{n=M+1}^{M+N} \chi(n) = \sum_{\substack{n=M+1 \\ (n,r)=1}}^{M+N} \chi^*(n)$$

$$= \sum_{n=M+1}^{M+N} \chi^*(n) \sum_{k|(n,r)} \mu(k)$$

$$= \sum_{k|r} \mu(k) \sum_{\substack{M < n \leq M+N \\ k|n}} \chi^*(n)$$

$$= \sum_{k|r} \mu(k)\chi^*(k) \sum_{M/k < m \leq (M+N)/k} \chi^*(m).$$

By the case already treated, we know that the inner sum above has absolute value not exceeding $d^{1/2} \log d$, and hence the given sum has absolute value not more than $2^{\omega(r)} d^{1/2} \log d$. But $2^{\omega(r)} \leq d(r) \ll r^{1/2} \leq (q/d)^{1/2}$, so we have proved

Theorem 9.18 (The Pólya–Vinogradov inequality) *Let χ be a non-principal character modulo q. Then for any integers M and N with $N > 0$,*

$$\sum_{n=M+1}^{M+N} \chi(n) \ll \sqrt{q} \log q.$$

In (9.16) we saw that the implicit constant can be taken to be 1 when χ is primitive. With a little more care it can be seen that the implicit constant can be taken to be 1 for all non-principal characters. The above estimate is important in many contexts, but we confine ourselves to two applications at this point.

Corollary 9.19 *Let χ be a non-principal character modulo p, and let n_χ be the least positive integer n such that $\chi(n) \neq 1$. Then $n_\chi \ll_\varepsilon p^{\frac{1}{2\sqrt{e}}+\varepsilon}$.*

Proof Suppose that $\chi(n) = 1$ for all $n \leq y$. Then $\chi(n) = 1$ whenever n is composed entirely of primes $q \leq y$. Hence, in the notation of Section 7.1, if $y \leq x < y^2$, then

$$\sum_{n \leq x} \chi(n) = \psi(x, y) + \sum_{y < q \leq x} \chi(q)[x/q]$$

where q denotes a prime. Thus

$$\left| \sum_{n \leq x} \chi(n) \right| \geq \psi(x, y) - \sum_{y < q \leq x} [x/q] = [x] - 2 \sum_{y < q \leq x} [x/q]$$

$$= x \left(1 - 2 \log \frac{\log x}{\log y} \right) + O \left(\frac{x}{\log x} \right).$$

If $x = p^{1/2}(\log p)^2$, then the sum on the left is $o(x)$, while if $y > x^{1/\sqrt{e}+\varepsilon}$, then the lower bound on the right is $\gg \varepsilon x$. Thus $n_\chi \ll_\varepsilon x^{1/\sqrt{e}+\varepsilon}$. $\square$

Corollary 9.20 *The number of primitive roots modulo p in the interval $[M + 1, M + N]$ is*

$$\frac{\varphi(p-1)}{p} N + O\left(p^{1/2+\varepsilon}\right).$$

Since the number of primitive roots in an interval of length p is exactly $\varphi(p - 1)$, the above asserts that primitive roots are roughly uniformly distributed into subintervals of length N provided that $N > p^{1/2+\varepsilon}$.

Proof Let $q_1, q_2, \ldots, q_r$ be the distinct prime factors of $p - 1$, and put $q = \prod_{i=1}^{r} q_i$. Then n is a primitive root modulo p if and only if $(\text{ind } n, q) = 1$. For $1 \leq i \leq r$ put

$$\chi_i(n) = e \left(\frac{\text{ind } n}{q_i} \right).$$

Then

$$\frac{1}{q_i} \sum_{a=1}^{q_i} \chi_i(n)^a = \begin{cases} 1 & \text{if } q_i \mid \text{ind } n, \\ 0 & \text{otherwise.} \end{cases}$$

Thus

$$\prod_{i=1}^{r}\left(\chi_0(n) - \frac{1}{q_i}\sum_{a_i=1}^{q_i}\chi_i(n)^{a_i}\right) = \begin{cases} 1 & \text{if } n \text{ is a primitive root } (\mathrm{mod} \ p), \\ 0 & \text{otherwise.} \end{cases}$$

The left-hand side above is

$$\prod_{i=1}^{r}\left((1 - 1/q_i)\chi_0(n) - \frac{1}{q_i}\sum_{a_i=1}^{q_i-1}\chi_i^{a_i}(n)\right) = \sum_{d|q}\frac{\varphi(q/d)}{q/d}\frac{\mu(d)}{d}\sum_{\substack{\chi \\ \mathrm{ord}\chi=d}}\chi(n).$$

Thus the number of primitive roots in the interval $[M+1, M+N]$ is

$$\frac{1}{q}\sum_{d|q}\varphi(q/d)\mu(d)\sum_{\substack{\chi \\ \mathrm{ord}\chi=d}}\sum_{n=M+1}^{M+N}\chi(n). \tag{9.17}$$

The only character of order $d = 1$ is the principal character χ_0, which gives us the main term

$$\frac{\varphi(q)}{q}((1 - 1/p)N + O(1)) = \frac{\varphi(p-1)}{p}N + O(1).$$

A character of order $d > 1$ is non-principal, and for such characters the innermost sum in (9.17) is $\ll p^{1/2}\log p$. Since there are $\varphi(d)$ such characters, the contribution in (9.17) of $d > 1$ is

$$\ll \frac{\varphi(q)}{q}p^{1/2}\log p\sum_{d|(p-1)}|\mu(d)| \ll 2^{\omega(p-1)}p^{1/2}\log p \ll p^{1/2+\varepsilon}.$$

This gives the stated result.   □

Suppose that χ is a non-principal character modulo q. Further insights into the Pólya–Vinogradov inequality may be gained by considering the sum $f_\chi(\alpha) = \sum_{0<n\le q\alpha}\chi(n)$ as a function of the real variable α, for $0 \le \alpha \le 1$. We extend the domain of $f_\chi(\alpha)$ by periodicity, and compute its Fourier coefficients:

$$\widehat{f}_\chi(k) = \int_0^1 f_\chi(\alpha)e(-k\alpha)\,d\alpha = \sum_{n=1}^q \chi(n)\int_{n/q}^1 e(-k\alpha)\,d\alpha.$$

The nature of this integral depends on whether $k = 0$ or not. In the former case we find that

$$\widehat{f}_\chi(0) = \sum_{n=1}^q \chi(n)\left(1 - \frac{n}{q}\right) = \frac{-1}{q}\sum_{n=1}^q n\chi(n),$$

while for $k \ne 0$ we have

$$\widehat{f}_\chi(k) = \sum_{n=1}^q \chi(n)\frac{1 - e(-kn/q)}{-2\pi i k} = \frac{1}{2\pi i k}\sum_{n=1}^q \chi(n)e(-kn/q) = \frac{c_\chi(-k)}{2\pi i k}.$$

It is convenient to restrict to primitive characters, since then $c_\chi(-k) = \overline{\chi}(-k)\tau(\chi)$ by Theorem 9.5. Since $f_\chi(\alpha)$ is a function of bounded variation it follows that

$$f_\chi(\alpha) = \frac{-1}{q} \sum_{n=1}^q n\chi(n) + \frac{\tau(\chi)}{2\pi i} \sum_{k \neq 0} \frac{\overline{\chi}(-k)}{k} e(k\alpha) \qquad (9.18)$$

at points of continuity of f_χ, with the understanding that the sum is calculated as the limit of the symmetric partial sums $\sum_{-K}^K$. If $\chi(-1) = 1$, then $f_\chi(\alpha)$ is an odd function and the contributions of k and of $-k$ can be combined to form a sine series. If $\chi(-1) = -1$, then $f_\chi(\alpha)$ is an even function, and the two terms merge to form a cosine series. In this case it is interesting to note that if we take $\alpha = 0$ then we obtain another proof of (9.9). Among other possible values of α that might be considered, the possibility $\alpha = 1/2$ is particularly striking. If $\chi(-1) = 1$ then $f_\chi(1/2) = 0$ by symmetry, so in continuing we suppose that $\chi(-1) = -1$. Note that if q is odd then $1/2$ is not of the form n/q, and hence $f_\chi(\alpha)$ is continuous at $1/2$. On the other hand, there is no primitive character modulo 2 and hence if q is even then $4 \mid q$. In this case we can solve the equation $n/q = 1/2$ by taking $n = q/2$, but then $q/2$ is even, so that $(q/2, q) > 1$, and hence $\chi(q/2) = 0$. Hence $f_\chi(\alpha)$ is continuous at $1/2$ in all cases, and we deduce that

$$\sum_{0 < n \leq q/2} \chi(n) = \frac{-1}{q} \sum_{n=1}^q n\chi(n) - \frac{\tau(\chi)}{\pi i} \sum_{k=1}^\infty \frac{\overline{\chi}(k)}{k}(-1)^k.$$

As we already discovered by taking $\alpha = 0$, the first term on the right is $\tau(\chi)L(1, \overline{\chi})/(\pi i)$. But

$$\sum_{k=1}^\infty \frac{\chi(k)(-1)^k}{k^s} = (2^{1-s}\chi(2) - 1)L(s, \chi)$$

for any character χ and any s with positive real part, so we have proved

Theorem 9.21 *Let χ be a primitive character modulo q such that $\chi(-1) = -1$. Then*

$$\sum_{1 \leq n \leq q/2} \chi(n) = (2 - \chi(2)) \frac{\tau(\chi)}{\pi i} L(1, \overline{\chi}).$$

In the special case that χ is a quadratic character we know the exact value of the Gauss sum, and hence we can say more.

Corollary 9.22 *If d is a quadratic discriminant with $d < 0$, then*

$$\sum_{1 \leq n \leq |d|/2} \left(\frac{d}{n}\right) > 0.$$

On taking $\alpha = (M + N)/q$ and then $\alpha = M/q$, and differencing, we see that

$$\sum_{n=M+1}^{M+N} \chi(n) = \frac{\tau(\chi)}{2\pi i} \sum_{k \neq 0} \frac{\overline{\chi}(-k)}{k} e(kM/q)(e(kN/q) - 1) + O(1).$$

Since $e(kN/q) - 1 \sim 2\pi i k N/q$ when $|k|$ is small compared with N/q, for rough heuristics we think of the above as being approximately

$$\frac{\tau(\chi)N}{q} \sum_{0 < |k| \leq N/q} \overline{\chi}(-k)e(kM/q).$$

Here a sum over an interval of length N reflects – approximately – to form a sum over an interval of length N/q. Further examples of this sort of phenomenon will emerge when we consider approximate functional equations of $\zeta(s)$ and of $L(s, \chi)$.

The Fourier expansion (9.18) is also useful in deriving quantitative estimates. We know not only that $\text{Var}_{[0,1]} f_\chi = \varphi(q)$, but (by Theorems 2.10 and 3.1) also that this variation is reasonably well distributed in subintervals, in the sense that $\text{Var}_{[\alpha,\beta]} f_\chi \ll \varphi(q)(\beta - \alpha)$ when $\beta - \alpha > q^{-1+\varepsilon}$. We apply Theorem D.2 to $f_\chi(\alpha)$, and divide the range of integration $(0, 1)$ into K intervals of length $1/K$, throughout each of which the integrand has a constant order of magnitude. Thus we see that

$$f_\chi(\alpha) = \frac{-1}{q} \sum_{n=1}^{q} n\chi(n) + \frac{\tau(\chi)}{2\pi i} \sum_{0 < |k| \leq K} \frac{\overline{\chi}(-k)}{k} e(k\alpha) + O\left(\frac{\varphi(q)}{K} \log 2K\right)$$

$$(9.19)$$

for $K \leq q^{1-\varepsilon}$. This can be used to obtain sharper constants in the Pólya–Vinogradov inequality; see Exercise 9.4.9.

We can also show that the estimate provided by the Pólya–Vinogradov inequality is in general not far from the truth.

Theorem 9.23 *Suppose that χ is a non-principal character modulo q. Then*

$$\max_{M,N} \left| \sum_{n=M+1}^{M+N} \chi(n) \right| \geq \frac{|\tau(\chi)|}{\pi}.$$

Proof Clearly

$$\left| \sum_{M=1}^{q} e(M/q) \sum_{n=M+1}^{M+N} \chi(n) \right| \leq \sum_{M=1}^{q} \left| \sum_{n=M+1}^{M+N} \chi(n) \right| \leq q \max_{M} \left| \sum_{n=M+1}^{M+N} \chi(n) \right|.$$

Here the sum on the left is

$$\sum_{n=1}^{N}\sum_{M=1}^{q} e(M/q)\chi(M+n) = \sum_{n=1}^{N} e(-n/q)\sum_{M=1}^{q} \chi(M)e(M/q).$$

By (9.14) this is

$$e\left(\frac{-(N+1)}{2q}\right)\frac{\sin \pi N/q}{\sin \pi/q}\tau(\chi).$$

If q is even, then we may take $N = q/2$, and then the quotient of sines is $= 1/(\sin \pi/q) \ge q/\pi$, while if q is odd, then we may take $N = (q-1)/2$, in which case the quotient of sines is

$$\frac{\cos \frac{\pi}{2q}}{\sin \frac{\pi}{q}} = \frac{1}{2\sin \frac{\pi}{2q}} \ge \frac{q}{\pi}.$$

The stated lower bound now follows by combining these estimates.   □

If χ is primitive modulo q, then the lower bound of Theorem 9.23 is $\sqrt{q}/\pi$. Further lower bounds of this nature can be derived by using Parseval's identity (4.4) for the finite Fourier transform; see Exercise 9.4.8. In addition to the lower bound above, which applies to all characters, for a sparse subset of characters we can obtain a better lower bound.

Theorem 9.24   (Paley) *There is a positive constant c such that*

$$\max_{M,N}\sum_{n=M+1}^{M+N}\left(\frac{d}{n}\right) > c\sqrt{d}\log\log d$$

for infinitely many positive quadratic discriminants d.

Proof   Let χ be a primitive character modulo q such that $\chi(-1) = 1$. By taking $M = k - h - 1$ and $N = 2h + 1$ in (9.15) we see that

$$\sum_{n=k-h}^{k+h}\chi(n) = \frac{1}{\tau(\overline{\chi})}\sum_{a=1}^{q}\overline{\chi}(a)e(ak/q)\frac{\sin \pi a(2h+1)/q}{\sin \pi a/q}.$$

Let h be the integer closest to $q/3$. Then the sine in the numerator is approximately $\sin 2\pi a/3$ when a is small. We shall choose χ so that $\chi(a) = \left(\frac{a}{3}\right)_L$ when a is small. Thus these two factors are strongly correlated. We would take $k = 0$ except for the need to dampen the effects of the larger values of a. To this end we sum over k, for $-K \le k \le K$ and divide by $2K + 1$. Thus by (9.14),

$$\frac{1}{2K+1} \sum_{k=-K}^{K} \sum_{n=k-h}^{k+h} \chi(n)$$

$$= \frac{1}{\tau(\overline{\chi})} \sum_{a=1}^{q} \overline{\chi}(a) \frac{\sin \pi a(2h+1)/q}{\sin \pi a/q} \frac{\sin \pi (2K+1)a/q}{(2K+1)\sin \pi a/q}. \qquad (9.20)$$

Here the last factor is approximately 1 if $\|a/q\| \le 1/K$, and decreases as $\|a/q\|$ becomes larger. Thus, despite its complicated appearance, the expression above is effectively

$$\frac{2q}{\pi \tau(\overline{\chi})} \sum_{a=1}^{A} \frac{\overline{\chi}(a) \sin 2\pi a/3}{a}$$

where $A = q/K$. To make this precise we observe that

$$\sin \pi (2h+1)a/q = \sin 2\pi a/3 + O(\|a/q\|)$$

and that

$$\frac{\sin \pi (2K+1)a/q}{(2K+1)\sin \pi a/q} = \begin{cases} 1 + O(K^2\|a/q\|^2) & (\|a/q\| \le 1/K), \\ O(K^{-1}\|a/q\|^{-1}) & (\|a/q\| > 1/K). \end{cases}$$

Thus the right-hand side of (9.20) is

$$= \frac{2}{\tau(\overline{\chi})} \sum_{a=1}^{q/K} \overline{\chi}(a) \left(\frac{1}{\pi a/q} + O\left(\frac{a}{q}\right) \right) \left(\sin 2\pi a/3 + O\left(\frac{a}{q}\right) \right)$$

$$\times \left(1 + O\left(\frac{K^2 a^2}{q^2}\right) \right) + O\left(\frac{1}{\sqrt{q}} \sum_{q/K < a \le q/2} \frac{q^2}{Ka^2} \right)$$

$$= \frac{2q}{\pi \tau(\overline{\chi})} \sum_{a=1}^{q/K} \frac{\overline{\chi}(a) \sin 2\pi a/3}{a} + O(\sqrt{q}). \qquad (9.21)$$

Now let y be a large parameter, and suppose that

$$q \equiv 5 \pmod 8,$$

$$\left(\frac{q}{p} \right)_L = \left(\frac{p}{3} \right)_L \qquad (3 < p \le y). \qquad (9.22)$$

Thus by the Chinese Remainder Theorem, q is restricted to certain residue classes modulo $Q = 8 \prod_{3 < p \le y} p$. Now let q be the least positive number that satisfies these constraints. Then q is square-free, and hence q is a quadratic discriminant, so we may take $\chi(n) = \left(\frac{q}{n}\right)_K$. Also, $q < Q$. By the Prime Number Theorem in the form of (6.13) we see that $\log Q = (1 + o(1))y$. Let K be the least integer such that $K > q/y$. Then by (9.22), $\chi(a) = \left(\frac{a}{3}\right)_L$ for $1 \leq a \leq q/K$, $(a, 3) = 1$. Thus $\sum_{1 \leq a \leq u} \chi(a) \sin 2\pi a/3 = u/\sqrt{3} + O(1)$, so the main term in (9.21) is

$$\frac{2\sqrt{q}}{\pi\sqrt{3}}(\log y + O(1)) \geq \left(\frac{2}{\pi\sqrt{3}} + o(1)\right)\sqrt{q}\log\log q.$$

This completes the proof.   □

In the two preceding theorems we have seen that the character sum can be large when N is comparable to q. For shorter sums we would expect the sum to be smaller, and indeed one would conjecture that if χ is a non-principal character modulo q, then

$$\sum_{n=M+1}^{M+N} \chi(n) \ll_\varepsilon N^{1/2}q^\varepsilon \tag{9.23}$$

for any $\varepsilon > 0$. Although our present knowledge falls far short of this, we now show that some improvement of the Pólya–Vinogradov inequality is possible, at least in some situations. Our approach depends on the Riemann hypothesis for curves over a finite field, in the form of the following character sum estimate, which we derive from the exposition of Schmidt (1976).

Lemma 9.25 (Weil) *Suppose that $d|(p - 1)$ with $d > 1$ and that χ is a character modulo p of order d. Suppose further that $e_j \geq 1$ $(1 \leq j \leq k)$, that $d \nmid e_j$ for some j with $1 \leq j \leq k$ and that the $c_1, c_2, \ldots, c_k$ are distinct modulo p. Then*

$$\left|\sum_{n=1}^{p} \chi\left((n + c_1)^{e_1}(n + c_2)^{e_2}\cdots(n + c_k)^{e_k}\right)\right| \leq (k - 1)p^{1/2}.$$

Proof Let $f(x) = (x + c_1)^{e_1}(x + c_2)^{e_2}\cdots(x + c_k)^{e_k}$. Then, by Lemma 4B of Schmidt (1976), $f(x)$ cannot satisfy $f(x) \equiv g(x)^d \pmod{p}$ identically where g is a polynomial with integer coefficients. The lemma then follows from Theorem 2C' ibidem.   □

Lemma 9.26 *Suppose that χ is a non-principal character modulo p and let*

$$S_{h,r} = \sum_{n=1}^{p}\left|\sum_{m=1}^{h}\chi(m + n)\right|^{2r}.$$

Then $S_{h,r} \ll r^{2r}\left(h^r p + h^{2r}p^{1/2}\right)$ for positive integers r.

Proof Clearly we may suppose that $h \leq p$. Let d denote the order of χ. Then $d > 1$ and

$$S_{h,r} = \sum_{m_1,\ldots,m_{2r}} \sum_{n=1}^{p} \chi((n+m_1)\cdots(n+m_r)(n+m_{r+1})^{d-1}\cdots(n+m_{2r})^{d-1}).$$

For a given $2r$–tuple $m_1, \ldots, m_{2r}$ let $c_1 < c_2 < \cdots < c_k$ be the distinct values of the m_j, and let a_l and b_l denote the number of occurrences of c_l amongst the $m_1, \ldots, m_r$ and $m_{r+1}, \ldots, m_{2r}$ respectively. Let $e_l = a_l + (d-1)b_l$. Then $(n+m_1)\cdots(n+m_r)(n+m_{r+1})^{d-1}\cdots(n+m_{2r})^{d-1} = (n+c_1)^{e_1}\cdots(n+c_k)^{e_k}$. Note that $e_1 + \cdots + e_k = r + r(d-1) = rd$. If there is an l such that $d \nmid e_l$, then by Lemma 9.25 the sum over n is bounded by $(k-1)p^{\frac{1}{2}}$, and so the total contribution to $S_{h,r}$ from such $2r$–tuples is

$$\leq 2rh^{2r}p^{\frac{1}{2}}.$$

On the other hand, if $d|e_l$ for every l, then $kd \leq e_1 + \cdots e_k = rd$ and so $k \leq r$. The number of choices of $m_1, \ldots, m_{2r}$ with $m_l \in \{c_1, \ldots, c_k\}$ is at most k^{2r} and the number of choices for $c_1, \ldots, c_k$ is $\binom{h}{k}$. Thus the total contribution to $S_{h,r}$ from these terms is bounded by

$$\sum_{k \leq r} k^{2r} \binom{h}{k} p \ll r^{2r} h^r p.$$

$\square$

Our main result takes the following form.

Theorem 9.27 (Burgess) *For any odd prime p and any positive integer r we have*

$$\sum_{n=M+1}^{M+N} \chi(n) \ll r N^{1-\frac{1}{r}} p^{\frac{r+1}{4r^2}} (\log p)^{\alpha_r}$$

where $\alpha_r = 1$ when $r = 1$ or 2 and $\alpha_r = \frac{1}{2r}$ otherwise.

Suppose that $\delta > 1/4$. If $N > p^{\delta}$, then the bound above is $o(N)$ if r is chosen suitably large in terms of δ. Thus any interval of length N contains both quadratic residues and quadratic non-residues. In addition the reasoning used to derive Corollary 9.19 applies here, so we see that the least positive quadratic non-residue modulo p is $\ll_{\varepsilon} p^{\frac{1}{4\sqrt{e}}+\varepsilon}$.

Proof When $r = 1$ or $N > p^{5/8}$ the bound is weaker than the Pólya–Vinogradov Inequality (Theorem 9.18), and when $r > 2$ and $N > p^{1/2}$ the stated bound is weaker than the case $r = 2$. Also, when $N \leq p^{\frac{r+1}{4r}}$ the bound is worse than trivial. Hence we may suppose that

$$p > p_0, \quad r \geq 2, \quad \text{and} \quad p^{\frac{r+1}{4r}} < N \leq \begin{cases} p^{5/8} & \text{when } r = 2, \\ p^{1/2} & \text{when } r > 2. \end{cases} \quad (9.24)$$

Let $S(M, N)$ denote the sum in question. Then

$$S(M, N) = \sum_{n=M+1}^{M+N} \chi(n + ab) + S(M, ab) - S(M + N, ab).$$

Let

$$\mathcal{M}(y) = \max_{\substack{M,N \\ N \leq y}} |S(M, N)|.$$

Then

$$S(M, N) = \sum_{n=M+1}^{M+N} \chi(n + ab) + 2\theta \mathcal{M}(ab)$$

where $|\theta| \leq 1$. We sum this over $a \in [1, A]$ and $b \in [1, B]$. Thus

$$AB S(M, N) = \sum_{n,a,b} \chi(n + ab) + 2AB\theta_1 \mathcal{M}(AB).$$

We suppose that

$$A < p \quad (9.25)$$

and then define $v(\ell)$ to be the number of pairs a, n with $a \in [1, A]$, $n \in [M + 1, M + N]$ and $n \equiv a\ell \pmod{p}$. Thus

$$\left| \sum_{n,a,b} \chi(n + ab) \right| = \left| \sum_{\ell=1}^{p} \sum_{\substack{n,a \\ n \equiv a\ell \pmod{p}}} \chi(a) \sum_{b} \chi(\ell + b) \right|$$

$$\leq \sum_{\ell=1}^{p} v(\ell) \left| \sum_{b} \chi(\ell + b) \right|.$$

By Hölder's inequality,

$$\left(\sum_{\ell=1}^{p} v(\ell) \left| \sum_{b} \chi(\ell + b) \right| \right)^{2r} \leq \left(\sum_{\ell=1}^{p} v(\ell)^{\frac{2r}{2r-1}} \right)^{2r-1} \sum_{\ell=1}^{p} \left| \sum_{b} \chi(\ell + b) \right|^{2r}$$

and

$$\left(\sum_{\ell=1}^{p} v(\ell)^{\frac{2r}{2r-1}} \right)^{2r-1} \leq \left(\sum_{\ell=1}^{p} v(\ell) \right)^{2r-2} \sum_{\ell=1}^{p} v(\ell)^2.$$

Clearly

$$\sum_{\ell=1}^{p} v(\ell) = AN.$$

We show below that if

$$AN < \frac{1}{2}p, \quad 1 \le A \le N, \tag{9.26}$$

then

$$\sum_{\ell=1}^{p} v(\ell)^2 \ll AN \log p. \tag{9.27}$$

Assuming this, we take $A = \left[\frac{1}{10}Np^{-1/(2r)}\right]$, $B = \left[p^{1/(2r)}\right]$. Then (9.24) gives (9.25) and (9.26). Thus from Lemma 9.26 with $h = B$ we see that

$$\sum_{n,a,b} \chi(n + ab) \ll rN^{2-\frac{1}{r}} p^{\frac{r+1}{4r^2}} (\log p)^{\frac{1}{2r}}.$$

Hence there is an absolute constant C such that

$$|S(M, N)| \le CrN^{1-\frac{1}{r}} p^{\frac{r+1}{4r^2}} (\log p)^{\frac{1}{2r}} + 2\mathcal{M}(N/10). \tag{9.28}$$

Choose M_1, N_1 with $N_1 \le N$ so that $|S(M_1, N_1)| = \mathcal{M}(N)$. If (9.24) fails because $N_1 \le p^{\frac{r+1}{4r}}$, then (9.28) with $M = M_1$, $N = N_1$ is trivial. Thus we have

$$\mathcal{M}(N) \le N^{1-\frac{1}{r}}\lambda + 2\mathcal{M}(N/10) \tag{9.29}$$

where

$$\lambda = Crp^{\frac{r+1}{4r^2}} (\log p)^{\frac{1}{2r}}.$$

Moreover (9.29) is also trivial when $N \le p^{\frac{r+1}{4r}}$. We apply (9.29) repeatedly with N replaced by $[N/10]$, $\left[[N/10]/10\right]$, and so on. Thus

$$\mathcal{M}(N) \le N^{1-\frac{1}{r}}\lambda \sum_{k=0}^{K} 2^k 10^{-k(1-\frac{1}{r})} + 2^{K+1}\mathcal{M}(10^{-K-1}N).$$

The trivial bound $\mathcal{M}(10^{-K-1}N) \ll 10^{-K}N$ with a judicious choice of K suffices to give

$$\mathcal{M}(N) \ll N^{1-\frac{1}{r}}\lambda$$

which completes the proof, apart from the need to establish (9.27) with (9.26). Clearly

$$\sum_{\ell} v(\ell)^2$$

is the number of choices of a, n, a', n', ℓ with $a, a' \in [1, A]$, $n, n' \in [1, N]$, $M + n \equiv a\ell \pmod{p}$, $M + n' \equiv a'\ell \pmod{p}$. Since $1 \le a, a' \le A < p$, by elimination of l we see that this is the number of solutions of $(a - a')M \equiv a'n - an' \pmod{p}$ with a, n, a', n' as before. Given any such pair a, a', choose k so that $k \equiv (a - a')M \pmod{p}$ and $|k| < p/2$. We have $1 \le a'n, an' \le AN \le \frac{1}{10}N^2 p^{-\frac{1}{2r}} < p/2$ in all cases. Thus $a'n - an' = k$. Given any one pair $n = n_0$, $n' = n_0'$ satisfying this equation we have, in general, $n = n_0 + \frac{a}{(a,a')}h$, $n' = n_0' + \frac{a'}{(a,a')}h$. Moreover $|h| \le \frac{N(a,a')}{\max\{a,a'\}}$. Therefore the total number of possible pairs n, n' is at most $1 + \frac{2N(a,a')}{\max\{a,a'\}}$. Hence

$$\sum_{\ell} v(\ell)^2 \ll A^2 + \sum_{1 \le a \le a' \le A} \frac{N(a, a')}{a'}$$

$$\ll A^2 + \sum_{d \le A} \sum_{1 \le b \le b' \le A/d} \frac{N}{b'}$$

$$\ll A^2 + AN \log 2A.$$

and so we have (9.27). $\qquad\qquad\qquad\qquad\qquad\qquad\qquad\qquad\qquad\square$

9.4.1   Exercises

1. Let χ be a non-principal character modulo q, and suppose that $(a, q) = 1$. Choose $\bar{a}$ so that $a\bar{a} \equiv 1 \pmod{q}$.
   (a) Explain why

   $$\overline{\chi}(a) \sum_{n=M+1}^{M+N} \chi(an + b) = \sum_{n=M+\bar{a}b+1}^{M+\bar{a}b+N} \chi(n).$$

   (b) Show that

   $$\sum_{n=M+1}^{M+N} \chi(an + b) \ll \sqrt{q} \log q.$$

2. With reference to the proof of Theorem 9.21, show that $2^{\omega(r)} \le c\sqrt{r}$ for all positive integers r where $c = 4/\sqrt{6}$, and that equality holds only when $r = 6$.

3. Show that if χ is a character modulo q with $\chi(-1) = -1$, then

$$\sum_{n=1}^{q} n^2 \chi(n) = q \sum_{n=1}^{q} n \chi(n).$$

4. (a) Let c_n and $f(n)$ have period q. Show that

$$\sum_{n=1}^{q} c_n f(n) = \sum_{n=1}^{q} c_n \frac{1}{q} \sum_{k=1}^{q} \widehat{f}(k) e(kn/q) = \frac{1}{q} \sum_{k=1}^{q} \widehat{f}(k)\widehat{c}(-k).$$

(b) Suppose that $1 \leq N \leq q$ and set $f(n) = 1$ for $M + 1 \leq n \leq M + N$, and $f(n) = 0$ for other residues (mod q). Show that $\widehat{f}(0) = N$ and by (9.14) or otherwise that

$$\widehat{f}(k) = e(-(2M + N + 1)k/q)\frac{\sin \pi k N/q}{\sin \pi k/q}$$

for $k \not\equiv 0 \pmod{q}$.

(c) By subtracting $\widehat{c}(0)N/q$ from both sides and applying the triangle inequality, show that

$$\left| \sum_{n=M+1}^{M+N} c_n - \frac{N}{q} \sum_{n=1}^{q} c_n \right| \leq \frac{1}{q} \sum_{k=1}^{q-1} \frac{|\widehat{c}(k)|}{\sin \pi k/q}.$$

5. (a) Suppose that a function f is concave upwards. Explain why

$$f(x) \leq \frac{1}{2\delta} \int_{x-\delta}^{x+\delta} f(u)\,du$$

for $\delta > 0$.

(b) Take $f(u) = \csc \pi u$, $x = k/q$, and $\delta = 1/(2q)$, and sum over k to see that

$$\sum_{k=1}^{q-1} \frac{1}{\sin \pi k/q} < q \int_{1/(2q)}^{1-1/(2q)} \frac{1}{\sin \pi u}\,du.$$

(c) Note that $\csc v$ has the antiderivative $\log(\csc v - \cot v)$, and hence deduce that the integral above is

$$= \frac{q}{\pi} \log \frac{1 + \cos \frac{\pi}{2q}}{1 - \cos \frac{\pi}{2q}}.$$

(d) By means of the inequalities $1 - \theta^2/2 \leq \cos \theta \leq 1$ deduce that the above is

$$< \frac{q}{\pi} \log \frac{16q^2}{\pi^2} = \frac{2q}{\pi} \log \frac{4q}{\pi}.$$

(e) Note that this is $< q \log q$ if $q > \exp((\log 4/\pi)/(1 - 2/\pi)) = 1.944\ldots$.

6. Let c_n be a sequence with period q and finite Fourier transform $\widehat{c}(k)$.

(a) Show that

$$\sum_{M=1}^{q}\left|\sum_{n=M+1}^{M+N} c_n - \frac{N}{q}\sum_{n=1}^{q} c_n\right|^2 = \frac{1}{q}\sum_{k=1}^{q-1} |\widehat{c}(k)|^2 \frac{\sin^2 \pi Nk/q}{\sin^2 \pi k/q}$$

for $1 \le N \le q$.

(b) Suppose that $c_n = 1$ for $0 < n < q$ and that $c_0 = 0$. Show that $\widehat{c}(0) = q - 1$ and that $\widehat{c}(k) = -1$ for $0 < k < q$. Deduce that

$$\sum_{k=1}^{q-1} \frac{\sin^2 \pi Nk/q}{\sin^2 \pi k/q} = (q - N)N$$

for $0 \le N \le q$.

(c) Take $q = 2N$ and write $k = 2n - 1$ to deduce that

$$\sum_{n=1}^{N} \frac{1}{\left(N \sin \pi \frac{2n-1}{2N}\right)^2} = 1.$$

Let N tend to infinity to show that $\sum_{n=1}^{\infty}(2n - 1)^{-2} = \pi^2/8$, and hence that $\zeta(2) = \pi^2/6$.

7. (a) Show that if χ is a primitive character modulo q, $q > 1$, then

$$\sum_{M=1}^{q}\left|\sum_{n=M+1}^{M+N} \chi(n)\right|^2 \le Nq$$

for $1 \le N \le q$.

(b) Show that if $\chi \ne \chi_0 \pmod{p}$, then

$$\sum_{M=1}^{p}\left|\sum_{n=M+1}^{M+N} \chi(n)\right|^2 = N(p - N)$$

for $1 \le N \le p$.

8. Let $f_\chi(\alpha) = \sum_{0 < n \le q\alpha} \chi(n)$. Show that if χ is a primitive character modulo q, then

$$\int_0^1 |f_\chi(\alpha) - a_\chi|^2 \, d\alpha = \frac{q}{12} \prod_{p|q}\left(1 - \frac{1}{p^2}\right)$$

where $a_\chi = 0$ if $\chi(-1) = 1$, and

$$a_\chi = \frac{-1}{q}\sum_{n=1}^{q} n\chi(n) = -iL(1, \overline{\chi})\tau(\chi)/\pi$$

if $\chi(-1) = -1$.

9. (a) Show that

$$\sum_{d|q} \frac{\log p}{p-1} \ll \log\log 3q.$$

(b) Recall Exercise 2.1.16, and show that

$$\sum_{\substack{k \le K \\ (k,q)=1}} \frac{1}{k} = \frac{\varphi(q)}{q} \log K + O\left(\frac{\varphi(q)}{q} \log\log q\right) + O\left(\frac{2^{\omega(q)}}{K}\right)$$

for $1 \le K \le q$.

(c) Suppose that χ is a primitive character modulo q, $q > 1$. Use Theorem D.2 to show that

$$\sum_{n=M+1}^{M+N} \chi(n) = \frac{\tau(\chi)}{2\pi i} \sum_{0 < |k| \le K} \frac{\overline{\chi}(-k)}{k} e(kM/q)(e(kN/q) - 1)$$

$$+ O\left(\frac{\varphi(q)}{K} \log 2K\right)$$

when $K < q^{1-\varepsilon}$.

(d) By taking $K = q^{1/2} \log q$ show that if χ is a primitive character modulo q, $q > 1$, then

$$\left| \sum_{n=M+1}^{M+N} \chi(n) \right| \le \frac{\varphi(q)}{\pi q} q^{1/2} \log q + O\left(q^{1/2} \log\log 3q\right).$$

10. (Bernstein 1914a,b) Let χ be a primitive character (mod q), with $q > 1$. Show that

$$\sum_{|n| \le q} (1 - |n|/q)\chi(n)e(n\alpha) \ll \sqrt{q}$$

uniformly in α.

9.5 Notes

Section 9.2. That the sum in (9.6) vanishes when $(n, q) > 1$ was proved by de la Vallée Poussin (1896), in a complicated way. We follow the simpler argument that Schur showed Landau (1908, pp. 430–431).

The evaluation of the sum c_χ is found in Hasse (1964, pp. 449–450). Our derivation follows that of Montgomery & Vaughan (1975). A different proof has been given by Joris (1977).

Section 9.3. Let $\zeta_K(s) = \sum_{\mathfrak{a}} N(\mathfrak{a})^{-s}$ be the Dedekind zeta function of the algebraic number field K. Here the sum is over all ideals $\mathfrak{a}$ in the ring $\mathcal{O}_K$ of integers in K. In case K is a quadratic extension of $\mathbb{Q}$, then the discriminant

d of K is a quadratic discriminant, $K = \mathbb{Q}(\sqrt{d})$, and $\zeta_K(s) = \zeta(s)L(s, \chi_d)$. In other words, the number of ideals of norm n is $\sum_{k|n} \chi_d(k)$.

Section 9.4. Concerning the constant that can be taken in Theorem 9.18, see Landau (1918), Cochrane (1987), Hildebrand (1988a,b), and Granville & Soundararajan (2005). Granville & Soundararajan (2005) also show that in the case of a cubic character, the sum in Theorem 9.18 is $\ll \sqrt{q}(\log q)^\theta$ where θ is an absolute constant, $\theta < 1$.

On the assumption of the Generalized Riemann Hypothesis for all Dirichlet characters, Montgomery & Vaughan (1977) have shown that

$$\sum_{n=M+1}^{M+N} \chi(n) \ll q^{1/2} \log \log q.$$

See Granville & Soundararajan (2005) for a much simpler proof. Paley's lower bound, Theorem 9.24 above, shows that the above is essentially best-possible. Nevertheless, it is known that one can do better a good deal of the time. In fact in Montgomery & Vaughan (1979) it is shown that for each $\theta \in (0, 1)$ there is a $c(\theta) > 0$ such that if $P > P_0(\theta)$, then for at least $\theta\pi(P)$ primes $p \leq P$ we have

$$\max_N \left| \sum_{n=1}^{N} \left(\frac{n}{p} \right) \right| \leq c(\theta)p^{1/2},$$

and if $q > P_0(\theta)$, then for at least $\theta\varphi(q)$ of the non-principal characters modulo q we have

$$\max_N \left| \sum_{n=1}^{N} \chi(n) \right| \leq c(\theta)q^{1/2}.$$

Walfisz (1942) and Chowla (1947) showed that there exist infinitely many primitive quadratic characters χ for which $L(1, \chi) \gtrsim e^{C_0} \log \log q$. In view of Theorem 9.21, this provides an alternative approach for proving estimates similar to Paley's Theorem 9.24. For recent developments concerning large $L(1, \chi)$, see Vaughan (1996), Montgomery & Vaughan (1999), and Granville & Soundararajan (2003).

Lemma 9.25 is a consequence of Weil's proof of the Riemann Hypothesis for curves over finite fields, and originally depended on considerable machinery from algebraic geometry. Later Stepanov used constructs from transcendence theory to estimate complete character sums, and subsequently Bombieri used Stepanov's ideas to give a proof of Weil's theorem that depends only on the Riemann–Roch theorem. Schmidt (1976) gives an exposition of this more elementary approach that even avoids the Riemann–Roch theorem. Friedlander & Iwaniec (1992) showed that the Pólya–Vinogradov inequality can be sharpened, in the direction of Burgess' estimates, without using Weil's estimates. The proof of Theorem 9.27 above is developed from one of Iwaniec appearing in Friedlander (1987), with a further wrinkle from Friedlander & Iwaniec (1993).

Burgess first (1957) treated the Legendre symbol and then (1962a, b) generalized his method to deal with arbitrary Dirichlet characters having cube-free conductor. Burgess' extension to composite moduli involves an extra new idea that does not extend well when the conductor is divisible by higher powers of primes. For some progress in this direction see Burgess (1986).

9.6 References

Apostol, T. M. (1970). Euler's φ-function and separable Gauss sums, *Proc. Amer. Math. Soc.* **24**, 482–485.

Baker, R. C. & Montgomery, H. L. (1990). Oscillations of quadratic L-functions, *Analytic Number Theory* (Urbana, 1989), Prog. Math. 85. Boston: Birkhäuser, pp. 23–40.

Bernstein, S. N. (1914a). Sur la convergence absolue des séries trigonométriques, *C. R. Acad, Sci. Paris* **158**, 1661–1663.

(1914b). Ob absoliutnoi skhodimosti trigonometricheskikh riadov, *Soobsch. Khar'k. matem. ob-va* (2) **14**, 145–152; 200–201.

Burgess, D. A. (1957). The distribution of quadratic residues and non-residues, *Mathematika* **4**, 106–112.

(1962a). On character sums and primitive roots, *Proc. London Math. Soc.* (3) **12**, 179–192.

(1962b). On character sums and L-series, *Proc. London Math. Soc.* (3) **12**, 193–206.

(1986). The character sum estimate with $r = 3$, *J. London Math. Soc.* (2) **33**, 219–226.

Chowla, S. (1947). On the class-number of the corpus $P(\sqrt{-k})$, *Proc. Nat. Inst. Sci. India* **13**, 197–200.

Chowla, S. & Mordell, L. J. (1961). Note on the nonvanishing of $L(1)$, *Proc. Amer. Math. Soc.* **12**, 283–284.

Cochrane, T. (1987). On a trigonometric inequality of Vinogradov, *J. Number Theory* **27**, 9–16.

Conway, J. H. (1997). *The Sensuous Quadratic Form*, Carus monograph 26. Washington: Math. Assoc. Amer.

Friedlander, J. B. (1987). Primes in arithmetic progressions and related topics, *Analytic Number Theory and Diophantine Problems* (Stillwater, 1984), Prog. Math. 70, Boston: Birkhäuser, pp. 125–134.

Friedlander, J. B. & Iwaniec, H. (1992). A mean-value theorem for character sums, *Michigan Math. J.* **39**, 153–159.

(1993). Estimates for character sums, *Proc. Amer. Math. Soc.* **119**, 365–372.

(1994). A note on character sums, *The Rademacher legacy to mathematics* (University Park, 1992), Contemp. Math. 166, Providence: Amer. Math. Soc., pp. 295–299.

Fujii, A., Gallagher, P. X., & Montgomery, H. L. (1976). Some hybrid bounds for character sums and Dirichlet L-series, *Topics in Number Theory* (Proc. Colloq.

Debrecen, 1974), Colloq. Math. Soc. Janos Bolyai 13. Amsterdam: North-Holland, pp. 41–57.

Granville, A. & Soundararajan, K. (2003). The distribution of values of $L(1, \chi_d)$, *Geom. Funct. Anal.* **13**, 992–1028; *Errata* **14** (2004), 245–246.

(2006). *Large character sums: pretentious characters and the Pólya-Vinogradov inequality*, to appear, 24 pp.

Hasse, H. (1964). *Vorlesungen über Zahlentheorie*, Second Edition, Grundl. Math. Wiss. 59. Berlin: Springer-Verlag.

Hildebrand, A. (1988a). On the constant in the Pólya–Vinogradov inequality, *Canad. Math. Bull.* **31**, 347–352.

(1988b). Large values of character sums, *J. Number Theory* **29**, 271–296.

Joris, H. (1977). On the evaluation of Gaussian sums for non-primitive characters, *Enseignement Math.* (2) **23**, 13–18.

Landau, E. (1908). Nouvelle démonstration pour la formule de Riemann sur le nombre des nombres premiers inférieurs à une limite donnée, et démonstration d'une formule plus générale pour le cas des nombres premiers d'une progression arithmétique, *Ann. École Norm. Sup.* (3) **25** 399–448; *Collected Works*, Vol. 4. Essen: Thales Verlag, 1986, pp. 87–130.

(1918). Abschätzungen von Charaktersummen, Einheiten und Klassenzahlen, *Nachr. Akad. Wiss.* Göttingen, 79–97; *Collected Works*, Vol. 7. Essen: Thales Verlag, 1986, pp. 114–132.

Martin, G. (2006). Inequities in the Shanks–Rényi prime number race, 32 pp., to appear.

Mattics, L. E. (1984). Advanced problem 6461, *Amer. Math. Monthly* **91**, 371.

Montgomery, H. L. (1976). Distribution questions concerning a character sum, *Topics in Number Theory* (Proc. Colloq. Debrecen, 1974), Colloq. Math. Soc. Janos Bolyai 13. Amsterdam: North-Holland, pp. 195–203.

(1980). An exponential polynomial formed with the Legendre symbol, *Acta Arith.* **37**, 375–380.

Montgomery, H. L. & Vaughan, R. C. (1975). The exceptional set in Goldbach's problem, *Acta Arith.* **27**, 353–370.

(1977). Exponential sums with multiplicative coefficients, *Invent. Math.* **43**, 69–82.

(1979). Mean values of character sums, *Canad. J. Math.* **31**, 476–487.

(1999). Extreme values of Dirichlet L-functions at 1, *Number Theory in Progress*, Vol. 2 (Zakopane–Kościelisko, 1997). Berlin: de Gruyter, pp. 1039–1052.

Mordell, L. J. (1933). The number of solutions of some congruences in two variables, *Math. Z.* **37**, 193–209.

Paley, R. E. A. C. (1932). A theorem of characters, *J. London Math. Soc.* **7**, 28–32.

Pólya, G. (1918). Über die Verteilung der quadratischen Reste und Nichtreste, *Nachr. Akad. Wiss. Göttingen*, 21–29.

Schmidt, W. M. (1976). *Equations over finite fields. An elementary approach*, Lecture Notes Math. 536, Berlin: Springer-Verlag.

Schur, I. (1918). Einige Bemerkungen zu der vorstehenden Arbeit des Herrn G. Pólya: Über die Verteilung der quadratischen Reste und Nichtreste, *Nachr. Akad. Wiss. Göttingen*, 30–36.

de la Vallée Poussin, C. J. (1896). Recherches analytiques sur la théorie des nombres premiers, I–III, *Ann. Soc. Sci. Bruxelles* **20**, 183–256, 281–362, 363–397.

Vaughan, R. C. (1996). Small values of Dirichlet L-functions at 1, *Analytic Number Theory*. (Allerton Park, 1995), Vol. 2, Prog. Math. 139, Boston: Birkhäuser, pp. 755–766.

Vinogradov, I. M. (1918). Sur la distribution des résidus et des nonrésidus des puissances, *J. Soc. Phys. Math. Univ. Permi*, 18–28.

  (1919). Über die Verteilung der quadratischen Reste und Nichtreste, *J. Soc. Phys. Math. Univ. Permi*, 1–14.

Vorhauer, U. M. A. (2006). *A note on comparative prime number theory*, to appear.

Walfisz, A. (1942). On the class-number of binary quadratic forms, *Trudy Tbliss. Mat. Inst.* **11**, 57–71.

10

Analytic properties of the zeta function and L-functions

10.1 Functional equations and analytic continuation

In Section 1.3 we saw that the zeta function can be analytically continued to the half-plane $\sigma > 0$. We now derive an important formula for the Riemann zeta function, one that serves to define the zeta function throughout the complex plane. From this formula we see that the zeta function is analytic at all points except for $s = 1$, and we find that $\zeta(s)$ is related to $\zeta(1 - s)$. In preparation for this we first use the Poisson summation formula to establish a corresponding functional equation for theta functions.

Theorem 10.1 *For arbitrary real α, and complex numbers z with $\Re z > 0$,*

$$\sum_{n=-\infty}^{\infty} e^{-\pi(n+\alpha)^2 z} = z^{-1/2} \sum_{k=-\infty}^{\infty} e(k\alpha)e^{-\pi k^2/z}, \tag{10.1}$$

and

$$\sum_{n=-\infty}^{\infty} (n + \alpha)e^{-\pi(n+\alpha)^2 z} = -iz^{-3/2} \sum_{k=-\infty}^{\infty} ke(k\alpha)e^{-\pi k^2/z} \tag{10.2}$$

where the branch of $z^{1/2}$ is determined by $1^{1/2} = 1$.

Proof We can obtain (10.2) from (10.1) by differentiating with respect to α, since the differentiated series are uniformly convergent for α in a compact set. As for (10.1), we note that if $g(u) = f(u + \alpha)$, then $\widehat{g}(t) = \widehat{f}(t)e(t\alpha)$. (Conventions governing the definition of the Fourier transform $\widehat{f}$ are established in Appendix D.) We apply the Poisson summation formula (Theorem D.3) to $g(u)$, where $f(u) = e^{-\pi u^2 z}$, and it remains only to demonstrate that $\widehat{f}(t) = z^{-1/2}e^{-\pi t^2/z}$. Writing

$$-\pi x^2 z - 2\pi itx = -\pi(x + it/z)^2 z - \pi t^2/z,$$

we see that

$$\widehat{f}(t) = e^{-\pi t^2/z} \int_{-\infty}^{+\infty} e^{-\pi(x+it/z)^2 z} \, dx.$$

We consider this integral to be a contour integral in the complex plane. We note that the integrand tends to 0 very rapidly as $|\Re x|$ tends to infinity with $|\Im x|$ bounded. Hence by Cauchy's theorem we may translate the path of integration to the line $x - it/z$, $-\infty < x < +\infty$, and we find that the above integral is $\int_{-\infty}^{+\infty} e^{-\pi x^2 z} \, dx$. We now turn the path of integration through an angle $-\frac{1}{2}\arg z$ and again apply Cauchy's theorem. After reparametrizing, we see that our integral is $z^{-1/2} \int_{-\infty}^{+\infty} e^{-\pi x^2} \, dx = z^{-1/2}$. This completes the proof. $\square$

Theorem 10.2 *For any complex number s, except $s = 0$ and $s = 1$, and any non-zero complex number z with $\Re z \geq 0$,*

$$\zeta(s)\Gamma(s/2)\pi^{-s/2} = \pi^{-s/2} \sum_{n=1}^{\infty} n^{-s} \Gamma(s/2, \pi n^2 z)$$

$$+ \pi^{(s-1)/2} \sum_{n=1}^{\infty} n^{s-1} \Gamma((1-s)/2, \pi n^2/z) \quad (10.3)$$

$$+ \frac{z^{(s-1)/2}}{s-1} - \frac{z^{s/2}}{s}.$$

Here $\Gamma(s, a)$ is the incomplete gamma function,

$$\Gamma(s, a) = \int_{a}^{\infty} e^{-w} w^{s-1} \, dw, \quad (10.4)$$

and we may take the path of integration to be the ray $w = a + u, 0 \leq u < \infty$, so that

$$\Gamma(s, a) = \int_{0}^{\infty} e^{-u-a}(u+a)^{s-1} \, du.$$

Now $(u+a)^{s-1} \ll |a|^{\sigma-1}$ uniformly for $\Re a \geq 0, |a| \geq \varepsilon > 0$, and $|\sigma| \leq C$, so that $n^{-s}\Gamma(s/2, \pi n^2 z) \ll n^{-2}$ uniformly for $\Re z \geq 0, |z| \geq \varepsilon, |s| \leq C$. Thus the two sums on the right are uniformly convergent for s in any compact set, and hence by a theorem of Weierstrass they represent entire functions. The last two terms have simple poles at 1 and 0, respectively. As for the left-hand side, we note that $\Gamma(s/2)$ has a pole at $s = 0$, and never vanishes, so it follows that $\zeta(s)$ is analytic for all $s \neq 1$. If we simultaneously replace s by $1 - s$ and z by $1/z$, then the two sums on the right in (10.3) are exchanged, and the last two terms are also exchanged, so that the value of the right-hand side is invariant. These observations may be summarized as follows:

Corollary 10.3 *The function*

$$\xi(s) = \frac{1}{2}s(s-1)\zeta(s)\Gamma(s/2)\pi^{-s/2} \tag{10.5}$$

is entire, and $\xi(s) = \xi(1-s)$ for all s.

This is the functional equation of the zeta function, first proved by Riemann in 1860. Since $\zeta(s) \neq 0$ for $\sigma \geq 1$, it follows that $\xi(s) \neq 0$ for $\sigma \geq 1$, and by the functional equation that $\xi(s) \neq 0$ for $\sigma \leq 0$. The zeros of $\zeta(s)$ in the *critical strip* $0 < \sigma < 1$ coincide precisely with those of $\xi(s)$. As $\Gamma(s/2)$ has simple poles at $s = 0, -2, -4, -6, \ldots$, the zeta function has simple zeros at $s = -2, -4, -6, \ldots$. These are the *trivial zeros* of the zeta function. The only other zeros of the zeta function are the *non-trivial zeros*, in the critical strip. The generic non-trivial zero is denoted $\rho = \beta + i\gamma$. By the Schwarz reflection principle, $\xi(\overline{s}) = \overline{\xi(s)}$; hence in particular $\xi\left(\frac{1}{2} - it\right) = \overline{\xi}\left(\frac{1}{2} + it\right)$. But the functional equation gives $\xi\left(\frac{1}{2} - it\right) = \xi\left(\frac{1}{2} + it\right)$, so it follows that $\xi\left(\frac{1}{2} + it\right)$ is real for all real t. Similarly, if ρ is a zero of $\xi(s)$ then so also are $\overline{\rho}$, $1 - \rho$, and $1 - \overline{\rho}$. The as yet unproved *Riemann Hypothesis* (RH) asserts that all non-trivial zeros of the zeta function have real part $1/2$; that is, all the zeros of $\xi(s)$ lie on the *critical line* $\sigma = 1/2$. We shall find it instructive to explore a number of consequences of this famous conjecture, in Chapter 13.

Proof of Theorem 10.2 By Euler's integral formula (Theorem C.2) for $\Gamma(s/2)$ we see that if $\sigma > 0$, then

$$\Gamma(s/2) = \int_0^\infty e^{-x} x^{s/2-1}\, dx. \tag{10.6}$$

By the linear change of variables $x = \pi n^2 u$ it follows that

$$n^{-s}\Gamma(s/2)\pi^{-s/2} = \int_0^\infty e^{-\pi n^2 u} u^{s/2-1}\, du.$$

We assume that $\sigma > 1$ and sum over n to find that

$$\zeta(s)\Gamma(s/2)\pi^{-s/2} = \sum_{n=1}^\infty \int_0^\infty e^{-\pi n^2 u} u^{s/2-1}\, du$$

$$= \int_0^\infty \left(\sum_{n=1}^\infty e^{-\pi n^2 u}\right) u^{s/2-1}\, du. \tag{10.7}$$

Here the exchange of integration and summation is permitted by absolute convergence. Suppose, for the present, that $\Re z > 0$. We may consider the integral above to be a contour integral in the complex plane, and by Cauchy's theorem we may replace the path of integration by the ray from 0 that passes through z. We now consider separately the integral from 0 to z, and the integral from

z to ∞. We call these integrals $\int_1$, $\int_2$, respectively. By reversing the steps we made in passing from (10.6) to (10.7) we see immediately that

$$\int_2 = \pi^{-s/2} \sum_{n=1}^{\infty} n^{-s} \Gamma(s/2, \pi n^2 z).$$

To treat $\int_1$ we let

$$\vartheta(u) = \sum_{-\infty}^{+\infty} e^{-\pi n^2 u} \qquad (10.8)$$

for $\Re u > 0$. Then the sum in the integrand in (10.7) is $(\vartheta(u) - 1)/2$. Thus

$$\int_1 = \frac{1}{2} \int_0^z \vartheta(u) u^{s/2-1}\, du - \frac{1}{2} \int_0^z u^{s/2-1}\, du.$$

Here the second integral is $\frac{2}{s} z^{s/2}$. By Theorem 10.1 we know that $\vartheta(u) = u^{-1/2} \vartheta(1/u)$. Hence the first term above is

$$\frac{1}{2} \int_0^z \vartheta(1/u) u^{s/2-3/2}\, du = \int_0^z \left(\sum_{n=1}^{\infty} e^{-\pi n^2/u} \right) u^{s/2-3/2}\, du + \frac{1}{2} \int_0^z u^{s/2-3/2}\, du.$$

Here the second integral is $\frac{2}{s-1} z^{(s-1)/2}$. By the change of variable $v = 1/u$ we see that the first term above is

$$\int_{1/z}^{\infty} \left(\sum_{n=1}^{\infty} e^{-\pi n^2 v} \right) v^{(1-s)/2-1}\, dv.$$

We exchange the order of summation and integration, and make the linear change of variables $x = \pi n^2 v$, to see that this is

$$\pi^{(s-1)/2} \sum_{n=1}^{\infty} n^{s-1} \Gamma((1-s)/2, \pi n^2/z).$$

Hence

$$\int_1 = \frac{z^{(s-1)/2}}{s-1} - \frac{z^{s/2}}{s} + \pi^{(s-1)/2} \sum_{n=1}^{\infty} n^{s-1} \Gamma((1-s)/2, \pi n^2/z),$$

so we have the desired identity for $\sigma > 1$. But, as already noted, the two sums represent entire functions, so the right-hand side of (10.3) is analytic for all s except for simple poles at $s = 1$ and $s = 0$. Hence by the uniqueness of analytic continuation the identity (10.3) holds for all s except at the poles. $\qquad\square$

The functional equation of Corollary 10.3 can also be expressed asymmetrically:

Corollary 10.4 *For all $s \neq 1$,*

$$\zeta(s) = \zeta(1-s) 2^s \pi^{s-1} \Gamma(1-s) \sin \frac{\pi s}{2}. \qquad (10.9)$$

Proof By the reflection principle (C.6) and the duplication formula (C.9), we see that

$$\frac{\Gamma\left(\frac{1-s}{2}\right)}{\Gamma\left(\frac{s}{2}\right)} = \frac{1}{\pi}\Gamma\left(\frac{1-s}{2}\right)\Gamma\left(1 - \frac{s}{2}\right)\sin\frac{\pi s}{2} = \pi^{-1/2}2^s\Gamma(1-s)\sin\frac{\pi s}{2}.$$

Thus the stated identity follows from Corollary 10.3.   □

By Stirling's formula, we can describe $|\zeta(s)|$ in terms of $|\zeta(1-s)|$.

Corollary 10.5 *Suppose that $A > 0$ is fixed. Then*

$$|\zeta(s)| \asymp \tau^{1/2-\sigma}|\zeta(1-s)|$$

uniformly for $|\sigma| \le A$ and $|t| \ge 1$. Here $\tau = |t| + 4$, as usual.

Proof Since the above is invariant when s is replaced by $1-s$, we may suppose that $-A \le \sigma \le 1/2$. We may also suppose that $t \ge 1$, since $|\zeta(\sigma - it)| = |\zeta(\sigma + it)|$. We consider the factors on the right-hand side of (10.9). By Stirling's formula as formulated in (C.18), we see that

$$|\Gamma(1-s)| \asymp \left|(1-s)^{1/2-s}\right| = |1 - s|^{1/2-\sigma}\exp(t\arg(1-s)).$$

But $\arg(1-s) = -\arctan t/(1-\sigma) = -\pi/2 + O(1/t)$ and $|1 - s| \sim t$, so $|\Gamma(1-s)| \asymp t^{1/2-\sigma}\exp(-\pi t/2)$. On the other hand, $\sin z = (e^{iz} - e^{-iz})/(2i)$, so $|\sin \pi s/2| \asymp \exp(\pi t/2)$, and we obtain the stated result.   □

Let σ be fixed, and let $\mu(\sigma)$ denote the infimum of those exponents μ such that $\zeta(\sigma + it) \ll \tau^\mu$. This is the *Lindelöf μ-function*. By Corollary 1.17 we know that $\mu(\sigma) = 0$ for $\sigma \ge 1$ and that $\mu(\sigma) \le 1 - \sigma$ for $0 < \sigma \le 1$. By Corollary 10.5 we see that $\mu(\sigma) = \mu(1-\sigma) + 1/2 - \sigma$. Hence in particular, $\mu(\sigma) = 1/2 - \sigma$ for $\sigma \le 0$. For $0 < \sigma < 1$ the value of $\mu(\sigma)$ is at present unknown, but the *Lindelöf Hypothesis* (LH) asserts that $\zeta(1/2 + it) \ll_\varepsilon \tau^\varepsilon$, which is to say that $\mu(1/2) = 0$. From this it follows that

$$\mu(\sigma) = \begin{cases} 0 & \text{for } \sigma \ge 1/2, \\ 1/2 - \sigma & \text{for } \sigma \le 1/2. \end{cases} \tag{10.10}$$

Three different proofs that LH implies the above are found in Exercises 10.1. 18–20. Also, from Exercises 10.1.20 and 10.1.21 we see that LH is equivalent to a certain assertion concerning the distribution of the zeros of $\zeta(s)$. Since this assertion is visibly weaker than RH, it is evident that RH implies LH. In Chapter 13 we shall show that RH implies a quantitative form of LH.

Concerning special values of the zeta function, we observe first that since $\zeta(s) \sim 1/(s - 1)$ for s near 1, it follows from Corollary 10.4 that

$$\zeta(0) = -1/2. \tag{10.11}$$

In addition, we note that Corollary B.3 asserts that

$$\zeta(2k) = \frac{(-1)^{k-1}2^{2k-1}B_{2k}}{(2k)!}\pi^{2k} \tag{10.12}$$

for each positive integer k. Hence by taking $s = 1 - 2k$ in Corollary 10.4 we deduce that

$$\zeta(1 - 2k) = \frac{-B_{2k}}{2k} \tag{10.13}$$

for positive integers k. An alternative proof of this is found in Appendix B. We may also determine the value of $\zeta'(0)$, as follows. Let $f(s) = (s - 1)\zeta(s)$. By Corollary 1.16 we know that $f(s) = 1 + C_0(s - 1) + \cdots$ for s near 1. On multiplying both sides of (10.9) by $s - 1$ we see that $f(s) = -\zeta(1 - s)2^s\pi^{s-1}\Gamma(2 - s)\sin\pi s/2$. On differentiating both sides and setting $s = 1$ we discover that $C_0 = 2\zeta'(0) - 2\zeta(0)\log 2\pi + 2\zeta(0)\Gamma'(1)$. But $\zeta(0) = -1/2$ and $\Gamma'(1) = -C_0$, so we find that

$$\zeta'(0) = -\frac{1}{2}\log 2\pi. \tag{10.14}$$

Our treatment of the zeta function extends readily to L-functions.

Theorem 10.6 *For z with $\Re z > 0$ let*

$$\vartheta_0(z, \chi) = \sum_{n=-\infty}^{\infty} \chi(n)e^{-\pi n^2 z/q},$$

$$\vartheta_1(z, \chi) = \sum_{n=-\infty}^{\infty} n\chi(n)e^{-\pi n^2 z/q}.$$

If χ is a primitive character modulo q, then

$$\vartheta_0(z, \chi) = \frac{\tau(\chi)}{q^{1/2}}z^{-1/2}\vartheta_0(1/z, \overline{\chi}),$$

$$\vartheta_1(z, \chi) = \frac{\tau(\chi)}{iq^{1/2}}z^{-3/2}\vartheta_1(1/z, \overline{\chi})$$

where the branch of $z^{1/2}$ is determined by $1^{1/2} = 1$.

Though both these functions are defined for all χ, we note that if $\chi(-1) = -1$, then $\vartheta_0(z, \chi) = 0$ for all z, while if $\chi(-1) = 1$, then $\vartheta_1(z, \chi) = 0$ identically. Thus $\vartheta_0(z, \chi)$ is of interest when $\chi(-1) = 1$, and $\vartheta_1(z, \chi)$ is useful when $\chi(-1) = -1$.

Proof Since χ is periodic with period q, it follows that

$$\vartheta_0(z, \chi) = \sum_{a=1}^{q} \chi(a) \sum_{m=-\infty}^{\infty} e^{-\pi(mq+a)^2 z/q}.$$

By (10.1) with $\alpha = a/q$ and z replaced by qz we see that the above is

$$= (qz)^{-1/2} \sum_{a=1}^{q} \chi(a) \sum_{k=-\infty}^{\infty} e^{-\pi k^2/(qz)} e(ak/q)$$

$$= (qz)^{-1/2} \sum_{k=-\infty}^{\infty} e^{-\pi k^2/(qz)} \sum_{a=1}^{q} \chi(a) e(ak/q).$$

Since χ is primitive, we know by Theorem 9.7 that the inner sum on the right is $\tau(\chi)\overline{\chi}(k)$ for all k. This gives the identity for ϑ_0. The identity for ϑ_1 is proved similarly, using (10.2).   □

In order to unify our formulæ we find it convenient to put

$$\kappa = \kappa(\chi) = \begin{cases} 0 & \text{if } \chi(-1) = 1, \\ 1 & \text{if } \chi(-1) = -1. \end{cases} \tag{10.15}$$

In this notation, the formulæ of Theorem 10.6 read

$$\vartheta_\kappa(z, \chi) = \frac{\varepsilon(\chi)}{z^{1/2+\kappa}} \vartheta_\kappa(1/z, \overline{\chi}) \tag{10.16}$$

where

$$\varepsilon(\chi) = \frac{\tau(\chi)}{i^\kappa \sqrt{q}}. \tag{10.17}$$

Suppose that χ is primitive. Some of our results concerning Gauss sums can be reformulated in terms of $\varepsilon(\chi)$. Firstly, from Theorem 9.7 we see that $|\varepsilon(\chi)| = 1$. Secondly, by Theorems 9.5 and 9.7 we see that $\varepsilon(\chi)\varepsilon(\overline{\chi}) = 1$. Finally, if χ is not only primitive but also quadratic, then $\varepsilon(\chi) = 1$, by Theorem 9.17.

In the same way that Theorem 10.2 was derived from (10.8), the following is an immediate consequence of (10.16).

Theorem 10.7 *Let χ be a primitive character modulo q with $q > 1$. Then for any complex numbers s and z with $\Re z \geq 0$,*

$$L(s, \chi)\Gamma((s + \kappa)/2)(q/\pi)^{(s+\kappa)/2}$$

$$= (q/\pi)^{(s+\kappa)/2} \sum_{n=1}^{\infty} \chi(n)n^{-s}\Gamma((s + \kappa)/2, \pi n^2 z/q) \tag{10.18}$$

$$+ \varepsilon(\chi)(q/\pi)^{(1-s+\kappa)/2} \sum_{n=1}^{\infty} \overline{\chi}(n)n^{s-1}\Gamma((1 - s + \kappa)/2, \pi n^2/(qz)).$$

As was the case with the zeta function, the above is first proved for $\sigma > 1$. Since each term of the series is entire, and since the series are locally uniformly convergent, the right-hand side is an entire function of s, and this provides an analytic continuation of $L(s, \chi)$ to the entire complex plane. If in the above we replace χ by $\overline{\chi}$, s by $1 - s$, and z by $1/z$, and then multiply both sides by $\varepsilon(\chi)$ then the right-hand side above is unchanged, and thus we obtain a functional equation for $L(s, \chi)$, as follows.

Corollary 10.8 *Let χ be a primitive character modulo q with $q > 1$. The function*

$$\xi(s, \chi) = L(s, \chi)\Gamma((s + \kappa)/2)(q/\pi)^{(s+\kappa)/2} \tag{10.19}$$

is entire, and $\xi(s, \chi) = \varepsilon(\chi)\xi(1 - s, \overline{\chi})$ for all s.

Let χ be a primitive character modulo q, $q > 1$. We already know that $L(s, \chi) \neq 0$ for $\sigma > 1$. Since the gamma function has no zeros, it follows that $\xi(s, \chi) \neq 0$ in this half-plane. By the functional equation, $\xi(s, \chi) \neq 0$ also for $\sigma < 0$, and hence $L(s, \chi) \neq 0$ for $\sigma < 0$ except that $L(s, \chi)$ must have simple zeros where the gamma factor has simple poles, which is to say at $-\kappa, -\kappa - 2, -\kappa - 4, \dots$. These are the *trivial zeros* of $L(s, \chi)$. Zeros $\rho = \beta + i\gamma$ of $L(s, \chi)$ in the *critical strip* $0 \leq \beta \leq 1$ are called *non-trivial*. The conjecture that these latter zeros all lie on the *critical line* $\sigma = 1/2$ is the *Generalized Riemann Hypothesis* (GRH). If ρ is a non-trivial zero of $L(s, \chi)$, then by the functional equation $1 - \rho$ is a zero of $L(s, \overline{\chi})$. Consequently $1 - \overline{\rho}$ is a zero of $L(s, \chi)$, since in general $\overline{L(s, \chi)} = L(\overline{s}, \overline{\chi})$. The pair of zeros $\rho, 1 - \overline{\rho}$ are symmetrically placed with respect to the critical line. Of course, if $\beta = 1/2$ then $\rho = 1 - \overline{\rho}$. For complex characters there is no symmetry about the real axis, but if χ is quadratic then $\overline{\chi} = \chi$, and so if ρ is a zero then so also are $\overline{\rho}$, $1 - \rho$, and $1 - \overline{\rho}$.

The functional equation of an L-function can also be expressed asymmetrically.

Corollary 10.9 *Suppose that χ is a primitive character (mod q) with $q > 1$. Then for all s,*

$$L(s, \chi) = \varepsilon(\chi)L(1 - s, \overline{\chi})2^s \pi^{s-1} q^{1/2-s}\Gamma(1 - s)\sin\frac{\pi}{2}(s + \kappa).$$

Proof When $\kappa = 0$ we proceed as in the proof of Corollary 10.4. When $\kappa = 1$ we use the reflection formula (C.6) and the duplication formula (C.9) to see that

$$\frac{\Gamma(1 - s/2)}{\Gamma((s + 1)/2)} = \frac{1}{\pi}\Gamma(1 - s/2)\Gamma(1/2 - s/2)\sin\pi(s + 1)/2$$

$$= 2^s \pi^{-1/2}\Gamma(1 - s)\sin\frac{\pi}{2}(s + 1).$$

This, with the identity $\xi(s, \chi) = \varepsilon(\chi)\xi(1 - s, \overline{\chi})$, gives the stated result. $\quad\square$

By the same method used to prove Corollary 10.5 we obtain

Corollary 10.10 *Let χ be a primitive character* (mod q) *with $q > 1$, and suppose that $A > 0$ is fixed. Then*

$$|L(s, \chi)| \asymp (q\tau)^{1/2-\sigma} |L(1 - s, \overline{\chi})|$$

uniformly for $|\sigma| \le A$ and $|t| \ge 1$. If $-A \le \sigma \le 1/2$ and $|t| \le 1$, then

$$L(s, \chi) \ll q^{1/2-\sigma} |L(1 - s, \overline{\chi})|.$$

Let χ be a character modulo q. If χ is imprimitive, then χ is induced by a primitive character χ^* modulo d, for some $d|q$, and

$$L(s, \chi) = L(s, \chi^*) \prod_{p|q} \left(1 - \frac{\chi^*(p)}{p^s}\right). \tag{10.20}$$

If $p|d$, then $\chi^*(p) = 0$, and thus in the above product we may confine our attention to those primes $p|q$ such that $p \nmid d$. For such a prime, the factor $1 - \chi^*(p)/p^s$ is an entire function whose zeros form an arithmetic progression on the imaginary axis. Thus $L(s, \chi)$ has all the zeros of $L(s, \chi^*)$, and if there are primes $p|q$ such that $p \nmid d$, then $L(s, \chi)$ has additional zeros on the imaginary axis. Such zeros constitute a finite union of arithmetic progressions. In the special case $\chi = \chi_0$, we have

$$L(s, \chi_0) = \zeta(s) \prod_{p|q} \left(1 - \frac{1}{p^s}\right).$$

Thus $L(s, \chi_0)$ has a pole at $s = 1$ with residue $\varphi(q)/q$, it has all the zeros of $\zeta(s)$, and it also has zeros of the form $2\pi i k / \log p$ where k takes integral values and $p|q$.

10.1.1 Exercises

1. Let $\vartheta(u)$ be defined as in (10.8). Show that $\vartheta'(1) = -\vartheta(1)/4$.
2. Let f be an even function in $L^1(\mathbb{R})$, let $\beta > 1$, suppose that $f(x) = O(x^{-\beta})$ as $x \to \infty$, and that $\widehat{f}(u) = O(u^{-\beta})$ as $u \to \infty$. Show that

$$2\zeta(s) \int_0^\infty f(x)x^{s-1}\,dx = 2\sum_{n=1}^\infty n^{-s} \int_n^\infty f(x)x^{s-1}\,dx$$

$$+ 2\sum_{n=1}^\infty n^{s-1} \int_n^\infty \widehat{f}(u)u^{-s}\,du$$

$$- f(0)/s + \widehat{f}(0)/(s - 1)$$

for $1 - \beta < \sigma < \beta$.

3. (Heilbronn 1938; cf. Weil 1967)

(a) Show that for $c > 1, x > 0$,
$$\frac{1}{2\pi i} \int_{c-i\infty}^{c+i\infty} \zeta(s)\Gamma(s/2)(\pi x)^{-s/2} \, ds = 2 \sum_{n=1}^{\infty} e^{-\pi n^2 x}.$$

(b) With $\vartheta(x)$ defined as in (10.8), use the functional equation of the zeta function to show that $\vartheta(x) = x^{-1/2}\vartheta(1/x)$ for $x > 0$.

4. (Lavrik 1965)

(a) Suppose that $\Re z > 0$, that $\sigma_0 > \max(0, -\sigma)$, and that $s \neq 0, s \neq -1$, $s \neq -2, \ldots$. By pulling the contour to the left and summing the residues, show that
$$\frac{1}{2\pi i} \int_{\sigma_0-i\infty}^{\sigma_0+i\infty} \Gamma(w+s)z^{-w} \frac{dw}{w} = \Gamma(s) - \sum_{k=0}^{\infty} \frac{(-1)^k z^{s+k}}{k!(s+k)}.$$

(b) Show that if $\sigma > 0$, then the right-hand side above is $\Gamma(s, z)$.

(c) Argue that both sides are entire functions of s, and hence that the identity
$$\Gamma(s, z) = \frac{1}{2\pi i} \int_{\sigma_0-i\infty}^{\sigma_0+i\infty} \Gamma(w+s)z^{-w} \frac{dw}{w}$$
holds for all complex s.

(d) Show that if $\sigma_0 > \max(0, (1-\sigma)/2)$, then
$$\pi^{-s/2} \sum_{n=1}^{\infty} n^{-s}\Gamma(s/2, \pi n^2 z)$$
$$= \frac{1}{2\pi i} \int_{\sigma_0-i\infty}^{\sigma_0+i\infty} \zeta(s+2w)\Gamma(w+s/2)\pi^{-w-s/2}z^{-w} \frac{dw}{w}.$$

(e) Suppose now that $s \neq 0$ and $s \neq 1$. Explain why the integrand has poles at $w = 0$, $w = (1-s)/2$, $w = -s/2$, and nowhere else.

(f) Show that when the contour is pulled to the left, the pole at $w = 0$ contributes $\zeta(s)\Gamma(s/2)\pi^{-s/2}$, the pole at $w = (1-s)/2$ contributes $z^{(s-1)/2}/(s-1)$, and the pole at $-s/2$ contributes $-z^{s/2}/s$.

(g) Suppose the contour is pulled to the left to an abscissa $\sigma_1 < \min(0, -\sigma/2)$. By means of the identity $\zeta(s)\Gamma(s/2)\pi^{-s/2} = \zeta(1-s)$ $\Gamma((1-s)/2)\pi^{(s-1)/2}$ and the change of variable $w \mapsto -w$, show that the expression is $\pi^{(s-1)/2} \sum_{n=1}^{\infty} n^{s-1}\Gamma((1-s)/2, \pi n^2/z)$. Thus demonstrate that Theorem 10.2 can be derived from Corollary 10.3.

5. Suppose that α is real, that $\Re z > 0$ and that χ is a primitive character (mod q).

(a) Show that
$$\sum_{n=-\infty}^{\infty} \chi(n)e^{-\pi(n+\alpha)^2 z/q} = \frac{\tau(\chi)}{q^{1/2}}z^{-1/2} \sum_{k=-\infty}^{\infty} \overline{\chi}(k)e(k\alpha/q)e^{-\pi k^2/(qz)}.$$

(b) By differentiating with respect to α, or otherwise, show that

$$\sum_{n=-\infty}^{\infty} \chi(n)(n+\alpha)e^{-\pi(n+\alpha)^2 z/q} = \frac{\tau(\chi)}{iq^{1/2}}z^{-3/2}\sum_{k=-\infty}^{\infty} \overline{\chi}(k)ke(k\alpha/q)e^{-\pi k^2/(qz)}.$$

6. Let α and β be real numbers, and suppose that $\Re z > 0$, and put

$$\vartheta_0(z;\alpha,\beta) = \sum_{n=-\infty}^{\infty} e(n\beta)e^{-\pi(n+\alpha)^2 z}.$$

(a) Show that if $f(x) = e(\beta x)e^{-\pi(x+\alpha)^2 z}$, then $\widehat{f}(t) = e(-\alpha\beta)z^{-1/2}$.
(b) Show that $\vartheta_0(z;\alpha,\beta) = e(-\alpha\beta)z^{-1/2}\vartheta(1/z,-\beta,\alpha)$.
(c) Without using the result of (b), show that $\vartheta_0(z;\alpha,\beta) = \vartheta_0(z;-\alpha,-\beta)$.

7. Show that

$$\sum_{n=-\infty}^{\infty}(1-2\pi n^2 x)e^{-\pi n^2 x} > \sum_{n=-\infty}^{\infty}(2\pi(n+1/2)^2 x - 1)e^{-\pi(n+1/2)^2 x} > 0$$

for all $x > 0$.

8. Use the functional equation of the zeta function in any convenient form to show that

$$\zeta(1-s) = \zeta(s)2^{1-s}\pi^{-s}\Gamma(s)\cos\frac{\pi s}{2}.$$

9. Show that if k is a positive integer, then

$$\zeta'(-2k) = \frac{(-1)^k (2k)!\zeta(2k+1)}{2^{2k+1}\pi^{2k}}.$$

10. Let $\vartheta(x)$ be defined as in (10.8). Show that

$$\zeta(s)\Gamma(s/2)\pi^{-s/2} = \frac{1}{s(s-1)} + \frac{1}{2}\int_1^{\infty}(x^{s/2}+x^{(1-s)/2})(\vartheta(x)-1)\frac{dx}{x}$$

for all s except $s = 1$ or $s = 0$.

11. (Walfisz 1931, p. 454) Show that

$$\sum_{\substack{a=1 \\ (a,b)=1}}^{\infty}\sum_{b=1}^{\infty}\frac{1}{a^2 b^2} = \frac{5}{2}.$$

12. (Mallik 1977) Let χ be a primitive quadratic character.
(a) Show that $\xi'(1/2,\chi) = 0$.
(b) Show that if $L(1/2,\chi) \neq 0$, then sgn $L'(1/2,\chi) = -$sgn $L(1/2,\chi)$.

13. Let χ be a primitive character modulo q, and let θ be a real number such that $e^{2i\theta} = \varepsilon(\chi)$. Thus $e^{i\theta}$ is one of the square roots of $\varepsilon(\chi)$. Show that $\xi(1/2+it,\chi)e^{-i\theta}$ is real for all real t.

14. Let χ be a primitive character modulo q with $q > 1$, and suppose that $\chi(-1) = 1$.

(a) For each positive integer k, show that

$$L(2k, \chi) = \frac{(-1)^{k-1}2^{2k-1}\pi^{2k}\tau(\chi)}{(2k)!\,q} \sum_{a=1}^{q} \overline{\chi}(a)B_{2k}(a/q).$$

(b) For positive integers k, deduce that

$$L(1 - 2k, \chi) = \frac{-q^{2k-1}}{2k} \sum_{a=1}^{q} \chi(a)B_{2k}(a/q).$$

15. Let χ be a primitive character modulo q with $q > 1$, and suppose that $\chi(-1) = -1$.

(a) For each non-negative integer k, show that

$$L(2k + 1, \chi) = \frac{i(-1)^{k}2^{2k}\pi^{2k+1}\tau(\chi)}{(2k + 1)!\,q} \sum_{a=1}^{q} \overline{\chi}(a)B_{2k+1}(a/q).$$

(b) Show that when $k = 0$, the above is consistent with the formula of Theorem 9.9.

(c) For non-negative integers k, deduce that

$$L(-2k, \chi) = \frac{-q^{2k}}{2k + 1} \sum_{a=1}^{q} \overline{\chi}(a)B_{2k+1}(a/q).$$

16. (a) Let p_1 and p_2 be distinct primes. Show that $(\log p_1)/(\log p_2)$ is irrational.

(b) Let χ be a character modulo q. Show that all zeros of $L(s, \chi)$ on the imaginary axis are simple, except possibly for zeros at the point $s = 0$.

(c) Let a positive integer m and a primitive character χ^* be given. Show that there is a character χ induced by χ^* such that $L(s, \chi)$ has a zero at $s = 0$ of exact multiplicity m.

17. (Landau 1907) (a) Let χ denote the character modulo 5 such that $\chi(2) = i$. Show that $L(1, \chi) = (-1 - 3i)\pi\tau(\chi)/25$.

(b) With χ as above, show that $L(2, \chi^2) = 4\sqrt{5}\pi^2/125$.

(c) Let χ be as above. By using Exercise 9.2.9, or otherwise, show that $\tau(\chi)^2 = (-1 - 2i)\sqrt{5}$.

(d) With χ as above, show that

$$\frac{L(1, \chi)^2}{L(2, \chi^2)} = 1 + i/2.$$

(e) Let χ denote a non-principal character modulo q. Show that

$$\sum_{n=1}^{\infty} 2^{\omega(n)}\chi(n)n^{-s} = \frac{L(s, \chi)^2}{L(2s, \chi^2)}$$

for $\sigma > 1/2$.

(f) Let $\varepsilon_n = 1$ if $n \equiv 1 \pmod 5$, $\varepsilon_n = -1$ if $n \equiv -1 \pmod 5$, and $\varepsilon_n = 0$ otherwise. Show that

$$\sum_{n=1}^{\infty} \frac{\varepsilon_n 2^{\omega(n)}}{n} = 1.$$

18. Suppose throughout that $0 < \delta \leq 1/2$. (a) Let $\alpha(s) = \sum_{n=1}^{\infty} a_n n^{-s}$ be a Dirichlet series with abscissa of convergence σ_c. Show that if $\sigma_0 > \max(\delta, \sigma_c)$, then

$$\sum_{n \leq x} a_n((x/n)^\delta - (n/x)^\delta) = \frac{\delta}{\pi i} \int_{\sigma_0 - i\infty}^{\sigma_0 + i\infty} \alpha(w) \frac{x^w}{(w - \delta)(w + \delta)} \, dw$$

(b) By taking $\alpha(w) = \zeta(1/2 + it + w)$, and considering the residues arising from poles at $w = 1/2 - it$ and at $w = \delta$, show that

$$\zeta(1/2 + \delta + it) = x^{-\delta} \sum_{n \leq x} n^{-1/2 - it}((x/n)^\delta - (n/x)^\delta)$$

$$+ \frac{\delta x^{-\delta}}{\pi} \int_{-\infty}^{\infty} \zeta(1/2 + it + iu) \frac{x^{iu}}{u^2 + \delta^2} \, du$$

$$- \frac{2\delta x^{1/2 - \delta - it}}{(1/2 - it - \delta)(1/2 - it + \delta)}$$

$$= T_1 + T_2 + T_3,$$

     say.

(c) Show that

$$T_1 \ll \left(1 + x^{1/2 - \delta}\right) \min\left(\frac{1}{|\delta - 1/2|}, \log x\right).$$

(d) Let $M(T) = \max_{0 \leq t \leq T} |\zeta(1/2 + it)|$. Show that

$$T_2 \ll x^{-\delta} M(2\tau)$$

     uniformly for $0 < \delta \leq 1/2$.

(e) Show that $T_3 \ll x^{1/2 - \delta}/\tau^2$.

(f) By taking $x = M(2\tau)^2$, show that

$$\zeta(\sigma + it) \ll M(2\tau)^{2 - 2\sigma} \min\left(\frac{1}{|\sigma - 1|}, \log M(2\tau)\right)$$

     uniformly for $1/2 \leq \sigma \leq 1$.

(g) Show that if $M(T) \ll_\varepsilon T^\varepsilon$, then $\mu(\sigma) = 0$ for $\sigma \geq 1/2$.

(h) By Corollary 10.5, deduce that if $M(T) \ll_\varepsilon T^\varepsilon$, then $\mu(\sigma) = 1/2 - \sigma$ when $\sigma \leq 1/2$.

19. Let $M(\sigma, T) = \max_{1 \le t \le T} |\zeta(\sigma + it)|$. Suppose that $\sigma, \sigma_1, \sigma_2$ are fixed, $0 \le \sigma_1 < \sigma < \sigma_2 \le 1$. Let C denote the rectangular contour with vertices $\sigma_2 - \sigma - i\tau/2, \sigma_2 - \sigma + i\tau/2, \sigma_1 - \sigma + i\tau/2, \sigma_1 - \sigma - i\tau/2$.

    (a) Show that
    $$\zeta(\sigma + it) = \frac{1}{2\pi i} \int_C \zeta(s + w) \frac{x^w}{w(w+1)} \, dw.$$

    (b) Deduce that
    $$\zeta(\sigma + it) \ll M(\sigma_1, 2\tau) x^{\sigma_1 - \sigma} + M(\sigma_2, 2\tau) x^{\sigma_2 - \sigma}.$$

    (c) By choosing x suitably, show that
    $$M(\sigma, T) \ll M(\sigma_1, 2T)^{(\sigma_2 - \sigma)/(\sigma_2 - \sigma_1)} M(\sigma_2, 2T)^{(\sigma - \sigma_1)/(\sigma_2 - \sigma_1)}.$$

    (d) Deduce that
    $$\mu(\sigma) \le \frac{\sigma_2 - \sigma}{\sigma_2 - \sigma_1} \mu(\sigma_1) + \frac{\sigma - \sigma_1}{\sigma_2 - \sigma_1} \mu(\sigma_2).$$

    (e) Conclude that $\mu(\sigma) \le \frac{1}{2}(1 - \sigma)$ for $0 \le \sigma \le 1$.

    (f) Show that if $\mu(1/2) = 0$, then (10.10) holds for all σ.

20. (Backlund 1918) Assume the Lindelöf Hypothesis (LH) throughout, and suppose that δ is a small fixed positive number and that t is not the ordinate γ of a zero ρ of $\zeta(s)$.

    (a) Show that the number of zeros $\rho = \beta + i\gamma$ of $\zeta(s)$ in the rectangle $1/2 + \delta \le \beta \le 1, T - 1 \le \gamma \le T + 1$ is $o(\log T)$.

    (b) Show that
    $$\frac{\zeta'}{\zeta}(s) = \sum_\rho \frac{1}{s - \rho} + o(\log \tau)$$

    uniformly for $1/2 + 2\delta \le \sigma \le 2$ where the sum is over those zeros ρ for which $1/2 + \delta \le \beta \le 1, t - 1 \le \gamma \le t + 1$.

    (c) Show that if $\sigma_1 < \sigma_2$ and $t \ne \gamma$, then
    $$\int_{\sigma_1}^{\sigma_2} \frac{\sigma - \beta}{(\sigma - \beta)^2 + (t - \gamma)^2} \, d\sigma = \frac{1}{2} \log \frac{(\sigma_2 - \beta)^2 + (t - \gamma)^2}{(\sigma_1 - \beta)^2 + (t - \gamma)^2}.$$

    (d) Show that if $1/2 \le \sigma_1 \le 1$ and $t \ne \gamma$, then
    $$\int_{\sigma_1}^2 \frac{\sigma - \beta}{(\sigma - \beta)^2 + (t - \gamma)^2} \, d\sigma \ge 0.$$

    (e) Show that if t is not the ordinate of a zero, then
    $$\int_{\sigma_1}^2 \Re \frac{\zeta'}{\zeta}(\sigma + it) \, d\sigma \ge -\varepsilon \log \tau$$

    uniformly for $1/2 + 2\delta \le \sigma \le 2$.

(f) Show that $\mu(\sigma) = 0$ for $1/2 < \sigma \le 2$.

(g) Deduce that $\mu(\sigma) = 1/2 - \sigma$ for $-1 \le \sigma < 1/2$.

(h) Show that

$$\int_{\sigma_1}^{\sigma_2} \frac{t - \gamma}{(\sigma - \beta)^2 + (t - \gamma)^2} \, d\sigma = \arctan\frac{t - \gamma}{\sigma_2 - \beta} - \arctan\frac{t - \gamma}{\sigma_1 - \beta}.$$

(i) Deduce that

$$\left| \int_{\sigma_1}^{\sigma_2} \frac{t - \gamma}{(\sigma - \beta)^2 + (t - \gamma)^2} \, d\sigma \right| \le \pi.$$

(j) Conclude that $\arg \zeta(1/2 + 2\delta + it) = o(\log \tau)$.

21. (Backlund 1918; cf. Littlewood 1924) Suppose now that the number of zeros ρ of $\zeta(s)$ in a rectangle $1/2 + \delta \le \beta \le 1, t - 1 \le \gamma \le t + 1$ is $o(\log \tau)$ as $t \to \infty$, and put

$$f(s) = \frac{\zeta'}{\zeta}(s) - \sum_\rho \frac{1}{s - \rho}$$

where the sum is over the $o(\log \tau)$ zeros in such a rectangle.

(a) Explain why $f(s) \ll \log \tau$ in the disc $|s - 2 - it_0| \le 3/2 - 2\delta$.

(b) Explain why $f(s) = o(\log \tau)$ in the disc $|s - 2 - it_0| \le 1/2$.

(c) Use Hadamard's three circles theorem to show that $f(s) = o(\log \tau)$ for $|s - 2 - it_0| \le 3/2 - 3\delta$.

(d) Deduce that $\zeta(1/2 + 3\delta + it) \ll \tau^\varepsilon$.

(e) Suppose that our hypothesis concerning the number of zeros in a rectangle holds for every fixed positive δ. Deduce that $\mu(\sigma) = 0$ for $\sigma > 1/2$.

(f) By Exercise 19(d), conclude that $\mu(1/2) = 0$, i.e., that LH follows.

22. For $0 < \alpha \le 1$ and $\sigma > 1$ let $\zeta(s, \alpha) = \sum_{n=0}^\infty (n + \alpha)^{-s}$ be the Hurwitz zeta function.

(a) Show that

$$\zeta(s, \alpha)\Gamma(s) = \int_0^\infty \frac{x^{s-1} e^{-\alpha x}}{1 - e^{-x}} \, dx$$

for $\sigma > 1$.

(b) Let

$$I(s, \alpha) = \int_{C(r)} \frac{z^{s-1} e^{-\alpha z}}{1 - e^{-z}} \, dz$$

where $C(r)$ is a contour that runs by a straight line from $ir + \infty$ to ir, by a semicircle from ir through $-r$ to $-ir$, and then by a straight line from $-ir$ to $-ir + \infty$. Note that the value of $I(s, \alpha)$ is independent of r for $0 < r < 2\pi$. By letting $r \to 0$ show that $I(s, \alpha) = (e^{2\pi i s} - 1)$ $\zeta(s, \alpha)\Gamma(s)$ for $\sigma > 1$.

(c) By means of (C.6), show that

$$\zeta(s, \alpha) = \frac{\Gamma(1 - s)e^{-\pi i s}}{2\pi i} I(s, \alpha)$$

for $\sigma > 1$.

(d) Show that $I(s, \alpha)$ is an entire function of s. Deduce by the above that $\zeta(s, \alpha)$ is meromorphic.

(e) Show that $I(k, \alpha) = 0$ for $k = 2, 3, \ldots$.

(f) Show that $I(1, \alpha) = 2\pi i$.

(g) Deduce that $\zeta(s, \alpha)$ is analytic everywhere except for a simple pole at $s = 1$ with residue 1.

(h) Show that if k is an integer, then

$$I(k, \alpha) = \oint_{|z|=1} z^{k-2} \left(\frac{ze^{(1-\alpha)z}}{e^z - 1} \right) dz.$$

(i) By Exercise B.3, deduce that if k is a non-negative integer, then

$$I(-k, \alpha) = 2\pi i\, B_{k+1}(1 - \alpha)/(k + 1)!.$$

(j) By Theorem B.1, deduce that if k is a positive integer then

$$\zeta(1 - k, \alpha) = \frac{-B_k(\alpha)}{k}.$$

In particular, $\zeta(0, \alpha) = 1/2 - \alpha$.

23. (Lerch 1894; cf. Berndt 1985) Let α be fixed, $0 < \alpha \leq 1$. (a) Show that

$$\zeta(s, \alpha) - \zeta(s) = \alpha^{-s} + \sum_{n=1}^{\infty}((n + \alpha)^{-s} - n^{-s})$$

for $\sigma > 0$.

(b) Show that

$$(n + \alpha)^{-s} - n^{-s} + \alpha s n^{-s-1} = s(s + 1) \int_n^{n+\alpha} (n + \alpha - u)u^{-s-2}\, du.$$

(c) Deduce that

$$\zeta(s, \alpha) - \zeta(s) + \alpha s\zeta(s + 1) = \alpha^{-s} + \sum_{n=1}^{\infty}((n + \alpha)^{-s} - n^{-s} + \alpha s n^{-s-1})$$

for $\sigma > -1$, and that the series is locally uniformly convergent in this half-plane.

(d) Show that

$$\zeta'(s, \alpha) - \zeta'(s) + \alpha\zeta(s+1) + \alpha s\zeta'(s+1)$$

$$= -\alpha^{-s}\log\alpha + \sum_{n=1}^{\infty}\left(\frac{-\log(n+\alpha)}{(n+\alpha)^s} + \frac{\log n}{n^s} + \frac{\alpha}{n^{s+1}} - \frac{\alpha s\log n}{n^{s+1}}\right)$$

for $\sigma > -1$. (Here $\zeta'(s, \alpha)$ is meant to denote $\frac{\partial}{\partial s}\zeta(s, \alpha)$.)

(e) By Corollary 1.16, or otherwise, show that

$$\lim_{s\to 0}\zeta(s+1) + s\zeta'(s+1) = C_0.$$

(f) Deduce that

$$\zeta'(0, \alpha) - \zeta'(0) + \alpha C_0 = -\log a + \sum_{n=1}^{\infty}(-\log(n+\alpha) + \log n + \alpha/n).$$

By (10.14) and the definition (C.1) of the gamma function, conclude that

$$\zeta'(0, \alpha) = \log\frac{\Gamma(\alpha)}{\sqrt{2\pi}}.$$

24. (a) Let χ be a character modulo q. Show that

$$L(s, \chi) = q^{-s}\sum_{a=1}^{q}\chi(a)\zeta(s, a/q).$$

(b) Show that if χ is a non-principal character modulo q, then

$$L(0, \chi) = \frac{-1}{q}\sum_{a=1}^{q}\chi(a)a.$$

(c) Show that if χ is a non-principal character modulo q, then

$$L'(0, \chi) = L(0, \chi)\log q + \sum_{a=1}^{q}\chi(a)\log\Gamma(a/q).$$

25. Let $Q(x, y) = ax^2 + bxy + cy^2$ where a, b, c are real numbers, and put $d = b^2 - 4ac$. Suppose that Q is positive-definite, which is to say that $a > 0$ and $d < 0$. For z with $\Re z > 0$, put

$$\vartheta_Q(z) = \sum_{m,n\in\mathbb{Z}}e^{-2\pi Q(m,n)z/\sqrt{-d}}.$$

(a) Show that

$$\vartheta_Q(z) = \sum_{n}e^{-\pi zn^2\sqrt{-d}/(2a)}\sum_{m}e^{-2\pi a(m+bn/(2a))^2z/\sqrt{-d}}.$$

(b) Apply Theorem 10.1 to the inner sum, take the sum over n inside, and apply Theorem 10.1 a second time to show that $\vartheta_Q(z) = \vartheta_Q(1/z)/z$.

(c) For $\sigma > 1$ put

$$\zeta_Q(s) = \sum_{(m,n)\neq(0,0)} Q(m,n)^{-s}.$$

Show that if $\Re z \geq 0$, then

$$\zeta_Q(s)\Gamma(s)(-d)^{s/2}(2\pi)^{-s}$$

$$= (-d)^{s/2}(2\pi)^{-s}\sum_{(m,n)\neq(0,0)} Q(m,n)^{-s}\Gamma\left(s,\frac{2\pi Q(m,n)z}{\sqrt{-d}}\right)$$

$$+ (-d)^{(1-s)/2}(2\pi)^{s-1}\sum_{(m,n)\neq(0,0)} Q(m,n)^{s-1}\Gamma\left(1-s,\frac{2\pi Q(m,n)}{z\sqrt{-d}}\right)$$

$$+ \frac{z^{s-1}}{2(s-1)} - \frac{z^{-s}}{2s}.$$

(d) Deduce that $\zeta_Q(s)$ is a meromorphic function whose only singularity is a simple pole at $s = 1$ with residue $\pi/\sqrt{-d}$.
(e) Put $\xi_Q(s) = \zeta_Q(s)\Gamma(s)(-d)^{s/2}(2\pi)^{-s}$. Show that $\xi_Q(s) = \xi_Q(1-s)$ for all s except $s = 0, s = 1$.
(f) Show that $\zeta_Q(0) = -1/2$.
(g) Show that $\zeta_Q(-k) = 0$ for all positive integers k.

26. Let K be an algebraic number field. The *Dedekind zeta function* of K is defined to be $\zeta_K(s) = \sum_{\mathfrak{a}} N(\mathfrak{a})^{-s}$ for $\sigma > 1$, where the sum is over all integral ideals in the ring $\mathcal{O}_K$ of algebraic integers in K. This is a natural generalization of the Riemann zeta function, and indeed $\zeta_{\mathbb{Q}}(s) = \zeta(s)$. Since ideals in $\mathcal{O}_K$ factor uniquely into prime ideals, and since $N(\mathfrak{ab}) = N(\mathfrak{a})N(\mathfrak{b})$ for any pair $\mathfrak{a}, \mathfrak{b}$ of ideals, it follows that

$$\zeta_K(s) = \prod_{\mathfrak{p}}(1 - N(\mathfrak{p})^{-s})^{-1}$$

for $\sigma > 1$. Let d denote the discriminant of K. In the case that K is a quadratic field, by analysing how rational primes split in K it emerges that $\zeta_K(s) = \zeta(s)L(s,\chi_d)$ where $\chi_d(n) = \left(\frac{d}{n}\right)_K$ is the Kronecker symbol. Thus the functional equations of $\zeta(s)$ and of $L(s,\chi_d)$ give a functional equation for $\zeta_K(s)$ in this case. From now on, suppose that K is a complex quadratic field, which is to say that $K = \mathbb{Q}(\sqrt{d})$ where $d < 0$ is a fundamental quadratic discriminant. Let w denote the number of units in $\mathcal{O}_K$, which is to say that $w = 6$ if $d = -3$, $w = 4$ if $d = -4$, and $w = 2$ if $d < -4$. Let h be the class number of K. Then there are precisely h reduced positive definite binary quadratic forms of discriminant d, say $Q_1, Q_2, \ldots, Q_h$. As m and n run over integral values, $(m,n) \neq (0,0)$, the values $Q_i(m,n)$ run over the the values $N(\mathfrak{a})$ for ideals $\mathfrak{a}$ in the ith ideal class C_i, each value being taken exactly w times. Thus

$$\zeta_{Q_i}(s) = w \sum_{\mathfrak{a}\in C_i} N(\mathfrak{a})^{-s}$$

in the notation of the preceding exercise, and

$$\zeta_K(s) = \frac{1}{w} \sum_{i=1}^{h} \zeta_{Q_i}(s).$$

(a) For $\Re z > 0$, let

$$\vartheta_K(z) = \sum_{i=1}^{h} \vartheta_{Q_i}(z) = h + w \sum_{n=1}^{\infty} r(n)e^{-2\pi nz/\sqrt{-d}}$$

where $r(n) = r_K(n) = \sum_{k|n} \chi_d(k)$ is the number of ideals in $\mathcal{O}_K$ with norm n. Show that $\vartheta_K(z) = \vartheta_K(1/z)/z$.

(b) Show that if $\Re z \geq 0$, then

$$\zeta_K(s)\Gamma(s)(-d)^{s/2}(2\pi)^{-s}$$
$$= (-d)^{s/2}(2\pi)^{-s} \sum_{n=1}^{\infty} r(n)n^{-s}\Gamma\left(s, 2\pi nz/\sqrt{-d}\right)$$
$$+ (-d)^{(1-s)/2}(2\pi)^{s-1} \sum_{n=1}^{\infty} r(n)n^{s-1}\Gamma\left(1-s, 2\pi n/(z\sqrt{-d})\right)$$
$$+ \frac{hz^{s-1}}{2w(s-1)} - \frac{hz^s}{2ws}.$$

(c) Deduce that $\zeta_K(s)$ is a meromorphic function whose only singularity is a simple pole at $s=1$ with residue $h\pi/(w\sqrt{-d})$.

(d) Put $\xi_K(s) = \zeta_K(s)\Gamma(s)(-d)^{s/2}(2\pi)^{-s}$. Show that $\xi_K(s) = \xi_K(1-s)$ for all s except $s=1$ and $s=0$.

(e) Show that $\zeta_K(0) = -h/(2w)$.

(f) Show that $\zeta_K(-k) = 0$ for all positive integers k.

(g) Show that $r(n^2) \geq 1$ for all positive integers n.

(h) Show that if $L(1/2, \chi_d) \geq 0$, then $h \gg (-d)^{1/4}\log(-d)$.

27. Let α be an arbitrary complex number and z a complex number with $\Re z > 0$. Let $f(u) = e^{-\pi(u+\alpha)^2 z}$. Show that $\widehat{f}(t) = z^{-1/2}e^{2\pi i t\alpha}e^{-\pi t^2/z}$. Deduce that the identities of Theorem 10.1 hold for arbitrary complex α.

28. *Grössencharaktere* for $\mathbb{Q}(\sqrt{-1})$, continued from Exercises 4.2.7 and 4.3.10. (a) By two applications of the preceding exercise, show that if z

and w are complex numbers with $\Re z > 0$, then

$$\sum_{a,b\in\mathbb{Z}} e^{-\pi(a^2+b^2)} e^{2\pi i(a+ib)w} = \frac{1}{z} \sum_{c,d\in\mathbb{Z}} e^{-\pi(c^2+d^2)/z} e^{2\pi i(c+id)w/z}.$$

(b) Differentiate both sides of the above m times with respect to w, and then set $w = 0$, to show that

$$\sum_{a,b} e^{-\pi(a^2+b^2)z}(a+ib)^m = \frac{1}{z^{m+1}} \sum_{c,d} e^{-\pi(c^2+d^2)/z}(c+id)^m.$$

(c) Explain why the above reduces to $0 = 0$ if $4 \nmid m$.

(d) Let χ_m and $L(s, \chi_m)$ be defined as before. Show that if m is a positive integer and $\Re z \geq 0$, then

$$L(s, \chi_m)\Gamma(s+2m)\pi^{-s}$$
$$= \frac{\pi^{-s}}{4} \sum_{(a,b)\neq(0,0)} \frac{\chi_m(a+ib)}{(a^2+b^2)^s}\Gamma(s+2m, \pi(a^2+b^2)z)$$
$$+ \frac{\pi^{s-1}}{4} \sum_{(a,b)\neq(0,0)} \frac{\chi_m(a+ib)}{(a^2+b^2)^{1-s}}\Gamma(1-s+2m, \pi(a^2+b^2)/z).$$

(e) Deduce that $L(s, \chi_m)$ is an entire function when m is a non-zero integer.

(f) For each positive integer m, put $\xi(s, \chi_m) = L(s, \chi_m)\Gamma(s+2m)\pi^{-s}$. Show that $\xi(s, \chi_m) = \xi(1-s, \chi_m)$ for all s.

(g) Show that if m is a positive integer, then $L(s, \chi_m)$ has simple zeros at $-2m, -2m-1, -2m-2, \ldots$, but no other zeros in the half-plane $\sigma < 0$.

(h) Show that $\xi(\sigma, \chi_m)$ is real for all real σ, and that $\xi(1/2 + it, \chi_m)$ is real for all real t.

10.2 Products and sums over zeros

If $P(z)$ is a polynomial, then we may express $P(z)$ as a product over its zeros z_i,

$$P(z) = c(z - z_1)(z - z_2)\cdots(z - z_n).$$

The question arises whether a more general entire function may be similarly represented as a product over its zeros, say

$$f(z) = c\prod_n \left(1 - \frac{z}{z_n}\right). \tag{10.21}$$

This is an issue that was addressed by Weierstrass and Hadamard. Rather than derive their extensive theory, we establish only a simple part of it that suffices for our purposes. We do not quite achieve a formula of the type (10.21) for the zeta function, but we obtain a serviceable substitute.

Lemma 10.11 *Suppose that $f(z)$ is an entire function with a zero of order K at 0, and that $f(z)$ vanishes at the non-zero numbers $z_1, z_2, z_3, \ldots$. Suppose also that there is a constant θ, $1 < \theta < 2$, such that*

$$\max_{|z| \leq R} |f(z)| \leq \exp(R^\theta)$$

for all sufficiently large R. Then there exist numbers $A = A(f)$ and $B = B(f)$, such that

$$f(z) = z^K e^{A+Bz} \prod_{k=1}^{\infty} \left(1 - \frac{z}{z_k}\right) e^{z/z_k}$$

for all z. Here the product is uniformly convergent for z in compact sets.

Proof We may suppose that $K = 0$, since if $K > 0$ then the function $f(z)/z^K$ does not vanish at the origin. Let $N_f(R)$ denote the number of zeros of $f(z)$ in the disc $|z| \leq R$. By Jensen's inequality (Lemma 6.1) we find that $N_f(R) \leq 8R^\theta$ for all sufficiently large R. Thus $\sum_{R < |z_k| \leq 2R} |z_k|^{-2} \leq 8R^{\theta-2}$, so by summing over dyadic blocks we see that $\sum_{k=1}^{\infty} |z_k|^{-2} < \infty$. (Alternatively, if more precision were desired, we could write this sum as $\int_0^\infty r^{-2} \, dN_{f(r)}$, and integrate by parts.) But $(1 - z)e^z = 1 + O(|z|^2)$ uniformly for $|z| \leq 1$, so the product

$$g(z) = \prod_{k=1}^{\infty} \left(1 - \frac{z}{z_k}\right) e^{z/z_k}$$

is uniformly convergent in compact regions, and hence represents an entire function. Thus $h(z) = f(z)/(f(0)g(z))$ is a non-vanishing entire function with $h(0) = 1$.

Next we derive an upper bound for $M_h(R)$. To this end we write the product above in three parts,

$$g(z) = \prod_{k \in \mathcal{K}_1} \prod_{k \in \mathcal{K}_2} \prod_{k \in \mathcal{K}_3} = P_1(z)P_2(z)P_3(z),$$

where $|z_k| \leq R/2$ for $k \in \mathcal{K}_1$, $R/2 < |z_k| \leq 3R$ for $k \in \mathcal{K}_2$, and $|z_k| > 3R$ for $k \in \mathcal{K}_3$. Suppose that $R \leq |z| \leq 2R$. If $|z_k| \leq R/2$, then $|1 - z/z_k| \geq |z/z_k| - 1 \geq 1$, and hence

$$|P_1(z)| \geq \prod_{k \in \mathcal{K}_1} e^{-2R/|z_k|}.$$

Now

$$\sum_{k \in \mathcal{K}_1} \frac{1}{|z_k|} \ll R^{\theta-1}.$$

Thus

$$|P_1(z)| \geq e^{-cR^\theta}$$

for all large R. Since card $K_2 \leq 72R^\theta$, it follows that there is an r, $R \leq r \leq 2R$, for which $|r - |z_k|| \geq 1/R^2$ for all k. If r is chosen in this way and $|z| = r$, then

$$|1 - z/z_k| \geq \frac{|r - |z_k||}{|z_k|} \geq \frac{1}{27R^3}$$

for all $k \in K_2$. Hence

$$|P_2(z)| \geq e^{-cR^\theta \log R}$$

when $|z| = r$. Finally,

$$|P_3(z)| \geq \prod_{k \in K_3} e^{-cR^2/|z_k|^2} \geq e^{-cR^\theta}$$

for $|z| \leq 2R$. Hence we see that for each large R there is an r, $R \leq r \leq 2R$, for which $|g(z)| \geq e^{-cR^\theta \log R}$ when $|z| = r$. Thus $|h(z)| \leq e^{cR^\theta \log R}$ for such z, and hence by the maximum modulus principle

$$M_h(R) \leq e^{cR^\theta \log R}.$$

Now put $j(z) = \log h(z)$ with $j(0) = 0$. Then $\Re j(z) \leq cR^\theta \log R$ for all large R, so that by the Borel–Carathéodory lemma (Lemma 6.2),

$$j(z) \ll R^\theta \log R$$

for all large R. But $\theta < 2$, so $j(z)$ must be a polynomial of degree at most 1, say $j(z) = A + Bz$, and the proof is complete. $\qquad\square$

In order to apply our lemma to $\xi(s)$ we need an upper bound for $|\xi(s)|$. From Corollary 1.17 we see that $\zeta(s) \ll |s|^{1/2}$ when $\sigma \geq 1/2$ and $|s| \geq 2$. Thus by Stirling's formula (Theorem C.1) it follows that

$$\xi(s) \ll \exp(|s| \log |s|) \tag{10.22}$$

when $\sigma \geq 1/2$ and $|s| \geq 2$. In view of the functional equation found in Corollary 10.3, this same upper bound therefore holds for all s with $|s| \geq 2$. Since

$$\xi(s) = (s - 1)\zeta(s)\Gamma(1 + s/2)\pi^{-s/2}, \tag{10.23}$$

it follows from (10.11) that $\xi(0) = 1/2$. Thus by Lemma 10.11 we obtain

Theorem 10.12 *Let $\xi(s)$ be defined as in Corollary* 10.3. *There is a constant B such that*

$$\xi(s) = \frac{1}{2}e^{Bs} \prod_\rho \left(1 - \frac{s}{\rho}\right) e^{s/\rho} \tag{10.24}$$

for all s. Here the product is extended over all zeros ρ of $\xi(s)$.

All known zeros of the zeta function are simple, and it is plausible to conjecture that they all are. In the (unlikely) event that a multiple zero is encountered, the associated factor in the above product is to be repeated as many times as the multiplicity.

Thus far we have remarked upon the zeros of $\xi(s)$ without having proved that they exist. However, from (10.24) we see that if $\xi(s)$ had at most finitely many zeros then there would be a constant $C > 0$ such that $\xi(s) \ll \exp(C|s|)$ for all large s. On the contrary, by Stirling's formula we find that $\xi(\sigma) = \exp\left(\frac{1}{2}\sigma \log \sigma + O(\sigma)\right)$ as $\sigma \to \infty$, so it is evident that $\xi(s)$ has infinitely many zeros. Concerning the density of the zeros, the following estimate is useful.

Theorem 10.13   *For $T \geq 0$, let $N(T)$ denote the number of zeros $\rho = \beta + i\gamma$ of $\xi(s)$ in the rectangle $0 < \beta < 1$, $0 < \gamma \leq T$. Any zeros with $\gamma = T$ should be counted with weight $1/2$. Then*

$$N(T+1) - N(T) \ll \log(T+2).$$

Proof   We apply Jensen's inequality (Lemma 6.1) to $\xi(s)$, on a disc with centre $2 + i(T + 1/2)$ and radius $R = 11/6$. By taking $r = 7/4$, it follows from the estimates of Corollary 1.17 that the number of zeros ρ in the rectangle $1/2 \leq \beta \leq 1$, $T \leq \gamma \leq T + 1$ is $\ll \log(T+2)$. (Alternatively, we could appeal to Theorem 6.8.) But ρ is a zero if and only if $1 - \overline{\rho}$ is a zero, so the rectangle $0 \leq \beta \leq 1/2$, $T \leq \gamma \leq T + 1$ contains the same number of zeros as the former one. Thus we have the result.   □

By summing the above over integral values of T, we deduce that $N(T) \ll T \log T$. Alternatively, this same upper bound follows from (10.22) by means of Jensen's inequality. Hence $\sum_\rho |\rho|^{-A} < \infty$ for all $A > 1$. With a little more work we could show that $\sum 1/|\rho| = \infty$ (see Exercise 10.1), and indeed that $N(T) \asymp T \log T$ for all large T (see Exercise 10.4). A much more precise asymptotic formula for $N(T)$ will be derived in Chapter 14.

We recall that the *logarithmic derivative* of a function $f(z)$ is defined to be $f'(z)/f(z)$. Since $f'(z)/f(z) = \frac{d}{dz} \log f(z)$, it follows that the logarithmic derivative of a product is the sum of the logarithmic derivatives of the factors. Although $\log f(z)$ is multiple-valued, the ambiguity involves only an additive constant, so $f'(z)/f(z)$ is a well-defined single-valued analytic function wherever $f(z)$ is analytic and non-zero. If f has a zero at a of multiplicity m, then f'/f has a simple pole at a with residue m. If f has a pole at a of multiplicity m then f'/f has a simple pole at a with residue $-m$. Hence if f is meromorphic then f'/f is meromorphic with only simple poles, which occur at the zeros and poles of f.

By taking logarithmic derivatives in the definition (10.5) of $\xi(s)$ we find that

$$\frac{\xi'}{\xi}(s) = \frac{1}{s} + \frac{1}{s-1} + \frac{\zeta'}{\zeta}(s) + \frac{1}{2}\frac{\Gamma'}{\Gamma}(s/2) - \frac{1}{2}\log \pi. \qquad (10.25)$$

By taking logarithmic derivatives in the functional equation of Corollary 10.3 we see that

$$\frac{\xi'}{\xi}(s) = -\frac{\xi'}{\xi}(1-s). \qquad (10.26)$$

By logarithmically differentiating the asymmetric form (10.9) of the functional equation, we discover that

$$\frac{\zeta'}{\zeta}(s) = -\frac{\zeta'}{\zeta}(1-s) + \log 2\pi - \frac{\Gamma'}{\Gamma}(1-s) + \frac{\pi}{2}\cot\frac{\pi s}{2}. \qquad (10.27)$$

By taking logarithmic derivatives of both sides of the identity (10.24) we obtain

Corollary 10.14 *Let B be defined as in Theorem 10.12. Then*

$$\frac{\xi'}{\xi}(s) = B + \sum_{\rho}\left(\frac{1}{s-\rho} + \frac{1}{\rho}\right) \qquad (10.28)$$

and

$$\frac{\zeta'}{\zeta}(s) = B + \frac{1}{2}\log\pi - \frac{1}{s-1} - \frac{1}{2}\frac{\Gamma'}{\Gamma}(s/2+1) + \sum_{\rho}\left(\frac{1}{s-\rho} + \frac{1}{\rho}\right).$$

$$\qquad (10.29)$$

Moreover,

$$B = -\frac{1}{2}\sum_{\rho}\left(\frac{1}{1-\rho} + \frac{1}{\rho}\right) = -\sum_{\rho}\Re\frac{1}{\rho} = \frac{-C_0}{2} - 1 + \frac{1}{2}\log 4\pi$$

$$= -0.0230957\ldots. \qquad (10.30)$$

In the above, it is to be understood that if $\xi(s)$ has a multiple zero ρ, then the summand arising from ρ is to be repeated as many times as the multiplicity.

Proof The second identity follows from the first by means of (10.25). As for (10.30), we observe first by taking $s = 0$ in (10.28) that $B = \frac{\xi'}{\xi}(0)$. Also, by taking $s = 1$ in (10.28) we find that $\frac{\xi'}{\xi}(1) = B + \sum_{\rho}(1/(1-\rho) + 1/\rho)$. By (10.26), this is $-B$, so we obtain the first identity in (10.30). Since B is real, we may write

$$B = -\frac{1}{2}\sum_{\rho}\left(\Re\frac{1}{1-\rho} + \Re\frac{1}{\rho}\right).$$

However, $\sum_{\rho}\Re 1/(1-\rho)$ and $\sum_{\rho}\Re 1/\rho$ are absolutely convergent, so these two sums may be written separately, above. Since $1 - \rho$ runs over zeros of the zeta function as ρ does, the two sums are equal, and we obtain the second identity in (10.30). By logarithmically differentiating the fundamental identity $s\Gamma(s) = \Gamma(s+1)$ we see that $1/s + \frac{\Gamma'}{\Gamma}(s) = \frac{\Gamma'}{\Gamma}(s+1)$. Hence (10.25) may be rewritten as

$$\frac{\xi'}{\xi}(s) = \frac{1}{s-1} + \frac{\zeta'}{\zeta}(s) + \frac{1}{2}\frac{\Gamma'}{\Gamma}(s/2+1) - \frac{1}{2}\log \pi.$$

We obtain the third identity in (10.30) by taking $s = 0$ in the above, in view of (10.11), (10.14), and (C.12). □

In order to extend our theory to include L-functions, we need an upper bound for $|L(s, \chi)|$ that corresponds to the bound for the zeta function provided by Corollary 1.17.

Lemma 10.15 *Let χ be a non-principal character modulo q, and suppose that $\delta > 0$ is fixed. Then*

$$L(s, \chi) \ll (1 + (q\tau)^{1-\sigma}) \min\left(\frac{1}{|\sigma - 1|}, \log q\tau\right)$$

uniformly for $\delta \leq \sigma \leq 2$.

Landau noted that an estimate relating to the zeta function often has a 'q-analogue' in which n^{-it} is replaced by $\chi(n)$ and τ is replaced by q. In the above we have a 'hybrid' of the two, with $\chi(n)n^{-it}$ and $q\tau$ throughout.

Proof Let $S(u, \chi) = \sum_{0 < n \leq u} \chi(n)$. Then for $\sigma > 0$,

$$L(s, \chi) = \sum_{n \leq x} \chi(n)n^{-s} + \int_x^\infty u^{-s}\, dS(u, \chi)$$

$$= \sum_{n \leq x} \chi(n)n^{-s} + S(u, \chi)u^{-s}\Big|_x^\infty - \int_x^\infty S(u, \chi)\, du^{-s}$$

$$= \sum_{n \leq x} \chi(n)n^{-s} - S(x, \chi)x^{-s} + s\int_x^\infty S(u, \chi)u^{-s-1}\, du.$$

This is analogous to Theorem 1.12. To estimate the sum we use (1.29). For the remaining terms we use the trivial estimate $S(u, \chi) \ll q$. The stated estimate then follows by taking $x = q\tau$. □

Now suppose that χ is a primitive character modulo q, $q > 1$. By Stirling's formula we see that $\xi(s, \chi) \ll q^{1/2+\sigma} \exp(|s| \log |s|)$ when $\sigma \geq 1/2$ and $|s| \geq 2$. By the functional equation of Corollary 10.8, it follows that

$$\xi(s, \chi) \ll \exp(|s| \log q|s|) \tag{10.31}$$

for all s with $|s| \geq 2$. Hence by Lemma 10.11 we obtain

Theorem 10.16 *Let χ be a primitive character modulo q, $q > 1$, and let $\xi(s, \chi)$ be defined as in Corollary 10.8. There is a constant $B(\chi)$ such that*

$$\xi(s, \chi) = \xi(0, \chi)e^{B(\chi)s} \prod_{\rho} \left(1 - \frac{s}{\rho}\right) e^{s/\rho} \qquad (10.32)$$

for all s. Here the product is extended over all zeros ρ of $\xi(s, \chi)$.

We expect that the zeros of $\xi(s, \chi)$ are all simple, but if a multiple zero is encountered, then the factor that it contributes to the above product is to be repeated as many times as its multiplicity. In analogy to Theorem 10.13, we have

Theorem 10.17 *Let χ be a character modulo q. The number of zeros $\rho = \beta + i\gamma$ of $L(s, \chi)$ in the rectangle $0 \le \beta \le 1$, $T \le \gamma \le T + 1$ is $\ll \log q(|T| + 2)$.*

Proof First suppose that χ is primitive. We apply Jensen's inequality (Lemma 6.1) to $L(s, \chi)$, on a disc with centre $2 + i(T + 1/2)$ and radius $R = 11/6$. By taking $r = 7/4$, it follows from the estimates of Lemma 10.15 that the number of zeros ρ in the rectangle $1/2 \le \beta \le 1$, $T \le \gamma \le T + 1$ is $\ll \log q(T + 2)$. But $L(\rho, \chi) = 0$ if and only if $L(1 - \overline{\rho}, \chi) = 0$ (except possibly for a trivial zero at $s = 0$ if $\chi(-1) = 1$), so the rectangle $0 \le \beta \le 1/2$, $T \le \gamma \le T + 1$ contains the same number of zeros as (or at most one more than) the former one. Thus we have the result when χ is primitive.

Suppose now that χ is induced by a primitive character χ^* modulo r, with $r | q$. Then

$$L(s, \chi) = L(s, \chi^*) \prod_{\substack{p | q \\ p \nmid r}} \left(1 - \frac{\chi^*(p)}{p^s}\right).$$

Here each factor in the product has zeros forming an arithmetic progression on the imaginary axis with common difference $2\pi i / \log p$. Thus $L(s, \chi)$ has $\ll \log r(|T| + 2)$ zeros of $L(s, \chi^*)$, and additionally has $\ll \sum_{p | q} \log p \ll \log q$ zeros on the imaginary axis with imaginary part between T and $T + 1$. This completes the argument. $\qquad\square$

Suppose that χ is a primitive character modulo q. By taking logarithmic derivatives in the definition (10.18) of $\xi(s, \chi)$, we see that

$$\frac{\xi'}{\xi}(s, \chi) = \frac{L'}{L}(s, \chi) + \frac{1}{2}\frac{\Gamma'}{\Gamma}((s + \kappa)/2) + \frac{1}{2}\log\frac{q}{\pi}. \qquad (10.33)$$

By taking logarithmic derivatives in the functional equation of Corollary 10.8

we see that

$$\frac{\xi'}{\xi}(s, \chi) = -\frac{\xi'}{\xi}(1 - s, \overline{\chi}). \tag{10.34}$$

By logarithmically differentiating the asymmetric form of the functional equation found in Corollary 10.9, we discover that

$$\frac{L'}{L}(s, \chi) = -\frac{L'}{L}(1 - s, \overline{\chi}) - \log\frac{q}{2\pi} - \frac{\Gamma'}{\Gamma}(1 - s) + \frac{\pi}{2}\cot\frac{\pi}{2}(s + \kappa) \tag{10.35}$$

By taking logarithmic derivatives of both sides of the identity (10.31) we obtain

Corollary 10.18 *Let χ be a primitive character modulo q, $q > 1$, and let $B(\chi)$ be defined as in Theorem 10.16. Then*

$$\frac{\xi'}{\xi}(s, \chi) = B(\chi) + \sum_\rho \left(\frac{1}{s - \rho} + \frac{1}{\rho}\right) \tag{10.36}$$

and

$$\frac{L'}{L}(s, \chi) = B(\chi) - \frac{1}{2}\frac{\Gamma'}{\Gamma}((s + \kappa)/2) - \frac{1}{2}\log\frac{q}{\pi} + \sum_\rho \left(\frac{1}{s - \rho} + \frac{1}{\rho}\right). \tag{10.37}$$

Moreover,

$$\Re B(\chi) = -\frac{1}{2}\sum_\rho \left(\frac{1}{1 - \rho} + \frac{1}{\rho}\right) = -\sum_\rho \Re\frac{1}{\rho} \tag{10.38}$$

and

$$B(\chi) = \frac{-1}{2}\log\frac{q}{\pi} - \frac{L'}{L}(1, \overline{\chi}) + \frac{1}{2}C_0 + (1 - \kappa)\log 2. \tag{10.39}$$

As always, multiple zeros are counted multiply.

Proof The second identity follows from the first by means of (10.33). To obtain the first identity in (10.38), we take $s = 1$ in (10.36), and apply (10.34) to see that

$$B(\chi) + \sum_\rho \left(\frac{1}{1 - \rho} + \frac{1}{\rho}\right) = \frac{\xi'}{\xi}(1, \chi) = -\frac{\xi'}{\xi}(0, \overline{\chi}) = -B(\overline{\chi}) = -\overline{B(\chi)}.$$

From Theorem 10.17 we know that the number of zeros ρ of $\xi(s, \chi)$ with $|\rho| \le R$ is $\ll R \log qR$ for $R \ge 2$. Hence the sums $\sum_\rho \Re 1/(1 - \rho)$ and $\sum_\rho \Re 1/\rho$ are absolutely convergent. As the map $\rho \mapsto 1 - \overline{\rho}$ merely permutes zeros of

$\xi(s, \chi)$, the first of these two sums is unchanged if we replace ρ by $1 - \overline{\rho}$. Hence the two sums are equal, and we obtain the second part of (10.38).

To derive (10.39) we first take $s = 0$ in (10.36) to see that $B(\chi) = \frac{\xi'}{\xi}(0, \chi)$. By (10.34) it follows that $B(\chi) = -\frac{\xi'}{\xi}(1, \overline{\chi})$. The stated identity now follows by taking $s = 1$ in (10.33), in view of (C.11) and (C.14). $\qquad\qquad\square$

10.2.1 Exercises

1. Let f satisfy the hypotheses of Lemma 10.11, and suppose that

$$\sum_{k=1}^{\infty} \frac{1}{|z_k|} < \infty.$$

(a) Show that there are numbers A and B and a non-negative integer K such that $f(z) = z^K e^{A+Bz} g(z)$ where $g(z) = \prod_{k=1}^{\infty}(1 - z/z_k)$.

(b) Observe that for any complex number w, $|1 - w| \leq e^{|w|}$ and show that there is a number C such that $|g(z)| \leq e^{C|z|}$.

(c) Deduce that $\sum_{\rho} 1/|\rho| = \infty$ where the sum is over all non-trivial zeros of the zeta function.

2. (a) Let B be the constant given in (10.30). Show that if $\rho = 1/2 + i\gamma$ is a zero of the zeta function on the critical line, then

$$|\gamma| \geq (-1/B - 1/4)^{1/2} = 6.5611\ldots.$$

(b) Let γ be given, and put $f(\beta) = \beta/(\beta^2 + \gamma^2)$. Show that if $0 \leq \beta \leq 1$, then $f(\beta) \geq \beta/(1 + \gamma^2)$. Deduce that if $0 \leq \beta \leq 1$, then $f(\beta) + f(1 - \beta) \geq f(0) + f(1)$.

(c) Show that if $\rho = \beta + i\gamma$ is a non-trivial zero of the zeta function with $\beta \neq 1/2$, then

$$|\gamma| \geq (-2/B - 1)^{1/2} = 9.2518\ldots.$$

3. (Landau 1903) Show that

$$\limsup_{m \to \infty} \left(\frac{1}{m!} \left| \sum_{n=1}^{\infty} \frac{\mu(n)(\log n)^m}{n} \right| \right)^{1/m} = \frac{1}{3}.$$

4. (a) Show that

$$\sum_{\rho} \Re \frac{1}{\sigma - \rho} = \frac{1}{2} \log \sigma + O(1)$$

for $\sigma \geq 2$, where the sum is over all non-trivial zeros of the zeta function.

(b) Deduce that

$$\sum_{\rho} \left(\Re\frac{1}{\sigma - \rho} - \frac{3}{4}\Re\frac{1}{2\sigma - \rho} \right) = \frac{1}{8}\log \sigma + O(1)$$

for $\sigma \geq 2$.

(c) Show that each summand above is $\leq 1/(\sigma - 1)$.

(d) Show that if $|\gamma| \geq 3\sigma$ and σ is large, then the summand arising from ρ in the sum above is ≤ 0.

(e) Conclude that $N(T) \gg T \log T$ when T is large.

5. Put $f(s) = \Re\left(\frac{1}{s+1} - \frac{3/4}{s+2} \right)$.

(a) Show that if $t \geq 2$, then

$$\sum_{\rho} f(1 + it - \rho) = \frac{1}{8}\log t + O(1)$$

where the sum is over all non-trivial zeros ρ of $\zeta(s)$.

(b) Show that $f(s) \leq 1$ when $\sigma \geq 0$.

(c) Show that if $0 \leq \sigma < 2$, then $f(s) \leq 0$ when

$$t^2 \geq \frac{(\sigma + 1)(\sigma + 2)(\sigma + 5)}{2 - \sigma}.$$

(d) Deduce that $f(s) \leq 0$ if $0 < \sigma < 1$ and $|t| \geq 6$.

(e) Show that $N(T + 6) - N(T - 6) \gg \log T$ for all $T > T_0$.

6. (a) Show that for s near 1 the Laurent expansion of $\frac{\zeta'}{\zeta}(s)$ begins

$$\frac{\zeta'}{\zeta}(s) = \frac{-1}{s - 1} + C_0 + \cdots.$$

(b) Deduce that

$$\frac{\zeta'}{\zeta}(1 - s) = \frac{1}{s} + C_0 + O(|s|)$$

for s near 0.

(c) Show that $\frac{\Gamma'}{\Gamma}(1) = -C_0$.

(d) Show that

$$\frac{\pi}{2}\cot\frac{\pi s}{2} = \frac{1}{s} + O(|s|)$$

for s near 0.

(e) Deduce by (10.27) that $\frac{\zeta'}{\zeta}(0) = \log 2\pi$.

(f) Use this to give a second proof that $\zeta'(0) = -\frac{1}{2}\log 2\pi$.

7. (Taylor 1945) (a) Show that if $\sigma > 1/2$, then $|\xi(s + 1/2)| > |\xi(s - 1/2)|$.

(b) Put $f(s) = \xi(s + 1/2) + \xi(s - 1/2)$. Show that all zeros of $f(s)$ have real part $1/2$.

(c) Assume RH. Show that if c is fixed, $c > 0$, then all zeros of $\xi(s + c) + \xi(s - c)$ have real part $1/2$.

8. (Vorhauer 2006) Let $B(\chi)$ denote the constant in Theorem 10.16.

   (a) Show that
   $$\frac{1 - \beta}{(1 - \beta)^2 + \gamma^2} + \frac{\beta}{\beta^2 + \gamma^2} \geq \frac{1}{1 + \gamma^2}$$
   uniformly for $0 \leq \beta \leq 1$.

   (b) Deduce that
   $$\Re B(\chi) \leq -\frac{1}{2} \sum_\gamma \frac{1}{1 + \gamma^2}.$$

   (c) Show that
   $$\frac{\xi'}{\xi}(2, \chi) = \frac{1}{2} \log q + O(1).$$

   (d) Show that
   $$\Re \frac{\xi'}{\xi}(2, \chi) = \sum_\rho \Re \frac{1}{2 - \rho}.$$

   (e) Show that
   $$\Re \frac{\xi'}{\xi}(2, \chi) = \frac{1}{2} \sum_\rho \Re \left(\frac{1}{2 - \rho} + \frac{1}{1 + \overline{\rho}} \right).$$

   (f) Show that
   $$\frac{2 - \beta}{(2 - \beta)^2 + \gamma^2} + \frac{1 + \beta}{(1 + \beta)^2 + \gamma^2} \leq \frac{3}{1 + \gamma^2}$$
   uniformly for $0 \leq \beta \leq 1$.

   (g) Conclude that
   $$\Re B(\chi) \leq \frac{-1}{6} \log q + O(1).$$

9. Let $K > 0$ be given, and put $E(z) = (1 - z) \exp \left(\sum_{k=1}^K z^k/k \right)$.

   (a) Show that
   $$E'(z) = -z^K \exp \left(\sum_{k=1}^K \frac{z^k}{k} \right).$$

   (b) Deduce that the power series coefficients of $E'(z)$ are all ≤ 0.

   (c) Write $E(z) = \sum_{m=0}^\infty A_m z^m$. Show that $A_0 = 1$, $A_m = 0$ for $1 \leq m \leq K$, $A_m < 0$ for $m > K$, and that $\sum_{m > K} A_m = -1$.

   (d) Show that if $|z| \leq r \leq 1$, then $|1 - E(z)| \leq 1 - E(r) \leq r^{K+1}$.

10.3 Notes

Section 10.1. The case $\alpha = 0$ of (10.1) was given by Poisson (1823). de la Vallée Poussin observed that the left-hand side of (10.1) has period 1 with respect to α, and then computed the Fourier coefficients of this function to obtain (10.1). This is rather similar to using the Poisson summation formula, as we have done. Theorem 10.1 is the basis for a very large class of functional equations and was first exploited systematically by Hecke. For the most general version see Tate's thesis, reproduced in Tate (1967). Riemann gave two proofs of Corollary 10.3. Riemann's second method involved using Theorem 10.1 to establish the formula of Exercise 10.1.10. This is the case $z = 1$ of Theorem 10.2, with the order of summation and integration reversed. Theorem 10.2 is due to Lavrik (1965), who derived it from Corollary 10.3 in the manner outlined in Exercise 10.1.4. For further proofs of the functional equation, see Titchmarsh (1986, Chapter 2).

The proof of Theorem 10.1 can be arranged so that one does not depend on the fact that $\int e^{-\pi x^2}\, dx = 1$. To see this, let c denote the value of this integral. Then the proof given establishes (10.1) with the factor c on the right-hand side. But if $z = 1$ and $\alpha = 0$ the two sides of (10.1) are visibly equal and positive, so it follows that $c = 1$.

The functional equation for $\zeta(s)$ was established by Riemann (1860), and that for $L(s, \chi)$ by de la Vallée Poussin (1896) although it was known in some special cases earlier. See the commentary of Landau (1909, p. 899).

Section 10.2 The product formula of Theorem 10.12 was established by Hadamard (1893). The constant $B(\chi)$ in Theorem 10.16 was long considered to be mysterious; the simple formula (10.39) for it is due to Vorhauer (2006).

10.4 References

Backlund, R. J. (1918). Über die Beziehung zwischen Anwachsen und Nullstellen der Zetafunktion, *Öfv. af finska vet. soc. förh.* **61A**, Nr. 9.

Berndt, B. C. (1985). The gamma function and the Hurwitz zeta-function, *Amer. Math. Monthly* **92**, 126–130.

Hadamard, J. (1893). Étude sur les propriétés des fonctions entières et en particulier d'une fonction considérée par Riemann, *J. Math. Pures Appl.* (4) **9**, 171–215.

Heilbronn, H. (1938). On Dirichlet series which satisfy a certain functional equation, *Quart J. Math. Oxford Ser.* **9**, 194–195.

Landau, E. (1903). Über die zahlentheoretische Funktion $\mu(k)$, *Sitzungsber. Kais. Akad. Wiss. Wien* **112**, 537–570; *Collected Works*, Vol. 2. Essen: Thales Verlag, 1986, pp. 60–93.

   (1907). Bemerkungen zu einer Arbeit des Herrn V. Furlan, *Rend. Circ. Mat. Palermo* **23**, 367–373; *Collected Works*, Vol. 3. Essen: Thales Verlag, 1986, pp. 316–322.

(1909). *Handbuch der Lehre von der Verteilung der Primzahlen*, Third edition. New York: Chelsea, 1974.

Lavrik, A. F. (1965). The abbreviated functional equation for the L-function of Dirichlet, *Izv. Akad. Nauk UzSSR Ser. Fiz.-Mat. Nauk* **9**, 17–22.

Lerch, M. (1894). Weitere Studien auf dem Gebiete der Malmstén'schen Reihen. Mit einem Briefe des Herrn Hermite, *Rozpravy* **3**, No. 28, 63 pp.

Littlewood, J. E. (1924). On the zeros of the Riemann Zeta-function, *Cambridge Philos. Soc. Proc.* **22**, 295–318.

Mallik, A. (1977). If $L(\frac{1}{2}, \chi) > 0$, then $L\left(\frac{1}{2}, \chi\right)$ cannot be a minimum, *Studia Sci. Math. Hungar.* **12**, 445–446.

Poisson, S. D. (1823). Suite de mémoire sur les intégrales définies et sur la sommation des séries, *l'École Royale, J. Polytechnique* **12**, 404–509.

Riemann, B. (1860). Ueber die Anzahl der Primzahlen unter einer gegebenen Grösse, *Monatsberichte der Königlichen Preussichen Akademie der Wissenschaften zu Berlin aus dem Jahre* 1859, 671–680; *Werke.* Leipzig: Teubner, 1876, pp. 3–47. Reprint: New York: Dover, 1953.

Tate, J. T. (1967). Fourier analysis in number fields, and Hecke's zeta-functions, *Algebraic Number Theory* (Brighton, 1965). Washington: Thompson, pp. 305–347.

Taylor, P. R. (1945). On the Riemann zeta function, *Quart. J. Math. Oxford Ser.* **16**, 1–21.

Titchmarsh, E. C. (1986). *The Theory of the Riemann Zeta-function*, Second Edition. Oxford: Oxford University Press.

de la Vallée Poussin, C. (1896). Recherches analytique sur la théorie des nombres premiers. Deuxième partie: Les fonctions de Dirichlet et les nombres premiers de la forme linéaire $Mx + N$, *Annales de la Société scientifique de Bruxelles*, **20**, 281–342.

Vorhauer, U. M. A. (2006). The Hadamard product formula for Dirichlet L-functions, to appear.

A. Walfisz (1931). Teilerprobleme, II, *Math. Z.* **34**, 448–472.

A. Weil (1967). Über die Bestimmung Dirichletscher Reihen durch Funktionalgleichungen, *Math. Ann.* **168**, 149–156.

11

Primes in arithmetic progressions: II

11.1 A zero-free region

For a given integer q, the primes not dividing q are distributed in the reduced residue classes modulo q. As there are no other obvious restrictions on the primes modulo q, we expect the primes to be uniformly distributed amongst the reduced residue classes. Let $\pi(x; q, a)$ denote the number of primes $p \leq x$ such that $p \equiv a \pmod{q}$. We anticipate that if $(a, q) = 1$, then

$$\pi(x; q, a) \sim \frac{x}{\varphi(q) \log x} \qquad \text{as } x \longrightarrow \infty.$$

This asymptotic estimate is the *Prime Number Theorem for arithmetic progressions*; it can readily be established by adapting the methods of Chapters 4 and 6. For many purposes, however, it is important to have a quantitative form of this, from which one can tell how large x should be, as a function of q, to ensure that $\pi(x; q, a)$ is near $\operatorname{li}(x)/\varphi(q)$. To obtain such an estimate we must first derive a zero-free region for the Dirichlet L-functions $L(s, \chi)$ that is explicit in its dependence on both q and t. For the most part our arguments are natural generalizations of the analysis in Chapter 6, but we shall encounter a new difficulty in connection with the possible existence of a real zero β near 1 of $L(s, \chi)$ when χ is a quadratic character.

The approximate partial fraction expansion of $\frac{\zeta'}{\zeta}(s)$ (cf. Lemma 6.4) depends on the upper bound for $|\zeta(s)|$ provided by Corollary 1.17. By using Lemma 10.15 in a similar manner, we now derive a corresponding approximate partial fraction formula for $\frac{L'}{L}(s, \chi)$. In order to formulate a unified result for both the principal and non-principal characters, it is convenient to employ the notation

$$E_0(\chi) = \begin{cases} 1 & \text{if } \chi = \chi_0, \\ 0 & \text{otherwise.} \end{cases} \tag{11.1}$$

Lemma 11.1 *If χ is a character* (mod q) *and* $5/6 \leq \sigma \leq 2$, *then*

$$-\frac{L'}{L}(s, \chi) = \frac{E_0(\chi)}{s-1} - \sum_{\rho} \frac{1}{s-\rho} + O(\log q\tau)$$

where the sum is over all zeros ρ of $L(s, \chi)$ for which $\left|\rho - \left(\frac{3}{2} + it\right)\right| \leq 5/6$.

Proof When χ is non-principal we apply Lemma 6.3 to the function

$$f(z) = L\left(z + \left(\frac{3}{2} + it\right), \chi\right)$$

with $R = 5/6$ and $r = 2/3$. By Lemma 10.15 we may take $M = Cq\tau$ for a suitable absolute constant C, and by the Euler product for $L(s, \chi)$ we see that

$$|f(0)| = \left|L\left(\tfrac{3}{2} + it, \chi\right)\right| = \prod_p \left|1 - \chi(p)p^{-\frac{3}{2}-it}\right|^{-1} \geq \prod_p \left(1 + p^{-3/2}\right)^{-1} \gg 1.$$

Now suppose that $\chi = \chi_0$. The zeros of the function $1 - p^{-s}$ form an arithmetic progression on the imaginary axis. Hence by (4.22), the zeros of $L(s, \chi_0)$ are the zeros of $\zeta(s)$ together with the union of several arithmetic progressions on the imaginary axis. Since these latter zeros all lie at a distance $\geq 3/2$ from the point $\frac{3}{2} + it$, none of them is included in the sum over ρ. Moreover, by taking logarithmic derivatives of both sides of (4.22) we see that

$$\frac{L'}{L}(s, \chi_0) = \frac{\zeta'}{\zeta}(s) + \sum_{p|q} \frac{\log p}{p^s - 1}.$$

But $(\log p)/(p^s - 1) \ll 1$ for $\sigma \geq 5/6$, so the sum over p is $\ll \omega(q) \ll \log q$ by Theorem 2.10. Hence we obtain the stated identity by appealing to Lemma 6.4. $\qquad\square$

The generalization of Lemma 6.5 is straightforward.

Lemma 11.2 *If $\sigma > 1$, then*

$$\Re\left(-3\frac{L'}{L}(\sigma, \chi_0) - 4\frac{L'}{L}(\sigma + it, \chi) - \frac{L'}{L}(\sigma + 2it, \chi^2)\right) \geq 0.$$

Proof By the Dirichlet series expansion (4.25) for $\frac{L'}{L}(s, \chi)$ we see that the left-hand side above is

$$\Re \sum_{\substack{n=1 \\ (n,q)=1}}^{\infty} \frac{\Lambda(n)}{n^{\sigma}}(3 + 4\chi(n)n^{-it} + \chi(n)^2 n^{-2it}).$$

The quantity $\chi(n)n^{-it}$ is unimodular when $(n, q) = 1$, so for such n there is a real number θ_n such that $\chi(n)n^{-it} = e^{i\theta_n}$. Thus the above is

$$\sum_{\substack{n=1 \\ (n,q)=1}}^{\infty} \frac{\Lambda(n)}{n^\sigma}(3 + 4\cos\theta_n + \cos 2\theta_n).$$

This is non-negative because $3 + 4\cos\theta + \cos 2\theta = 2(1 + \cos\theta)^2 \geq 0$ for all θ. $\qquad\square$

The groundwork laid above enables us to establish a variant of Theorem 6.6 for Dirichlet L-functions.

Theorem 11.3 *There is an absolute constant $c > 0$ such that if χ is a Dirichlet character modulo q, then the region*

$$R_q = \{s : \sigma > 1 - c/\log q\tau\}$$

contains no zero of $L(s, \chi)$ unless χ is a quadratic character, in which case $L(s, \chi)$ has at most one, necessarily real, zero $\beta < 1$ in R_q.

A zero lying in R_q, as described above, is called *exceptional*. No exceptional zero is known, and indeed it may be conjectured that if χ is quadratic, then $L(\sigma, \chi) > 0$ for all $\sigma > 0$. We give further study to exceptional zeros in the next section.

Proof The case $\chi = \chi_0$ is immediate from (4.22) and Theorem 6.6, so we may assume that χ is non-principal. Also, the Euler product (4.21) for $L(s, \chi)$ is absolutely convergent when $\sigma > 1$, and hence $L(s, \chi) \neq 0$ for such s. Thus it suffices to consider a zero $\rho_0 = \beta_0 + i\gamma_0$ of $L(s, \chi)$ with $12/13 \leq \beta_0 \leq 1$. We consider several cases, the first of which parallels the proof of Theorem 6.6 most closely. $\qquad\square$

CASE I. COMPLEX χ. If $\sigma > 1$ and ρ is a zero of an L-function, then $\Re(s - \rho) > 0$ and hence $\Re(1/(s - \rho)) > 0$. Thus by Lemma 11.1, if $0 < \delta \leq 1$, then

$$-\Re\frac{L'}{L}(1 + \delta, \chi_0) \leq \frac{1}{\delta} + c_1\log q,$$

$$-\Re\frac{L'}{L}(1 + \delta + i\gamma_0, \chi) \leq \frac{-1}{1 + \delta - \beta_0} + c_1\log q(|\gamma_0| + 4), \qquad (11.2)$$

$$-\Re\frac{L'}{L}(1 + \delta + 2i\gamma_0, \chi^2) \leq c_1\log q(2|\gamma_0| + 4)$$

for some absolute constant c_1. The hypothesis that χ is complex is needed for this last inequality, to ensure that $\chi^2 \neq \chi_0$ in the appeal to Lemma 11.1. We multiply both sides of the first inequality by 3, the second by 4, and sum all three. By Lemma 11.2, the resulting left-hand side is non-negative. That is,

$$\frac{3}{\delta} - \frac{4}{1 + \delta - \beta_0} + c_2 \log q(|\gamma_0| + 4) \geq 0$$

for some constant c_2. If $\beta_0 = 1$, then letting $\delta \to 0^+$ gives an immediate contradiction, so it may be assumed that $\beta_0 < 1$. Then, on taking $\delta = 6(1 - \beta_0)$, it follows that

$$1 - \beta_0 \geq \frac{1}{14c_2 \log q(|\gamma_0| + 4)}.$$

Hence $\rho_0 \notin R_q$ if c is chosen sufficiently small.

This argument also applies with only small changes when χ is quadratic, provided that $|\gamma_0|$ is large. We can even allow $|\gamma_0|$ to be small, as long as it is large compared with $1 - \beta_0$. We now consider such a case.

CASE 2. QUADRATIC χ, $|\gamma_0| \geq 6(1 - \beta_0)$. By Theorem 4.9, $L(1, \chi) \neq 0$, so $\gamma_0 \neq 0$. Hence we can proceed as above, except that as $\chi^2 = \chi_0$ the third inequality in (11.2) must be replaced by the weaker inequality

$$-\Re\frac{L'}{L}(1 + \delta + 2i\gamma_0, \chi^2) \leq \frac{\delta}{\delta^2 + 4\gamma_0^2} + c_1 \log q(2|\gamma_0| + 4).$$

Again if $\beta_0 = 1$, then taking $\delta \to 0^+$ gives a contradiction. Thus it can be supposed that $\beta_0 < 1$. Since $|\gamma_0| \geq 6(1 - \beta_0)$, this implies that

$$-\Re\frac{L'}{L}(1 + \delta + 2i\gamma_0, \chi^2) \leq \frac{\delta}{\delta^2 + 144(1 - \beta_0)^2} + c_1 \log q(2|\gamma_0| + 4).$$

We combine this inequality with the first two inequalities in (11.2) and apply Lemma 11.2 with $\sigma = 1 + \delta = 1 + 6(1 - \beta_0)$ to see that

$$\frac{1}{1 - \beta_0} \left(\frac{3}{6} - \frac{4}{7} + \frac{6}{180} \right) + c_2 \log q(|\gamma_0| + 4) \geq 0.$$

The factor in large parentheses above is $-4/105 < -1/27$, so

$$1 - \beta_0 \geq \frac{1}{27c_2 \log q(|\gamma_0| + 4)}.$$

CASE 3. QUADRATIC χ, $0 < |\gamma_0| \leq 6(1 - \beta_0)$. Since $L(s, \chi)$ is real when s is real, it follows by the Schwarz reflection principle that $L(\beta_0 - i\gamma_0, \chi) = 0$. Hence by Lemma 11.1 we see that if $1 < \sigma \leq 2$, then

$$-\Re\frac{L'}{L}(\sigma, \chi) \leq -\Re\frac{1}{\sigma - \rho_0} - \Re\frac{1}{\sigma - \overline{\rho_0}} + c_1 \log 4q$$

$$= \frac{-2(\sigma - \beta_0)}{(\sigma - \beta_0)^2 + \gamma_0^2} + c_1 \log 4q$$

$$\leq \frac{-2(\sigma - \beta_0)}{(\sigma - \beta_0)^2 + 36(1 - \beta_0)^2} + c_1 \log 4q. \tag{11.3}$$

Rather than apply Lemma 11.2 we simply observe that if $\sigma > 1$, then

$$-\frac{L'}{L}(\sigma, \chi_0) - \frac{L'}{L}(\sigma, \chi) = \sum_{\substack{n=1 \\ (n,q)=1}}^{\infty} \frac{\Lambda(n)}{n^\sigma}(1 + \chi(n)) \geq 0. \qquad (11.4)$$

We put $\sigma = 1 + \delta = 1 + a(1 - \beta_0)$ and combine the first inequality in (11.2) and (11.3) in the above to deduce that

$$\frac{1}{1 - \beta_0}\left(\frac{1}{a} - \frac{2(a+1)}{(a+1)^2 + 36}\right) + c_2 \log 4q \geq 0.$$

The factor in large parentheses is $\sim -1/a$ as $a \to \infty$, so it is certainly possible to choose a value of a so that this factor is negative. Indeed, when $a = 13$ this factor is $-33/754 < -1/27$, and hence

$$1 - \beta_0 \geq \frac{1}{27c_2 \log 4q}.$$

(We note that our supposition that $\beta_0 \geq 12/13$ implies that $\sigma = 1 + 13(1 - \beta_0) \leq 2$, so that Lemma 11.1 is applicable.)

CASE 4. QUADRATIC χ, REAL ZEROS. If β_0 is a real zero of $L(s, \chi)$, then $\beta_0 < 1$ by Theorem 4.9. Suppose that $\beta_0 \leq \beta_1 < 1$ are two such zeros. Then by Lemma 11.1,

$$-\Re\frac{L'}{L}(\sigma, \chi) \leq -\frac{1}{\sigma - \beta_0} - \frac{1}{\sigma - \beta_1} + c_1 \log 4q$$
$$\leq -\frac{2}{\sigma - \beta_0} + c_1 \log 4q.$$

On combining the first part of (11.2) and the above in (11.4) with $\sigma = 1 + \delta = 1 + a(1 - \beta_0)$, we find that

$$\frac{1}{1 - \beta_0}\left(\frac{1}{a} - \frac{2}{a+1}\right) + c_2 \log 4q \geq 0.$$

On taking $a = 2$ we deduce that

$$1 - \beta_0 \geq \frac{1}{6c_2 \log 4q}.$$

This completes the proof. $\qquad\qquad\square$

In the same way that Theorem 6.7 was derived from Theorem 6.6, we now derive estimates for $\frac{L'}{L}(s, \chi)$ and $\log L(s, \chi)$ in a portion of the critical strip.

Theorem 11.4 *Let χ be a non-principal character modulo q, let c be the constant in* Theorem 3, *and suppose that $\sigma \geq 1 - c/(2 \log q\tau)$. If $L(s, \chi)$ has no exceptional zero, or if β_1 is an exceptional zero of $L(s, \chi)$ but $|s - \beta_1| \geq$*

$1/\log q$, *then*

$$\frac{L'}{L}(s, \chi) \ll \log q\tau, \qquad (11.5)$$

$$|\log L(s, \chi)| \le \log\log q\tau + O(1), \qquad (11.6)$$

and

$$\frac{1}{L(s, \chi)} \ll \log q\tau. \qquad (11.7)$$

Alternatively, if β_1 is an exceptional zero of $L(s, \chi)$ and $|s - \beta_1| \le 1/\log q$, then

$$\frac{L'}{L}(s, \chi) = \frac{1}{s - \beta_1} + O(\log q) \quad (s \neq \beta_1), \qquad (11.8)$$

$$|\arg L(s, \chi)| \le \log\log q + O(1) \quad (s \neq \beta_1), \qquad (11.9)$$

and

$$|s - \beta_1| \ll |L(s, \chi)| \ll |s - \beta_1|(\log q)^2. \qquad (11.10)$$

Proof If $\sigma > 1$, then by Corollary 1.11 we see that

$$\left|\frac{L'}{L}(s, \chi)\right| \le \sum_{n=1}^{\infty} \Lambda(n)n^{-\sigma} = -\frac{\zeta'}{\zeta}(\sigma) \ll \frac{1}{\sigma - 1}.$$

Hence (11.5) is obvious if $\sigma \ge 1 + 1/\log q\tau$. Let $s_1 = 1 + 1/\log q\tau + it$. Then

$$\frac{L'}{L}(s_1, \chi) \ll \log q\tau.$$

From this and Lemma 11.1 it follows that

$$\sum_\rho \frac{1}{s_1 - \rho} \ll \log q\tau \qquad (11.11)$$

where the sum is over those zeros of $L(s, \chi)$ for which $|\rho - (3/2 + it)| \le 5/6$. Hence

$$\sum_\rho \frac{1}{s - \rho} = \sum_\rho \left(\frac{1}{s - \rho} - \frac{1}{s_1 - \rho}\right) + O(\log q\tau). \qquad (11.12)$$

Suppose that $1 - c/(2\log q\tau) \le \sigma \le 1 + 1/\log q\tau$ and that $|s - \beta_1| \ge 1/\log q$ if $L(s, \chi)$ has an exceptional zero β_1. Since $|s - \rho| \asymp |s_1 - \rho|$ for all zeros ρ, it follows that

$$\frac{1}{s - \rho} - \frac{1}{s_1 - \rho} = \frac{1 + 1/\log q\tau - \sigma}{(s - \rho)(s_1 - \rho)} \ll \frac{1}{|s_1 - \rho|^2 \log q\tau} \ll \Re \frac{1}{s_1 - \rho}.$$

On summing this over ρ and appealing to (11.11) we find that

$$\sum_{\rho} \frac{1}{s - \rho} \ll \log q\tau, \tag{11.13}$$

and (11.5) follows by Lemma 11.1.

To derive (11.6) we first note that if $\sigma > 1$, then

$$|\log L(s, \chi)| \leq \sum_{n=2}^{\infty} \frac{\Lambda(n)}{\log n} n^{-\sigma} = \log \zeta(\sigma).$$

Since $\zeta(\sigma) \leq \sigma/(\sigma - 1)$ by Corollary 1.14, we see that (11.6) holds when $\sigma \geq 1 + 1/\log q\tau$. In particular, (11.6) holds at the point $s_1 = 1 + 1/\log q\tau + it$. To treat the remaining s it suffices to note that

$$\log L(s, \chi) - \log L(s_1, \chi) = \int_{s_1}^{s} \frac{L'}{L}(w, \chi)\,dw \ll |s_1 - s| \log q\tau \ll 1$$

by (11.5). The estimate (11.6) trivially implies (11.7) since $\log 1/|L(s, \chi)| = -\Re \log L(s, \chi)$.

Now suppose that $L(s, \chi)$ has an exceptional zero β_1 such that $|s - \beta_1| \leq 1/\log q$. Then $1 - c/(2\log 4q) \leq \sigma \leq 1 + 1/\log q$, so by Lemma 11.1,

$$\frac{L'}{L}(s, \chi) = \frac{1}{s - \beta_1} + {\sum_{\rho}}' \frac{1}{s - \rho} + O(\log q)$$

where $\sum_{\rho}'$ denotes a sum over all zeros ρ such that $|\rho - (3/2 + it)| \leq 5/6$ *except* for the exceptional zero β_1. The proof of (11.13) applies to $\sum_{\rho}'$, so we have (11.8). Proceeding as in the proof of (11.6), we find that

$$\log L(s, \chi) = \log \frac{s - \beta_1}{s_1 - \beta_1} + \log L(s_1, \chi) + O(1),$$

which implies that

$$\left| \log L(s, \chi) - \log \frac{s - \beta_1}{s_1 - \beta_1} \right| \leq |\log L(s_1, \chi)| + O(1) \leq \log \log q + O(1).$$

But $\arg(s - \beta_1) \ll 1$, $\arg(s_1 - \beta_1) \ll 1$, and $\log|s_1 - \beta_1| = -\log \log q + O(1)$, so we have (11.9) and (11.10). $\qquad\square$

Our methods yield not only a zero-free region, but also enable us to bound the number of zeros ρ of $L(s, \chi)$ that might lie near $1 + it$.

Theorem 11.5 *Let $n(r; t, \chi)$ denote the number of zeros ρ of $L(s, \chi)$ in the disc $|\rho - (1 + it)| \leq r$. Then $n(r; t, \chi) \ll r \log q\tau$ uniformly for $1/\log q\tau \leq r \leq 3/4$.*

Here the constraint $r \geq 1/\log q\tau$ is needed because $L(s, \chi)$ might have an exceptional zero. If $L(s, \chi)$ has no exceptional zero, then the bound holds uniformly for $0 \leq r \leq 3/4$, in view of the zero-free region of Theorem 11.3.

Proof In view of Theorem 6.8, we may suppose that χ is non-principal. Suppose first that $1/\log q\tau \leq r \leq 1/3$. Take $s_1 = 1 + r + it$. Then $\Re(s_1 - \rho)^{-1} \geq 0$ for all zeros ρ, and $\Re(s_1 - \rho)^{-1} \gg 1/r$ if ρ is counted by $n(r; t, \chi)$. Hence

$$\frac{1}{r}n(r; t, \chi) \ll \Re \sum_{\rho} \frac{1}{s_1 - \rho}$$

where the sum is over all zeros ρ such that $|\rho - (3/2 + it)| \leq 5/6$. By Lemma 11.1 we see that the above is $\ll \log q\tau$, since

$$\left| \frac{L'}{L}(s_1) \right| \leq -\frac{\zeta'}{\zeta}(1 + r) \asymp \frac{1}{r} \ll \log q\tau.$$

If $1/3 \leq r \leq 3/4$, then it suffices to apply Jensen's inequality to $L(s, \chi)$ on a disc with centre $3/2 + it$, with $R = 4/3$ and $r = 5/4$, in view of the estimates provided by Lemma 10.15. $\qquad\square$

11.1.1 Exercises

1. Let $S(x; q)$ denote the number of integers $n, 0 < n \leq x$, such that $(n, q) = 1$, and put $R(x; q) = S(x; q) - (\varphi(q)/q)x$.
   (a) Show that if $\sigma > 0$, $x > 0$, and $s \neq 1$, then

   $$L(s, \chi_0) - \sum_{n \leq x} \chi_0(n)n^{-s} + \frac{\varphi(q)}{q} \cdot \frac{x^{1-s}}{s - 1} - \frac{R(x; q)}{x^s} \mid s \int_x^{\infty} R(u; q)u^{-s-1} du.$$

   Show that this includes Theorem 1.12 as a special case.
   (b) Let $\delta > 0$ be fixed. Show that if $\sigma \geq \delta$, then

   $$L(s, \chi_0) = \frac{\varphi(q)}{q} \cdot \frac{x^{1-s}}{s - 1} + \sum_{n \leq x} \chi_0(n)n^{-s} + O(d(q)|s|x^{-\sigma}).$$

2. Suppose that δ is fixed, $0 < \delta < 1$. Show that

   $$\sum_{p \mid q} \frac{\log p}{p^s - 1} \ll (\log q)^{1-\delta}$$

   uniformly for $\sigma \geq \delta$. (This improves on the estimate used in the latter part of the proof of Lemma 11.1.)

3. (a) Show that if $\sigma > 0$, then

   $$\zeta(s) = \frac{1}{s - 1} + \frac{1}{2} - s \int_1^{\infty} (\{x\} - 1/2)x^{-s-1} dx.$$

(b) Show that if $f(x)$ is a monotonically decreasing function, then

$$\int_0^1 (x - 1/2)f(x)\,dx \le 0.$$

(c) Show that

$$\zeta(\sigma) > \frac{1}{\sigma - 1} + \frac{1}{2}$$

for $\sigma > 0$.

(d) Show that

$$-\zeta'(s) = \frac{1}{(s - 1)^2} + \int_1^\infty (\{x\} - 1/2)(1 - s\log x)x^{-s-1}\,dx$$

for $\sigma > 0$.

(e) Show that if $\sigma > 0$, then

$$\left|\zeta'(\sigma) + \frac{1}{(\sigma - 1)^2}\right| < \frac{1}{2}\int_1^\infty |1 - \sigma\log x|x^{-\sigma-1}\,dx = \frac{1}{e\sigma}.$$

(f) Justify the following chain of inequalities for $\sigma > 1$:

$$-\frac{\zeta'}{\zeta}(\sigma) < \frac{\frac{1}{(\sigma-1)^2} + \frac{1}{e\sigma}}{\frac{1}{\sigma-1} + \frac{1}{2}} = \frac{1}{\sigma - 1} \cdot \frac{1 + \frac{(\sigma-1)^2}{e\sigma}}{1 + \frac{\sigma-1}{2}} < \frac{1}{\sigma - 1}.$$

(g) Show that if χ_0 is the principal character (mod q), then

$$-\frac{L'}{L}(\sigma, \chi_0) < \frac{1}{\sigma - 1}$$

for $\sigma > 1$. (This improves on the first inequality in (11.2), in the proof of Theorem 11.3.)

4. Let χ be a character (mod q), and suppose that the order d of χ is odd.
   (a) Show that $\Re\chi(n) \ge -\cos\pi/d$ for all integers n.
   (b) Show that if $\sigma > 1$, then $\log|L(\sigma, \chi)| \ge -(\cos\pi/d)\log\zeta(\sigma)$.
   (c) Show that $L(1, \chi) \asymp L(1 + 1/\log q, \chi)$.
   (d) Show that $|L(1, \chi)| \gg (\log q)^{-\cos\pi/d}$.
   (e) Deduce in particular that if χ is a cubic character (mod q), then $|L(1, \chi)| \gg 1/\sqrt{\log q}$.

5. *Grössencharaktere* for $\mathbb{Q}(\sqrt{-1})$, continued from Exercise 10.1.28. For an ideal $\mathfrak{a} = (a + ib)$ in the ring $\mathcal{O}\{a + ib : a, b \in \mathbb{Z}\}$ of Gaussian integers, put $\chi_m(\mathfrak{a}) = e^{4mi\,\arg(a+ib)}$. The ideal $\mathfrak{a}$ is the set of (Gaussian integer) multiples of the number $a + ib$, but it can equally well be expressed as the set of Gaussian integer multiples of $(a + ib)i^k$ for $k = 0, 1, 2, 3$. Note that the stated value of $\chi_m(\mathfrak{a})$ is independent of the choice of k.

(a) Show that

$$L(s, \chi_m) = \prod_{\mathfrak{p}} \left(1 - \frac{\chi_m(\mathfrak{p})}{N(\mathfrak{p})^s}\right)^{-1}$$

for $\sigma > 1$, where the product is over all prime ideals $\mathfrak{p}$ in the ring.

(b) Let $\Lambda(\mathfrak{a}) = \log(a^2 + b^2)$ if $\mathfrak{a} = (a + ib)^k$ for some positive integer k and $a + ib$ is a Gaussian prime, and $\Lambda(\mathfrak{a}) = 0$ otherwise. Show that

$$\frac{L'}{L}(s, \chi_m) = -\sum_{\mathfrak{a}} \frac{\Lambda(\mathfrak{a})\chi_m(\mathfrak{a})}{N(\mathfrak{a})^s}$$

for $\sigma > 1$.

(c) Show that there is an absolute constant $c > 0$ such that $L(s, \chi_m) \neq 0$ for $\sigma > 1 - c/\log m\tau$ for every positive integer m.

11.2 Exceptional zeros

Although there is no known quadratic character χ for which $L(s, \chi)$ has an exceptional real zero, the possible existence of such zeros is a recurring issue in the theory in its current stage of development. The techniques of the preceding section do not seem to offer a means of eliminating exceptional zeros entirely, but nevertheless they may be used to show that such zeros occur at most rarely. To this end we introduce a variant of Lemma 11.5 that allows us to consider two different quadratic characters.

Lemma 11.6 (Landau) *Suppose that χ_1 and χ_2 are quadratic characters. If $\sigma > 1$, then*

$$-\frac{\zeta'}{\zeta}(\sigma) - \frac{L'}{L}(\sigma, \chi_1) - \frac{L'}{L}(\sigma, \chi_2) - \frac{L'}{L}(\sigma, \chi_1\chi_2) \geq 0.$$

Proof It suffices to express the left-hand side as a Dirichlet series and to note that

$$1 + \chi_1(n) + \chi_2(n) + \chi_1\chi_2(n) = (1 + \chi_1(n))(1 + \chi_2(n)) \geq 0$$

for all n. $\qquad\square$

Theorem 11.7 (Landau) *There is a constant $c > 0$ such that if χ_1 and χ_2 are quadratic characters modulo q_1 and q_2, respectively, and if $\chi_1\chi_2$ is non-principal, then $L(s, \chi_1)L(s, \chi_2)$ has at most one real zero β such that $1 - c/\log q_1q_2 < \beta < 1$.*

Proof Since any given L-function can have at most one such zero, if there are two zeros, then one of them, say β_1, is a zero of $L(s, \chi_1)$, and the other, β_2, is a zero of $L(s, \chi_2)$. We may assume that c is so small that $5/6 \le \beta_i < 1$. Also, we note that $\chi_1 \chi_2$ is a non-principal character (mod $q_1 q_2$). Hence by four applications of Lemma 11.1 we see that if $0 < \delta \le 1$, then

$$-\frac{\zeta'}{\zeta}(1+\delta) \le \frac{1}{\delta} + c_1 \log 4,$$

$$-\frac{L'}{L}(1+\delta, \chi_i) \le \frac{-1}{1+\delta-\beta_i} + c_1 \log q_i,$$

$$-\frac{L'}{L}(1+\delta, \chi_1\chi_2) \le c_1 \log q_1 q_2.$$

We sum these inequalities and apply Lemma 11.4 to see that

$$\frac{1}{\delta} - \frac{1}{1+\delta-\beta_1} - \frac{1}{1+\delta-\beta_2} + c_2 \log q_1 q_2 \ge 0.$$

Without loss of generality we may suppose that $\beta_1 \le \beta_2$. Then

$$\frac{1}{\delta} - \frac{2}{1+\delta-\beta_1} + c_2 \log q_1 q_2 \ge 0,$$

and by taking $\delta = 2(1 - \beta_1)$ we deduce that

$$1 - \beta_1 \ge \frac{1}{6c_2 \log q_1 q_2}.$$

$\square$

The following corollaries are immediate.

Corollary 11.8 (Landau) *There is a positive constant $c > 0$ such that $\prod_\chi L(s, \chi)$ has at most one zero in the region $\sigma > 1 - c/\log q\tau$. Here the product is over all Dirichlet characters χ (mod q). If such a zero exists then it is necessarily real and the associated character χ is quadratic.*

Corollary 11.9 (Landau) *For each positive number A there is a $c(A) > 0$ such that if $\{q_i\}$ is a strictly increasing sequence of natural numbers with the property that for each q_i there is a primitive quadratic character χ_i (mod q_i) for which $L(s, \chi_i)$ has a zero β_i satisfying*

$$\beta_i > 1 - \frac{c(A)}{\log q_i},$$

then

$$q_{i+1} > q_i^A.$$

Corollary 11.10 (Page) *There is a constant $c > 0$ such that for every $Q \geq 1$ the region $\sigma \geq 1 - c/\log Q\tau$ contains at most one zero of the function*

$$\prod_{q \leq Q} \prod_{\chi}^{*} L(s, \chi)$$

where $\prod_{\chi}^{*}$ denotes a product over all primitive characters χ (mod q). If such a zero exists, then it is necessarily real and the associated character χ is quadratic.

We now turn to the problem of showing that even an exceptional zero cannot be too close to 1. By taking $s = 1$ in (11.10) we see that this is equivalent to showing that $L(1, \chi)$ cannot be too small. Suppose that χ is a primitive quadratic character modulo q, and let $r(n) = \sum_{d|n} \chi(d)$. Then $r(n) \geq 0$ for all n and $r(n) \geq 1$ when n is a perfect square. Since $\sum_{n=1}^{\infty} r(n)n^{-s} = \zeta(s)L(s, \chi)$ for $\sigma > 1$, we find that

$$\sum_{n \leq x} r(n)n^{-s} = \frac{L(1, \chi)x^{1-s}}{1 - s} + \zeta(s)L(s, \chi) + \text{error terms}. \quad (11.14)$$

Here the error terms are small if x is sufficiently large in terms of q. Estimates of this kind can be derived from Corollary 1.15 by the method of the hyperbola, or else by employing an inverse Mellin transform. Suppose that $0 \leq s < 1$ in the above. We can give a lower bound for the left-hand side, which yields a lower bound for $L(1, \chi)$ if the second term on the right-hand side does not interfere. Since $\zeta(s) < 0$ for $0 < s < 1$ (cf. Corollary 1.14), this term is harmless if $L(s, \chi) \geq 0$. If this cannot be arranged, we may alternatively eliminate this term by taking two values of x and differencing. Since the method of the hyperbola leads to tedious details, we use an inverse Mellin transform to derive a more precise version of (11.14). To make the estimates easier we introduce an Abelian weighting of the sum. By (5.23) with x replaced by $1/x$ we see that

$$\sum_{n=1}^{\infty} r(n)e^{n/x} = \frac{1}{2\pi i} \int_{2-i\infty}^{2+i\infty} \zeta(s)L(s, \chi)\Gamma(s)x^s \, ds.$$

We move the contour of integration to the line $\Re s = -1/2$, which gives rise to residues at the poles at $s = 1$ and $s = 0$. Thus the above is

$$= L(1, \chi)x + \zeta(0)L(0, \chi) + \frac{1}{2\pi i} \int_{-1/2-i\infty}^{-1/2+i\infty} \zeta(s)L(s, \chi)\Gamma(s)x^s \, ds.$$

By Corollary 10.5 we know that $\zeta(-1/2 + it) \ll \tau$, by Corollary 10.10 we know that $L(-1/2 + it, \chi) \ll q\tau$, and by (C.19) we know that $\Gamma(-1/2 + it) \ll \tau^{-1}e^{-\pi\tau/2}$. Hence the integral is $\ll qx^{-1/2}$. By (10.11) we know that $\zeta(0) = -1/2$, and by Corollary 10.9 we know that $L(0, \chi) \geq 0$. (More precisely, $L(0, \chi) = 0$ if $\chi(-1) = 1$, and $L(0, \chi) \asymp q^{1/2}L(1, \chi)$ if $\chi(-1) = -1$.) Since the perfect squares on the left-hand side contribute an amount $\gg x^{1/2}$, we deduce that

$$x^{1/2} \ll xL(1, \chi) + qx^{-1/2}.$$

On taking $x = Cq$ with C a large constant we deduce that $L(1, \chi) \gg q^{-1/2}$. Now consider the possibility that χ is an imprimitive quadratic character. Then there is a primitive quadratic character χ^* modulo d, with $d|q$, that induces χ. Thus $L(1, \chi) = L(1, \chi^*) \prod_{p|q/d}(1 - \chi^*(p)/p) \geq L(1, \chi^*)\varphi(q/d)d/q \gg d^{-1/2}(\log\log 3q/d)^{-1} \gg q^{-1/2}$, by Theorem 2.9, so we have

Theorem 11.11   *If χ is a quadratic character modulo q, then $L(1, \chi) \gg q^{-1/2}$.*

By (11.10) the following corollary is immediate.

Corollary 11.12   *There is an absolute constant $c > 0$ such that if χ is a quadratic character modulo q and $L(s, \chi)$ has an exceptional zero β_1, then*

$$\beta_1 \leq 1 - \frac{c}{q^{1/2}(\log q)^2}.$$

By elaborating on the above argument we can obtain better lower bounds for $1 - \beta_1$. To facilitate this we first establish a convenient inequality that depends only on the analyticity and size of the relevant Dirichlet series in the immediate vicinity of the real axis.

Lemma 11.13   (Estermann) *Suppose that $f(s)$ is analytic for $|s - 2| \leq 3/2$, and that $|f(s)| \leq M$ for s in this disc. Suppose also that*

$$F(s) = \zeta(s)f(s) = \sum_{n=1}^{\infty} r(n)n^{-s}$$

for $\sigma > 1$, that $r(1) = 1$, and that $r(n) \geq 0$ for all n. If there is a $\sigma \in [19/20, 1)$ such that $f(\sigma) \geq 0$, then

$$f(1) \geq \frac{1}{4}(1 - \sigma)M^{-3(1-\sigma)}.$$

To put this in perspective, we recall that our proof in Chapter 4 that $L(1, \chi) \neq 0$ depended on Landau's theorem (Theorem 1.7). The above amounts to a quantitative elaboration of Landau's theorem, for if $f(1)$ were 0, then $F(s)$ would be analytic for $s > 1/2$, so by Landau's theorem the Dirichlet series would converge when $\sigma > 1/2$. This would imply that $F(\sigma) > 0$ for $\sigma > 1/2$. But $\zeta(\sigma) < 0$ for $1/2 < \sigma < 1$ (cf. Corollary 1.14), so it would follow that

$f(\sigma) < 0$ in this interval. Thus the hypothesis above that $f(\sigma) \geq 0$ implies – by Landau's theorem – that $f(1) > 0$. In the above we obtain not just this qualitative information but a quantitative lower bound for $f(1)$ in terms of the size of σ and the size of $f(s)$ in a surrounding disc.

Proof As in the proof of Landau's theorem we begin by expanding $F(s)$ in powers of $2 - s$,

$$F(s) = \sum_{k=0}^{\infty} b_k (2 - s)^k \qquad (11.15)$$

for $|s - 2| < 1$. By Cauchy's coefficient formula we know that

$$b_k = \frac{(-1)^k}{k!} F^{(k)}(2) = \frac{1}{k!} \sum_{n=1}^{\infty} r(n) n^{-2} (\log n)^k.$$

Thus $b_k \geq 0$ for all k, and $b_0 = \sum_{n=1}^{\infty} r(n) n^{-2} \geq 1$. For $|s - 2| < 1$ we may write

$$\frac{1}{s - 1} = \frac{1}{1 - (2 - s)} = \sum_{k=0}^{\infty} (2 - s)^k.$$

On multiplying this by $f(1)$ and subtracting from (11.15) we deduce that

$$F(s) - \frac{f(1)}{s - 1} = \sum_{k=0}^{\infty} (b_k - f(1))(2 - s)^k \qquad (11.16)$$

for $|s - 2| < 1$. But the left-hand side is analytic for $|s - 2| \leq 3/2$, so the series converges in this larger disc. In order to estimate the coefficients on the right-hand side we bound the left-hand side when s lies on the circle $|s - 2| = 3/2$. To this end, we note by (1.24) that

$$|\zeta(s)| = \left| 1 + \frac{1}{s - 1} + s \int_1^{\infty} \frac{[u] - u}{u^{s+1}} \, du \right|$$

$$\leq 1 + \frac{1}{|s - 1|} + \frac{|s|}{\sigma}.$$

The relation $|s - 2| = 3/2$ implies that $|s - 1| \geq 1/2$, that $|s| \leq 7/2$, and that $\sigma \geq 1/2$. Hence $|\zeta(s)| \leq 10$ for the s under consideration. Since $|f(1)/(s - 1)| \leq 2M$, it follows that the left-hand side of (11.16) has modulus $\leq 12M$ for $|s - 2| \leq 3/2$. By the Cauchy coefficient inequalities we deduce that $|b_k - f(1)| \leq 12M(2/3)^k$. We apply this bound for all $k > K$ where K is a parameter to be chosen later. Thus from (11.16) we see that if $1/2 < \sigma \leq 2$, then

$$\zeta(\sigma) f(\sigma) - \frac{f(1)}{\sigma - 1} \geq \sum_{k=0}^{K} (b_k - f(1))(2 - \sigma)^k - 12M \sum_{k > K} \left(\tfrac{2}{3}(2 - \sigma) \right)^k.$$

We observe that if $19/20 \leq \sigma < 1$, then $\frac{2}{3}(2 - \sigma) \leq 7/10$. We also recall that $b_0 \geq 1$ and that $b_k \geq 0$ for all k. Hence the above is

$$\geq 1 - f(1)\frac{1 - (2 - \sigma)^{K+1}}{1 - (2 - \sigma)} - 40M(7/10)^{K+1}.$$

On cancelling the common term $f(1)/(1 - \sigma)$ from both sides, and rearranging, we find that

$$1 \leq \frac{f(1)(2 - \sigma)^{K+1}}{1 - \sigma} + \zeta(\sigma)f(\sigma) + 40M(7/10)^{K+1},$$

a relation comparable to (11.14). To ensure that the last term on the right does not overwhelm the left-hand side, we take $K = [(\log 80M)/\log 10/7]$. Then the last term on the right is $\leq 1/2$. Since $\zeta(\sigma) < 0$ by Corollary 1.14, and $f(\sigma) \geq 0$ by hypothesis, it follows that

$$f(1) \geq \frac{1}{2}(1 - \sigma)(2 - \sigma)^{-K-1} \geq \frac{10}{21}(1 - \sigma)(2 - \sigma)^{-K}. \qquad (11.17)$$

But

$$(2 - \sigma)^K \leq (2 - \sigma)^{(\log 80M)/\log 10/7} = (80M)^{(\log(2-\sigma))/\log 10/7}$$
$$\leq 80^{(\log 21/20)/\log 10/7} M^{(\log(2-\sigma))/\log 10/7}.$$

Here the first factor is $< 13/7$. Since $\log(1 + \delta) \leq \delta$ for any $\delta \geq 0$, on taking $\delta = 1 - \sigma$ we see that $\log(2 - \sigma) \leq 1 - \sigma$. Also, $\log 10/7 > 1/3$ and it can certainly be supposed that $M \geq 1$, so the expression above is $< (13/7)M^{3(1-\sigma)}$. This with (11.17) gives the desired lower bound for $f(1)$. $\square$

We are now prepared to prove an important strengthening of Theorem 11.11.

Theorem 11.14   (Siegel) *For each positive number ε there is a positive constant $C(\varepsilon)$ such that if χ is a quadratic character modulo q, then*

$$L(1, \chi) > C(\varepsilon)q^{-\varepsilon}.$$

Proof We assume, as we may, that $\varepsilon \leq 1/5$. For the present we restrict our attention to primitive characters. We consider two cases, according to whether there exists a primitive quadratic character χ_1 such that $L(s, \chi_1)$ has a real zero β_1 in the interval $[1 - \varepsilon/4, 1)$, or not. Suppose first that there is no such zero. We take $f(s) = L(s, \chi)$, $\sigma = 1 - \varepsilon/4$. Then $f(\sigma) > 0$ and by Lemma 10.15 we may take $M \ll q^{1/2}$. Hence by Lemma 11.13, $f(1) \gg \varepsilon q^{-3\varepsilon/8}$. Thus there is a constant $C_1(\varepsilon) > 0$ such that $L(1, \chi) \geq C_1(\varepsilon)q^{-\varepsilon}$.

Now consider the contrary case, in which there is a primitive quadratic character χ_1 modulo q_1 such that $L(s, \chi_1)$ has a real zero $\beta_1 \geq 1 - \varepsilon/4$. Since $L(1, \chi_1) > 0$ there is a constant $C_2(\varepsilon) > 0$ such that $L(1, \chi_1) \geq C_2(\varepsilon)q_1^{-\varepsilon}$.

Now suppose that χ is a primitive quadratic character, $\chi \neq \chi_1$. We apply Lemma 11.13 with $f(s) = L(s, \chi)L(s, \chi_1)L(s, \chi\chi_1)$. To see that the Dirichlet series coefficients of $\zeta(s)f(s)$ are non-negative, we note first that if $g(s)$ is a Dirichlet series with non-negative coefficients, then $\exp g(s)$ is also a Dirichlet series with non-negative coefficients, since the power series coefficients of the exponential function are non-negative. Then it suffices to apply this observation with

$$g(s) = \log \zeta(s)f(s) = \sum_{n=1}^{\infty} \frac{\Lambda(n)}{\log n}(1 + \chi(n))(1 + \chi_1(n))n^{-s}.$$

In view of Lemma 10.15 we may take $M = C_3 q q_1$. On taking $\sigma = \beta_1$, we find that

$$f(1) \geq \frac{1}{4}(C_3 q q_1)^{-3(1-\beta_1)} \geq \frac{1}{4}(C_3 q q_1)^{-3\varepsilon/4} \geq C_4(\varepsilon)q^{-\varepsilon}.$$

Now

$$f(1) = L(1, \chi)L(1, \chi_1)L(1, \chi\chi_1) \ll L(1, \chi)(\log qq_1)^2$$

by Lemma 10.15, and hence we deduce that

$$L(1, \chi) \geq C_5(\varepsilon)q^{-2\varepsilon}. \tag{11.18}$$

We may assume that $C_5 \leq C_1$, so that (11.18) holds in either case.

We now extend to imprimitive characters. Suppose that χ is induced by a primitive character χ^* (mod d), so that $q = dr$ for some r. Then

$$L(1, \chi) = L(1, \chi^*)\prod_{p|r}\left(1 - \frac{\chi^*(p)}{p}\right) \geq L(1, \chi^*)\frac{\varphi(r)}{r} \geq C_5(\varepsilon)d^{-2\varepsilon}\frac{\varphi(r)}{r}.$$

By Theorem 2.9 the above is

$$\geq C_6(\varepsilon)(dr)^{-2\varepsilon} = C_6(\varepsilon)q^{-2\varepsilon},$$

and hence the proof is complete. $\qquad\square$

We are unable to compute the value of the constant $C(\varepsilon)$ in Siegel's theorem when $\varepsilon < 1/2$, because we have no way of estimating the size of the smallest possible q_1 when the second case arises in the proof. Such a constant is called 'non-effective.' This is our first encounter with a non-effective constant, so the distinction between effectively computable constants and non-effective constants arises here for the first time.

Corollary 11.15 *For any $\varepsilon > 0$ there is a positive number $C(\varepsilon)$ such that if χ is a quadratic character modulo q and β is a real zero of $L(s, \chi)$, then $\beta < 1 - C(\varepsilon)q^{-\varepsilon}$.*

Proof We may certainly suppose that $\beta > 1 - c/\log 4q > 1 - \frac{1}{\log q}$, where c is the number appearing in Theorem 11.3, so that β is an exceptional zero by the criterion following that theorem. By taking $s = 1$ in (10) we see that

$$L(1, \chi) \ll (1 - \beta)(\log q)^2$$

and the corollary follows easily from the theorem. $\qquad\square$

11.2.1 Exercises

1. Call a modulus q 'exceptional' if there is a primitive quadratic character $\chi \pmod q$ such that $L(s, \chi)$ has a real zero β such that $\beta > 1 - c/\log q$. Show that if c is sufficiently small, then the number of exceptional q not exceeding Q is $\ll \log\log Q$.
2. Use the last part of Theorem 4 to show that if $L(s, \chi)$ has an exceptional zero β_1, then $L'(\beta_1, \chi) \gg 1$.
3. (cf. Mahler 1934, Davenport 1966, Haneke 1973, Goldfeld & Schinzel 1975) Suppose that χ is a quadratic character, and put $r(n) = \sum_{d|n} \chi(d)$.
   (a) Show that
   $$\sum_{n \le y} \frac{\chi(n)}{n} = L(1, \chi) + O\left(q^{1/2}y^{-1}\log q\right).$$
   (b) Show that
   $$\sum_{n \le y} \frac{\chi(n)\log n}{n} = -L'(1, \chi) + O(q^{1/2}y^{-1}(\log qy)^2).$$
   (c) Verify that
   $$\sum_{n \le x} \frac{r(n)}{n} = \sum_{d \le y} \frac{\chi(d)}{d} \sum_{m \le x/d} \frac{1}{m} + \sum_{m \le x/y} \frac{1}{m} \sum_{d \le x/m} \frac{\chi(d)}{d}$$
   $$- \left(\sum_{d \le y} \frac{\chi(d)}{d}\right)\left(\sum_{m \le x/y} \frac{1}{m}\right)$$
   $$= \Sigma_1 + \Sigma_2 - \Sigma_3,$$
   say.
   (d) Show that
   $$\Sigma_1 = (\log x + C_0)L(1, \chi) + L'(1, \chi) + O\left(q^{1/2}y^{-1}(\log qy)^2\right) + O(yx^{-1}).$$
   (e) Show that
   $$\Sigma_2 = (\log x/y + C_0)L(1, \chi) + O(yx^{-1}\log q) + O\left(q^{1/2}y^{-1}\log q\right).$$

(f) Show that

$$\Sigma_3 = (\log x/y + C_0)L(1, \chi) + O(yx^{-1}\log q) + O\left(q^{1/2}y^{-1}(\log qx)^2\right).$$

(g) Show that

$$\sum_{n \le x} \frac{r(n)}{n} = (\log x + C_0)L(1, \chi) + L'(1, \chi) + O\left(q^{1/4}x^{-1/2}(\log qx)^{3/2}\right).$$

(h) Show that for each $c < 1/2$ there is a constant $q_0(c)$ such that if $q \ge q_0(c)$ and $L(1, \chi) < c/\log q$, then

$$L'(1, \chi) \asymp \sum_{n \le q} \frac{r(n)}{n}.$$

(i) Show that $L''(\sigma, \chi) \ll (\log q)^3$ for $\sigma \ge 1 - 1/\log q$.

(j) Show that there is an absolute constant $c > 0$ such that if $L(s, \chi)$ has an exceptional zero β_1 for which $\beta_1 \ge 1 - c/(\log q)^3$, then

$$L(1, \chi) \asymp (1 - \beta_1) \sum_{n \le q} \frac{r(n)}{n}.$$

4. Use Estermann's lemma (Lemma 11.13) to give a second proof that if $L(s, \chi)$ has an exceptional zero β_1, then $L(1, \chi) \gg 1 - \beta_1$ (cf. (11.10) of Theorem 11.4).

5. Use Estermann's lemma (Lemma 11.13) to give a second proof that if χ is a cubic character (mod q), then $L(1, \chi) \gg (\log q)^{-1/2}$ (cf. Exercise 11.1.4(e)).

6. (Tatuzawa 1951) Let χ_1 and χ_2 be distinct primitive quadratic characters, modulo q_1 and q_2, respectively, and suppose that $L(1, \chi_i) < C\varepsilon q_i^{-\varepsilon}$ for $i = 1, 2$ where $0 < \varepsilon \le 1$ and $C > 0$.

  (a) Show that $\min_{x>1} \frac{x}{\log x} = e$. By a change of variables, deduce that if $\varepsilon > 0$, then $\min_{x>1} x^\varepsilon / \log x = e\varepsilon$. Use this to show that $\min_{x>1} x^\varepsilon/(\log x)^2 = e^2\varepsilon^2/4$.

  (b) Explain why there exists a constant $c_1 > 0$ such that $L(1, \chi) \ge c_1/\log q$ whenever $L(s, \chi)$ has no exceptional zero. Let $C_1 = ec_1$. Show that if $C < C_1$, then $L(s, \chi_1)$ and $L(s, \chi_2)$ have exceptional zeros, say β_1 and β_2. (From now on, suppose that $C < C_1$.)

  (c) Explain why there is a positive constant c_2 such that $L(1, \chi) \ge c_2(1 - \beta)$ whenever β is an exceptional zero of $L(s, \chi)$. Let $C_2 = c_2/6$. Show that if $C < C_2$, then $\beta > 1 - \varepsilon/6$. Let $C_3 = c_2/20$. Show that if $C < C_3$, then $\beta > 19/20$. (From now on, suppose that $C < C_i$ for $i = 1, 2, 3$.)

  (d) Explain why there is a constant $c_3 > 0$ such that at most one of $L(s, \chi_1)$, $L(s, \chi_2)$ has a zero in the interval $[1 - c_3/\log q_1q_2, 1]$.

  (e) Show that $L(s, \chi_1)L(s, \chi_2)$ has a zero β that satisfies the three inequalities $\beta \ge 19/20$, $\beta \ge 1 - \varepsilon/6$, $\beta \le 1 - c_3/\log q_1q_2$.

(f) Let $f(s) = L(s, \chi_1)L(s, \chi_2)L(s, \chi_1\chi_2)$. Show that there is an absolute constant $c_4 > 0$ such that $f(1) \geq c_4(\log q_1 q_2)^{-1}(q_1 q_2)^{-\varepsilon/2}$.

(g) Explain why there is a constant $c_5 > 0$ such that $L(1, \chi_1\chi_2) \leq c_5 \log q_1 q_2$.

(h) Show that $C \geq c_4^{1/2} c_5^{-1/2} e/4$.

(i) Conclude that there is a positive effectively computable absolute C such that if $0 < \varepsilon \leq 1$, then the inequality $L(1, \chi) > C\varepsilon q^{-\varepsilon}$ holds for all primitive quadratic characters, with at most one exception.

7. (Fekete & Pólya 1912, Pólya & Szegö 1925, p. 44, Heilbronn 1937) Let $S_1(x, \chi) = \sum_{1 \leq n \leq x} \chi(n)$.

(a) Show that if χ is a quadratic character such that $S_1(x, \chi) \geq 0$ for all $x \geq 1$, then $L(\sigma, \chi) > 0$ for all $\sigma > 0$.

(b) Let $\chi_d(n) = \left(\frac{d}{n}\right)$. Show that the hypothesis above holds for $d = -3, -4, -7, -8$, but not for $d = 5, 8$.

(c) For $k > 1$ let $S_k(N, \chi) = \sum_{n=1}^{N} S_{k-1}(n, \chi)$. Show that

$$S_k(N, \chi) = \sum_{n=1}^{N} \binom{N - n + k - 1}{k - 1} \chi(n).$$

(d) Let $\Delta f(x) = f(x + 1) - f(x)$ and $\Delta_k f(x) = \Delta(\Delta_{k-1} f(x))$. Show that $\Delta_k f(x) = \sum_{r=0}^{k}(-1)^r \binom{k}{r} f(x + k - r)$, and that if $f^{(k)}(x)$ is continuous then

$$\Delta_k f(x) = \int_{x}^{x+1} \int_{u_1}^{u_1+1} \cdots \int_{u_{k-1}}^{u_{k-1}+1} f^{(k)}(u_k)\, du_k du_{k-1} \cdots du_1.$$

(e) Show that if $\sigma > 0$, then $(-1)^k \Delta_k(x^{-\sigma}) > 0$ for all $x > 0$.

(f) Show that $L(s, \chi) = (-1)^k \sum_{n=1}^{\infty} S_k(n, \chi)\Delta_k(n^{-s})$ for $\sigma > 0$.

(g) Show that if χ is a quadratic character and k is an integer such that $S_k(N, \chi) \geq 0$ for all integers $N \geq 1$, then $L(\sigma, \chi) > 0$ for all $\sigma > 0$.

(h) For $\chi_5(n) = \left(\frac{5}{n}\right)$ and $\chi_8(n) = \left(\frac{8}{n}\right)$ find the least k such that the hypothesis above is satisfied.

(i) Let $P(z, \chi) = \sum_{n=1}^{\infty} \chi(n)z^n$ for $|z| < 1$. Show that $P(z, \chi)(1 - z)^{-k} = \sum_{n=1}^{\infty} S_k(n, \chi)z^n$ for $|z| < 1$.

(j) Show that if χ is a quadratic character for which $S_k(N, \chi) \geq 0$ for all positive integers N, then $P(z, \chi) > 0$ for $0 < z < 1$.

(k) Show that $\sum_{n=1}^{12} \left(\frac{n}{163}\right)(7/10)^n = -0.0483$, and that $\sum_{n=13}^{\infty}(7/10)^n = 0.0323$. Deduce that $P(0.7, \chi_{-163}) < 0$, and hence that for any k there is an N for which $S_k(N, \chi_{-163}) < 0$.

8. S. Chowla (1972) conjectured that for any primitive quadratic character χ^* there is a character χ induced by χ^* such that $S_1(x, \chi) \geq 0$ for all $x \geq 1$ (in the notation of the preceding exercise). Show that Chowla's conjecture implies that $L(\sigma, \chi) > 0$ when χ is a quadratic character and $\sigma > 0$. See also Rosser (1950).

9. (Bateman & Chowla 1953) Suppose that k is a positive integer such that

$$\sum_{1 \leq n \leq x} \frac{\lambda(n)}{n} \left(1 - \frac{n}{x}\right)^k \geq 0 \qquad (11.19)$$

for all $x \geq 1$. (It is not known whether there is such a k.) (a) Show that if χ is a quadratic character, then

$$\sum_{1 \leq n \leq x} \frac{\chi(n)}{n} \left(1 - \frac{n}{x}\right)^k \geq \sum_{1 \leq n \leq x} \frac{\lambda(n)}{n} \left(1 - \frac{n}{x}\right)^k$$

for all $x \geq 1$.

(b) Show that if there is a k such that (11.19) holds for all $x \geq 1$, then $L(\sigma, \chi) > 0$ when χ is a quadratic character and $\sigma > 0$.

11.3 The Prime Number Theorem for arithmetic progressions

The various inequalities for zeros of Dirichlet L-functions established above are motivated by a desire to imitate for primes in arithmetic progressions the quantitative form of the Prime Number Theorem achieved in Theorem 6.9. For $(a, q) = 1$ we set

$$\pi(x; q, a) = \sum_{\substack{p \leq x \\ p \equiv a \, (q)}} 1, \quad \vartheta(x; q, a) = \sum_{\substack{p \leq x \\ p \equiv a \, (q)}} \log p, \quad \psi(x; q, a) = \sum_{\substack{n \leq x \\ n \equiv a \, (q)}} \Lambda(n),$$

$$(11.20)$$

and correspondingly for any Dirichlet character χ we put

$$\pi(x, \chi) = \sum_{p \leq x} \chi(p), \quad \vartheta(x, \chi) = \sum_{p \leq x} \chi(p) \log p, \quad \psi(x, \chi) = \sum_{n \leq x} \chi(n) \Lambda(n).$$

$$(11.21)$$

By multiplying both sides of (4.27) by $\Lambda(n)$, and summing over $n \leq x$, we see that

$$\psi(x; q, a) = \frac{1}{\varphi(q)} \sum_{\chi} \overline{\chi}(a) \psi(x, \chi), \qquad (11.22)$$

and similarly for $\pi(x; q, a)$ and $\vartheta(x; q, a)$. We deal with $\psi(x, \chi)$ in much the same way that we dealt with $\psi(x)$ in Chapter 6.

Theorem 11.16   *There is a constant $c_1 > 0$ such that if $q \le \exp(2c_1\sqrt{\log x})$, then*

$$\psi(x, \chi) = E_0(\chi)x + O\left(x \exp\left(-c_1\sqrt{\log x}\right)\right) \qquad (11.23)$$

when $L(s, \chi)$ has no exceptional zero, but

$$\psi(x, \chi) = -\frac{x^{\beta_1}}{\beta_1} + O\left(x \exp\left(-c_1\sqrt{\log x}\right)\right) \qquad (11.24)$$

when $L(s, \chi)$ has an exceptional zero β_1. Here $E_0(\chi) = 1$ if $\chi = \chi_0$, and $E_0(\chi) = 0$ otherwise.

Proof   By Theorems 4.8 and 5.2 we see that

$$\psi(x, \chi) = \frac{-1}{2\pi i} \int_{\sigma_0-iT}^{\sigma_0+iT} \frac{L'}{L}(s, \chi)\frac{x^s}{s}\, ds + R$$

where $\sigma_0 > 1$ and

$$R \ll \sum_{x/2 < n < 2x} \Lambda(n) \min\left(1, \frac{x}{T|x-n|}\right) + \frac{(4x)^{\sigma_0}}{T} \sum_{n=1}^{\infty} \frac{\Lambda(n)}{n^{\sigma_0}}$$

by Corollary 5.3. As in the proof of Theorem 6.9 we suppose that $2 \le T \le x$ and set $\sigma_0 = 1 + 1/\log x$. Thus

$$R \ll \frac{x}{T}(\log x)^2,$$

as before. As in the proof of Theorem 6.9, we let C denote a closed contour that consists of line segments joining the points $\sigma_0 - iT$, $\sigma_0 + iT$, $\sigma_1 + iT$, $\sigma_1 - iT$, but now the choice of σ_1 is a little more complicated, since we want to ensure that C does not pass too closely to an exceptional zero.

CASE 1. *There is no exceptional zero.* In this case we take $\sigma_1 = 1 - c/(5\log qT)$ where c is the constant in Theorem 11.3. If χ is non-principal, then the integrand is analytic on and inside C, but if $\chi = \chi_0$, then it has a pole at $s = 1$ with residue x. Hence

$$\frac{-1}{2\pi i}\int_C \frac{L'}{L}(s, \chi)\frac{x^s}{s}\, ds = E_0(\chi)x. \qquad (11.25)$$

We estimate the integrals from $\sigma_0 + iT$ to $\sigma_1 + iT$, from $\sigma_1 + iT$ to $\sigma_1 - iT$, and from $\sigma_1 - iT$ to $\sigma_0 - iT$ as in the proof of Theorem 6.9, using the estimate (11.5) of Theorem 11.4. Thus we find that

$$\psi(x, \chi) - E_0(\chi)x \ll x(\log x)^2 \left(\frac{1}{T} + \exp\left(\frac{-c\log x}{5\log qT}\right)\right). \qquad (11.26)$$

CASE 2. *There is an exceptional zero β_1, and it satisfies $\beta_1 \ge 1 - c/(4\log qT)$.* In this case we take $\sigma_1 = 1 - c/(3\log qT)$. The integrand in (11.25) now has a pole inside $\mathcal{C}$ at β_1, so the left-hand side of (11.25) has the value $-x^{\beta_1}/\beta_1$. Otherwise, the estimates proceed as before, and we find that

$$\psi(x, \chi) = -\frac{x^{\beta_1}}{\beta_1} + O\left(x(\log x)^2\left(\frac{1}{T} + \exp\left(\frac{-c\log x}{5\log qT}\right)\right)\right). \quad (11.27)$$

CASE 3. *There is an exceptional zero β_1, but it satisfies $\beta_1 < 1 - c/(4\log qT)$.* We proceed exactly as in Case 1, and so we obtain (11.26). To pass to (11.27) it suffices to note that

$$\frac{x^{\beta_1}}{\beta_1} \ll x\exp\left(\frac{-c\log x}{5\log qT}\right)$$

in the current case.

We have established (11.26) if there is no exceptional zero, and (11.27) if there is one. To complete our argument, we need only observe that if $c_1 = \sqrt{c/20}$, if $q \le \exp(2c_1\sqrt{\log x})$, and if $T = \exp(2c_1\sqrt{\log x})$, then (11.26) gives (11.23) and (11.27) gives (11.24). $\quad\square$

We are now in a position to prove

Corollary 11.17 (Page) *Let c_1 be the same constant as in Theorem 11.16. If $(a, q) = 1$, then*

$$\psi(x; q, a) = \frac{x}{\varphi(q)} + O\left(x\exp\left(-c_1\sqrt{\log x}\right)\right) \quad (11.28)$$

when there is no exceptional character modulo q, and

$$\psi(x; q, a) = \frac{x}{\varphi(q)} - \frac{\chi_1(a)x^{\beta_1}}{\varphi(q)\beta_1} + O\left(x\exp\left(-c_1\sqrt{\log x}\right)\right) \quad (11.29)$$

when there is an exceptional character χ_1 modulo q and β_1 is the concomitant zero.

Proof If $q \le \exp\left(2c_1\sqrt{\log x}\right)$, then we have only to insert the estimates of Theorem 11.16 into (11.22). If q is larger, then the stated estimates are still valid, but are worse than trivial. To see this, note first that the largest term in $\psi(x; q, a)$ is $\le \log x$, and the number of terms is $\le x/q + 1$, so it is immediate that

$$\psi(x; q, a) \le (x/q + 1)\log x \ll x\exp(-c_1\sqrt{\log x})$$

when $q \ge \exp(2c_1\sqrt{\log x})$. $\quad\square$

Presumably, exceptional zeros do not exist. However, if such a zero does exist, then we have a second main term in (11.29) that is bigger than the error term when $x < \exp(c_1^2/(1 - \beta_1)^2)$. If β_1 is extremely close to 1, then one might have $\beta_1 \geq 1 - 1/\log x$, and in such a situation the second main term is of the same order of magnitude as the first main term, since

$$x - \frac{x^{\beta_1}}{\beta_1} = (\beta_1 - 1)x^{\beta_1}/\beta_1 + (\log x) \int_{\beta_1}^{1} x^{\sigma} \, d\sigma \asymp (1 - \beta_1)x \log x. \quad (11.30)$$

Thus if $1 - \beta_1$ is small compared with $1/\log x$, then the main term is nearly doubled if $\chi_1(a) = -1$, and it is nearly annihilated if $\chi_1(a) = 1$. Unfortunately, the upper bound provided by the Brun–Titchmarsh theorem (Theorem 3.9) is not quite strong enough to refute such a possibility.

The constants c and c_1 in Theorems 11.3, 11.4, 11.16 and Corollary 11.17 are effectively computable. However, if we are willing to accept non-effective constants, then by Siegel's theorem (Theorem 11.14), or more precisely by its corollary (Corollary 11.15), we can eliminate the second main term, provided that q is more sharply limited.

Corollary 11.18 *Let c_1 be the same constant as in Theorem 11.16. For any positive A there is an $x_0(A)$ such that if $q \leq (\log x)^A$, then*

$$\psi(x, \chi) = E_0(\chi)x + O\left(x \exp\left(-c_1\sqrt{\log x}\right)\right) \quad (11.31)$$

for $x \geq x_0(A)$.

Proof Suppose that χ is quadratic and that $L(s, \chi)$ has an exceptional zero β_1. Then

$$x^{\beta_1} = x \exp(-(1 - \beta_1) \log x) \leq x \exp(-C(\varepsilon)q^{-\varepsilon} \log x)$$

by Siegel's theorem (Corollary 11.15). Since $q \leq (\log x)^A$, the above is

$$\leq x \exp(-C(\varepsilon)(\log x)^{1-A\varepsilon}).$$

In order to reach (11.31) we need to take ε a little smaller than $1/(2A)$, say $\varepsilon = 1/(3A)$. Then the above is

$$\leq x \exp\left(-c_1\sqrt{\log x}\right)$$

provided that $x \geq x_0 = \exp((c_1/C(\varepsilon))^6)$. $\qquad\qquad\square$

The constraint $q \leq (\log x)^A$ can be rewritten as $x \geq \exp(q^{1/A})$. This implies the constraint $x \geq x_0(A)$ if q is sufficiently large, say $q \geq q_0(A)$. We note also that the implicit constant in (11.31) is absolute. If we were to allow the implicit constant to depend on A, e.g. to be as large as $\exp((c_1/C(\varepsilon))^3)$, then we would obtain an estimate

$$\psi(x, \chi) \ll_A x \exp\left(-c_1\sqrt{\log x}\right)$$

that is valid for all q and all $x \geq \exp(q^{1/A})$, though of course the implicit constant is so large that the bound is worse than the trivial $\psi(x, \chi) \ll x$ when $x < x_0$. By applying (11.22) and (11.28), we obtain

Corollary 11.19 (The Siegel–Walfisz theorem) *Let c_1 be the constant in Theorem 11.16, and suppose that A is given, $A > 0$. If $q \leq (\log x)^A$ and $(a, q) = 1$, then*

$$\psi(x; q, a) = \frac{x}{\varphi(q)} + O_A\left(x \exp\left(-c_1\sqrt{\log x}\right)\right).$$

Pertaining to $\vartheta(x; q, a)$ and $\pi(x; q, a)$ we have estimates similar to those of Corollary 11.17.

Corollary 11.20 *Let c_1 be the constant in Theorem 11.16. If $(a, q) = 1$, then*

$$\vartheta(x; q, a) = \frac{x}{\varphi(q)} + O\left(x \exp\left(-c_1\sqrt{\log x}\right)\right) \qquad (11.32)$$

and

$$\pi(x; q, a) = \frac{\operatorname{li}(x)}{\varphi(q)} + O\left(x \exp\left(-c_1\sqrt{\log x}\right)\right) \qquad (11.33)$$

when there is no exceptional character modulo q, but

$$\vartheta(x; q, a) = \frac{x}{\varphi(q)} - \frac{\chi_1(a)x^{\beta_1}}{\varphi(q)\beta_1} + O\left(x \exp\left(-c_1\sqrt{\log x}\right)\right) \quad (11.34)$$

and

$$\pi(x; q, a) = \frac{\operatorname{li}(x)}{\varphi(q)} - \frac{\chi_1(a)\operatorname{li}\left(x^{\beta_1}\right)}{\varphi(q)} + O\left(x \exp\left(-c_1\sqrt{\log x}\right)\right) \quad (11.35)$$

when there is an exceptional character χ_1 modulo q and β_1 is the concomitant zero.

Proof Since

$$0 \leq \psi(x; q, a) - \vartheta(x; q, a) \leq \psi(x) - \vartheta(x) \ll x^{1/2},$$

the assertions concerning $\vartheta(x; q, a)$ follow immediately from Corollary 11.17. As for $\pi(x; q, a)$, we write

$$\pi(x; q, a) = \int_{2-}^{x} \frac{1}{\log u} \, d\vartheta(u; q, a) = \frac{\operatorname{li}(x)}{\varphi(q)} + \int_{2-}^{x} \frac{1}{\log u} \, d(\vartheta(u; q, a) - u/\varphi(q)).$$

This last integral we integrate by parts (as in the proof of Theorem 6.9), and find that it is

$$\frac{\vartheta(u;q,a) - u/\varphi(q)}{\log u}\Big|_{2-}^{x} - \int_{2}^{x} \frac{\vartheta(u;q,a) - u/\varphi(q)}{u(\log u)^2}\, du.$$

If there is no exceptional zero, then the numerator in the integrand is $\ll u \exp(-c_1\sqrt{\log u}) \ll x \exp(-c_1\sqrt{\log x})$, so we obtain (11.33). If there is an exceptional character χ_1, then the main term is reduced by $\chi_1(a)/\varphi(q)$ times the amount

$$\int_{2}^{x} \frac{1}{\log u}\, d\frac{u^{\beta_1}}{\beta_1} = \int_{2}^{x} \frac{u^{\beta_1-1}}{\log u}\, du = \int_{2^{\beta_1}}^{x^{\beta_1}} \frac{1}{\log v}\, dv = \operatorname{li}(x^{\beta_1}) + O(1).$$

The error term is still treated in the same way, so we obtain (11.35).   □

By arguing in the same manner from Corollary 11.19, we obtain

Corollary 11.21 *Let c_1 be the constant in Theorem* 11.16, *and suppose that A is given, $A > 0$. If $q \leq (\log x)^A$ and $(a, q) = 1$, then*

$$\vartheta(x;q,a) = \frac{x}{\varphi(q)} + O_A\bigl(x \exp\bigl(-c_1\sqrt{\log x}\bigr)\bigr) \tag{11.36}$$

and

$$\pi(x;q,a) = \frac{\operatorname{li}(x)}{\varphi(q)} + O_A\bigl(x \exp\bigl(-c_1\sqrt{\log x}\bigr)\bigr). \tag{11.37}$$

11.3.1   Exercises

1. Suppose that χ is a character modulo q. Explain why

$$\psi(x,\chi) = \sum_{\substack{a=1 \\ (a,q)=1}}^{q} \chi(a)\psi(x;q,a).$$

2. Suppose that $\exp(2c_1\sqrt{\log x}) \leq q \leq x$. Show that there is a positive constant c_2 such that

$$\psi(x,\chi) = E_0(\chi)x + O\left(x \exp\left(\frac{-c_2\log x}{\log q}\right)\right)$$

if $L(s,\chi)$ has no exceptional zero, and that

$$\psi(x,\chi) = -\frac{x^{\beta_1}}{\beta_1} + \left(x \exp\left(\frac{-c_2\log x}{\log q}\right)\right)$$

if $L(s,\chi)$ has the exceptional zero β_1.

3. Show that if $q \leq \exp(2c_1\sqrt{\log x})$, then

$$\vartheta(x,\chi) = E_0(\chi)x + O\bigl(x \exp\bigl(-c_1\sqrt{\log x}\bigr)\bigr)$$

when $L(s, \chi)$ has no exceptional zero, and that

$$\vartheta(x, \chi) = -\frac{x^{\beta_1}}{\beta_1} + O\big(x \exp\big(-c_1\sqrt{\log x}\big)\big)$$

when $L(s, \chi)$ has an exceptional zero β_1.

4. Suppose that $q \le \exp(c_1\sqrt{\log x})$, and put $x_0 = \exp\big(\big(\frac{\log q}{2c_1}\big)^2\big)$.

   (a) Explain why $\pi(x_0; \chi) \ll x_0 \le x^{1/4}$.

   (b) Treat $\pi(x, \chi) - \pi(x_0, \chi)$ as in the proof of Corollary 11.20 to show that

$$\pi(x, \chi) \ll x \exp\big(-c_1\sqrt{\log x}\big)$$

   if $L(s, \chi)$ has no exceptional zero, and that

$$\pi(x, \chi) = -\operatorname{li}(x^{\beta_1}) + O\big(x \exp\big(-c_1\sqrt{\log x}\big)\big)$$

   if $L(s, \chi)$ has the exceptional zero β_1.

5. Suppose that A is given, $A > 0$. Show that if $q \le (\log x)^A$, then

$$\vartheta(x, \chi) = E_0(x)x + O\big(x \exp\big(-c_1\sqrt{\log x}\big)\big),$$

and that

$$\pi(x, \chi) = E_0(\chi)\operatorname{li}(x) + O\big(x \exp\big(-c_1\sqrt{\log x}\big)\big).$$

By analogy with (11.20) we set

$$\Lambda(x; q, a) = \sum_{\substack{n \le x \\ n \equiv a(q)}} \lambda(n), \quad M(x; q, a) = \sum_{\substack{n \le x \\ n \equiv a(q)}} \mu(n). \qquad (11.38)$$

Here it is no longer natural to restrict to $(a, q) = 1$. Correspondingly, if χ is a character modulo q, we put

$$\Lambda(x, \chi) = \sum_{n \le x} \chi(n)\lambda(n), \quad M(x, \chi) = \sum_{n \le x} \chi(n)\mu(n). \qquad (11.39)$$

6. Let c_1 be the constant of Theorem 11.16, suppose that $q \le \exp(2c_1\sqrt{\log x})$ and that χ is a character modulo q. Show that

$$\Lambda(x, \chi) \ll x \exp\big(-c_1\sqrt{\log x}\big)$$

when $L(s, \chi)$ has no exceptional zero, and that

$$\Lambda(x, \chi) = \frac{L(2\beta_1, \chi_0)x^{\beta_1}}{L'(\beta_1, \chi)\beta_1} + O\big(x \exp\big(-c_1\sqrt{\log x}\big)\big)$$

when $L(s, \chi)$ has an exceptional zero β_1. (Note that in this latter case, the result of Exercise 11.1.2 is useful.)

7. Let c_1 be the constant of Theorem 11.16, suppose that $q \leq \exp(2c_1\sqrt{\log x})$ and that χ is a character modulo q. Show that

$$M(x, \chi) \ll x \exp\left(-c_1\sqrt{\log x}\right)$$

when $L(s, \chi)$ has no exceptional zero, and that

$$M(x, \chi) = \frac{x^{\beta_1}}{L'(\beta_1, \chi)\beta_1} + O\left(x \exp\left(-c_1\sqrt{\log x}\right)\right)$$

when $L(s, \chi)$ has an exceptional zero β_1.

8. Let c_1 be the constant in Theorem 11.16, and suppose that A is given, $A > 0$. Show that if $q \leq (\log x)^A$ and χ is a character modulo q, then

$$\Lambda(x, \chi) \ll_A \exp\left(-c_1\sqrt{\log x}\right),$$

and that

$$M(x, \chi) \ll_A x \exp\left(-c_1\sqrt{\log x}\right).$$

9. Show that if $(a, q) = 1$, then

$$\Lambda(x; q, a) = \frac{1}{\varphi(q)} \sum_{\chi} \overline{\chi}(a)\Lambda(x, \chi),$$

and that

$$M(x; q, a) = \frac{1}{\varphi(q)} \sum_{\chi} \overline{\chi}(a)M(x, \chi).$$

10. Let c_1 be the constant in Theorem 11.16. Show that if $(a, q) = 1$, then

$$\Lambda(x; q, a) \ll x \exp\left(-c_1\sqrt{\log x}\right)$$

if there is no exceptional χ modulo q, and that

$$\Lambda(x; q, a) = \frac{\chi_1(a)L(2\beta_1, \chi_0)x^{\beta_1}}{\varphi(q)L'(\beta_1, \chi_1)\beta_1} + O\left(x \exp\left(-c_1\sqrt{\log x}\right)\right)$$

if there is an exceptional character χ_1 modulo q with associated zero β_1.

11. Suppose that $(a, q) = d$, and write $a = db, q = dr$.
   (a) Show that $\Lambda(x; q, a) = \lambda(d)\Lambda(x/d; r, b)$.
   (c) Show that

$$\Lambda(x; q, a) \ll \frac{x}{d} \exp\left(-c_1\sqrt{\log x/d}\right)$$

if no L-function modulo r has an exceptional zero, and that

$$\Lambda(x; q, a) = \frac{\lambda(d)\chi_1(b)L(2\beta_1, \chi_0)(x/d)^{\beta_1}}{\varphi(r)L'(\beta_1, \chi_1)\beta_1} + O\left(\frac{x}{d} \exp\left(-c_1\sqrt{\log x/d}\right)\right)$$

if there is an exceptional character χ_1 modulo r with associated zero β_1. Here χ_0 is the principal character modulo r.

(d) Show that if $q \le (\log x)^A$, then

$$\Lambda(x; q, a) \ll_A x \exp\left(-c_1\sqrt{\log x}\right)$$

for all a.

12. Suppose that $(a, q) = 1$. Show that

$$M(x; q, a) \ll x \exp\left(-c_1\sqrt{\log x}\right)$$

if there is no exceptional character χ modulo q, and that

$$M(x; q, a) = \frac{\chi_1(a)x^{\beta_1}}{\varphi(q)L'(\beta_1, \chi_1)\beta_1} + O\left(x \exp\left(-c_1\sqrt{\log x}\right)\right)$$

if there is an exceptional character χ_1 modulo q with associated zero β_1.

13. Suppose that $d = (a, q)$, and write $q = dr$, $a = bd$.

(a) Show that if d is not square-free, then $M(x; q, a) = 0$.

(b) Explain why one does not expect that $M(x; q, a) = \mu(d)M(x/d; r, b)$ is true in general.

(c) Show instead that

$$M(x; q, a) = \mu(d) \sum_{\substack{k|d \\ (k,r)=1}} \mu(k)M(x/(dk); r, b\overline{k})$$

where $k\overline{k} \equiv 1 \pmod{r}$.

(d) Show that $M(x; q, a) \ll x/q$ in any case.

(e) Deduce that $M(x; q, a) \ll x \exp(-c\sqrt{\log x})$ if there is no exceptional character modulo r, and that

$$M(x; q, a) = \frac{\mu(d)\chi_1(b)(x/d)^{\beta_1}}{\varphi(r)L'(\beta_1, \chi_1)\beta_1} \prod_{\substack{p|d \\ p\nmid r}} \left(1 - \frac{\chi_1(p)}{p^{\beta_1}}\right) + O\left(x \exp\left(-c\sqrt{\log x}\right)\right)$$

if there is an exceptional character χ_1 with associated zero β_1.

(f) Show that if $q \le (\log x)^A$, then $M(x; q, a) \ll_A x \exp(-c\sqrt{\log x})$ for all a.

14. *Grössencharaktere* for $\mathbb{Q}(\sqrt{-1})$, continued from Exercise 11.1.5. Put $\psi(x, \chi_m) = \sum_{N(\mathfrak{a}) \le x} \Lambda(\mathfrak{a})\chi_m(\mathfrak{a})$. Show that if $1 \le m \le \exp(\sqrt{\log x})$, then $\psi(x, \chi_m) \ll x \exp(-c\sqrt{\log x})$ where $c > 0$ is a suitable absolute constant.

11.4   Applications

The fundamental estimates of the preceding section can be applied to a wide variety of counting problems, of which the following are representative examples.

Theorem 11.22   (Walfisz) *Let $A > 0$ be fixed, and let $R(n)$ denote the number of ways of writing n as a sum of a prime and a square-free number. Then*

$$R(n) = c(n)\mathrm{li}(n) + O\left(n/(\log n)^A\right)$$

where

$$c(n)=\prod_{p\nmid n}\left(1-\frac{1}{p(p-1)}\right)=\left(\prod_{p\mid n}\left(1+\frac{1}{p^2-p-1}\right)\right)\left(\prod_{p}\left(1-\frac{1}{p(p-1)}\right)\right).$$

Proof   Clearly

$$R(n) = \sum_{p<n}\mu(n-p)^2$$

$$= \sum_{p<n}\sum_{d^2\mid(n-p)}\mu(d)$$

by (2.4). Here the divisibility relation is equivalent to asserting that $p \equiv n \pmod{d^2}$. Hence on inverting the order of summations we see that the above is

$$= \sum_{d\le\sqrt{n}}\mu(d)\pi(n-1;d^2,n).$$

If $(d, n) > 1$, then the summand is $O(1)$, and hence such $d \le \sqrt{n}$ contribute an amount that is $O(\sqrt{n})$. We now restrict our attention to those d for which $(d, n) = 1$. For small d, say $d \le y = (\log x)^A$ we can apply the Siegel–Walfisz theorem (Corollary 11.19). Thus we see that

$$\sum_{\substack{d\le y\\(d,n)=1}}\mu(d)\pi(n-1;d^2,n) = \mathrm{li}(x)\sum_{\substack{d\le y\\(d,n)=1}}\frac{\mu(d)}{\varphi(d^2)} + O\left(xy\exp\left(-c\sqrt{\log x}\right)\right).$$

Since $\varphi(d^2) = d\varphi(d)$, we see that the sum in the main term is

$$\sum_{\substack{d=1\\(d,n)=1}}^{\infty}\frac{\mu(d)}{d\varphi(d)} + O\left(\sum_{d>y}\frac{1}{d\varphi(d)}\right) = \prod_{p\nmid n}\left(1-\frac{1}{p(p-1)}\right) + O(1/y)$$

by (1.31). To treat $d > y$ we could appeal to the Brun–Titchmarsh theorem (Theorem 3.9), but the moduli d^2 are increasing so rapidly that the trivial estimate $\pi(x; q, a) \ll 1 + x/q$ is enough:

$$\sum_{y<d<\sqrt{n}} \pi(n-1; d^2, n) \ll \sum_{y<d<\sqrt{n}} \frac{n}{d^2} \ll \frac{n}{y}.$$

On combining our estimates we obtain the stated result. $\qquad\square$

In some situations, as below, we find it fruitful to use the Prime Number Theorem for arithmetic progressions in conjunction with sieve estimates.

Theorem 11.23 *Let $N(x)$ denote the number of integers $n \le x$ for which $(n, \varphi(n)) = 1$. Then*

$$N(x) \sim \frac{e^{-C_0} x}{\log\log\log x}$$

as $x \to \infty$.

Proof We note that $(n, \varphi(n)) = 1$ if and only if n has the following two properties: (i) n is square-free, and (ii) there do not exist prime factors p, p' of n such that $p' \equiv 1 \pmod{p}$. Let $p(n)$ denote the least prime factor of n. We shall show that if $p(n)$ is small compared with $\log\log x$ then n is unlikely to have the property (ii). We also show that n is likely to have both properties (i) and (ii) if $p(n)$ is large compared with $\log\log x$. Thus $N(x)$ is approximately the number of integers $n \le x$ for which $p(n) > \log\log x$.

Let $A_p(x)$ denote the number of $n \le x$ that satisfy (i) and (ii) and for which $p(n) = p$. Thus

$$N(x) = \sum_{p \le x} A_p(x).$$

We begin by estimating $A_p(x)$ when $p \le \log\log x$. Let p be given, and suppose that n is an integer such that $p(n) = p$ and for which (ii) holds. Write $n = pm$; then m is relatively prime to all prime numbers $< p$ and also to all primes $\equiv 1 \pmod{p}$. Thus by the sieve estimate (3.20) we see that

$$A_p(x) \ll \frac{x}{p} \left(\prod_{p'<p} \left(1 - \frac{1}{p'}\right) \right) \prod_{\substack{p' \le x/p \\ p' \equiv 1(p)}} \left(1 - \frac{1}{p'}\right).$$

Here the first product is $\asymp 1/\log p$ by Mertens' estimate (Theorem 2.7(e)). By Theorem 4.12(d) we know that the second product is $\asymp (\log x)^{-1/(p-1)}$ for any fixed prime p. To derive a bound that is uniform in p we appeal to the Siegel–Walfisz theorem (Corollary 11.19), by which we see that $\pi(u; p, 1) \asymp$

$u/(p \log u)$ uniformly for $u \geq e^p$. Hence by integrating by parts we deduce that

$$\sum_{\substack{e^p \leq p' \leq x/p \\ p' \equiv 1(p)}} \frac{1}{p'} \asymp \frac{1}{p}(\log \log x/p - \log p) \asymp \frac{\log \log x}{p}$$

uniformly for $p \leq \log \log x$. Hence there is a constant $c > 0$ such that in this range,

$$A_p(x) \ll \frac{x}{p \log p} \exp(-c(\log \log x)/p).$$

Now it is not hard to show that the number of integers $n \leq x$ such that $p(n) = p$ is $\asymp x/(p \log p)$ uniformly for $p \leq x/2$. Hence the exponential above reflects the relative improbability that n satisfies condition (ii). On summing, we find that

$$\sum_{\frac{1}{2}U < p \leq U} A_p(x) \ll \frac{x}{(\log U)^2} \exp(-c(\log \log x)/U).$$

We take $U = 2^{-k} \log \log x$ and sum over k to see that

$$\sum_{p \leq \log \log x} A_p(x) \ll \frac{x}{(\log \log \log x)^2}.$$

We now consider n for which $p(n)$ is large, say $p(n) \geq y$ where y, to be chosen later, is somewhat larger than $\log \log x$. Let $\Phi(x, y)$ denote the number of integers $n \leq x$ composed entirely of prime numbers $> y$. By the sieve of Eratosthenes (Theorem 3.1) and Mertens' estimate (Theorem 2.7(e)) we see that

$$\sum_{y < p \leq x} A_p(x) \leq \Phi(x, y) = \frac{e^{-C_0}x}{\log y} + O\left(\frac{x}{(\log y)^2}\right) + O\left(e^{y/\log y}\right).$$

To derive a corresponding lower bound for the left-hand side we start with the numbers counted by $\Phi(x, y)$ and then delete those that do not satisfy (i) or (ii). If n does not satisfy (i), then there is a prime number p such that $p^2 | n$. The number of such $n \leq x$ is not more than $[x/p^2] \leq x/p^2$. Hence the total number of n counted in $\Phi(x, y)$ for which (i) fails is not more than $x \sum_{p>y} p^{-2} \ll x/(y \log y)$. Similarly, if n does not satisfy (ii), then there exist primes p, p' with $pp' | n$ such that $p' \equiv 1 \pmod{p}$. If p and p' are given, then the number of $n \leq x$ for which $pp' | n$ is $\leq x/(pp')$. Hence the total number of n counted in $\Phi(x, y)$ for which (ii) fails is not more than

$$x \sum_{y \leq p \leq \sqrt{x}} \frac{1}{p} \sum_{\substack{p' \leq x/p \\ p' \equiv 1(p)}} \frac{1}{p'}. \tag{11.40}$$

By the Brun–Titchmarsh inequality (Theorem 3.9) we see that

$$\sum_{\substack{U < p' \leq 2U \\ p' \equiv 1(p)}} \frac{1}{p'} \ll \frac{1}{p \log 2U/p}$$

uniformly for $U \geq p$. We take $U = 2^k p$ and sum over k to see that the inner sum in (11.40) is $\ll (\log \log 4x/p^2)/p$. Hence the expression (11.40) is

$$\ll x(\log \log x) \sum_{p > y} \frac{1}{p^2} \ll \frac{x \log \log x}{y \log y}.$$

On combining our estimates we see that

$$\sum_{y \leq p \leq x} A_p(x) \geq \frac{e^{C_0} x}{\log y} - O\left(\frac{x}{(\log y)^2}\right) - O\left(e^{y/\log y}\right)$$

$$- O\left(\frac{x}{y \log y}\right) - O\left(\frac{x \log \log x}{y \log y}\right).$$

In order that the last error term above is of a smaller order of magnitude than the main term, it is necessary to choose y so that $y/\log \log x \to \infty$. Thus there is necessarily a remaining range $\log \log x < p \leq y$ to be treated. By using the sieve (i.e., (3.20)) as in our treatment of small p we see that the number of integers $n \leq x$ for which $p(n) = p$ is $\ll x/(p \log p)$, uniformly for $p \leq \sqrt{x}$. Hence $A_p(x) \ll x/(p \log p)$, and consequently

$$\sum_{U \leq p \leq 2U} A_p(x) \ll \frac{x}{(\log U)^2}.$$

We put $U = 2^k \log \log x$ and sum over $1 \leq k \leq K$ where $K \ll \log \frac{y}{\log \log x}$ to see that

$$\sum_{\log \log x \leq p \leq y} A_p(x) \ll \frac{x}{(\log \log \log x)^2} \log \frac{y}{\log \log x}.$$

In order that this is a smaller order of magnitude than the main term, it is necessary to take $y \leq (\log \log x)^{(1+\varepsilon)}$ with $\varepsilon \to 0$ as $x \to \infty$. By taking y to be of this form with ε tending to 0 slowly, we obtain the stated result. □

11.4.1 Exercises

1. Let $R(n)$ be defined as in Theorem 11.22.
   (a) Show that if there is a primitive quadratic character $\chi_1 \pmod{q_1}$, $q_1 \leq \exp(\sqrt{\log x})$, for which $L(s, \chi_1)$ has a real zero $\beta_1 > 1 - c(\log x)^{-1/2}$, then

$$R(n) = c(n)\text{li}(n) - \chi_1(n)c_1(n)\text{li}(n^{\beta_1}) + O\left(n \exp\left(-c\sqrt{\log n}\right)\right)$$

where

$$c_1(n) = \sum_{\substack{d=1 \\ (d,n)=1 \\ q_1|d^2}}^{\infty} \frac{\mu(d)}{d\varphi(d)}.$$

(b) Show that $c_1(n) = 0$ if $8|q_1$.

(c) Show that if q_1 is odd, then

$$c_1(n) = \frac{\mu(q_1)c(q_1 n)}{q_1 \varphi(q_1)}.$$

(d) Show that if $4\|q_1$, then

$$c_1(n) = \frac{4\mu(q_1/2)c(q_1 n)}{q_1 \varphi(q_1)}.$$

2. In the proof of Theorem 11.23, specify ε as an explicit function of x to show that

$$N(x) = \frac{x}{\log\log\log x}\left(e^{-C_0} + O\left(\frac{\log\log\log\log x}{\log\log\log x}\right)\right).$$

3. Let a be a fixed non-zero integer. Show that the number of primes $p \le x$ such that $p + a$ is square-free is $c(a)\mathrm{li}(x) + O_A(x(\log x)^{-A})$ where $c(a)$ is defined as in Theorem 11.22.

4. Show that the appeal to the Siegel–Walfisz theorem in the proof of Theorem 11.23 can be replaced by an appeal to Page's theorem in conjunction with Corollary 11.12.

5. (Vaughan 1973) Let A and B be positive numbers. Show that

$$\sum_{p \le x}\left(\frac{\varphi(p-1)}{p-1}\right)^B = C\,\mathrm{li}(x) + O_{A,B}(x/(\log x)^A)$$

where

$$C = \prod_p\left(1 - \frac{1 - (1 - 1/p)^B}{p-1}\right).$$

6. (Erdős 1951)

(a) Let $r(n)$ denote the number of solutions of $p + 2^k = n$ with p prime and $k \ge 1$, and let $y = c\sqrt{\log x}$ where c is a sufficiently small positive constant. Define $q' = \prod_{2 < p \le y} p$. If there is a primitive character χ^* modulo q^* with $q^*|q'$ for which $L(s, \chi^*)$ has an exceptional zero, then let p be any prime divisor of q^* and define $q = q'/p$. Otherwise let $q = q'$. Prove that

$$\sum_{m \le x/q} r(qm) = \frac{x}{\varphi(q)\log 2} + O\left(\frac{x}{\varphi(q)\log x}\right).$$

(b) Show that $r(n) = \Omega(\log\log n)$.

11.5 Notes

Section 11.1. Theorem 11.3 is a combination of work by Gronwall (1913) and Titchmarsh (1930).

Section 11.2. Lemma 11.6, Theorem 11.7, and Corollaries 11.8, 11.9 originate in Landau (1918a, b), while Corollary 11.10 is from Page (1935). Theorem 11.11 can also be proved by appealing to the Dirichlet class number formula, which asserts that if d is a quadratic discriminant and $\chi_d(n) = \left(\frac{d}{n}\right)_K$ is the associated quadratic character, then

$$L(1, \chi_d) = \begin{cases} \dfrac{2\pi h}{w\sqrt{-d}} & (d < 0), \\[2ex] \dfrac{h \log \varepsilon}{\sqrt{d}} & (d > 0); \end{cases}$$

see Davenport (2000, Section 6). If $d < 0$, then $\chi_d(-1) = -1$, $\mathbb{Q}(\sqrt{d})$ is an imaginary quadratic field with class number h, and w denotes the number of roots of unity in the field (which is to say that $w = 6$ if $d = -3$, $w = 4$ if $d = -4$, and $w = 2$ otherwise). If $d > 0$, then $\chi_d(-1) = 1$, $\mathbb{Q}(\sqrt{d})$ is a real quadratic field with class number h and fundamental unit ε. Since $\varepsilon \gg \sqrt{d}$, it follows that if χ is a quadratic character with $\chi(-1) = 1$, then $L(1, \chi) \gg (\log q)/q^{1/2}$.

Corollary 11.12 has been sharpened by Davenport (1966), Haneke (1973), and by Goldfeld & Schinzel (1975).

Section 11.3. Let $h(d)$ denote the number of equivalence classes of primitive binary quadratic forms of discriminant d. Gauss (1801, Section 303) conjectured that $h(d) \to \infty$ as $d \to -\infty$. (The behaviour for $d > 0$ is quite different – the heuristics of Cohen & Lenstra (1984a, b) predict that $h(p) = 1$ for a positive proportion of primes $p \equiv 1 \pmod 4$.) For Gauss, the generic binary quadratic form was written $ax^2 + 2bxy + cy^2$, which is to say that the middle coefficient is even. Put $\Delta = b^2 - ac$. In Gauss's notation, Landau (1903) found that if $\Delta < 0$, then the class number is 1 precisely when $\Delta = -1, -2, -3, -4, -7$. Binary quadratic forms $ax^2 + bxy + cy^2$ with $d = b^2 - 4ac$ correspond, when d is a fundamental quadratic discriminant, to ideals in the ring $\mathcal{O}_K$ of integers in the quadratic number field $K = \mathbb{Q}(\sqrt{d})$. In this notation, $h(d) = 1$ if and only if $\mathcal{O}_K$ is a unique factorization domain. The problem of determining all $d < 0$ for which $h(d) = 1$ is now solved, but historically it was enormously more difficult than the class number 1 problem settled by Landau. Landau (1918b) recorded Hecke's observation that if $d < 0$ is a quadratic discriminant and $L(s, \chi_d) > 0$ for $1 - c/\log|d| < s < 1$, then $h(d) \gg_c |d|^{1/2}/\log|d|$. In view of Dirichlet's class number formula (4.36), we have obtained Hecke's result – by a different method – in Theorem 11.4. Thus we have a good lower bound for $h(d)$ when $d < 0$, except for those d for which $L(s, \chi_d)$ has an exceptional real zero. Deuring (1933) showed that if $h(d) = 1$ has infinitely many solutions with $d < 0$, then the Riemann Hypothesis is true. Mordell (1934) showed that the same conclusion can be derived from the weaker hypothesis that $h(d)$ does not tend to infinity as $d \to -\infty$. Heilbronn (1934) found that instead of arguing from a hypothetical zero ρ of the zeta function with $\beta > 1/2$ one could just as well argue from an exceptional zero of a quadratic L-function, and thus proved Gauss's conjecture that $h(d) \to \infty$ as $d \to -\infty$. Landau (1935) put Heilbronn's theorem in a quantitative form: $h(d) > |d|^{3/8-\varepsilon}$ as $d \to -\infty$. Through a different arrangement of the technical details, Siegel (1935) sharpened Landau's argument to show that $h(d) > |d|^{1/2-\varepsilon}$, which by (4.36) is the case $d < 0$ of Theorem 11.14. To achieve his result, Siegel first generalized to algebraic number fields the formula (found in Exercise 10.1.10) that Riemann used to prove the functional equation for $\zeta(s)$. Then Siegel applied this to the quartic number field $K = \mathbb{Q}(\sqrt{d_1}, \sqrt{d_2})$ whose Dedekind zeta function is $\zeta_K(s) = \zeta(s)L(s, \chi_{d_1})L(s, \chi_{d_2})L(s, \chi_{d_1 d_2})$. It is now recognized that Siegel's formula arises through the choice of the kernel in a Mellin transform, and that many other choices work just as well; see Goldfeld(1974). Our exposition is based on that of Estermann (1948).

It is easy to show that the complex quadratic field of discriminant $d < 0$ has unique factorization in the nine cases $d = -3, -4, -7, -8, -11, -19,$ $-43, -67, -163$. Heilbronn & Linfoot (1934) showed that there could exist at most one more such discriminant. The 'problem of the tenth discriminant' was solved first by Heegner (1952). However, Heegner's paper contained many assertions for which proofs were not provided, and Heegner also used results from Weber's *Algebra* which were known not to be trustworthy. Consequently, for many years Heegner's paper was thought to be incorrect. Baker (1966) proved a fundamental lower bound for linear forms in logarithms of algebraic numbers, which by means of a result of Gel'fond & Linnik (1948) reduced the class number 1 problem to a finite calculation. Meanwhile, Stark (1967) showed that there is no tenth discriminant by translating Heegner's argument into parallel language where it could be checked. After a reexamination of Heegner's work, Deuring (1968), Birch (1969), and Stark (1969) all concluded that Heegner's paper was after all correct. Gel'fond & Linnik reduced the class number problem to a question concerning linear forms in three logarithms, which Baker treated successfully. However, with a small modification of their argument, Gel'fond & Linnik could have reduced the problem to linear forms in two logarithms, which Gel'fond had already treated. Thus one could say that Gel'fond & Linnik 'should' have solved the problem in 1948.

Baker (1971) and Stark (1971b, 1972) reduced the complete determination of complex quadratic fields with $h(d) = 2$ to a finite calculation which was provided by Bundschuh & Hock (1969), Ellison *et al.* (1971), Montgomery & Weinberger (1973), and by Stark (1975).

The effective determination of all quadratic discriminants $d < 0$ for which $h(d)$ takes specific larger values became possible only with the addition of further ideas. Goldfeld (1976) showed that a zero at $s = 1/2$ of the L-function of an elliptic curve would be useful if it is of sufficiently high multiplicity. In particular, if (i) the Birch–Swinnerton-Dyer conjectures are true, and if (ii) there exist elliptic curves of arbitrarily high rank, then $h(d) \gg_A (\log |d|)^A$ for arbitrarily large A, with an effectively computable implicit constant. Although these conjectures remain unproved, Gross & Zagier (1986) were able to establish enough to give an effective lower bound for $h(d)$ tending to infinity. For accounts of this, see Zagier (1984), Goldfeld (1985), Coates (1986), and finally Oesterlé (1988), who developed the Goldfeld and Gross–Zagier work to show that

$$h(d) \geq \frac{1}{55}(\log |d|) \prod_{\substack{p|d \\ p<|d|}} \left(1 - \frac{[2\sqrt{p}]}{p+1}\right).$$

By means of this inequality, Arno (1992), Wagner (1996), and Arno, Robinson & Wheeler (1998) treated progressively larger collections of class numbers. Most recently, Watkins (2004) settled the complete determination of all discriminants $d < 0$ for which $h(d) \leq 100$.

With regard to Corollary 11.17, Page (1935) states the final conclusion in a less precise form in which the term corresponding to the exceptional zero is replaced by $O(x^{\beta_1}/\phi(q))$.

The deduction of Corollaries 11.18 and 11.19 from Siegel's theorem was first recorded by Walfisz (1936).

Section 11.4. Theorem 11.22 is due to Walfisz (1936). In a weaker form it occurs first in Estermann (1931), and is given in a somewhat refined form but without the benefit of Siegel's theorem in Page (1935). For similar theorems see see Mirsky (1949).

Theorem 11.23 is due to Erdős (1948).

11.6 References

Arno, S. (1992). The imaginary quadratic fields of class number 4, *Acta Arith.* **60**, 321–334.

Arno, S., Robinson, M. L., & Wheeler, F. S. (1998). Imaginary quadratic fields with small class number, *Acta Arith.* **83**, 295–330.

Baker, A. (1966). Linear forms in the logarithms of algebraic numbers, I, *Mathematika* **13**, 204–216.

(1971). Imaginary quadratic fields with class number 2, *Ann. of Math.* (2) **94**, 139–152.

Bateman, P. T. & Chowla, S. (1953).The equivalence of two conjectures in the theory of numbers, *J. Indian Math. Soc.* (N.S.) **17**, 177–181.

Birch, B. J. (1969). Weber's class invariants, *Mathematika* **16**, 283–294.

Buell, D. A. (1999). The last exhaustive computation of class groups of complex quadratic number fields, *Number Theory* (Ottawa, 1996), CRM Proc. Lecture Notes 19, Providence: Amer. Math. Soc., pp. 35–53.

Bundschuh, P. & Hock, A. (1969). Bestimmung aller imaginär-quadratischen Zahlkörper der Klassenzahl Eins mit Hilfe eines Satzes von Baker, *Math. Z.* **111**, 191–204.

Coates, J. (1986). The work of Gross and Zagier on Heegner points and the derivatives of *L*-series, *Seminar Bourbaki*, Vol. 1984/1985, *Astérisque* No. 133–134, 55–72.

Chowla, S. (1972). On *L*-series and related topics, *Proc. Number Theory Conf.* (Boulder, 1972), Boulder: University of Colorado, pp. 41–42.

Cohen, H. & Lenstra, H. (1984a). Heuristics on class groups, *Number Theory* (New York, 1982). Lecture Notes in Math. 1052. Berlin: Springer-Verlag, pp. 26–36.

(1984b). Heuristics on class groups of number fields, *Number Theory* (Noordwijkerhout, 1983). Lecture Notes in Math. 1068. Berlin: Springer-Verlag, pp. 33–62.

Davenport, H. (1966). Eine Bemerkung über Dirichlets *L*-Funktionen, *Nachr. Akad. Wiss. Göttingen Math.-Phys.* Kl. II, 203–212; *Collected Works*, Vol. 4. London: Academic Press, 1977, pp. 1816–1825.

(2000). *Multiplicative Number Theory*, Third edition, Graduate Texts in Math. 74. New York: Springer-Verlag.

Deuring, M. (1933). Imaginäre quadratische Zahlkörper mit der Klassenzahl 1, *Math. Z.* **37**, 405–415.

(1968). Imaginäre quadratische Zahlkörper mit der Klassenzahl Eins, *Invent. Math.* **5**, 169–179.

Ellison, W. J., Pesek, J., Stall, D. S. & Lunnon, W. F. (1971). A postscript to a paper of A. Baker, *Bull. London Math. Soc.* **3**, 75–78.

Erdős, P. (1948). Some asymptotic formulas in number theory, *J. Indian Math. Soc.* (N. S.) **12**, 75–78.

(1951). On some problems of Bellman and a theorem of Romanoff, *J. Chinese Math. Soc.* (N. S.) **1**, 409–421.

Estermann, T. (1931). On the representations of a number as the sum of a prime and a quadratfrei number, *J. London Math. Soc.* **6**, 219–221.

(1948). On Dirichlet's *L* functions, *J. London Math. Soc.* **23**, 275–279.

Fekete, M. & Pólya, G. (1912). Über ein Problem von Laguerre, *Rend. Circ. Mat. Palermo* **34**, 1–32.

Gauss, C. F. (1801). *Disquisitiones Arithmeticae*, Leipzig: Fleischer.

Gel'fond, A. O. & Linnik, Yu. V. (1948). On Thue's method in the problem of effectiveness in quadratic fields, *Dokl. Akad. Nauk SSSR* **61**,773–776.

Goldfeld, D. M. (1974). A simple proof of Siegel's theorem, *Proc. Nat. Acad. Sci. U.S.A.* **71**, 1055.

(1975). On Siegel's zero, *Ann. Scuola Norm. Sup. Pisa Cl. Sci.* (4) **2**, 571–583.

(1976). The class number of quadratic fields and the conjectures of Birch and Swinnerton-Dyer, *Ann. Scuola Norm. Sup. Pisa Cl. Sci.* (4) **3**, 624–663.

(1985). Gauss' class number problems for imaginary quadratic fields, *Bull. Amer. Math. Soc.* **13**, 23–37.

(2004). The Gauss class number problem for imaginary quadratic fields, *Heegner Points and Rankin L-series*, Math. Sci. Res. Inst. Publ. 49. Cambridge: Cambridge University Press, 25–36.

Goldfeld, D. M. & Schinzel, A. (1975). On Siegel's zero, *Ann. Scuola Norm. Sup. Pisa Cl. Sci.* (4) **2**, 571–583.

Gronwall, T. H. (1913). Sur les séries de Dirichlet correspondant à des caractères complexes, *Rend. Circ. Mat. Palermo* **35**, 145–159.

Gross, B. H. & Zagier, D. B. (1986). Heegner points and derivatives of *L*-series, *Invent. Math.* **84**, 225–320.

Haneke, W. (1973). Über die reellen Nullstellen der Dirichletschen *L*-Reihen, *Acta Arith.* **22**, 391–421; *Corrigendum*, **31** (1976), 99–100.

Heegner, K. (1952). Diophantische Analysis und Modulfunktionen, *Math. Z.* **56**, 227–253.

Heilbronn, H. (1934). On the class-number in imaginary quadratic fields, *Quart. J. Math. Oxford Ser.* **5**, 150–160.

(1937). On real characters, *Acta Arith.* **2**, 212–213.

Heilbronn, H. & Linfoot, E. (1934). On the imaginary quadratic corpora of class-number one, *Quart. J. Math. Oxford Ser.* **5**, 293–301.

Landau, E. (1903). Über die Klassenzahl der binären quadratischen Formen von negativer Discriminante, *Math. Ann.* **56**, 671–676; *Collected Works*, Vol. 1. Essen: Thales Verlag, 1985, pp. 354–359.

(1918a). Über imaginär-quadratische Zahlkörper mit gleicher Klassenzahl, *Nachr. Akad. Wiss. Göttingen*, 277–284; *Collected Works*, Vol. 7. Essen: Thales Verlag, 1986, pp. 142–160.

(1918b). Über die Klassenzahl imaginär-quadratischer Zahlkörper, *Nachr. Akad. Wiss. Göttingen*, 285–295; *Collected Works*, Vol. 7. Essen: Thales Verlag, pp. 150–160.

(1935). Bemerkungen zum Heilbronnschen Satz, *Acta Arith.* **1**, 1–18; *Collected Works*, Vol. 9. Essen: Thales Verlag, 1987, pp. 265–282.

Mahler, K. (1934). On Hecke's theorem on the real zeros of the *L*-functions and the class number of quadratic fields, *J. London Math. Soc.* **9**, 298–302.

Mirsky, L. (1949). The number of representations of an integer as the sum of a prime and a *k*-free integer, *Amer. Math. Monthly* **56**, 17–19.

Montgomery, H. L. & Weinberger, P. J. (1973). Notes on small class numbers, *Acta Arith.* **24**, 529–542.

Mordell, L. J. (1934). On the Riemann Hypothesis and imaginary quadratic fields with given class number, *J. London Math. Soc.* **9**, 405–415.

Oesterlé, J. (1988). Le problème de Gauss sur le nombre de classes, *Enseignement Math.* (2) **34**, 43–67.

Page, A. (1935). On the number of primes in an arithmetic progression, *Proc. London Math. Soc.* (2) **39**, 116–141.

Pólya, G. & Szegö, G. (1925). *Aufgaben und Lehrsätze aus der Analysis*, Vol. 2, Grundl. Math. Wiss. 20. Berlin: Springer.

Rosser, J. B. (1950). Real roots of real Dirichlet *L*-series, *J. Research Nat. Bur. Standards* **45**, 505–514.

Siegel, C. L. (1935). Über die Classenzahl quadratischer Zahlkörper, *Acta Arith.* **1**, 83–86.

(1968). Zum Beweis des Starkschen Satzes, *Invent. Math.* **5**, 180–191.

Stark, H. M. (1967). A complete determination of the complex quadratic fields of class-number one, *Michigan Math. J.* **14**, 1–27.

(1969). On the "gap" in a theorem of Heegner, *J. Number Theory* **1**, 16–27.

(1971a). Recent advances in determining all complex quadratic fields of a given class-number, *Number Theory Institute* (Stony Brook, 1969), Proc. Sympos. Pure Math. 20. Providence: Amer. Math. Soc., pp. 401–414.

(1971b). A transcendence theorem for class-number problems, *Ann. of Math.* (2) **94**, 153–173.

(1972). A transcendence theorem for class-number problems, II, *Ann. of Math.* (2) **96**, 174–209.

(1973). Class-numbers of complex quadratic fields, *Modular Functions of One Variable*, I (Proc. Internat. Summer School, Univ. Antwerp, Antwerp, 1972), Lecture Notes in Math. 320. Berlin: Springer-Verlag, pp. 153–174.

(1975). On complex quadratic fields with class-number two, *Math. Comp.* **29**, 289–302.

Tatuzawa, T. (1951). On a theorem of Siegel, *Japan. J. Math.* **21**, 163–178.

Titchmarsh, E. C. (1930). A divisor problem, *Rend. Circ. Mat. Palermo* **54**, 414–429; Correction, **57** (1933), 478–479.

Vaughan, R. C. (1973). Some applications of Montgomery's sieve, *J. Number Theory* **5**, 64–79.

Wagner, C. (1996). Class number 5, 6 and 7, *Math. Comp.* **65**, 785–800.

Walfisz, A. (1936). Zur additiven Zahlentheorie. II, *Math. Z.* **40**, 592-607.

Watkins, M. (2004). Class numbers of imaginary quadratic fields, *Math. Comp.* **73**, 907–938.

Zagier, D. (1984). *L*-series of elliptic curves, the Birch–Swinnerton-Dyer conjecture, and the class number problem of Gauss, *Notices Amer. Math. Soc.* **31**, 739–743.

12

Explicit formulæ

12.1 Classical formulæ

When we proved the Prime Number Theorem, we confined the contour of integration to the zero-free region. If we pull the contour further to the left, then we encounter a number of poles that leave residues, and thus we can express the error term in the Prime Number Theorem as a sum over the zeros of $\zeta(s)$. Let $\psi_0(x) = (\psi(x^+) + \psi(x^-))/2$. By applying Perron's formula (Theorem 5.1) to the Dirichlet series $-\frac{\zeta'}{\zeta}(s) = \sum_n \Lambda(n) n^{-s}$, we see that

$$\psi_0(x) = \lim_{T \to \infty} \frac{-1}{2\pi i} \int_{\sigma_0 - iT}^{\sigma_0 + iT} \frac{\zeta'}{\zeta}(s) \frac{x^s}{s} \, ds.$$

Here the integrand has a pole at $s = 1$, at zeros ρ, at $s = 0$, and at the trivial zeros $-2k$. Since x^s decays very rapidly as $\sigma \to -\infty$, it is reasonable to expect that we can pull the contour to the left, and thus show that the above is

$$= x - \lim_{T \to \infty} \sum_{\substack{\rho \\ |\gamma| \le T}} \frac{x^\rho}{\rho} - \frac{\zeta'}{\zeta}(0) + \sum_{k=1}^{\infty} \frac{x^{-2k}}{2k}. \tag{12.1}$$

Here $\frac{\zeta'}{\zeta}(0) = \log 2\pi$ by (10.11) and (10.14), and the sum over the trivial zeros is

$$-\frac{1}{2} \log(1 - 1/x^2),$$

which is continuous and tends to 0 as $x \to \infty$. In order to give a rigorous proof of the above, we first establish estimates for $\frac{\zeta'}{\zeta}(s)$.

Lemma 12.1 *We have*

$$\frac{\zeta'}{\zeta}(s) = \frac{-1}{s-1} + \sum_{\substack{\rho \\ |\gamma - t| \le 1}} \frac{1}{s - \rho} + O(\log \tau) \tag{12.2}$$

uniformly for $-1 \le \sigma \le 2$.

397

Here the first term on the right is significant only for $|t| \leq 1$. We could prove the above by the same method that we used to prove Lemma 6.4, but we find it instructive to argue instead from Corollary 10.14.

Proof   By combining (10.29) and Theorem C.1, it is immediate that

$$\frac{\zeta'}{\zeta}(s) = \frac{-1}{s-1} + \sum_{\rho} \left(\frac{1}{s-\rho} + \frac{1}{\rho} \right) - \frac{1}{2} \log \tau + O(1).$$

On applying this at $\sigma + it$ and at $2 + it$, and differencing, it follows that

$$\frac{\zeta'}{\zeta}(s) = \frac{-1}{s-1} + \sum_{\rho} \left(\frac{1}{s-\rho} - \frac{1}{2+it-\rho} \right) + O(1).$$

By Theorem 10.13 it is clear that

$$\sum_{\substack{\rho \\ |\gamma - t| \leq 1}} \frac{1}{2 + it - \rho} \ll \sum_{\substack{\rho \\ |\gamma - t| \leq 1}} 1 \ll \log \tau.$$

Now suppose that n is a positive integer, and consider those zeros ρ for which $n \leq |\gamma - t| \leq n + 1$. Since

$$\frac{1}{s-\rho} - \frac{1}{2+it-\rho} = \frac{2-\sigma}{(s-\rho)(2+it-\rho)} \ll \frac{1}{n^2},$$

it follows that such zeros contribute an amount

$$\ll \frac{N(t+n+1) - N(t+n) + N(t-n) - N(t-n-1)}{n^2} \ll \frac{\log(\tau+n)}{n^2}.$$

On summing over n we obtain the stated estimate.   □

Lemma 12.2   *For each real number $T \geq 2$ there is a T_1, $T \leq T_1 \leq T + 1$, such that*

$$\frac{\zeta'}{\zeta}(\sigma + iT_1) \ll (\log T)^2$$

uniformly for $-1 \leq \sigma \leq 2$.

Proof   By Theorem 10.13, there is a $T_1 \in [T, T+1]$ such that $|T_1 - \gamma| \gg 1/\log T$ for all zeros ρ. Since each summand in (12.2) is $\ll \log T$, and there are $\ll \log T$ summands, the estimate is immediate.   □

The next lemma is useful in Chapter 14, but we establish it here since it is a also an immediate corollary of Lemma 12.1.

Lemma 12.3   *For any real number t,*

$$\arg \zeta(\sigma + it) \ll \log \tau$$

uniformly for $-1 \leq \sigma \leq 2$.

The function $\log \zeta(s)$ has a branch point at $s = 1$, and also at zeros ρ of the zeta function. To obtain a single branch of the logarithm, we remove from the complex plane the interval $(-\infty, 1]$, and also intervals of the form $(-\infty + i\gamma, \beta + i\gamma]$. What remains is simply connected, and in this region we take that branch of $\log \zeta(s)$ for which $\log \zeta(s) \to 0$ as $\sigma \to \infty$. This is the branch of the logarithm that we have expanded as a Dirichlet series, for $\sigma > 1$ (cf. Corollary 1.11). Thus, if t is not the ordinate of a zero, we define $\arg \zeta(s) = \Im \log \zeta(s)$ by continuous variation from $\infty + it$ to $\sigma + it$, which is to say that

$$\arg \zeta(s) = -\int_{\sigma}^{\infty} \Im \frac{\zeta'}{\zeta}(\alpha + it) \, d\alpha.$$

If t is the ordinate of a zero then we set $\arg \zeta(s) = (\arg \zeta(\sigma + it^+) + \arg \zeta(\sigma + it^-))/2$.

Proof Suppose that $-1 \le \sigma \le 2$, and that t is not the ordinate of a zero. Then

$$\arg \zeta(\sigma + it) = \arg \zeta(2 + it) - \int_{\sigma}^{2} \Im \frac{\zeta'}{\zeta}(\alpha + it) \, d\alpha.$$

Here $\arg \zeta(2 + it) \ll 1$ uniformly in t, by Corollary 1.11. Thus by Lemma 12.1, the right-hand side above is

$$-\sum_{|\gamma - t| \le 1} \int_{\sigma}^{2} \Im \frac{1}{\alpha + it - \rho} \, d\alpha + O(\log \tau).$$

Here the summand is

$$\arctan \frac{\sigma - \beta}{t - \gamma} - \arctan \frac{2 - \beta}{t - \gamma}.$$

If $t > \gamma$, then this lies between $-\pi$ and 0, while if $t < \gamma$, then the above lies between 0 and π. Thus in any case the quantity is bounded, and by Theorem 10.13 the number of summands is $\ll \log \tau$, so we have the result when t is not the ordinate of a zero. Since the ordinates of zeros have no finite limit point, we obtain the same bound when t is the ordinate of a zero, since in that case $\arg \zeta(s) = (\arg \zeta(\sigma + it^+) + \arg \zeta(\sigma - it^-))/2$. □

Lemma 12.4 *Let $\mathcal{A}$ denote the set of those points $s \in \mathbb{C}$ such that $\sigma \le -1$ and $|s + 2k| \ge 1/4$ for every positive integer k. Then*

$$\frac{\zeta'}{\zeta}(s) \ll \log(|s| + 1)$$

uniformly for $s \in \mathcal{A}$.

Proof   We recall (10.27), in which the first two terms are bounded for $s \in \mathcal{A}$. Also,

$$\frac{\Gamma'}{\Gamma}(1 - s) \ll \log(|s| + 1)$$

by Theorem C.1. Finally

$$\cot \frac{\pi s}{2} = i + \frac{2i}{e^{i\pi s} - 1} \ll 1$$

since s is bounded away from even integers, so we have the result. $\square$

We are now in a position to prove the explicit formula (12.1) in a quantitative form.

Theorem 12.5   *Let c be a constant, $c > 1$, suppose that $x \geq c$, that $T \geq 2$, and let $\langle x \rangle$ denote the distance from x to the nearest prime power, other than x itself. Then*

$$\psi_0(x) = x - \sum_{\substack{\rho \\ |\gamma| \leq T}} \frac{x^\rho}{\rho} - \log 2\pi - \frac{1}{2}\log(1 - 1/x^2) + R(x, T) \quad (12.3)$$

where

$$R(x, T) \ll (\log x) \min\left(1, \frac{x}{T\langle x \rangle}\right) + \frac{x}{T}(\log xT)^2. \quad (12.4)$$

Since $\langle x \rangle > 0$ for all x, we obtain (12.1) by letting $T \to \infty$ in the above. Moreover, if $n_1 < n_2$ are two consecutive prime powers, then from the above we see that $\sum_{|\gamma| \leq T} x^\rho/\rho$ converges uniformly for x in an interval of the form $[n_1 + \delta, n_2 - \delta]$. This sum, of course, cannot be uniformly convergent for x in a neighbourhood of a prime power, since $\psi_0(x)$ has jump discontinuities at such points, but we see from the above that it is boundedly convergent in the neighbourhood of a prime power. The sum over ρ is also convergent when $x = 1$, but it is not boundedly convergent near 1, since $\log(1 - 1/x^2) \to -\infty$ as $x \to 1^+$.

Proof   Let T_1 be the number supplied by Lemma 12.2. Then by Theorem 5.2 and its Corollary 5.3, with $\sigma_0 = 1 + 1/\log x$, we see that

$$\psi_0(x) = \frac{-1}{2\pi i} \int_{\sigma_0 - iT_1}^{\sigma_0 + iT_1} \frac{\zeta'}{\zeta}(s) \frac{x^s}{s} \, ds + R_1$$

where

$$R_1 \ll \sum_{\substack{x/2 < n < 2x \\ n \neq x}} \Lambda(n) \min\left(1, \frac{x}{T|x - n|}\right) + \frac{x}{T}\sum_{n=1}^{\infty} \frac{\Lambda(n)}{n^{\sigma_0}}.$$

Here the second sum is $-\frac{\zeta'}{\zeta}(\sigma_0) \asymp 1/(\sigma_0 - 1) = \log x$. In the first sum, the terms for which $x + 1 \le n < 2x$ contribute an amount

$$\ll \sum_{x+1 \le n < 2x} \frac{x \log x}{T(n-x)} \ll \frac{x}{T}(\log x)^2.$$

The terms for which $x/2 < n \le x - 1$ are handled similarly. Finally, any terms for which $x - 1 < n < x + 1$ contribute an amount

$$\ll (\log x) \min\left(1, \frac{x}{T\langle x \rangle}\right),$$

so

$$R_1 \ll (\log x) \min\left(1, \frac{x}{T\langle x \rangle}\right) + \frac{x}{T}(\log x)^2.$$

Let K denote an odd positive integer, and let $\mathcal{C}$ denote the contour consisting of line segments connecting $\sigma_0 - iT_1, -K - iT_1, -K + iT_1, \sigma_0 + iT_1$. Then by Cauchy's residue theorem,

$$\psi_0(x) = x - \sum_{\substack{\rho \\ |\gamma| < T_1}} \frac{x^\rho}{\rho} + \sum_{1 \le k < K/2} \frac{x^{-2k}}{2k} - \frac{\zeta'}{\zeta}(0) + R_1 + R_2$$

where

$$R_2 = \frac{-1}{2\pi i} \int_{\mathcal{C}} \frac{\zeta'}{\zeta}(s) \frac{x^s}{s} \, ds.$$

Since $|\sigma \pm iT_1| \ge T$, we see by Lemma 12.2 that

$$\int_{-1 \pm iT_1}^{\sigma_0 \pm iT_1} \frac{\zeta'}{\zeta}(s) \frac{x^s}{s} \, ds \ll \frac{(\log T)^2}{T} \int_{-1}^{\sigma_0} x^\sigma \, d\sigma \ll \frac{x(\log T)^2}{T \log x} \ll \frac{x(\log T)^2}{T}.$$

Similarly, since $(\log |\sigma \pm iT_1|)/|\sigma \pm iT_1| \ll (\log T)/T$, we see by Lemma 12.4 that

$$\int_{-K \pm iT_1}^{-1 \pm iT_1} \frac{\zeta'}{\zeta}(s) x^s \, ds \ll \frac{\log T}{T} \int_{-\infty}^{-1} x^\sigma \, d\sigma \ll \frac{\log T}{xT \log x} \ll \frac{\log T}{T}.$$

As $|-K + it| \ge K$, by Lemma 12.4 we also see that

$$\int_{-K-iT_1}^{-K+iT_1} \frac{\zeta'}{\zeta}(s) \frac{x^s}{s} \, ds \ll \frac{\log KT}{K} x^{-K} \int_{-T_1}^{T_1} 1 \, dt \ll \frac{T \log KT}{K x^K}.$$

This tends to 0 as $K \to \infty$, so we obtain the stated result. $\qquad\square$

Let $\psi_0(x, \chi) = (\psi(x^+, \chi) + \psi(x^-, \chi))/2$. Not surprisingly, our treatment of $\psi_0(x)$ extends readily to provide explicit formulæ for $\psi_0(x, \chi)$.

Lemma 12.6 *Let χ be a primitive character modulo q with $q > 1$. Then*

$$\frac{L'}{L}(s, \chi) = \sum_{\substack{\rho \\ |\gamma - t| \leq 1}} \frac{1}{s - \rho} + O(\log q\tau) \qquad (12.5)$$

uniformly for $-1 \leq \sigma \leq 2$.

Proof By combining (10.37) and Theorem C.1, it is immediate that

$$\frac{L'}{L}(s, \chi) = B(\chi) + \sum_{\rho} \left(\frac{1}{s - \rho} + \frac{1}{\rho} \right) + O(\log q\tau).$$

On applying this at $\sigma + it$ and $2 + it$, and differencing, it follows that

$$\frac{L'}{L}(s, \chi) = \sum_{\rho} \left(\frac{1}{s - \rho} - \frac{1}{2 + it - \rho} \right) + O(\log q\tau).$$

By Theorem 10.17 it is clear that

$$\sum_{\substack{\rho \\ |\gamma - t| \leq 1}} \frac{1}{2 + it - \rho} \ll \sum_{\substack{\rho \\ |\gamma - t| \leq 1}} 1 \ll \log q\tau.$$

Now suppose that n is a positive integer, and consider those zeros ρ for which $n \leq |\gamma - t| \leq n + 1$. Since

$$\frac{1}{s - \rho} - \frac{1}{2 + it - \rho} = \frac{2 - \sigma}{(s - \rho)(2 + it - \rho)} \ll \frac{1}{n^2},$$

it follows that such zeros contribute an amount

$$\ll \frac{\log q + \log(|t + n| + 2) + \log(|t - n| + 2)}{n^2} \ll \frac{\log q(\tau + n)}{n^2}.$$

On summing over n we obtain the stated estimate. $\qquad\square$

Lemma 12.7 *Let χ be a primitive character modulo q, and suppose that $T \geq 2$. Then there is a T_1, $T \leq T_1 \leq T + 1$, such that*

$$\frac{L'}{L}(\sigma \pm iT_1, \chi) \ll (\log qT)^2$$

uniformly for $-1 \leq \sigma \leq 2$.

Proof By Theorem 10.17, there is a $T_1 \in [T, T + 1]$ such that both $|T_1 - \gamma| \gg 1/\log qT$ and $|T_1 + \gamma| \gg 1/\log qT$ for all zeros ρ of $L(s, \chi)$. Since each summand in (12.5) is $\ll \log qT$, and there are $\ll \log qT$ summands, the estimate is immediate. $\qquad\square$

Lemma 12.8 *Let χ be a primitive character modulo q, $q > 1$. Then*

$$\arg L(s, \chi) \ll \log q\tau$$

uniformly for $-1 \leq \sigma \leq 2$.

Proof Suppose that $-1 \le \sigma \le 2$, and that t is not the ordinate of a zero. Then

$$\arg L(\sigma + it, \chi) = \arg L(2 + it, \chi) - \int_\sigma^2 \Im \frac{L'}{L}(\alpha + it, \chi) \, d\alpha.$$

Here $\arg L(2 + it, \chi) \ll 1$ uniformly in t, by Theorem 4.8. Thus by Lemma 12.6, the right-hand side above is

$$-\sum_{|\gamma - t| \le 1} \int_\sigma^2 \Im \frac{1}{\alpha + it - \rho} \, d\alpha + O(\log q\tau).$$

Here the summand is

$$\arctan \frac{\sigma - \beta}{t - \gamma} - \arctan \frac{2 - \beta}{t - \gamma}.$$

If $t > \gamma$, then this lies between $-\pi$ and 0, while if $t < \gamma$, then the above lies between 0 and π. Thus in any case the quantity is bounded, and by Theorem 10.17 the number of summands is $\ll \log \tau$, so we have the result when t is not the ordinate of a zero. Since the ordinates of zeros have no finite limit point, we obtain the same bound when t is the ordinate of a zero, since in that case $\arg L(s, \chi) = (\arg L(\sigma + it^+, \chi) + \arg L(\sigma - it^-, \chi))/2$. $\qquad\square$

Lemma 12.9 *Let χ be a primitive character modulo q with $q > 1$, put $\kappa = 0$ or 1 according as $\chi(-1) = 1$ or -1, and let $\mathcal{A}(\kappa)$ denote the set of points $s \in \mathbb{C}$ such that $\sigma \le -1$ and $|s + 2n - \kappa| \ge 1/4$ for each positive integer n. Then*

$$\frac{L'}{L}(s, \chi) \ll \log(2q|s|)$$

uniformly for $s \in \mathcal{A}(\kappa)$.

Proof By (10.35) and Theorem C.1 we see that

$$\frac{L'}{L}(s, \chi) = \frac{\pi}{2} \cot \frac{\pi}{2}(s + \kappa) + O(\log q) + O(\log(|s| + 2)).$$

Here

$$\cot \frac{\pi}{2}(s + \kappa) = i + \frac{2i}{e^{i\pi(s+\kappa)} - 1} \ll 1$$

since s is bounded away from integers with the parity of κ. $\qquad\square$

Theorem 12.10 *Let c be a constant, $c > 1$. Suppose that $x \ge c$, that $T \ge 2$, and that χ is a primitive character modulo q with $q > 1$. Then*

$$\psi_0(x, \chi) = -\sum_{\substack{\rho \\ |\gamma| \le T}} \frac{x^\rho}{\rho} - \frac{1}{2} \log(x - 1)$$

$$- \frac{\chi(-1)}{2} \log(x + 1) + C(\chi) + R(x, T; \chi) \qquad (12.6)$$

where

$$C(\chi) = \frac{L'}{L}(1, \overline{\chi}) + \log\frac{q}{2\pi} - C_0 \tag{12.7}$$

and

$$R(x, T; \chi) \ll (\log x)\min\left(1, \frac{x}{T\langle x\rangle}\right) + \frac{x}{T}(\log qxT)^2. \tag{12.8}$$

Here $\langle x\rangle$ denotes the distance from x to the nearest prime power, other than x itself.

Proof Put $\sigma_0 = 1 + 1/\log x$. By arguing as in the proof of Theorem 12.5, we see that

$$\psi_0(x, \chi) = \frac{-1}{2\pi i}\int_{\sigma_0 - iT_1}^{\sigma_0 + iT_1} \frac{L'}{L}(s, \chi)\frac{x^s}{s}\,ds + R_1$$

where

$$R_1 \ll (\log x)\min\left(1, \frac{x}{T\langle x\rangle}\right) + \frac{x}{T}(\log x)^2.$$

Let K be chosen so that $K - \kappa$ is an odd positive integer, and let $\mathcal{C}$ denote the contour consisting of the line segments connecting $\sigma_0 - iT_1$, $-K - iT_1$, $-K + iT_1$, $\sigma_0 + iT_1$ where T_1 is chosen as in Lemma 12.7. Since K and κ have opposite parity, the line segment from $-K - iT_1$ to $-K + iT_1$ lies in the region $\mathcal{A}(\kappa)$ of Lemma 12.9. Thus by Cauchy's residue theorem,

$$\psi_0(x, \chi) = -\sum_{\substack{\rho \\ |\gamma| < T_1}}\frac{x^\rho}{\rho} + \sum_{1 \le k < (K+\kappa)/2}\frac{x^{\kappa - 2k}}{2k - \kappa} + E + R_1 + R_2$$

where $\kappa = 0$ if $\chi(-1) = 1$ and $\kappa = 1$ if $\chi(-1) = -1$, E is the residue of

$$-\frac{L'}{L}(s, \chi)\frac{x^s}{s}$$

at $s = 0$, and

$$R_2 = \frac{-1}{2\pi i}\int_{\mathcal{C}}\frac{L'}{L}(s, \chi)\frac{x^s}{s}\,ds.$$

By proceeding as in the latter part of the proof of Theorem 12.5, but using now Lemma 12.7 and Lemma 12.9 in place of Lemma 12.2 and Lemma 12.4, we see that

$$R_2 \ll \frac{x}{T}(\log qT)^2 + \frac{T\log qK}{Kx^K}.$$

This last term tends to 0 as $K \to \infty$. Put

$$R_3 = -\sum_{\substack{\rho \\ T<|\gamma|<T_1}} \frac{x^\rho}{\rho}.$$

Then $R(x, T) = R_1 + R_2 + R_3$, and $R_3 \ll xT^{-1} \log qT$ by Theorem 10.17.

It remains to compute the residue E. By logarithmic differentiation of the functional equation in the asymmetric form of Corollary 10.9, we find that

$$\frac{L'}{L}(s, \chi) = -\frac{L'}{L}(1 - s, \overline{\chi}) - \log\frac{q}{2\pi} - \frac{\Gamma'}{\Gamma}(1 - s) + \frac{\pi}{2}\cot\frac{\pi}{2}(s + \kappa) \tag{12.9}$$

If $\chi(-1) = -1$, then $\frac{L'}{L}(s, \chi)$ is analytic at $s = 0$, so

$$E = -\frac{L'}{L}(0, \chi) = \frac{L'}{L}(1, \overline{\chi}) + \log\frac{q}{2\pi} - C_0,$$

in view of (C.11). Since $\cot z$ is an odd function, its Laurent expansion about $z = 0$ is of the form $\cot z = 1/z + \sum_{k=1}^{\infty} c_k z^{2k-1}$. Hence if $\chi(-1) = 1$, we see by (12.8) that the Laurent expansion of $\frac{L'}{L}(s, \chi)$ begins

$$\frac{L'}{L}(s, \chi) = \frac{1}{s} - \frac{L'}{L}(1, \overline{\chi}) - \log\frac{q}{2\pi} + C_0 + \cdots$$

Hence

$$E = -\log x + \frac{L'}{L}(1, \overline{\chi}) + \log\frac{q}{2\pi} - C_0$$

in this case.

Finally, we note that

$$\sum_{k=1}^{\infty} \frac{x^{-2k}}{2k} = -\frac{1}{2}\log(1 - x^{-2}), \qquad \sum_{k=1}^{\infty} \frac{x^{1-2k}}{2k-1} = \frac{1}{2}\log\frac{x+1}{x-1}.$$

This completes the proof. □

By letting $T \to \infty$ we immediately obtain

Corollary 12.11 *Suppose that χ is a primitive character modulo q, $q > 1$, and that $x > 1$. Then*

$$\psi_0(x, \chi) = -\sum_\rho \frac{x^\rho}{\rho} - \frac{1}{2}\log(x - 1) - \frac{\chi(-1)}{2}\log(x + 1) + C(\chi). \tag{12.10}$$

By Theorem 11.4 we see that $C(\chi) \ll \log q$ if $L(s, \chi)$ has no exceptional zero, and that

$$C(\chi) = \frac{1}{1 - \beta_1} + O(\log q)$$

if $L(s, \chi)$ has the exceptional zero β_1. In this latter case, the sum over ρ includes a large term due to $\rho = 1 - \beta_1$. This, however, is largely cancelled by $C(\chi)$, since

$$-\frac{x^{1-\beta_1} - 1}{1 - \beta_1} = -\frac{\log x}{1 - \beta_1} \int_0^{1-\beta_1} x^\sigma \, d\sigma \ll x^{1-\beta_1} \log x. \qquad (12.11)$$

This is quite small compared with the contribution $-x^{\beta_1}/\beta_1$ made by $\rho = \beta_1$, not to mention the contributions of other zeros with $\beta \geq 1/2$.

In principle, we could derive an explicit formula for $\psi_0(x, \chi)$ when χ is imprimitive, by taking into account the contributions made by zeros on the imaginary axis. However, we find it simpler to pass from $\psi_0(x, \chi^*)$ to $\psi_0(x, \chi)$ by elementary reasoning. Suppose that χ is a character modulo q induced by the primitive character χ^* modulo d, where $d | q$. (The possibility that $d = 1$ is not excluded here.) Then

$$\psi_0(x, \chi^*) - \psi_0(x, \chi) = \sum_{\substack{p|q \\ p\nmid d}} \sum_{\substack{k \\ 1 < p^k \leq x}} \chi^*\left(p^k\right) \log p$$

$$\ll \sum_{\substack{p|q \\ p\nmid d}} \left[\frac{\log x}{\log p}\right] \log p \qquad (12.12)$$

$$\leq \omega(q/d) \log x$$

$$\ll (\log q/d)(\log x).$$

Note that the distinction between $\psi_0(x, \chi)$ and $\psi(x, \chi)$ can be dropped at this point:

$$\psi(x, \chi) = \psi_0(x, \chi^*) + O((\log 2q)(\log x)). \qquad (12.13)$$

This estimate, though somewhat crude, suffices for most purposes.

The explicit formulæ that we have established thus far arise from Perron's formula. We may similarly derive other explicit formulæ using other kernels in the inverse Mellin transform. Examples of such formulæ are found in Exercises 12.1.5–10. In some cases it may not be so easy to apply complex variable techniques, but for such weighted sums over primes we may use the formulæ above, with integration by parts. For example, from Theorem 12.5 we see that

$$\sum_{n \leq x} w(n)\Lambda(n) = \int_{2-}^{x} w(u)d\psi(u)$$

$$= \int_2^x w(u) \, du - \sum_{\substack{\rho \\ |\gamma| \leq T}} \int_2^x w(u)u^{\rho-1} \, du + \text{smaller terms}.$$

To facilitate the estimation of these 'smaller terms' it is useful to record a little more information concerning the error terms in the truncated explicit formula.

Theorem 12.12 *Suppose that c is a constant, c > 1, and let χ be a character modulo q. For x ≥ c and T ≥ 2 there exist functions $E_1(x, \chi)$ and $E_2(x, T, \chi)$ with the following properties:*

$$\psi(x, \chi) = E_0(\chi)x - \sum_{\substack{\rho \\ |\gamma| \le T}} \frac{x^\rho}{\rho} + E_1(x, \chi) + E_2(x, T, \chi); \quad (12.14)$$

$$\int_c^x 1 \, |d E_1(u, \chi)| \ll (\log xq)^2; \quad (12.15)$$

$$E_2(x, T, \chi) \ll \log x + \frac{x}{T}(\log xTq)^2; \quad (12.16)$$

$$\int_c^x |E_2(u, T, \chi)| \, du \ll \frac{x^2}{T}(\log xTq)^2. \quad (12.17)$$

Proof Suppose first that χ is non-principal. Thus χ is induced by a primitive character χ^* (mod d) where $1 < d \le q$. Put

$$E_1(x, \chi) = \psi_0(x, \chi) - \psi_0(x, \chi^*) - \frac{1}{2}\log(x - 1)$$

$$- \frac{\chi(-1)}{2}\log(x + 1) + C(\chi^*), \quad (12.18)$$

$$E_2(x, T, \chi) = \psi(x, \chi) - \psi_0(x, \chi) + R(x, T; \chi^*) \quad (12.19)$$

where $R(x, T; \chi^*)$ is defined by taking $\chi = \chi^*$ in (12.6). Thus (12.6) gives (12.14). By (12.12) we see that

$$\int_c^x 1 \, |d(\psi_0(u, \chi) - \psi_0(u, \chi^*))| \ll \sum_{\substack{p|q \\ p \nmid d}} \left\lceil \frac{\log x}{\log p} \right\rceil \log p \ll (\log x)(\log q).$$

Thus we have (12.15). It is also clear that (12.8) gives (12.16). To obtain (12.17), we note that

$$\int_c^x \min\left(1, \frac{u}{T\langle u \rangle}\right) du \le \frac{x}{T} \sum_{p^k \le 2x} \left(1 + \int_{x/T}^x \frac{1}{u} du\right) \ll \frac{x^2 \log T}{T \log x}.$$

Since $\psi(x, \chi) - \psi_0(x, \chi) = 0$ except for jump discontinuities at the prime powers, this term makes no contribution to the integral (12.17). Thus we have (12.17).

Now suppose that χ is principal. Put

$$E_1(x, \chi_0) = \psi(x, \chi_0) - \psi_0(x) - \log 2\pi - \frac{1}{2}\log(1 - 1/x^2),$$

$$E_2(x, T, \chi_0) = \psi(x, \chi_0) - \psi_0(x, \chi_0) + R(x, T)$$

where $R(x, T)$ is defined by (12.3). Then the desired assertions follow from (12.3) and (12.4) in the same way as in the former case, so the proof is complete. □

12.1.1 Exercises

1. Suppose that $|s - 1| \geq 1$. Show that

$$\log \zeta(s) = \sum_{\substack{\rho \\ |\gamma - t| \leq 1}} \log(s - \rho) + O(\log \tau)$$

uniformly for $-1 \leq \sigma \leq 2$, where $\log \zeta(s)$ is defined by continuous variation along the ray from $\sigma + it$ to $\infty + it$, with $\log \zeta(\infty + it) = 0$, and $|\Im \log(s - \rho)| < \pi$.

2. (a) By using the Brun–Titchmarsh inequality, show that

$$\sum_{x+1 \leq n \leq 2x} \frac{\Lambda(n)}{n - x} \ll (\log x)(\log \log x).$$

   (b) Let R_1 be defined as in the proof of Theorem 12.5. Show that

$$R_1 \ll (\log x) \min \left(1, \frac{x}{T \langle x \rangle} \right) + \frac{x}{T} (\log x)(\log \log x).$$

3. Let δ be a small positive number. For a given $T \geq 4$, let $\mathcal{S} = \{t \in [T, T+1] : \min_\gamma |t - \gamma| \geq \delta / \log T\}$, and for $T \leq t \leq T + 1$ define

$$f(t) = \log T + \sum_{T-1 \leq \gamma \leq T+2} \frac{1}{|t - \gamma|}$$

   where the sum is over ordinates γ of zeros of the zeta function.
   (a) Show that if $T \leq t \leq T + 1$, then

$$\max_{-1 \leq \sigma \leq 2} \left| \frac{\zeta'}{\zeta}(s) \right| \ll f(t).$$

   (b) Show that meas $\mathcal{S} \asymp 1$ whenever δ is a sufficiently small positive constant.
   (c) Show that

$$\int_{\mathcal{S}} f(t) \, dt \ll (\log T) \log \log T.$$

   (d) Deduce that for every $T \geq 4$ there is a $T_1 \in [T, T+1]$ such that

$$\max_{-1 \leq \sigma \leq 2} \left| \frac{\zeta'}{\zeta}(\sigma + iT_1) \right| \ll (\log T) \log \log T.$$

4. Show that if $s \neq 1$, and $\zeta(s) \neq 0$, then

$$\sum_{n \leq x} \frac{\Lambda(n)}{n^s} = \frac{x^{1-s}}{1 - s} - \frac{\zeta'}{\zeta}(s) - \sum_\rho \frac{x^{\rho-s}}{\rho - s} + \sum_{k=1}^{\infty} \frac{x^{-2k-s}}{2k + s}$$

where it is understood that the term $n = x$ is counted with weight $1/2$ if x is a prime power, and the sum over ρ is calculated as $\lim_{T\to\infty}\sum_{|\gamma|\le T}$.

5. (cf. Ingham 1932, p. 81) By (12.1) we know that

$$\sum_\rho \frac{x^\rho}{\rho} = x - \psi_0(x) - \log 2\pi - \frac{1}{2}\log(1 - 1/x^2)$$

for $x > 1$. Show that if $0 < x < 1$, then

$$\sum_\rho \frac{x^\rho}{\rho} = \sum_{n\le 1/x} \frac{\Lambda(n)}{n} + \log x + C_0 + x + \frac{1}{2}\log\frac{1-x}{1+x}.$$

6. (de la Vallée Poussin 1896) Show that if $x > 1$, then

$$\sum_{n\le x}\Lambda(n)(x - n) = \frac{1}{2}x^2 - \sum_\rho \frac{x^{\rho+1}}{\rho(\rho + 1)} - (\log 2\pi)x + \frac{\zeta'}{\zeta}(-1)$$

$$-\sum_{k=1}^\infty \frac{x^{-2k+1}}{2k(2k - 1)}.$$

7. Show that if $x > 1$, then

$$\sum_{n\le x}\Lambda(n)\log x/n = x - \sum_\rho \frac{x^\rho}{\rho^2} - (\log 2\pi)\log x - \left(\frac{\zeta'}{\zeta}\right)'(0) - \frac{1}{4}\sum_{k=1}^\infty \frac{x^{-2k}}{k^2}.$$

8. (Hardy & Littlewood 1918; Wigert 1920) (a) Let k be a non-negative integer. Show that for s near $-k$, the Laurent expansion of $\Gamma(s)$ begins

$$\Gamma(s) = \frac{(-1)^k}{k!(s + k)} + \frac{(-1)^k}{k!}\frac{\Gamma'}{\Gamma}(k + 1) + \cdots .$$

(b) Let k be a positive integer. Show that for s near $-2k$, the Laurent expansion of $\frac{\zeta'}{\zeta}(s)$ begins

$$\frac{\zeta'}{\zeta}(s) = \frac{1}{s + 2k} - \frac{\zeta'}{\zeta}(2k + 1) + \log 2\pi - \frac{\Gamma'}{\Gamma}(2k + 1) + \cdots .$$

(c) Show that if $\Re z > 0$, then

$$\sum_{n=1}^\infty \Lambda(n)e^{-n/z} = z - \sum_\rho \Gamma(\rho)z^\rho - e^{-1/z}\log 2\pi + (-1 + \cosh 1/z)\log z$$

$$+ \sum_{k=1}^\infty (-1)^k \frac{\zeta'}{\zeta}(k + 1)\frac{z^{-k}}{k!} - \sum_{k=0}^\infty \frac{\Gamma'}{\Gamma}(2k + 2)\frac{z^{-2k-1}}{(2k + 1)!}.$$

9. Suppose that $a > 0$, that $x \geq 1$, and that x is not of the form $e^{2a^2 k}$ where k is a positive integer. Show that

$$\frac{1}{\sqrt{2\pi} \, a} \sum_{n=1}^{\infty} \Lambda(n) \exp\left(\frac{-(\log x/n)^2}{2a^2}\right)$$

$$= e^{a^2/2} x - \sum_{\rho} e^{a^2 \rho^2/2} x^{\rho} + \sum_{0 < k < \frac{\log x}{2a^2}} e^{2a^2 k^2} x^{-2k}$$

$$-\frac{1}{2\pi} \exp\left(\frac{-(\log x)^2}{2a^2}\right) \int_{-\infty}^{\infty} \frac{\zeta'}{\zeta}(-(\log x)/a^2 + it) e^{-a^2 t^2/2} \, dt.$$

12.2 Weil's explicit formula

In order to see better the relationship between a sum over zeros and a corresponding sum over primes, we now derive an explicit formula that applies to a general class of kernels. (The next theorem is not used later, and can be omitted on a first reading.)

Theorem 12.13 (Weil) *Let $F(x)$ be a measurable function such that*

$$\int_{-\infty}^{\infty} e^{(\frac{1}{2}+\delta_0)2\pi |x|} |F(x)| \, dx < \infty, \tag{12.20}$$

and

$$\int_{-\infty}^{\infty} e^{(\frac{1}{2}+\delta_0)2\pi |x|} |dF(x)| < \infty \tag{12.21}$$

where $\delta_0 > 0$ is fixed. Suppose that $F(x) = \frac{1}{2}(F(x^-) + F(x^+))$ for all x, and that $F(x) + F(-x) = 2F(0) + O(|x|)$. Put

$$\Phi(s) = \int_{-\infty}^{\infty} F(x) e^{-(s-1/2)2\pi x} \, dx$$

for $-\delta_0 < \sigma < 1 + \delta_0$. Let χ be a primitive character modulo q. Then

$$\lim_{T \to \infty} \sum_{|\gamma| \leq T} \Phi(\rho) = E_0(\chi)(\Phi(0) + \Phi(1)) + \frac{1}{2\pi}\left(\log q/\pi + \frac{\Gamma'}{\Gamma}(1/4 + \kappa/2)\right) F(0)$$

$$-\frac{1}{2\pi} \sum_{n=1}^{\infty} \frac{\Lambda(n)}{n^{1/2}}\left(\chi(n) F\left(\frac{-1}{2\pi} \log n\right) + \overline{\chi}(n) F\left(\frac{1}{2\pi} \log n\right)\right)$$

$$+ \int_0^{\infty} \frac{e^{-(1+2\kappa)\pi x}}{1 - e^{-4\pi x}} (2F(0) - F(x) - F(-x)) \, dx. \tag{12.22}$$

Here $E_0(\chi) = 1$ if $\chi = \chi_0$, $E_0(\chi) = 0$ otherwise, and $\kappa = 0$ if $\chi(-1) = 1$, $\kappa = 1$ if $\chi(-1) = -1$.

We note that if $\rho = 1/2 + i\gamma$, then

$$\Phi(\rho) = \int_{-\infty}^{\infty} F(x)e(-\gamma x)\,dx = \widehat{F}(\gamma).$$

The values of Γ'/Γ can be evaluated explicitly; from Appendix C we see that

$$\frac{\Gamma'}{\Gamma}(1/4) = -C_0 - 3\log 2 - \pi/2$$

and

$$\frac{\Gamma'}{\Gamma}(3/4) = -C_0 - 3\log 2 + \pi/2.$$

Here C_0 is Euler's constant. Since $\int |dfg| \le \int |f|\,|dg| + \int |g|\,|df|$, from (12.20) and (12.21) we see that $e^{a|x|}F(x)$ is of bounded variation for any a, $0 \le a \le (1/2 + \delta_0)2\pi$. Hence $F(x) \ll \exp(-(1/2 + \delta_0)2\pi|x|)$, and $\Phi(s)$ is analytic in the strip $-\delta_0 < \sigma < 1 + \delta_0$. For $|t| \le 1$ we note that $\phi(s) \ll 1$. For $|t| \ge 1$ we integrate by parts to see that

$$\Phi(s) = \frac{1}{2\pi it} \int_{-\infty}^{\infty} e(-tx)\,d\left(F(x)\exp((1-2\sigma)\pi x)\right);$$

hence $\Phi(s) \ll 1/(|t|+1)$ uniformly for $-\delta_0 \le \sigma \le 1 + \delta_0$. In these estimates, and in the proof below, implicit constants may depend on F and on δ_0.

Proof We note that

$$\sum_{|\gamma| \le T_1} \Phi(\rho) = \frac{1}{2\pi i} \int_C \Phi(s)\frac{\xi'}{\xi}(s,\chi)\,ds$$

where C is the closed polygonal contour with vertices $-\delta_1 + iT_1$, $-\delta_1 - iT_1$, $1 + \delta_1 - iT_1$, $1 + \delta_1 + iT_1$. Here $0 < \delta_1 < \delta_0$, and T_1 is chosen so that $|T - T_1| \le 1$, and so that

$$\frac{\xi'}{\xi}(\sigma \pm iT_1, \chi) \ll (\log qT)^2$$

uniformly for $-1 \le \sigma \le 2$. Thus

$$\sum_{|\gamma| \le T} \Phi(\rho) = \frac{1}{2\pi i}\left(\int_{1+\delta_1-iT}^{1+\delta_1+iT} + \int_{-\delta_1+iT}^{-\delta_1-iT}\right)\Phi(s)\frac{\xi'}{\xi}(s,\chi)\,ds + O\left(\frac{(\log T)^2}{T}\right).$$

By the functional equation for $\xi(s,\chi)$, we see that

$$\frac{\xi'}{\xi}(s,\chi) = -\frac{\xi'}{\xi}(1-s, \overline{\chi}).$$

Hence the integral above is

$$\frac{1}{2\pi i} \int_{1+\delta_1-iT}^{1+\delta_1+iT} \Phi(s)\frac{\xi'}{\xi}(s,\chi) + \Phi(1-s)\frac{\xi'}{\xi}(s,\overline{\chi})\,ds. \qquad (12.23)$$

From (10.25) and (10.33) we see that

$$\frac{\xi'}{\xi}(s,\chi) = E_0(\chi)\left(\frac{1}{s}+\frac{1}{s-1}\right) + \frac{1}{2}\log\frac{q}{\pi} + \frac{1}{2}\frac{\Gamma'}{\Gamma}((s+\kappa)/2) + \frac{L'}{L}(s,\chi). \qquad (12.24)$$

For $1 < \sigma < 1+\delta_0$,

$$\Phi(s)\frac{L'}{L}(s,\chi) = -\Phi(s)\sum_{n=1}^{\infty}\Lambda(n)\chi(n)n^{-s} \qquad (12.25)$$

$$= -\sum_{n=1}^{\infty}\Lambda(n)\chi(n)n^{-1/2}\int_{-\infty}^{\infty}F\left(x-\frac{1}{2\pi}\log n\right)e^{-(s-1/2)2\pi x}\,dx,$$

and similarly

$$\Phi(1-s)\frac{L'}{L}(s,\overline{\chi}) = -\sum_{n=1}^{\infty}\Lambda(n)\overline{\chi}(n)n^{-1/2}$$

$$\times \int_{-\infty}^{\infty}F\left(-x+\frac{1}{2\pi}\log n\right)e^{-(s-1/2)2\pi x}\,dx. \qquad (12.26)$$

From the estimate $F(x) \ll e^{-(1/2+\delta_0)2\pi|x|}$ we see that

$$\sum_{n}\Lambda(n)n^{-1/2}\int_{-\infty}^{\infty}\left|F\left(x-\tfrac{1}{2\pi}\log n\right)\right|\left|e^{-(1/2+\delta_1)2\pi x}\right|dx$$

$$\ll \sum_{n=1}^{\infty}\Lambda(n)n^{-1/2}\left(\int_{(\log n)/(2\pi)}^{\infty}e^{-(1+\delta_0+\delta_1)2\pi x}n^{1/2+\delta_0}\,dx\right.$$

$$\left. + \int_{-\infty}^{(\log n)/(2\pi)}e^{(\delta_0-\delta_1)2\pi x}n^{-1/2-\delta_0}\,dx\right)$$

$$\ll \sum_{n}\Lambda(n)n^{-1-\delta_1} \ll 1.$$

A similar calculation relates to the second term (12.26), and hence for $s = 1+\delta_1+it$,

$$\Phi(s)\frac{L'}{L}(s,\chi) + \Phi(1-s)\frac{L'}{L}(s,\overline{\chi}) = \int_{-\infty}^{\infty}H(x)e(-tx)\,dx = \widehat{H}(t)$$

where

$$H(x) = -\sum_{n=1}^{\infty} \frac{\Lambda(n)}{n^{1/2}} \left(\chi(n)F\left(x - \frac{\log n}{2\pi}\right) \right.$$
$$\left. + \overline{\chi}(n)F\left(-x + \frac{\log n}{2\pi}\right) \right) e^{-(1/2+\delta_1)2\pi x}.$$

Now $H(x)$ is of bounded variation, since

$$\text{Var} H \le \sum_n \frac{\Lambda(n)}{n^{1/2}} \text{Var}\left(F\left(x - \frac{\log n}{2\pi}\right) e^{-(1/2+\delta_1)2\pi x} \right)$$
$$+ \sum_n \frac{\Lambda(n)}{n^{1/2}} \text{Var}\left(F\left(-x + \frac{\log n}{2\pi}\right) e^{-(1/2+\delta_1)2\pi x} \right)$$
$$= 2\left(\sum_n \Lambda(n)n^{-1-\delta_1} \right) \text{Var}\left(F(x)e^{-(1/2+\delta_1)2\pi x} \right) \ll 1.$$

Moreover, $H(x) = (H(x^+) + H(x^-))/2$, and thus by the Fourier integral theorem,

$$\lim_{T\to\infty} \int_{-T}^{T} \widehat{H}(t)\,dt = H(0).$$

That is,

$$\lim_{T\to\infty} \frac{1}{2\pi i} \int_{1+\delta_1-iT}^{1+\delta_1+iT} \Phi(s)\frac{L'}{L}(s,\chi) + \Phi(1-s)\frac{L'}{L}(s,\overline{\chi})\,ds$$
$$= \frac{-1}{2\pi} \sum_n \frac{\Lambda(n)}{n^{1/2}} \left(\chi(n)F\left(\frac{-\log n}{2\pi}\right) + \overline{\chi}(n)F\left(\frac{\log n}{2\pi}\right) \right).$$

The remaining terms from (12.24) contribute to the integral (12.23) an amount

$$\frac{1}{2\pi i} \int_{1+\delta_1-iT}^{1+\delta_1+iT} G(s)\,ds.$$

where

$$G(s) = \left(E_0(\chi)\left(\frac{1}{s} + \frac{1}{s-1} \right) + \frac{1}{2}\log\frac{q}{\pi} + \frac{1}{2}\frac{\Gamma'}{\Gamma}\left(\frac{s+\kappa}{2} \right) \right)(\Phi(s) + \Phi(1-s))$$

By Cauchy's theorem this is

$$\frac{1}{2\pi i} \int_{1/2-iT}^{1/2+iT} G(s)\,ds + E_0(\chi)(\Phi(0) + \Phi(1)) + O\left(\frac{\log^2 qT}{T} \right).$$

To treat this latter integral we note that

$$\frac{1}{2\pi i} \int_{1/2-iT}^{1/2+iT} \left(\frac{1}{s} + \frac{1}{s-1} \right) (\Phi(s) + \Phi(1-s))\,ds$$

$$= \frac{-4i}{\pi} \int_{-T}^{T} \frac{t}{1+4t^2} \left(\Phi\left(\frac{1}{2}+it\right) + \Phi\left(\frac{1}{2}-it\right) \right) dt = 0.$$

Now $\Phi(1/2+it) = \widehat{F}(t)$, and hence

$$\frac{1}{2\pi i} \int_{1/2-iT}^{1/2+iT} \frac{1}{2}(\log q/\pi)(\Phi(s) + \Phi(1-s))\,ds$$

$$= \frac{\log q/\pi}{4\pi} \int_{-T}^{T} \widehat{F}(t) + \widehat{F}(-t)\,dt \longrightarrow \frac{F(0)}{2\pi} \log q/\pi$$

as T tends to infinity. Thus to complete the proof of the theorem it suffices to establish

Lemma 12.14   *Let $a > 0$ and $b > 0$ be fixed. If $J \in L^1(\mathbb{R})$, J is of bounded variation on $\mathbb{R}$, and if $J(x) = J(0) + O(|x|)$, then*

$$\lim_{T \to \infty} \int_{-T}^{T} \frac{\Gamma'}{\Gamma}(a \pm ibt)\widehat{J}(t)\,dt$$

$$= \frac{\Gamma'}{\Gamma}(a)J(0) + \frac{2\pi}{b} \int_{0}^{\infty} \frac{e^{-2\pi ax/b}}{1 - e^{-2\pi x/b}}(J(0) - J(\mp x))\,dx. \quad (12.27)$$

If G and J are in $L^1(\mathbb{R})$, then

$$\int_{-\infty}^{\infty} G(t)\widehat{J}(t)\,dt = \int_{-\infty}^{\infty} \widehat{G}(x)J(x)\,dx,$$

since both sides are

$$\int_{-\infty}^{\infty} \int_{-\infty}^{\infty} G(t)J(x)e(-tx)\,dx\,dt.$$

We cannot apply this with $G(t) = \frac{\Gamma'}{\Gamma}(a \pm ibt)$, since this function is not in $L^1(\mathbb{R})$. Nevertheless, the right-hand side of (12.27) is a linear functional of J, which thus serves as a surrogate for the Fourier transform of $\frac{\Gamma'}{\Gamma}(a \pm ibt)$, at least when the test function J is sufficiently well-behaved.

Proof   It suffices to consider the $+$ sign on the left-hand side of (12.27), for if $K(x) = J(-x)$ then $\widehat{K}(t) = \widehat{J}(-t)$. We suppose first that $J(0) = 0$. The integral with respect to t on the left-hand side of (12.27) is

$$\int_{-\infty}^{\infty} J(x) \left(\int_{-T}^{T} \frac{\Gamma'}{\Gamma}(a + ibt)e(-xt)\,dt \right) dx.$$

Since $\frac{\Gamma'}{\Gamma}(a+ibt) \ll \log(|t|+2)$, the inner integral above is $\ll T \log T$, uniformly in x. Put $\delta = T^{-2/3}$. The contribution to the above by those x for which $|x| \le \delta$ is

$$\ll \int_{-\delta}^{\delta} |x| T \log T \, dx \ll \delta^2 T \log T = T^{-1/3} \log T.$$

For $|x| \ge \delta$ we appeal to Theorem C.5 to estimate the inner integral. The error term in Theorem C.5 contributes an amount

$$\ll \int_{\delta}^{\infty} \min(x,1) T^{-1} x^{-2} \, dx \ll T^{-1} \log T.$$

By integrating by parts we see that

$$\int_{\delta}^{\infty} J(x)\frac{e(-xT)}{x} \, dx = \frac{J(\delta)e(-\delta T)}{2\pi i \delta T} - \frac{1}{2\pi i T}\int_{\delta}^{\infty} J(x)\frac{e(-xT)}{x^2}\,dx$$

$$+ \frac{1}{2\pi i T}\int_{\delta}^{\infty}\frac{e(-xT)}{x}\,dJ(x)$$

$$\ll \frac{1}{T} + \frac{1}{T}\int_{\delta}^{\infty}\min(x,1)x^{-2}\,dx + \frac{1}{\delta T}\int_{\delta}^{\infty}|dJ|$$

$$\ll T^{-1/3},$$

and similarly for the three related terms. Hence

$$\int_{-T}^{T}\frac{\Gamma'}{\Gamma}(a+ibt)\widehat{J}(t)\,dt = \frac{-2\pi}{b}\int_{-\infty}^{-\delta}\frac{e^{2\pi ax/b}}{1-e^{2\pi x/b}}J(x)\,dx + O\left(T^{-1/3}\log T\right).$$

On the right-hand side we see that $\int_{-\delta}^{0}\cdots \ll \delta$, so that

$$\lim_{T\to\infty}\int_{-T}^{T}\frac{\Gamma'}{\Gamma}(a+ibt)\widehat{J}(t)\,dt = \frac{-2\pi}{b}\int_{0}^{\infty}\frac{e^{-2\pi ax/b}}{1-e^{-2\pi x/b}}J(-x)\,dx$$

provided that $J(0)=0$. To obtain the general case we apply the above to the function $K(x) = J(x) - J(0)e^{-\pi x^2/A}$ where $A>0$ is large. Then $\widehat{K}(t) = \widehat{J}(t) - J(0)\sqrt{A}e^{-\pi At^2}$, and hence

$$\lim_{T\to\infty}\int_{-T}^{T}\frac{\Gamma'}{\Gamma}(a+ibt)\widehat{K}(t)\,dt = \lim_{T\to\infty}\int_{-T}^{T}\frac{\Gamma'}{\Gamma}(a+ibt)\widehat{J}(t)\,dt$$

$$- J(0)\sqrt{A}\int_{-\infty}^{\infty}\frac{\Gamma'}{\Gamma}(a+ibt)e^{-\pi At^2}\,dt.$$

This last integral is

$$\int_{-\infty}^{\infty}\left(\frac{\Gamma'}{\Gamma}(a)+O(|t|)\right)e^{-\pi At^2}\,dt = \frac{\Gamma'}{\Gamma}(a)A^{-1/2}+O(A^{-1}).$$

On the other hand,

$$-2\pi \int_0^\infty \frac{e^{-2\pi ax/b}}{1 - e^{-2\pi x/b}} K(-x)\,dx$$

$$= 2\pi \int_0^\infty \frac{e^{-2\pi ax/b}}{1 - e^{-2\pi x/b}} (J(0) - J(-x))\,dx$$

$$+ 2\pi J(0) \int_0^\infty \frac{e^{-2\pi ax/b}}{1 - e^{-2\pi x/b}} \left(e^{-\pi x^2/A} - 1\right)dx.$$

Now $e^{-\alpha} = 1 + O(\alpha)$ for $\alpha \geq 0$, and hence this last integral is

$$\ll \int_0^1 x A^{-1}\,dx + \int_1^\infty e^{-2\pi ax/b} x^2 A^{-1}\,dx \ll A^{-1}.$$

On combining these estimates, we see that (12.29) holds apart from an error term $O(A^{-1/2})$, and we obtain the result since A can be arbitrarily large. □

12.3 Notes

Section 12.1. Let $\Pi(x) = \sum_{n \leq x} \Lambda(n)/\log n$. Riemann (1859) gave a heuristic proof that if $x > 1$, and x is not a prime power, then

$$\Pi(x) = \mathrm{Li}(x) - \sum_\rho \mathrm{Li}(x^\rho) - \log 2 + \int_x^\infty \frac{du}{(u^2 - 1)u \log u}.$$

Here the sum over the zeros is conditionally convergent, and it is to be understood that it is computed as the limit, as $T \to \infty$, of the sum over those zeros for which $|\gamma| \leq T$. The above formula was first proved rigorously by von Mangoldt (1895), and additional proofs were subsequently given by Landau (1908a, b). For further discussion of the explicit formula in the form given by Riemann, see Edwards (1974, Chapter 1). von Mangoldt (1895) also proved the explicit formula (12.1). Landau (1909, Section 89) was the first to show that the limit in (12.1) is attained uniformly for x in a compact interval not containing a prime power. Cramér (1918) showed that (12.1) can be derived from the above. von Koch (1910) and Landau (1912) estimated the error term that arises when the explicit formula is truncated, as in Theorem 12.5. The explicit formula for $\psi_0(x, \chi)$ was first established by Landau (1908b), but with not so much attention to the constant term. In the customary form of this explicit formula (cf. Davenport (2000, p. 117)), the constant term is expressed in terms of the constant $B(\chi)$ that arises in the Hadamard product formula for $\xi(s, \chi)$. Our presentation, which avoids this, is that of Vorhauer (2006).

Section 12.2. Although many specific explicit formulæ were derived by various authors for a variety of purposes, it was Guinand (1942) who first suggested that it would be possible to specify a general class of such formulæ. Guinand (1948) did this assuming the Riemann Hypothesis, but it seems that he imposed RH only in order to obtain a wider class of test functions. Theorem 12.13 is a special case of the main result of Weil (1952), who treats general L-functions associated with *Grössencharaktere* χ, which are representations of the group of idèle-classes of an algebraic number field k into the multiplicative group of non-zero complex numbers. Weil also showed that a necessary and sufficient condition for the Riemann hypothesis to hold for L is that the right-hand side corresponding to (12.22) is non-negative for all functions F of a certain class. Gallagher (1987) widened the class of test functions in Guinand's formula and gave several applications. See also Besenfelder (1977a, b), Yoshida (1982), Jorgenson, Lang & Goldfeld (1994), and Bombieri & Lagarias (1999).

12.4 References

Barner, K. (1981). On A. Weil's explicit formula, *J. Reine Angew. Math.* **323**, 139–152.
Besenfelder, H.-J. (1977a). Die Weilsche "Explizite Formel" und temperierte Distributionen, *J. Reine Angew. Math.* **293–294**, 228–257.
(1977b). Zur Nullstellenfreiheit der Riemannschen Zeta-funktion auf der Geraden $\sigma = 1$, *J. Reine Angew. Math.* **295**, 116–119.
Besenfelder, H.-J. & Palm, G. (1997). Einige Äquivalenzen zur Riemannschen Vermutung, *J. Reine Angew. Math.* **293–294**, 109–115.
Bombieri, E. & Lagarias, J. C. (1999). Complements to Li's criterion for the Riemann hypothesis, *J. Number Theory* **77**, 274–287.
Cramér, H. (1918). Über die Herleitung der Riemannschen Primzahlformel, *Arkiv för Mat. Astr. Fys.* **13**, no. 24, 7 pp.
Davenport, H. (2000). *Multiplicative Number Theory*, Third Edition, Graduate Texts Math. 74. New York: Springer-Verlag.
Edwards, H. M. (1974). *Riemann's Zeta Function*, Pure and Applied Math. 58. New York: Academic Press.
Gallagher, P. X. (1987). Applications of Guinand's formula, *Analytic number theory and Diophantine problems* (Stillwater, 1984), Progress in Math. 70. Boston: Birkhäusen, pp. 135–157.
Guinand, A. P. (1937). A class of self-reciprocal functions connected with summation formulæ, *Proc. London Math. Soc.* (2) **43**, 439–448.
(1938). Summation formulæ and self-reciprocal functions, *Quart. J. Math. Oxford Ser.* **9**, 53–67.
(1939a). Finite summation formulæ, *Quart. J. Math.* **10**, 38–44.
(1939b). Summation formulæ and self-reciprocal functions (II), *Quart. J. Math.* **10**, 104–118.

(1939c). A formula for $\zeta(s)$ in the critical strip, *J. London Math. Soc.* **14**, 97–100.

(1941). On Poisson's summation formula, *Ann. of Math.* (2) **42**, 591–603.

(1942). Summation formulæ and self-reciprocal functions (III), *Quart. J. Math.* **13**, 30–39.

(1948). A summation formula in the theory of prime numbers, *Proc. London Math. Soc.* **50**, 107–119.

Hardy, G. H. & Littlewood, J. E. (1918). Contributions to the theory of the Riemann zeta-function and the theory of the distribution of primes, *Acta Math.* **41**, 119–196; *Collected Papers,* Vol. 2. Oxford: Clarendon Press, 1967, pp. 20–97.

Ingham, A. E. (1932). *The Distribution of Prime Numbers*, Cambridge Tract No. 30. Cambridge: Cambridge University Press.

Jorgenson, J., Lang, S., & Goldfeld, D. (1994). *Explicit Formulas*. Lecture Notes in Math. 1593. Berlin: Springer-Verlag.

von Koch, H. (1910). Contributions à la théorie des nombres premiers, *Acta Math.* **33**, 293–320.

Landau, E. (1908a). Neuer Beweis der Riemannschen Primzahlformel, *Sitzungsber. Königl. Preuß. Akad. Wiss. Berlin*, 737–745; *Collected Works*, Vol. 4, Essen: Thales Verlag, 1986, pp. 11–19.

(1908b). Nouvelle démonstration pour la formule de Riemann sur le nombre des nombres premiers inférieurs à une limite donnée, et démonstration d'une formule plus générale pour le cas des nombres premiers d'une progression arithmétique, *Ann. l'École Norm. Sup.* (3) **25**, 399–442; *Collected Works*, Vol. 4, Essen: Thales Verlag, 1986, pp. 87–130.

(1909). *Handbuch der Lehre von der Verleilung der Primzahlen*. Leipzig: Teubner. Reprint: New York: Chelsea, 1953.

(1912). Über einige Summen, die von den Nullstellen der Riemannschen Zetafunktion abhängen, *Acta Math.* **35**, 271–294; *Collected Works*, Vol. 5. Essen: Thales Verlag, 1986, pp. 62–85.

von Mangoldt, H. (1895). Zu Riemann's Abhandlung "Ueber die Anzahl der Primzahlen unter einer gegebenen Grösse", *J. Reine Angew. Math.* **114**, 255–305.

Riemann, B. (1859). Ueber die Anzahl der Primzahlen unter einer gegebenen Grösse, *Monatsber. Kgl. Preuss. Akad. Wiss. Berlin*, 671–680; *Werke*, Leipzig: Teubner, 1876, pp. 3–47. Reprint: New York: Dover, 1953.

de la Vallée Poussin, C. J. (1896). Recherches analytiques sur la théorie des nombres premiers, I–III, *Ann. Soc. Sci. Bruxelles* **20**, 183–256, 281–362, 363–397.

Vorhauer, U. M. A. (2006). *The Hadamard product formula for Dirichlet L-functions*, to appear.

Weil, A. (1952). Sur les "formules explicites" de la théorie des nombres premiers, *Comm. Sém. Math. Univ. Lund [Medd. Lunds Univ. Mat. Sem.]*, Tome Supplementaire, 252–265.

Wigert, S. (1920). Sur la théorie de la fonction $\zeta(s)$ de Riemann, *Ark. Mat.* **14**, 1–17.

Yoshida H. (1992). On Hermitian forms attached to zeta functions, *Zeta functions in geometry* (Tokyo, 1990), Adv. Stud. Pure Math. 21. Tokyo: Kinokuniya , 281–325.

13

Conditional estimates

13.1 Estimates for primes

From the explicit formula for $\psi_0(x)$ we see that the contribution to the error term $\psi_0(x) - x$ made by a typical zero $\rho = \beta + i\gamma$ is $-x^\rho/\rho$. This has absolute value $\asymp x^\beta/|\gamma|$, which diminishes as $|\gamma|$ increases, but it depends much more sensitively on the value of β. We recall that if ρ is a zero, then so also is $1 - \rho$. Since at least one of these has real part $\geq 1/2$, we see that the Riemann Hypothesis represents the best of all possible worlds, in the sense that the error term in the Prime Number Theorem is smallest when the Riemann Hypothesis is true. By Theorem 10.13 we find that

$$\sum_{\substack{\rho \\ |\gamma| \leq T}} \frac{1}{|\rho|} \ll \sum_{1 \leq n \leq T} \frac{\log 2n}{n} \ll (\log T)^2. \tag{13.1}$$

Thus by taking $T = x$ in Theorem 12.5, we obtain

Theorem 13.1 *Assume RH. Then for $x \geq 2$,*

$$\psi(x) = x + O\left(x^{1/2}(\log x)^2\right), \tag{13.2}$$

$$\vartheta(x) = x + O\left(x^{1/2}(\log x)^2\right), \tag{13.3}$$

$$\pi(x) = \mathrm{li}(x) + O\left(x^{1/2}\log x\right). \tag{13.4}$$

In Chapter 15 we shall show that these estimates for the error term are within a factor $(\log x)^2$ of being best possible, which is not surprising since each zero individually contributes an amount of the order $x^{1/2}$.

Proof The second assertion follows from the first by Corollary 2.5. By integration by parts we find that

$$\pi(x) = \int_2^x \frac{1}{\log u}\, du + \frac{\vartheta(x) - x}{\log x} + \frac{2}{\log 2} + \int_2^x \frac{\vartheta(u) - u}{u(\log u)^2}\, du, \tag{13.5}$$

and so the third assertion follows from the second. $\qquad \square$

419

The factor $(\log x)^2$ in (13.2) can be avoided if we take smoother weights. For example, put

$$\psi_1(x) = \sum_{n \le x} (x - n)\Lambda(n). \qquad (13.6)$$

Then we have the explicit formula

$$\psi_1(x) = \frac{x^2}{2} - \sum_{\rho} \frac{x^{\rho+1}}{\rho(\rho+1)} - \frac{\zeta'}{\zeta}(0)x + \frac{\zeta'}{\zeta}(-1) + O\left(x^{-1/2}\right) \qquad (13.7)$$

for $x \ge 2$. Assuming RH, it follows easily that

$$\psi_1(x) = \frac{1}{2}x^2 + O\left(x^{3/2}\right). \qquad (13.8)$$

Assuming RH, we can also describe more precisely the relationships between the three standard prime-counting functions $\psi(x)$, $\vartheta(x)$, and $\pi(x)$.

Theorem 13.2 *Assume RH. Then*

$$\vartheta(x) = \psi(x) - x^{1/2} + O\left(x^{1/3}\right), \qquad (13.9)$$

and

$$\pi(x) - \operatorname{li}(x) = \frac{\vartheta(x) - x}{\log x} + O\left(\frac{x^{1/2}}{(\log x)^2}\right). \qquad (13.10)$$

Proof By an easy elaboration on Corollary 2.5, we see that

$$\vartheta(x) = \psi(x) - \psi\left(x^{1/2}\right) + O\left(x^{1/3}\right).$$

Hence (13.9) follows immediately from (13.2). To obtain (13.10), put

$$\vartheta_1(x) = \sum_{p \le x}(x - p)\log p = \int_2^x \vartheta(u)\,du.$$

By (13.8) and (13.9) it follows that $\vartheta_1(x) = x^2/2 + O\left(x^{3/2}\right)$. By integration by parts we see that the final integral in (13.5) is

$$\left[\frac{\vartheta_1(u) - u^2/2}{u(\log u)^2}\right]_2^x + \int_2^x \frac{\vartheta_1(u) - u^2/2}{(u \log u)^2}(1 + 2/\log u)\,du$$

$$\ll \frac{x^{1/2}}{(\log x)^2} + \int_2^x u^{-1/2}(\log u)^{-2}\,du$$

$$\ll \frac{x^{1/2}}{(\log x)^2}.$$

Thus (13.10) follows from (13.5). $\qquad\qquad \square$

As for primes in short gaps, we see from (13.4) that

$$\pi(x+h) - \pi(x) = \int_x^{x+h} \frac{1}{\log u}\, du + O\left(x^{1/2}\log x\right).$$

Here the main term on the right is larger than the error term if $h \geq Cx^{1/2}(\log x)^2$. We can do slightly better than this by counting primes between x and $x+h$ with a smoother weight.

Theorem 13.3 (Cramér) *There is a constant $C > 0$ such that if the Riemann Hypothesis is true, then for every $x \geq 2$ the interval $(x, x + Cx^{1/2}\log x)$ contains at least $x^{1/2}$ prime numbers.*

Proof Let h be a parameter to be determined, and put $w(u) = 1 - |u - x|/h$ when $|u - x| \leq h$, and $w(u) = 0$ otherwise. Then by three applications of (13.7) we see that

$$\sum_n \Lambda(n)w(n) = \frac{1}{h}(\psi_1(x+h) - 2\psi_1(x) + \psi_1(x-h))$$

$$= h - \frac{1}{h}\sum_\rho \frac{(x+h)^{\rho+1} - 2x^{\rho+1} + (x-h)^{\rho+1}}{\rho(\rho+1)} + O\left(\frac{1}{hx}\right).$$

$$(13.11)$$

Assuming RH, we note that the summand here is obviously

$$\ll \frac{x^{3/2}}{\gamma^2}. \qquad (13.12)$$

Moreover, if $\gamma > x/h$, then the three terms in the numerator may have quite different arguments, in which case the above estimate is the best that we can assert in general. On the other hand, if γ is smaller, then some cancellation must occur in the numerator. To see this, note that the summand may be written

$$\int_{x-h}^{x+h} (h - |x - u|)u^{\rho-1}\, du \ll h^2 x^{-1/2} \qquad (13.13)$$

assuming RH. This improves on (13.12) when $|\gamma| < x/h$. We use this estimate for the size of the summand together with Theorem 10.13 to see that the sum in (13.11) is $\ll hx^{1/2}\log x/h$. Hence if $h = Cx^{1/2}\log x$, then

$$\sum_{x-h<n<x+h} \Lambda(n) \geq \frac{h}{2}.$$

To complete the proof it remains to estimate the contribution made by higher powers of primes on the left-hand side. The number of squares in this interval is $\ll \log x$, so the squares of the primes contribute an amount that is $\ll (\log x)^2$. For each $k > 2$ there is at most one k^{th} power in the interval. Moreover, if p^k is in the interval, then $k \ll \log x$. Hence the higher powers contribute an amount $\ll (\log x)^2$, and the proof is complete.   □

Although Cramér's theorem is highly non-trivial, and is significantly stronger than anything that we know how to prove unconditionally, it is nevertheless disappointing that it falls so far short of what we conjecture to be true, namely that for every $\varepsilon > 0$ the interval $[x, x + x^\varepsilon]$ contains a prime, for all $x > x_0(\varepsilon)$. In order to understand the weakness in our approach, write

$$\psi(x + h) - \psi(x) - h = -\sum_\rho \frac{(x + h)^\rho - x^\rho}{\rho} + \cdots. \qquad (13.14)$$

The contribution of zeros with $|\gamma| > x/h$ can be attenuated by employing a smoother weight, but no amount of smoothing will eliminate the smaller zeros. However, if $|\gamma| \le x/h$ then the argument of $(x + h)^\rho$ is near that of x^ρ, so there is some significant cancellation in the numerators above. Indeed,

$$\frac{(x + h)^\rho - x^\rho}{\rho} = \int_x^{x+h} u^{\rho-1} \, du \ll hx^{-1/2}$$

if $0 \le h \le x$ and $\beta = 1/2$. Taking this a step further, we see that the above is

$$= hx^{\rho-1} + O(h^2 |\gamma| x^{\beta-2}).$$

Thus the left-hand side of (13.14) bears a passing resemblance to

$$-hx^{-1/2} \sum_{|\gamma| \le x/h} x^{i\gamma}, \qquad (13.15)$$

if we assume RH. Here the sum has $\asymp xh^{-1} \log x/h$ terms, and with sums of independent random variables in mind, we might guess that the above sum is $\ll (x/h)^{1/2+\varepsilon}$, which suggests

Conjecture 13.4 *If $2 \le h \le x$, then*

$$\psi(x + h) - \psi(x) = h + O_\varepsilon\left(h^{1/2} x^\varepsilon\right).$$

Although we expect there to be considerable cancellation in (13.15), any such cancellation that might occur among the contributions of the zeros is discarded in the proof of Theorem 13.3. Thus it seems that if we are to argue through zeta zeros to obtain an improvement of Theorem 13.3, then we need not just RH but also some deeper information concerning the distribution of the γ — more precisely that the numbers $\gamma \log x$ are approximately uniformly distributed modulo 2π. Although we cannot demonstrate that the desired cancellation occurs for all x, we can show that there is considerable cancellation in mean square.

Theorem 13.5 *Assume RH. Then for $X \geq 2$,*

$$\int_X^{2X} (\psi(x) - x)^2\, dx \ll X^2.$$

Note that if we were to use the pointwise bound of Theorem 13.1 to bound the left-hand side above, then we would obtain an estimate that is larger than the above by a factor $(\log X)^4$. From the above we see that $\psi(x) = x + O(x^{1/2})$ on average.

Proof Take $T = X$ in the explicit formula of Theorem 12.5. Then

$$\psi(x) = x - \sum_{|\gamma| \leq X} \frac{x^\rho}{\rho} + R(x)$$

where

$$\int_X^{2X} R(x)^2\, dx \ll X(\log X)^4 + \sum_{X/2 < p^k < 3X} \left(\log p^k\right)^2 \left(1 + \int_1^\infty u^{-2}\, du\right)$$

$$\ll X(\log X)^4.$$

On the other hand, the sum over zeros contributes

$$\int_X^{2X} \left| \sum_{|\gamma| \leq X} \frac{x^\rho}{\rho} \right|^2 dx = \sum_{\substack{\gamma_1, \gamma_2 \\ |\gamma_i| \leq X}} \frac{1}{\rho_1 \overline{\rho_2}} \int_X^{2X} x^{1 + i(\gamma_1 - \gamma_2)}\, dx$$

$$\ll X^2 \sum_{\gamma_1, \gamma_2} \frac{1}{|\rho_1 \rho_2|\, |2 + i(\gamma_1 - \gamma_2)|}.$$

To complete the proof it suffices to show that

$$\sum_{\gamma_1, \gamma_2} \frac{1}{|\gamma_1 \gamma_2|(1 + |\gamma_1 - \gamma_2|)} < \infty. \tag{13.16}$$

In view of the symmetry of zeros about the real axis, we may confine our attention to $\gamma_1 > 0$. For each such zero, we consider γ_2 in various ranges. By Theorem 10.13, the sum over $\gamma_2 < -\gamma_1$ is

$$\sum_{\substack{\gamma_2 \\ \gamma_2 < -\gamma_1}} \frac{1}{|\gamma_2|(1 + |\gamma_1 - \gamma_2|)} \ll \sum_{\substack{\gamma_2 \\ \gamma_2 < -\gamma_1}} \frac{1}{\gamma_2^2} \ll \sum_{n > \gamma_1} \frac{\log n}{n^2} \ll \frac{\log \gamma_1}{\gamma_1}.$$

Similarly, the sum over those γ_2 for which $|\gamma_2| \leq \frac{1}{2}\gamma_1$ is

$$\ll \frac{1}{\gamma_1} \sum_{\substack{\gamma_2 \\ 0 < \gamma_2 \leq \gamma_1}} \frac{1}{\gamma_2} \ll \frac{1}{\gamma_1} \sum_{1 \leq n \leq \gamma_1} \frac{\log n}{n} \ll \frac{(\log \gamma_1)^2}{\gamma_1}.$$

The sum over those γ_2 for which $\frac{1}{2}\gamma_1 < \gamma_2 < \frac{3}{2}\gamma_1$ is

$$\ll \frac{1}{\gamma_1} \sum_{\substack{\gamma_2 \\ |\gamma_2 - \gamma_1| \leq \gamma_1/2}} \frac{1}{1 + |\gamma_1 - \gamma_2|} \ll \frac{\log \gamma_1}{\gamma_1} \sum_{1 \leq n \leq \gamma_1} \frac{1}{n} \ll \frac{(\log \gamma_1)^2}{\gamma_1},$$

and finally the sum over $\gamma_2 \geq \frac{3}{2}\gamma_1$ is

$$\ll \sum_{\substack{\gamma_2 \\ \gamma_2 \geq \frac{3}{2}\gamma_1}} \frac{1}{\gamma_2^2} \ll \sum_{n > \gamma_1} \frac{\log n}{n^2} \ll \frac{\log \gamma_1}{\gamma_1}.$$

We sum these estimates, multiply by $1/\gamma_1$, and sum over γ_1 to see that the expression (13.16) is

$$\ll \sum_{\gamma_1 > 0} \frac{(\log \gamma_1)^2}{\gamma_1^2} \ll \sum_{n=1}^{\infty} \frac{(\log n)^3}{n^2} < \infty.$$

This completes the proof. $\qquad\qquad\qquad\qquad\qquad\qquad\qquad\qquad\qquad\qquad\square$

The oscillations of $x^{i\gamma} = e^{i\gamma \log x}$ become slower as x increases, since $\frac{d}{dx} \log x = 1/x \to 0$ as $x \to \infty$. However, with the change of variable $x = e^u$ we have $x^{i\gamma} = e^{i\gamma u}$, which is a periodic function of u. Put

$$f(u) = \frac{\psi(e^u) - e^u}{e^{u/2}}. \tag{13.17}$$

Assuming RH, the explicit formula of Theorem 12.5 gives

$$f(u) = -\sum_{\rho} \frac{e^{i\gamma u}}{\rho} + o(1)$$

as $u \to \infty$. This provides a kind of Fourier expansion of $f(u)$. Since

$$\int_U^{U+1} |f(u)|^2 \, du = \int_{e^U}^{e^{U+1}} (\psi(x) - x)^2 \frac{dx}{x^2} \asymp e^{-2U} \int_{e^U}^{e^{U+1}} (\psi(x) - x)^2 \, dx,$$

Theorem 13.5 is equivalent (assuming RH) to the estimate

$$\int_U^{U+1} |f(u)|^2 \, du \ll 1. \tag{13.18}$$

By averaging $|f(u)|^2$ over a longer interval we obtain not just an upper bound, but an asymptotic formula.

Theorem 13.6 *Assume RH, and let $f(u)$ be defined as in (13.17). Then*

$$\lim_{U \to \infty} \frac{1}{U} \int_0^U |f(u)|^2 \, du = \sum_{\text{distinct } \gamma} \frac{m_\rho^2}{|\rho|^2}$$

where m_ρ denotes the multiplicity of the zero ρ.

Proof Since the explicit formula for $\psi_0(x)$ is uniformly convergent in intervals free of prime powers, and is boundedly convergent in a neighbourhood of a prime power, it follows that

$$\frac{1}{U} \int_1^U |f(u)|^2 \, du$$

$$= \lim_{T \to \infty} \sum_{\substack{\gamma_1, \gamma_2 \\ |\gamma_i| \leq T}} \frac{1}{\rho_1 \overline{\rho_2} U} \int_1^U e^{i(\gamma_1 - \gamma_2)u} \, du \, + o(1)$$

$$= \left(1 - \frac{1}{U}\right) \sum_{\substack{\gamma_1, \gamma_2 \\ \gamma_1 = \gamma_2}} \frac{1}{|\rho_1|^2} + O\left(\sum_{\substack{\gamma_1, \gamma_2 \\ \gamma_1 \neq \gamma_2}} \frac{1}{|\gamma_1 \gamma_2|} \min\left(1, \frac{1}{U|\gamma_1 - \gamma_2|}\right)\right) + o(1).$$

Here the sum over $\gamma_1 \neq \gamma_2$ is finite already when $U = 1$, in view of (13.16). Since each term in this sum tends to 0 as $U \to \infty$, it follows that

$$\lim_{U \to \infty} \frac{1}{U} \int_1^U |f(u)|^2 \, du = \sum_{\substack{\gamma_1, \gamma_2 \\ \gamma_1 = \gamma_2}} \frac{1}{|\rho_1|^2}.$$

Suppose that $\rho = 1/2 + i\gamma$ is a zero, and that its multiplicity is m_ρ. Then the equation $\gamma_i = \gamma$ has m_ρ solutions for $i = 1$ and for $i = 2$. Thus there are m_ρ^2 pairs (γ_1, γ_2) such that $\gamma_1 = \gamma_2 = \gamma$, so we have the result. $\qquad\square$

We now return to the distribution of primes in arithmetic progressions.

Theorem 13.7 *Let q be given, and suppose that GRH holds for all L-functions modulo q. Then for $x \geq 2$,*

$$\psi(x, \chi) = E_0(\chi)x + O\left(x^{1/2}(\log x)(\log qx)\right), \tag{13.19}$$

$$\vartheta(x, \chi) = E_0(\chi)x + O\left(x^{1/2}(\log x)(\log qx)\right), \tag{13.20}$$

$$\pi(x, \chi) = E_0(\chi)\text{li}(x) + O\left(x^{1/2} \log qx\right) \tag{13.21}$$

where $E_0(\chi) = 1$ or 0 according as $\chi = \chi_0$ or not.

Proof For χ_0 these relations follow from Theorem 1 and (12.14). Suppose that χ is non-principal, and that χ^* is a primitive character that induces χ. Thus χ^* is a character modulo d for some $d|q$, $1 < d \leq q$. By taking $T = x$ in the explicit formula for $\psi(x, \chi^*)$, and appealing to Theorem 10.17, we see that

$$\psi(x, \chi^*) \ll x^{1/2}(\log qx)(\log x),$$

and then by (12.14) we have (13.19). By the triangle inequality, $|\psi(x, \chi) - \vartheta(x, \chi)| \leq \psi(x) - \vartheta(x)$. From Corollary 2.5 we know that this latter quantity is $\ll x^{1/2}$, so (13.20) follows from (13.19). On inserting (13.20) into the identity

$$\pi(x, \chi) = \frac{\vartheta(x, \chi)}{\log x} + \int_2^x \frac{\vartheta(u, \chi)}{u(\log u)^2} \, du,$$

we obtain (13.21). $\qquad\square$

Corollary 13.8 *Let q be given, and assume GRH for all L-functions modulo q. Suppose that $(a, q) = 1$. Then for $x \geq 2$,*

$$\psi(x; q, a) = \frac{x}{\varphi(q)} + O\big(x^{1/2}(\log x)^2\big), \qquad (13.22)$$

$$\vartheta(x; q, a) = \frac{x}{\varphi(q)} + O\big(x^{1/2}(\log x)^2\big), \qquad (13.23)$$

$$\pi(x; q, a) = \frac{\text{li}(x)}{\varphi(q)} + O\big(x^{1/2} \log x\big). \qquad (13.24)$$

Note that trivially,

$$0 \leq \psi(x; q, a) \leq (\log x) \sum_{\substack{0 < n \leq x \\ n \equiv a\,(q)}} 1 \leq (\log x)(1 + x/q).$$

Thus we see that the bound (13.22) is worse than trivial if $q > x^{1/2}$. However, if q is smaller, say $q \leq x^\theta$ with $\theta < 1/2$, then (13.22) provides a form of the Prime Number Theorem for arithmetic progressions with a much better error term than we were able to prove unconditionally (cf. Corollary 11.17).

Proof In view of the remarks above, we may assume that $q \leq x^{1/2}$. By (11.22) we see that

$$\psi(x; q, a) - \frac{x}{\varphi(q)} = \frac{\psi(x, \chi_0) - x}{\varphi(q)} + \frac{1}{\varphi(q)} \sum_{\chi \neq \chi_0} \overline{\chi}(a)\psi(x, \chi). \qquad (13.25)$$

Thus by the triangle inequality,

$$\left|\psi(x; q, a) - \frac{x}{\varphi(q)}\right| \leq \frac{|\psi(x, \chi_0) - x|}{\varphi(q)} + \frac{1}{\varphi(q)} \sum_{\chi \neq \chi_0} |\psi(x, \chi)|, \qquad (13.26)$$

and so (13.22) follows from (13.19). The other relations are proved similarly. $\qquad \square$

Since $L(s, \chi)$ has $\asymp \log q$ zeros with $\gamma \ll 1$, we expect (assuming GRH) that $\psi(x, \chi)$ is usually about $(x \log q)^{1/2}$ in size. Thus the estimates of Theorem 13.7 are close to what we presume would be best possible. On the right-hand side of (13.25), we have $\varphi(q)$ terms. With sums of independent random variables in mind, we would expect therefore that the right-hand side of (13.25) is usually $\ll (x(\log q)/\varphi(q))^{1/2}$. Since we are unable to prove that there is cancellation in (13.25), we have no recourse but to use the triangle inequality, as in (13.26). However, we conjecture that a lot has been lost at this point.

Conjecture 13.9 If $(a, q) = 1$ and $q \leq x$, then

$$\psi(x; q, a) = \frac{x}{\varphi(q)} + O_\varepsilon\big(x^{1/2+\varepsilon}/q^{1/2}\big).$$

Although we are unable to confirm our speculations concerning cancellation in (13.25) for any individual a, we can show that such cancellation must occur on average.

Corollary 13.10 *Assume GRH for all L-functions modulo q. If $2 \leq q \leq x$, then*

$$\sum_{\substack{a=1 \\ (a,q)=1}}^{q} (\psi(x;q,a) - x/\varphi(q))^2 \ll x(\log x)^4.$$

Proof We claim that

$$\sum_{\substack{a=1 \\ (a,q)=1}}^{q} \left| \sum_{\chi} c(\chi)\chi(a) \right|^2 = \varphi(q) \sum_{\chi} |c(\chi)|^2 \qquad (13.27)$$

for arbitrary complex numbers $c(\chi)$. To understand why this holds, expand the left-hand side and take the sum over a inside, to see that it is

$$= \sum_{\chi_1} \sum_{\chi_2} c(\chi_1)\overline{c(\chi_2)} \sum_{\substack{a=1 \\ (a,q)=1}}^{q} \chi_1(a)\overline{\chi_2}(a).$$

By the basic orthogonality property of Dirichlet characters (cf (4.14)), the inner sum here is $\varphi(q)$ if $\chi_1 = \chi_2$, and is 0 otherwise, and this gives (13.27). By taking $c(\chi) = (\psi(x, \overline{\chi}) - E_0(\chi)x)/\varphi(q)$, it follows by (11.22) that

$$\sum_{\substack{a=1 \\ (a,q)=1}}^{q} (\psi(x;q,a) - x/\varphi(q))^2 = \frac{1}{\varphi(q)} \sum_{\chi} |\psi(x, \chi) - E_0(\chi)x|^2,$$

The stated estimate now follows from (13.19). □

For non-principal χ let $n(\chi)$ denote the least character non-residue of χ, which is to say the least positive integer n such that $\chi(n) \neq 1$ and $\chi(n) \neq 0$. Since

$$\psi(x, \chi_0) = \psi(x) + O((\log q)(\log x)) \asymp x$$

for $x \geq C(\log q)(\log \log q)$, it follows by taking $x = C(\log q)^2(\log \log q)^2$ in (13.19) that $n(\chi) \ll (\log q)^2(\log \log q)^2$. As was the case with Cramér's theorem (Theorem 13.3), we can do slightly better by using a weighted sum of primes.

Theorem 13.11 *Let χ be a non-principal character modulo q, and assume that $L(s, \chi) \neq 0$ for $\sigma > 1/2$. Then $n(\chi) \ll (\log q)^2$.*

Proof   By taking $k = 1$ in (5.17)–(5.19), we see that

$$\sum_{n \leq x} \chi(n)\Lambda(n)(x - n) = \frac{-1}{2\pi i} \int_{\sigma_0 - i\infty}^{\sigma_0 + i\infty} \frac{L'}{L}(s, \chi) \frac{x^{s+1}}{s(s+1)} ds.$$

On pulling the contour to the line $\sigma = 1/4$, we see that the above is

$$-\sum_\rho \frac{x^{\rho+1}}{\rho(\rho + 1)} - \frac{x^{5/4}}{2\pi} \int_{-\infty}^\infty \frac{L'}{L}(1/4 + it, \chi) \frac{x^{it}}{(1/4 + it)(5/4 + it)} dt.$$

By Theorem 10.17, the sum over ρ is $\ll x^{3/2} \log q$. By Theorem 10.17 with Lemma 12.7, we see that $\frac{L'}{L}(1/4 + it, \chi) \ll \log q\tau$. Hence the second term above is $\ll x^{5/4} \log q$. Thus

$$\sum_{n \leq x} \chi(n)\Lambda(n)(x - n) \ll x^{3/2} \log q. \qquad (13.28)$$

On the other hand,

$$\sum_{n \leq x} \chi_0(n)\Lambda(n)(x - n) = \sum_{n \leq x} \Lambda(n)(x - n) + O(x(\log x)(\log q)) \gg x^2$$

$$(13.29)$$

if $x \geq C(\log q)(\log \log q)$. If $\chi(n) = \chi_0(n)$ for all prime powers $n \leq x$, then the left-hand sides of (13.28) and (13.29) are equal. However, the right-hand sides are inconsistent if we take $x = C(\log q)^2$, so we obtain the stated result.   □

Weaker hypotheses concerning the zeros of $L(s, \chi)$ also imply bounds for $n(\chi)$. The argument here depends on a careful selection of the kernel in the inverse Mellin transform.

Theorem 13.12   *Let χ be a non-principal character* (mod q), *and suppose that δ is chosen, $1/\log q \leq \delta \leq 1/2$, so that $L(s, \chi) \neq 0$ for $1 - \delta < \sigma < 1$, $0 < |t| \leq \delta^2 \log q$. Then $n(\chi) < (A\delta \log q)^{1/\delta}$. Here A is a suitable absolute constant.*

Proof   First we show that if $1/\log q \leq R \leq 1$, then

$$\sum_{|\rho - 1| > R} \frac{1}{|\rho - 1|^2} \ll \frac{\log q}{R}. \qquad (13.30)$$

To see this, note that

$$\sum_{R < |\rho - 1| \leq 2R} \frac{1}{|\rho - 1|^2} \ll \frac{1}{R^2} n(2R; 0, \chi) \ll \frac{\log q}{R}$$

by Theorems 11.5 and 10.17. On replacing R by $2^k R$, and summing, we deduce that

$$\sum_{R<|\rho-1|\le 1} \frac{1}{|\rho - 1|^2} \ll \frac{\log q}{R}.$$

As for zeros farther from 1, we note by Theorem 10.17 that

$$\sum_{|\rho-1|>1} \frac{1}{|\rho - 1|^2} \ll \sum_{n=1}^{\infty} \frac{\log 2qn}{n^2} \ll \log q,$$

and so we have (13.30) for all $R \ge 1/\log q$.

Let x and y be parameters to be chosen later so that $2 < y \le x^{1/3}$. For $x/y^2 \le u \le xy^2$ set $w(u) = (2\log y - |\log(x/u)|)x/u$, and put $w(u) = 0$ otherwise. Then

$$\sum_n w(n)\chi(n)\Lambda(n) = \frac{-1}{2\pi i}\int_{\sigma_0-i\infty}^{\sigma_0+i\infty} \frac{L'}{L}(s,\chi)\left(\frac{y^{s-1} - y^{1-s}}{s-1}\right)^2 x^s\, ds \quad (13.31)$$

for $\sigma_0 > 1$. We move the contour to the abscissa $\sigma_0 = -1/2$, and find that the above is

$$= -\sum_\rho \left(\frac{y^{\rho-1} - y^{1-\rho}}{\rho - 1}\right)^2 x^\rho - (1-\kappa)(y-1/y)^2$$

$$\quad (13.32)$$

$$-\frac{1}{2\pi i}\int_{-1/2-i\infty}^{-1/2+i\infty} \frac{L'}{L}(s,\chi)\left(\frac{y^{s-1} - y^{1-s}}{s-1}\right)^2 x^s\, ds.$$

Here the second term arises because $L(s,\chi)$ has a trivial zero at $s=0$ if $\chi(-1) = 1$. Suppose that χ is induced by a primitive character $\chi^\star$. Then by (10.20) we see that

$$\frac{L'}{L}(s,\chi) = \frac{L'}{L}(s,\chi^\star) + \sum_{p|q} \frac{\chi^\star(p)\log p}{p^s - \chi^\star(p)}.$$

When $\sigma = -1/2$, the summand above is $\ll \log p$, and so by Lemma 12.9 we see that $\frac{L'}{L}(-1/2 + it, \chi) \ll \log q\tau$. Hence the last term in (13.32) is $\ll x^{-1/2}y^3 \log q$. If χ is imprimitive, then $L(s,\chi)$ may have infinitely many zeros on the imaginary axis. Such zeros are to be included in the sums in (13.30) and (13.32). If a zero ρ is real, then its contribution in (13.32) is negative. If ρ is a zero for which $\beta \le 1-\delta$, then its contribution to (13.32) is

$$\ll \frac{x^{1-\delta}y^{2\delta}}{|\rho - 1|^2}.$$

From (13.30) with $R = \delta$ we see that the total contribution of such zeros is

$$\ll x^{1-\delta}y^{2\delta}(\log q)/\delta.$$

If ρ is a zero for which $\beta > 1 - \delta$ and ρ is not real, then by hypothesis we have $|\gamma| \geq \delta^2 \log q$. The summand in (13.32) is $\ll x/|\rho - 1|$, so that from (13.30) with $R = \delta^2 \log q$ we see that such zeros contribute an amount $\ll x/\delta^2$. On combining these estimates we find that there is an absolute constant $c_1 > 0$ such that

$$\Re \sum_n w(n)\chi(n)\Lambda(n) \leq c_1\left(x^{1-\delta}y^{2\delta}\delta^{-1}\log q + x\delta^{-2}\right). \qquad (13.33)$$

If we replace χ by χ_0 in (13.31) and argue as in the proof of the Prime Number Theorem, we find that

$$\sum_n w(n)\chi_0(n)\Lambda(n) = 4(\log y)^2 x + O\left(x\exp\left(-c\sqrt{\log x}\right)\right) + O(y^2 \log q). \qquad (13.34)$$

Here the second error term reflects the possible contribution of zeros of $L(s, \chi_0)$ on the imaginary axis. If $\chi(n) = \chi_0(n)$ for all n for which $w(n) \neq 0$, then the left-hand side in (13.33) is identical with that in (13.34). Thus we wish to show that the right-hand sides cannot be equal, with a choice of x and y for which xy^2 is as small as possible. To this end, note that if $x = (C^3\delta \log q)^{1/\delta}$ and $y = C^{1/\delta}$, then the right-hand side of (13.33) is $\asymp (1 + 1/C)x/\delta^2$, while the right-hand side of (13.34) is $\asymp (\log C)^2 x/\delta^2$, uniformly for $C \geq 2$. Thus if C is a sufficiently large absolute constant, then the left-hand members of (13.33) and (13.34) cannot be identical, and we have the stated result. □

13.1.1 Exercises

1. Let $\Theta = \sup_\rho \beta$ where ρ runs over all non-trivial zeros of $\zeta(s)$. Show that

$$\psi(x) = x + O(x^\Theta(\log x)^2),$$
$$\vartheta(x) = x + O(x^\Theta(\log x)^2),$$
$$\pi(x) = = \mathrm{li}(x) + O(x^\Theta \log x).$$

2. Let $F(x)$ be as in the proof of Theorem 13.3. Suppose that $2 \leq \Delta \leq h \leq x$, and put $w(u) = 0$ for $u \leq x - \Delta$, $w(u) = (u - x + \Delta)/\Delta$ for $x - \Delta \leq u \leq x$, $w(u) = 1$ for $x \leq u \leq x + h$, $w(u) = (x + h + \Delta - u)/\Delta$ for $x + h \leq u \leq x + h + \Delta$, $w(u) = 0$ for $u \geq x + h + \Delta$.
(a) Show that

$$\sum_n \Lambda(n)w(n) = \frac{1}{\Delta}(F(x+h+\Delta) - F(x+h) - F(x) + F(x-\Delta))$$

$$= h + \Delta - \frac{1}{\Delta}\sum_\rho S(\rho) + O\left(\frac{1}{\Delta x}\right)$$

where

$$S(\rho) = \frac{(x+h+\Delta)^{\rho+1} - (x+h)^{\rho+1} - x^{\rho+1} + (x-\Delta)^{\rho+1}}{\rho(\rho+1)}.$$

(b) Show that if RH holds, then $S(\rho) \ll h\Delta x^{-1/2}$ for $|\gamma| \le x/h$, that $S(\rho) \le \Delta x^{1/2}/|\gamma|$ for $x/h \le |\gamma| \le x/\Delta$, and that $S(\rho) \ll x^{3/2}/\gamma^2$ for $\gamma| \ge x/\Delta$.

(c) Show that if RH holds, then

$$\psi(x+h) - \psi(x) = h + O\left(x^{1/2}(\log x)\log\frac{2h}{x^{1/2}\log x}\right)$$

uniformly for $x^{1/2}\log x \le h \le x$.

3. Assume RH. Show that

$$\int_2^X (\psi(x)-x)^2 \frac{dx}{x^2} \sim (\log X) \sum_\rho \frac{m_\rho^2}{|\rho|^2}$$

as $X \to \infty$.

4. Assume RH. Suppose that T is given, $T \ge 2$, and let $f(u)$ be defined as in (13.17). Show that

$$\lim_{U\to\infty} \frac{1}{U} \int_1^U \left| f(u) + \sum_{\substack{\rho \\ |\gamma|\le T}} \frac{e^{i\gamma u}}{\rho} \right|^2 du = \sum_{\substack{\rho \\ |\gamma|>T}} \frac{m_\rho^2}{|\rho|^2}.$$

5. Assume GRH for all L-functions modulo q. (a) Show that

$$\sum_{n\le x} \chi(n)\Lambda(n)(x-n) = E_0(\chi)x^2/2 + O\left(x^{3/2}\log q\right),$$

$$\sum_{p\le x} \chi(p)(\log p)(x-p) = E_0(\chi)x^2/2 + O\left(x^{3/2}\log q\right).$$

(b) Show that if $(a,q) = 1$, then

$$\sum_{\substack{n\le x \\ n\equiv a\,(q)}} \Lambda(n)(x-n) = \frac{x^2}{2\varphi(q)} + O\left(x^{3/2}\log q\right),$$

$$\sum_{\substack{p\le x \\ p\equiv a\,(q)}} (\log p)(x-p) = \frac{x^2}{2\varphi(q)} + O\left(x^{3/2}\log q\right).$$

(c) Deduce that if $(a,q) = 1$, then the least prime $p \equiv a \pmod q$ is $\ll \varphi(q)^2(\log q)^2$.

6. Assume Conjecture 13.9. Show that if $(a,q) = 1$, then there is a prime number $p \equiv a \pmod q$ such that $p \ll_\varepsilon q^{1+\varepsilon}$.

7. Let χ be a non-principal character, and let $n(\chi)$ denote the least positive integer n such that $\chi(n) \ne 1$, $\chi(n) \ne 0$. Show that $n(\chi)$ is a prime number.

8. (Montgomery 1971, p. 121) Let χ be a character modulo q, and let d denote the order of χ.

   (a) Show that

   $$\frac{1}{d}\sum_{k=1}^{d}\chi^k(n)e(-ak/d) = \begin{cases} 1 & \text{if } \chi(n) = e(a/d), \\ 0 & \text{otherwise.} \end{cases}$$

   (b) Assume that GRH holds for the $d-1$ L-functions $L(s, \chi^k)$ where $0 < k < d$. Show that for each dth root of unity $e(a/d)$ there is a prime p such that $\chi(p) = e(a/d)$, with $p \ll d^2(\log q)^2$.

9. (Montgomery 1971, p. 122) Let $\mathcal{P}(y)$ denote the set of those primes p such that $\left(\frac{n}{p}\right) = 1$ for all $n \leq y$, and let $P(y)$ be the product of all primes not exceeding y. Suppose that $2 \leq y \leq x$.

   (a) Explain why

   $$\sum_{\substack{x<p\leq 2x \\ p\in\mathcal{P}(y)}} \log p = 2^{-\pi(y)} \sum_{x<p\leq 2x} (\log p) \prod_{p_1\leq y} \left(1 + \left(\frac{p_1}{p}\right)\right).$$

   (b) For each $m|P(y)$, $m > 1$, let χ_m be the quadratic character determined by quadratic reciprocity so that $\chi_m(p) = \prod_{p_1|m} \left(\frac{p_1}{p}\right)$. Also, let $\chi_1(n) = 1$ for all n. Explain why the above is

   $$= 2^{-\pi(y)} \sum_{m|P(y)} (\vartheta(2x, \chi_m) - \vartheta(x, \chi_m)).$$

   (c) Assume GRH for all quadratic L-functions. Show that the above is

   $$= 2^{-\pi(y)}x(1 + o(1)) + O\left(x^{1/2}(\log x)^2\right).$$

   (d) Show that if $y = \frac{2}{3}(\log x)(\log\log x)$, then the above is positive, for all sufficiently large x.

   (e) Let $n_2(p)$ denote the least quadratic non-residue of p, which is to say the least positive integer n such that $\left(\frac{n}{p}\right) = -1$. Show that if GRH is true for all quadratic L-functions, then there exist infinitely many primes p such that $n_2(p) > \frac{2}{3}(\log p)(\log\log p)$.

10. (Littlewood 1924a; cf. Goldston 1982)

   (a) Show (unconditionally) that

   $$\psi(x) \leq x - \sum_{\rho} \frac{(x+h)^{\rho+1} - x^{\rho+1}}{h\rho(\rho+1)} + O(h)$$

   for $2 \leq h \leq x/2$.

   (b) Show (unconditionally) that

   $$\psi(x) \geq x - \sum_{\rho} \frac{x^{\rho+1} - (x-h)^{\rho+1}}{h\rho(\rho+1)} - O(h)$$

   for $2 \leq h \leq x/2$.

(c) Now, and in the following, assume RH. Show that

$$\sum_{\substack{\rho \\ |\gamma|>x/h}} \frac{(x+h)^{\rho+1} - x^{\rho+1}}{h\rho(\rho+1)} \ll x^{1/2} \log x / h.$$

(d) Show that if $|\gamma| \leq x/h$, then

$$\frac{(x+h)^{\rho+1} - x^{\rho+1}}{h\rho(\rho+1)} = \frac{x^{\rho}}{\rho} + O(x^{-1/2}h).$$

(e) Show that

$$\sum_{\substack{\rho \\ |\gamma|\leq x/h}} \frac{(x+h)^{\rho+1} - x^{\rho+1}}{h\rho(\rho+1)} = \sum_{\substack{\rho \\ |\gamma|\leq x/h}} \frac{x^{\rho}}{\rho} + O(x^{1/2} \log x / h).$$

(f) Show that

$$\psi(x) = x - \sum_{\substack{\rho \\ |\gamma|\leq\sqrt{x}/\log x}} \frac{x^{\rho}}{\rho} + O(x^{1/2} \log x).$$

13.2 Estimates for the zeta function

We now show that our estimates of $\zeta(s)$ and of $\frac{\zeta'}{\zeta}(s)$ can be improved if we assume RH. To this end, we begin with a useful explicit formula. For $x \geq 2$, $y \geq 2$, put

$$w(u) = w(x, y; u) = \begin{cases} 1 & \text{if } 1 \leq u \leq x; \\ 1 - \frac{\log u/x}{\log y} & \text{if } x \leq u \leq xy; \\ 0 & \text{if } u \geq xy. \end{cases}$$

Then by two applications of (5.20) we find that

$$\sum_{n\leq xy} w(n)\frac{\Lambda(n)}{n^s} = \frac{-1}{2\pi i \log y} \int_{\sigma_0-i\infty}^{\sigma_0+i\infty} \frac{\zeta'}{\zeta}(s+w)\frac{(xy)^w - x^w}{w^2} \, dw,$$

and on pulling the contour to the left we see that this is

$$= -\frac{\zeta'}{\zeta}(s) + \frac{(xy)^{1-s} - x^{1-s}}{(1-s)^2 \log y}$$

$$- \sum_{\rho} \frac{(xy)^{\rho-s} - x^{\rho-s}}{(\rho - s)^2 \log y} - \sum_{k=1}^{\infty} \frac{(xy)^{-2k-s} - x^{-2k-s}}{(2k+s)^2 \log y} \qquad (13.35)$$

provided that $s \neq 1$ and that $\zeta(s) \neq 0$. This much is true unconditionally, but from now on we assume RH, and show that the sum on the left provides a useful approximation to $-\frac{\zeta'}{\zeta}(s)$ when $\sigma > 1/2$.

Theorem 13.13 *Assume RH. Then*

$$\left|\frac{\zeta'}{\zeta}(s)\right| \le \sum_{n \le (\log \tau)^2} \frac{\Lambda(n)}{n^\sigma} + O((\log \tau)^{2-2\sigma}) \tag{13.36}$$

uniformly for $1/2 + 1/\log \log \tau \le \sigma \le 3/2$, $|t| \ge 1$.

Proof If $\sigma \ge 1/2$, then $|y^{\rho-s} - 1| \le 2$. Hence for $\sigma > 1/2$, the sum over ρ in (13.25) has absolute value not exceeding

$$\frac{2x^{1/2-\sigma}}{\log y} \sum_\rho \frac{1}{|s - \rho|^2}.$$

By (10.29) and (10.30) we see that

$$(\sigma - 1/2) \sum_\rho \frac{1}{(\sigma - 1/2)^2 + (t - \gamma)^2}$$

$$= \Re\frac{\zeta'}{\zeta}(s) + \frac{1}{2}\Re\frac{\Gamma'}{\Gamma}(s/2 + 1) - \frac{1}{2}\log \pi + \frac{\sigma - 1}{(\sigma - 1)^2 + t^2},$$

and by Theorem C.1 this is

$$= \Re\frac{\zeta'}{\zeta}(s) + \frac{1}{2}\log \tau + O(1).$$

On inserting this in (13.35), we find that

$$\frac{\zeta'}{\zeta}(s) = -\sum_{n \le xy} w(n)\frac{\Lambda(n)}{n^s} + \frac{\theta 2x^{1/2-\sigma}}{(\sigma - 1/2)\log y}\left|\Re\frac{\zeta'}{\zeta}(s)\right|$$

$$+ O\left(\frac{x^{1/2-\sigma}\log \tau}{(\sigma - 1/2)\log y}\right) + O\left(\frac{(xy)^{1-\sigma}}{\tau^2}\right) + O\left(\frac{y^{1-\sigma}}{\tau^2}\right) \tag{13.37}$$

where θ is a complex number satisfying $|\theta| \le 1$. Thus

$$\frac{\zeta'}{\zeta}(s) \ll \left|\sum_{n \le xy} w(n)\frac{\Lambda(n)}{n^s}\right| + \frac{x^{1/2-\sigma}\log \tau}{(\sigma - 1/2)\log y} + \frac{(xy)^{1-\sigma}}{\tau^2} + \frac{y^{1-\sigma}}{\tau^2} \tag{13.38}$$

provided that

$$\frac{2x^{1/2-\sigma}}{(\sigma - 1/2)\log y} \le c < 1. \tag{13.39}$$

We take

$$y = \exp\left(\frac{1}{\sigma - 1/2}\right), \qquad x = (\log \tau)^2/y.$$

Then the left-hand side of (13.39) is $2e(\log \tau)^{1-2\sigma}$, and so (13.39) holds with

$c = 2/e$ for $\sigma \geq 1/2 + 1/\log \log \tau$. We observe that

$$\sum_{n \leq xy} w(n) \frac{\Lambda(n)}{n^s} \ll \sum_{n \leq (\log \tau)^2} \frac{\Lambda(n)}{n^{1/2}} \ll \log \tau$$

uniformly for $\sigma \geq 1/2$. On inserting this in (13.38), we find that

$$\frac{\zeta'}{\zeta}(s) \ll \log \tau$$

uniformly for $\sigma \geq 1/2 + 1/\log \log \tau$, $|t| \geq 1$. We insert this on the right-hand side of (13.37) to obtain the stated estimate. □

Corollary 13.14 *Assume RH. Then*

$$\frac{\zeta'}{\zeta}(s) \ll ((\log \tau)^{2-2\sigma} + 1) \min \left(\frac{1}{|\sigma - 1|}, \log \log \tau \right)$$

uniformly for $1/2 + 1/\log \log \tau \leq \sigma \leq 3/2$, $|t| \geq 1$.

Proof By Chebyshev's estimate (Theorem 2.4) we know that

$$\sum_{U \leq n < eU} \frac{\Lambda(n)}{n^\sigma} \ll U^{1-\sigma}.$$

On summing this over $U = e^k$ for $0 \leq k \leq 2 \log \log \tau$, we obtain the stated bound from Theorem 13.13. □

Corollary 13.15 *Assume RH. Then*

$$|\log \zeta(s)| \leq \sum_{n \leq (\log \tau)^2} \frac{\Lambda(n)}{n^\sigma \log n} + O \left(\frac{(\log \tau)^{2-2\sigma}}{\log \log \tau} \right) \tag{13.40}$$

uniformly for $1/2 + 1/\log \log \tau \leq \sigma \leq 3/2$, $|t| \geq 1$.

Proof Since

$$\log \zeta(\sigma + it) = \log \zeta(3/2 + it) - \int_\sigma^{3/2} \frac{\zeta'}{\zeta}(\alpha + it) \, d\alpha,$$

it follows by the triangle inequality that

$$|\log \zeta(\sigma + it)| \leq |\log \zeta(3/2 + it)| + \int_\sigma^{3/2} \left| \frac{\zeta'}{\zeta}(\alpha + it) \right| d\alpha,$$

which by Corollary 13.13 is

$$\leq |\log \zeta(3/2 + it)| + \sum_{n \leq (\log \tau)^2} \frac{\Lambda(n)}{\log n} (n^{-\sigma} - n^{-3/2}) + O \left(\frac{(\log \tau)^{2-2\sigma}}{\log \log \tau} \right).$$

But

$$|\log \zeta(3/2 + it)| = \left| \sum_{n=1}^{\infty} \frac{\Lambda(n)}{\log n} n^{-3/2 - it} \right| \le \sum_{n=1}^{\infty} \frac{\Lambda(n)}{\log n} n^{-3/2},$$

so it follows that

$$|\log \zeta(\sigma + it)| \le \sum_{n \le (\log \tau)^2} \frac{\Lambda(n)}{\log n} n^{-\sigma} + \sum_{n > (\log \tau)^2} \frac{\Lambda(n)}{\log n} n^{-3/2} + O\left(\frac{(\log \tau)^{2 - 2\sigma}}{\log \log \tau} \right).$$

$$(13.41)$$

By the Chebyshev estimate $\psi(x) \ll x$ we see that

$$\sum_{U < n \le 2U} \frac{\Lambda(n)}{\log n} n^{-3/2} \ll U^{-1/2} (\log U)^{-1}.$$

By taking $U = (\log \tau)^2 2^k$, and summing over $k \ge 0$, we deduce that

$$\sum_{n > (\log \tau)^2} \frac{\Lambda(n)}{\log n} n^{-3/2} \ll (\log \tau)^{-1} (\log \log \tau)^{-1}.$$

Since this is majorized by the error term in (13.41), we have (13.40). $\qquad \square$

Corollary 13.16 *Assume RH. If $|t| \ge 1$, then*

$$|\log \zeta(s)| \le \log \frac{1}{\sigma - 1} + O(\sigma - 1) \qquad (13.42)$$

for $1 + 1/\log \log \tau \le \sigma \le 3/2$,

$$|\log \zeta(s)| \le \log \log \log \tau + O(1) \qquad (13.43)$$

for $1 - 1/\log \log \tau \le \sigma \le 1 + 1/\log \log \tau$, and

$$|\log \zeta(s)| \le \log \frac{1}{1 - \sigma} + O\left(\frac{(\log \tau)^{2 - 2\sigma}}{(1 - \sigma) \log \log \tau} \right) \qquad (13.44)$$

for $1/2 + 1/\log \log \tau \le \sigma \le 1 - 1/\log \log \tau$.

Proof To establish (13.42), we note that if $1 < \sigma \le 3/2$, then

$$|\log \zeta(s)| = \left| \sum_{n=1}^{\infty} \frac{\Lambda(n)}{\log n} n^{-s} \right| \le \sum_{n=1}^{\infty} \frac{\Lambda(n)}{\log n} n^{-\sigma} = \log \zeta(\sigma)$$

$$= \log \left(1/(\sigma - 1) + O(1) \right) = \log \frac{1}{\sigma - 1} + O(\sigma - 1).$$

As for (13.43), we note first that

$$\sum_{n \le z} \frac{\Lambda(n)}{n \log n} = \log \log z + O(1).$$

by Mertens' estimates (Theorem 2.7). Also, if $\sigma = 1 + O(1/\log z)$, then

$$n^{-\sigma} - n^{-1} = \int_1^\sigma n^{-\alpha}\, d\alpha \log n \ll |\sigma - 1| n^{-1} \log n$$

for $1 \le n \le z$, so that

$$\sum_{n \le z} \frac{\Lambda(n)}{\log n}(n^{-\sigma} - n^{-1}) \ll |\sigma - 1| \sum_{n \le z} \frac{\Lambda(n)}{n} \ll |\sigma - 1| \log z \ll 1.$$

On combining these estimates with $z = (\log \tau)^2$, we see that the sum in (13.40) is $\le \log \log \log \tau + O(1)$, which gives the desired estimate.

Concerning (13.44), we note that

$$\sum_{n \le z} \frac{\Lambda(n)}{\log n} n^{-\sigma} = \int_{2-}^z \frac{1}{u^\sigma \log u}\, d\psi(u)$$

$$= \int_2^z \frac{1}{u^\sigma \log u}\, du + \frac{\psi(z) - z}{z^\sigma \log z} + 2^{1-\sigma}/\log 2$$

$$+ \int_2^z \frac{\psi(u) - u}{u^{\sigma+1} \log u}\left(\sigma + \frac{1}{\log u}\right) du. \qquad (13.45)$$

By the change of variable $v = u^{1-\sigma}$, the first integral immediately above is $\mathrm{li}(z^{1-\sigma}) - \mathrm{li}(2^{1-\sigma})$. But

$$\mathrm{li}(z^{1-\sigma}) \ll \frac{z^{1-\sigma}}{(1-\sigma)\log z}$$

for $\sigma \le 1 - 1/\log z$, and

$$-\mathrm{li}(2^{1-\sigma}) = \int_{2^{1-\sigma}}^2 \frac{dv}{\log v} = \int_{2^{1-\sigma}}^2 \left(\frac{1}{v-1} + O(1)\right) dv$$

$$= -\log(2^{1-\sigma} - 1) + O(1) = \log \frac{1}{\sigma - 1} + O(1).$$

By Theorem 13.1, the second term in (13.45) is $\ll z^{1/2-\sigma}\log z$, and the final integral in (13.45) is

$$\ll \int_2^\infty u^{-\sigma-1/2} \log u\, du \ll (\sigma - 1/2)^{-2}.$$

On combining these estimates, we find that

$$\sum_{n \le z} \frac{\Lambda(n)}{n^\sigma \log n} = \log \frac{1}{1-\sigma} + O\left(\frac{z^{1-\sigma}}{(1-\sigma)\log z}\right),$$

uniformly for $1/2 < \sigma \le 1 - 1/\log z$. On taking $z = (\log \tau)^2$, the desired estimate now follows from (13.40). $\qquad\square$

From Corollary 13.16 we see that if RH holds, then

$$\frac{1}{\log\log\tau} \ll |\zeta(1+it)| \ll \log\log\tau$$

for $|t| \geq 1$. We can make this more precise by taking a little more care.

Corollary 13.17   *Assume RH. Then* $|\zeta(1+it)| \leq 2e^{C_0}\log\log\tau + O(1)$.

Proof   We observe that

$$\sum_{n\leq z}\frac{\Lambda(n)}{n\log n} = \sum_{p^k\leq z}\frac{\Lambda(n)}{n\log n} \leq \sum_{p\leq z}\sum_{k=1}^{\infty}\frac{1}{kp^k} = \log\prod_{p\leq z}\left(1-\frac{1}{p}\right)^{-1}$$
$$= C_0 + \log\log z + O(1/\log z)$$

by Mertens' estimate (Theorem 2.7). We take $z = (\log\tau)^2$, insert this in Corollary 13.15, and exponentiate to obtain the stated bound. $\square$

To complete the picture, we estimate $|\zeta(s)|$ and $\arg\zeta(s)$ when σ is near $1/2$. Of these estimates, the upper bound for $|\zeta(s)|$ is the most immediate.

Theorem 13.18   *Assume RH. There is an absolute constant $C > 0$ such that*

$$|\zeta(s)| < \exp\left(\frac{C\log\tau}{\log\log\tau}\right)$$

uniformly for $\sigma \geq 1/2$, $|t| \geq 1$.

Note that this is a quantitative form of the Lindelöf Hypothesis (LH).

Proof   Put $\sigma_1 = 1/2 + 1/\log\log\tau$. For $\sigma \geq \sigma_1$, the above is contained in Corollary 13.14. Suppose that $1/2 \leq \sigma \leq \sigma_1$. Since $\Re 1/(s-\rho) \geq 0$ for all zeros ρ, from Lemma 12.1 it follows that there is an absolute constant $A > 0$ such that

$$\Re\frac{\zeta'}{\zeta}(s) \geq -A\log\tau$$

uniformly for $1/2 \leq \sigma \leq 2$, $|t| \geq 1$. Hence

$$\log|\zeta(s)| = \log|\zeta(\sigma_1+it)| - \int_{\sigma}^{\sigma_1}\Re\frac{\zeta'}{\zeta}(\alpha+it)\,d\alpha$$
$$\leq \log|\zeta(\sigma_1+it)| + A(\sigma_1-\sigma)\log\tau.$$

Here the first member on the right-hand side is bounded by Corollary 13.15, and $0 \leq \sigma_1 - \sigma \leq 1/\log\log\tau$, so we have the stated bound. $\square$

To obtain the remaining estimates, we first establish two lemmas, which are of interest in their own right.

Lemma 13.19 *Assume RH. Then for $T \geq 4$,*

$$N(T + 1/\log\log T) - N(T) \ll \frac{\log T}{\log\log T}.$$

Proof Take $s = 1/2 + 1/\log\log T + iT$. Then $\frac{\zeta'}{\zeta}(s) \ll \log T$ by Corollary 13.14. Hence by Lemma 12.1 it follows that

$$\sum_{\substack{\rho \\ |\gamma - T| \leq 1}} \frac{1}{s - \rho} \ll \log T.$$

Here each summand has positive real part, and for $T \leq \gamma \leq T + 1/\log\log T$ the real part is $\geq \frac{1}{2}\log\log T$, so we obtain the stated bound. $\square$

By mimicking the proof of Lemma 12.1, we obtain

Lemma 13.20 *Assume RH. If $|\sigma - 1/2| \leq 1/\log\log\tau$, then*

$$\frac{\zeta'}{\zeta}(s) = \sum_{\substack{\rho \\ |\gamma - t| \leq 1/\log\log\tau}} \frac{1}{s - \rho} + O(\log\tau).$$

In applying the above, one is free to replace the condition $|\gamma - t| \leq 1/\log\log\tau$ by a different condition, say $|\gamma - t| \leq \delta$, provided that $\delta \asymp 1/\log\log\tau$. To see why this is so, note that a summand in one sum that is missing in the other has absolute value $\asymp \log\log\tau$, and that by Lemma 13.19 there are $\ll (\log\tau)/\log\log\tau$ such summands. Hence the total contribution made by terms in one sum but not the other is $\ll \log\tau$, and a discrepancy of this size may be absorbed in the error term.

Proof Put $\sigma_1 = 1/2 + 1/\log\log\tau$, and set $s_1 = \sigma_1 + it$. We apply Lemma 12.1 at s_1 and at s, and difference, to see that

$$\frac{\zeta'}{\zeta}(s) = \frac{\zeta'}{\zeta}(s_1) + \sum_{|\gamma - t| \leq 1} \left(\frac{1}{s - \rho} - \frac{1}{s_1 - \rho} \right) + O(\log\tau).$$

Here the first term on the right-hand side is $\ll \log\tau$, by Corollary 13.14. Let k be a positive integer, and consider zeros for which $k/\log\log\tau \leq |\gamma - t| \leq (k + 1)/\log\log\tau$. By the preceding lemma, there are $\ll (\log\tau)/\log\log\tau$ such zeros, each one of which contributes an amount $\ll (\log\log\tau)/k^2$ to the above sum. On summing over k we see that the contribution of zeros for which $|\gamma - t| > 1/\log\log\tau$ is $\ll \log\tau$. Finally, for the zeros with $|\gamma - t| \leq 1$, we observe that $|1/(s_1 - \rho)| \leq \log\log\tau$, and there are $\ll (\log\tau)/\log\log\tau$ such zeros, so we have the stated result. $\square$

If t is not the ordinate of a zero of the zeta function, then we define $\arg\zeta(s)$ by continuous variation along the ray $\alpha + it$ where α runs from σ to $+\infty$, and $\arg(+\infty + it) = 0$. If t is the ordinate of a zero, then we put $\arg \zeta(s) = (\arg \zeta(\sigma + it^+) + \arg \zeta(\sigma + it^-))/2$.

Theorem 13.21 *Assume RH. Then*

$$\arg \zeta(s) \ll \frac{\log \tau}{\log \log \tau}$$

uniformly for $\sigma \geq 1/2$, $|t| \geq 1$.

Proof We may assume that t is not the ordinate of a zero. Let σ_1 and s_1 be defined as in the preceding proof. If $\sigma \geq \sigma_1$, then the above follows from Corollary 13.16. Suppose now that $1/2 \leq \sigma \leq \sigma_1$. Then

$$\arg \zeta(s) = \arg \zeta(s_1) - \int_\sigma^{\sigma_1} \Im \frac{\zeta'}{\zeta}(\alpha + it)\, d\alpha.$$

Since $0 \leq \sigma_1 - \sigma \leq 1/\log \log \tau$, by Lemma 13.20 the right-hand side above is

$$= - \sum_{|\gamma - t| \leq 1/\log \log \tau} \int_\sigma^{\sigma_1} \Im \frac{1}{\alpha + it - \rho}\, d\alpha + O\left(\frac{\log \tau}{\log \log \tau}\right).$$

Here the summand is

$$\arctan \frac{\sigma - 1/2}{\gamma - t} - \arctan \frac{\sigma_1 - 1/2}{\gamma - t}.$$

If $\gamma > t$, then the above lies between 0 and $\pi/2$, while if $\gamma < t$, then it lies between $-\pi/2$ and 0. In either case, the contribution is bounded, and there are $\ll (\log \tau)/\log \log \tau$ summands by Lemma 13.19, so we have the result. $\square$

Although a lower bound for $|\zeta(s)|$ at all heights is out of the question, we can show, assuming RH, that there are heights for which a lower bound can be established.

Theorem 13.22 *Assume RH. There is an absolute constant C such that for every $T \geq 4$ there is a t, $T \leq t \leq T + 1$, such that*

$$|\zeta(s)| \geq \exp\left(\frac{-C \log T}{\log \log T}\right)$$

uniformly for $-1 \leq \sigma \leq 2$.

Proof By Corollary 10.5 we see that if $-1 \leq \sigma \leq 1/2$, then $|\zeta(s)| \gg |\zeta(1 - \sigma + it)|$. Thus we may restrict our attention to $1/2 \leq \sigma \leq 2$. Put $\sigma_1 = 1/2 + 1/\log \log T$. From Corollary 13.16 we have the desired lower bound for all heights, for $\sigma_1 \leq \sigma \leq 2$. For the remaining interval, $I = [1/2, \sigma_1]$, we show that

$$\int_T^{T+1} \log \frac{1}{\min_{\sigma \in I} |\zeta(s)|} \, dt \ll \frac{\log T}{\log \log T}. \tag{13.46}$$

Put $s_1 = \sigma_1 + it$. Then

$$\log |\zeta(s)| = \log |\zeta(s_1)| - \int_\sigma^{\sigma_1} \Re \frac{\zeta'}{\zeta}(\alpha + it) \, d\alpha.$$

By Corollary 13.16 and Lemma 13.20, this is

$$= -\int_\sigma^{\sigma_1} \sum_{\substack{\rho \\ |\gamma - t| \le \delta}} \Re \frac{1}{\alpha + it - \rho} \, d\alpha + O\left(\frac{\log T}{\log \log T}\right)$$

where $\delta = 1/\log \log T$. The summands are non-negative, so the above is

$$\ge -\int_{1/2}^{\sigma_1} \sum_{\substack{\rho \\ |\gamma - t| \le \delta}} \Re \frac{1}{\alpha + it - \rho} \, d\alpha + O\left(\frac{\log T}{\log \log T}\right).$$

Since this lower bound applies for all $\sigma \in I$, the above provides a lower bound for $\log \min_{\sigma \in I} |\zeta(s)|$. We note that

$$\int_{1/2}^{\sigma_1} \int_{\gamma-\delta}^{\gamma+\delta} \Re \frac{1}{\alpha + it - \rho} \, dt \, d\alpha = \int_0^\delta \int_{-\delta}^\delta \frac{x}{x^2 + y^2} \, dy \, dx$$
$$\le \int_{-\pi/2}^{\pi/2} \int_0^{2\delta} \frac{r \cos \theta}{r^2} \, r dr \, d\theta = 4\delta.$$

Hence

$$\int_T^{T+1} \int_{1/2}^{\sigma_1} \sum_{\substack{\rho \\ |\gamma - t| \le \delta}} \Re \frac{1}{\alpha + it - \rho} \, d\alpha \, dt \ll \sum_{\substack{\rho \\ T-1 \le \gamma \le T+2}} \delta \ll \frac{\log T}{\log \log T},$$

so we have (13.46), and the proof is complete. $\qquad\square$

By Theorem 5.2 and Corollary 5.3 with $\sigma_0 = 1 + 1/\log x$ and $1 \le T \le x$, we see that

$$M(x) = \frac{1}{2\pi i} \int_{\sigma_0 - iT}^{\sigma_0 + iT} \frac{x^s}{\zeta(s)s} \, ds + O\left(\frac{x \log x}{T}\right). \tag{13.47}$$

By Corollary 13.16 we see (assuming RH) that $|\zeta(1/2 + \varepsilon + it)| \gg_\varepsilon \tau^{-\varepsilon}$. Hence, by moving the contour to the abscissa $1/2 + \varepsilon$, we deduce that $M(x) \ll_\varepsilon x^{1/2+\varepsilon}$. This can be made more precise, by determining ε as a function of x, but in order to do so we need a lower bound for $|\zeta(s)|$ when $1/2 < \sigma \le 1/2 + 1/\log \log \tau$.

Theorem 13.23 *Assume RH. There is a constant $C > 0$ such that if $|t| \geq 1$, then*

$$\left|\frac{1}{\zeta(s)}\right| \leq \begin{cases} \exp\left(\frac{C\log\tau}{\log\log\tau}\right) & \text{for } \sigma \geq 1/2 + 1/\log\log\tau, \\ \exp\left(\frac{C\log\tau}{\log\log\tau}\log\frac{e}{(\sigma-1/2)\log\log\tau}\right) & \text{for } 1/2 < \sigma \leq 1/2 + 1/\log\log\tau. \end{cases}$$

Proof The first part follows from Corollary 13.14. Let σ_1 and s_1 be defined as in the proof of Lemma 13.20, and suppose that $1/2 < \sigma \leq \sigma_1$. Then

$$\log\zeta(s) = \log\zeta(s_1) - \int_\sigma^{\sigma_1} \frac{\zeta'}{\zeta}(\alpha + it)\,d\alpha.$$

Here the first term on the right is $\ll (\log\tau)/\log\log\tau$, by Corollary 13.16. By Lemma 13.19 we know that the sum in Lemma 13.20 has $\ll (\log\tau)/\log\log\tau$ terms. Since each term has absolute value $\leq 1/(\sigma - 1/2)$, it follows that

$$\frac{\zeta'}{\zeta}(\alpha + it) \ll \frac{\log\tau}{(\alpha - 1/2)\log\log\tau}$$

for $1/2 < \alpha \leq \sigma_1$. Hence

$$\log\zeta(s) \ll \left(1 + \log\frac{\sigma_1 - 1/2}{\sigma - 1/2}\right)\frac{\log\tau}{\log\log\tau},$$

which gives the stated bound. □

Theorem 13.24 *Assume RH. Then there is an absolute constant $C > 0$ such that*

$$M(x) \ll x^{1/2}\exp\left(\frac{C\log x}{\log\log x}\right)$$

for $x \geq 4$.

Proof Put $\sigma_1 = 1/2 + 1/\log\log x$, and let $\mathcal{C}$ denote the contour that passes by straight line segments from $\sigma_0 - ix$ to $\sigma_1 - ix$ to $\sigma_1 + ix$ to $\sigma_0 + ix$. Then

$$\int_{\sigma_0-ix}^{\sigma_0+ix} \frac{x^s}{\zeta(s)s}\,ds = \int_{\mathcal{C}} \frac{x^s}{\zeta(s)s}\,ds,$$

since the integrand is analytic in the rectangle enclosed by these contours. By the first case of Theorem 13.22 we see that

$$\int_{\sigma_1+ix}^{\sigma_0+ix} \frac{x^s}{\zeta(s)s}\,ds \ll \exp\left(\frac{C\log x}{\log\log x}\right)\int_{\sigma_1} \sigma_0 x^{\sigma-1}\,d\sigma \ll \exp\left(\frac{C\log x}{\log\log x}\right),$$

and the same estimate applies to the integral from $\sigma_1 - ix$ to $\sigma_0 - ix$. Similarly, by the second part of Theorem 13.22 we see that

$$\int_{\sigma_1-ix}^{\sigma_1+ix} \frac{x^s}{\zeta(s)s}\,ds \ll x^{\sigma_1}\int_0^x \exp\left(\frac{C\log\tau}{\log\log\tau}\log\frac{e\log\log x}{\log\log\tau}\right)\frac{dt}{\tau}.$$

By logarithmic differentiation we may confirm that the argument of the exponential is an increasing function of t for $0 \le t \le x$. Thus we obtain the stated bound by taking $T = x$ in (13.47). $\qquad\qquad\square$

13.2.1 Exercises

1. (a) Show (unconditionally) that

$$\Re\frac{\xi'}{\xi}(s) = \sum_\rho \Re\frac{1}{s - \rho}$$

whenever $\xi(s) \ne 0$.

(b) Show (unconditionally) that

$$\Re\frac{\xi'}{\xi}(1/2 + it) = 0$$

for all t such that $\xi(1/2 + it) \ne 0$.

(c) Assume RH. Show that

$$\Re\frac{\xi'}{\xi}(s) \begin{cases} > 0 & \text{if } \sigma > 1/2, \\ = 0 & \text{if } \sigma = 1/2 \text{ and } \xi(s) \ne 0, \\ < 0 & \text{if } \sigma < 1/2. \end{cases}$$

(d) Assume RH. Show that if $\xi'(s) = 0$, then $\Re s = 1/2$.

(e) Assume RH, and let t be any fixed real number. Show that $|\xi(\sigma + it)|$ is a strictly increasing function of σ for $1/2 \le \sigma < \infty$, and that $|\xi(\sigma + it)|$ is a strictly decreasing function of σ for $-\infty < \sigma \le 1/2$.

(f) Assume RH, and suppose that t is a fixed real number. Show that $(\sigma - 1/2)\Re\frac{\xi'}{\xi}(\sigma + it)$ is an increasing function of σ for $1/2 \le \sigma < \infty$.

(g) Assume RH. Show that if $1/2 < \sigma_2 \le \sigma_1$, then

$$|\xi(\sigma_2 + it)| \ge |\xi(\sigma_1 + it)| \cdot \left(\frac{\sigma_2 - 1/2}{\sigma_1 - 1/2}\right)^{(\sigma_1 - 1/2)\Re\frac{\xi'}{\xi}(\sigma_1 + it)}.$$

2. (a) Show (unconditionally) that if $\xi(s) \ne 0$, then

$$\frac{\xi''}{\xi}(s) - \left(\frac{\xi'}{\xi}(s)\right)^2 = -\sum_\rho \frac{1}{(s - \rho)^2}.$$

(b) Show (unconditionally) that if t is real, then $\xi'(1/2 + it) \in i\mathbb{R}$.

(c) Show (unconditionally) that if t is real, then $\xi''(1/2 + it) \in \mathbb{R}$.

(d) Show (unconditionally) that if t is real, then

$$\sum_\rho \frac{1}{(1/2 + it - \rho)^2}$$

is real.

(e) Assume RH. Show that if $\xi(1/2+it) \neq 0$, then

$$\frac{\xi''}{\xi}(1/2+it) > \left(\frac{\xi'}{\xi}\right)^2 (1/2+it).$$

(f) Assume RH. Show that if $\xi(1/2+it) \neq 0$ and $\xi'(1/2+it) = 0$, then $\operatorname{sgn} \xi''(1/2+it) = \operatorname{sgn} \xi(1/2+it)$.

(g) Assume RH. Show that if $\xi(1/2+it) \neq 0$ and $\xi'(1/2+it) = 0$, then

$$\operatorname{sgn} \frac{\partial^2}{\partial t^2}\xi(1/2+it) = -\operatorname{sgn} \xi(1/2+it).$$

(h) Assume RH. Suppose that $\xi(1/2+i\gamma) = \xi(1/2+i\gamma') = 0$, and that $\xi(1/2+it) \neq 0$ for $\gamma < t < \gamma'$. Show that $\xi'(1/2+it)$ has exactly one zero with $\gamma < t < \gamma'$, and that this zero is necessarily simple.

(i) Assume RH. In the above notation, show that the number of zeros of $\xi'(1/2+it)$ in the interval $[\gamma, \gamma')$, counting multiplicity, is the same as the number of zeros of $\xi(1/2+it)$ in the same interval.

(j) Assume RH. Let $N_1(T)$ denote the number of zeros of $\xi'(s)$ with imaginary part in the interval $[0, T]$. Show that $N_1(T) = N(T) + O(1)$.

3. Let χ be a primitive character modulo $q, q > 1$, and suppose that $L(s, \chi) \neq 0$ for $\sigma > 1/2$. Show that

$$\left|\frac{L'}{L}(s, \chi)\right| \leq \sum_{n \leq (\log q\tau)^2} \frac{\Lambda(n)}{n^\sigma} + O\left(\frac{(\log q\tau)^{2-2\sigma}}{\log\log\tau}\right)$$

uniformly for $1/2 + 1/\log\log q\tau \leq \sigma \leq 3/2$.

4. Let χ be a primitive character modulo $q, q > 1$, and suppose that $L(s, \chi) \neq 0$ for $\sigma > 1/2$. Show that

$$\frac{L'}{L}(s, \chi) \ll ((\log q\tau)^{2-2\sigma} + 1) \min\left(\frac{1}{|\sigma - 1|}, \log\log q\tau\right)$$

uniformly for $1/2 + 1/\log\log q\tau \leq \sigma \leq 3/2$.

5. Let χ be a primitive character modulo $q, q > 1$, and suppose that $L(s, \chi) \neq 0$ for $\sigma > 1/2$. Show that

$$|\log L(s, \chi)| \leq \sum_{n \leq (\log q\tau)^2} \frac{\Lambda(n)}{n^\sigma \log n} + O\left(\frac{(\log q\tau)^{2-2\sigma}}{\log\log q\tau}\right)$$

uniformly for $1/2 + 1/\log\log q\tau \leq \sigma \leq 3/2$.

6. Let χ be a primitive character modulo $q, q > 1$, and suppose that $L(s, \chi) \neq 0$ for $\sigma > 1/2$.

(a) Show that

$$|L(s, \chi)| \leq \log\frac{1}{\sigma - 1} + O(\sigma - 1)$$

uniformly for $1 + 1/\log \log q\tau \le \sigma \le 3/2$.

(b) Show that

$$|L(s, \chi)| \le \log \log q\tau + O(1)$$

uniformly for $1 - 1/\log \log q\tau \le \sigma \le 1 + 1/\log \log q\tau$.

(c) Show that

$$|L(s, \chi)| \le \log \frac{1}{1 - \sigma} + O\left(\frac{(\log q\tau)^{2-2\sigma}}{(1 - \sigma)\log \log q\tau}\right)$$

uniformly for $1/2 + 1/\log \log q\tau \le \sigma \le 1 - 1/\log \log q\tau$.

7. Let χ be a primitive character modulo $q, q > 1$, and suppose that $L(s, \chi) \ne 0$ for $\sigma > 1/2$. Show that $|L(1 + it, \chi)| \le 2e^{C_0} \log \log q\tau$.

8. Let χ be a primitive Dirichlet character modulo q with $q > 1$, and suppose that $L(s, \chi) \ne 0$ for $\sigma > 1/2$. Show that there is an absolute constant $C > 0$ such that

$$|L(s, \chi)| \le \exp\left(\frac{C \log q\tau}{\log \log q\tau}\right)$$

uniformly for $1/2 \le \sigma \le 3/2$.

9. Let χ be a primitive character modulo $q, q > 1$, and suppose that $L(s, \chi) \ne 0$ for $\sigma > 1/2$. Show that the number of zeros $\rho = 1/2 + i\gamma$ of $L(s, \chi)$ with $T \le \gamma \le T + 1/\log \log q\tau$ is $\ll (\log q\tau)/(\log \log q\tau)$ uniformly in T.

10. Let χ be a primitive character modulo $q, q > 1$, and suppose that $L(s, \chi) \ne 0$ for $\sigma > 1/2$. Show that if $|\sigma - 1/2| \le 1/\log \log q\tau$, then

$$\frac{L'}{L}(s, \chi) = \sum_{|\gamma - t| \le 1/\log \log q\tau} \frac{1}{s - \rho} + O(\log q\tau).$$

11. (Selberg 1946b, Section 5) Let χ be a primitive character modulo $q, q > 1$, and suppose that $L(s, \chi) \ne 0$ for $\sigma > 1/2$. Show that

$$\arg L(s, \chi) \ll \frac{\log q\tau}{\log \log q\tau}$$

uniformly for $\sigma \ge 1/2$.

12. Let χ be a character modulo q, and suppose that χ is induced by a primitive character χ^* where χ^* is a character modulo d for some $d|q$. Show that

$$\frac{L'}{L}(s, \chi) - \frac{L'}{L}(s, \chi^*) \ll ((\log q)^{1-\sigma} + 1) \min\left(\frac{1}{|\sigma - 1|}, \log \log q\right).$$

13. (Vorhauer 2006) Let χ be a primitive character modulo $q, q > 1$, and suppose that $L(s, \chi) \ne 0$ for $\sigma > 1/2$. Show that

$$\lim_{T \to \infty} \sum_{|r| \le T} \frac{1}{\rho} = \frac{1}{2} \log q + O(\log \log q).$$

14. (Axer 1911) Assume RH.

(a) Show that if $c = 1/4 + \varepsilon$, then

$$\int_{c-iT}^{c+iT} \left| \frac{\zeta(s)x^s}{\zeta(2s)s} \right| |ds| \ll x^{1/4+\varepsilon} T^{1/4+\varepsilon}.$$

(b) Let $Q(x)$ denote the number of square-free integers not exceeding x. Show that if RH is true, then

$$Q(x) = \frac{6}{\pi^2} x + O\left(x^{2/5+\varepsilon}\right).$$

(A better estimate is obtained in Exercise 16 below.)

15. Assume RH.

(a) Show that if $c = 1/2 + \varepsilon$, then

$$\int_{c-iT}^{c+iT} \left| \frac{\zeta(s)x^s}{\zeta(2s)s(s+1)} \right| |ds| \ll x^{1/4+\varepsilon} T^\varepsilon.$$

(b) Show that if RH is true, then

$$\sum_{n \leq x} \mu(n)^2 (1 - n/x) = \frac{3}{\pi^2} x + O\left(x^{1/4+\varepsilon}\right).$$

16. (Montgomery & Vaughan 1981)

(a) Show that

$$Q(x) = \sum_{\substack{d,m \\ d^2 m \leq x}} \mu(d).$$

Let Σ_1 denote the sum of the above terms for which $d \leq y$, and let Σ_2 denote the sum of the above terms for which $d > y$. Here y is a parameter to be determined later, $1 \leq y \leq x^{1/2}$.

(b) Put

$$S(x, y) = \sum_{d \leq y} \mu(d) B_1(x/d^2)$$

where $B_1(u) = u - 1/2$ is the first Bernoulli polynomial. Show that

$$\Sigma_1 = x \sum_{d \leq y} \frac{\mu(d)^2}{d} - \frac{1}{2} M(y) - S(x, y).$$

(c) Assume RH. Show that if $\sigma \geq 1/2 + 2\varepsilon$, then

$$\sum_{d \leq y} \frac{\mu(d)}{d^s} = \frac{1}{2\pi i} \int_{C_0} \frac{y^{w-s}}{\zeta(w)(w-s)} dw$$

where C_0 is a contour running from $\sigma_0 - i\infty$ to $\sigma_0 - iy$ to $1/2 + \varepsilon - iy$ to $1/2 + \varepsilon + iy$ to $\sigma_0 + iy$ to $\sigma_0 + i\infty$ and $\sigma_0 = 1 + 1/\log y$. Deduce that

$$\sum_{d \leq y} \frac{\mu(d)}{d^s} = \frac{1}{\zeta(s)} + O\left(y^{1/2 - \sigma + \varepsilon} \tau^\varepsilon\right).$$

(d) Put $f_y(s) = 1/\zeta(s) - \sum_{d \leq y} \mu(d)/d^s$. Show that

$$\Sigma_2 = \frac{1}{2\pi i} \int_{\sigma_1 - i\infty}^{\sigma_1 + i\infty} \zeta(s) f_y(2s) \frac{x^s}{s} \, ds$$

where $\sigma_1 = 1 + 1/\log x$.

(e) Show (unconditionally) that

$$\Sigma_2 = f_y(2) + \frac{1}{2\pi i} \int_{C_1} \zeta(s) f_y(2s) \frac{x^s}{s} \, ds$$

where C_1 is a contour running from $\sigma_1 - i\infty$ to $\sigma_1 - ix$ to $1/2 - ix$ to $1/2 + ix$ to $\sigma_1 + ix$ to $\sigma_1 + i\infty$.

(f) Assume RH. Show that $\Sigma_2 \ll x^{1/2 + \varepsilon} y^{-1/2}$.

(g) Note that the estimate $S(x, y) \ll y$ is trivial.

(h) Show that if RH is true, then

$$Q(x) = \frac{6}{\pi^2} x + O\left(x^{1/3 + \varepsilon}\right).$$

13.3 Notes

Section 13.1. Theorem 13.1 is due to von Koch (1901). Theorems 13.3 and 13.5 are due to Cramér (1921). The order of magnitude of the estimate in Theorem 13.5 is optimal, in view of Theorem 13.6, which is from Cramér (1922). Wintner (1941) showed (assuming RH) that the function $f(u)$ defined in (13.17) has a limiting distribution. That is, there is a weakly monotonic function $F(x)$ with $\lim_{x \to -\infty} F(x) = 0$, $\lim_{x \to +\infty} F(x) = 1$, such that

$$\lim_{U \to \infty} \frac{1}{U} \operatorname{meas}\{u \in [0, U] : f(u) \leq x\} = F(x)$$

whenever x is a point of continuity of F. The result of Exercise 13.1.4 is useful in this connection. If in addition to RH, the ordinates $\gamma > 0$ are linearly independent over the field $\mathbb{Q}$ of rational numbers, then this distribution function is the same as the distribution function of the random variable

$$X = 2 \sum_{\gamma > 0} \frac{\cos 2\pi X_\gamma}{\rho}$$

where the X_γ are independent random variables, each one uniformly distributed on $[0, 1]$. It can be shown (unconditionally) that the distribution function F_X of X satisfies the inequalities

$$\exp\left(-c_1\sqrt{x}e^{\sqrt{2\pi x}}\right) < 1 - F_X(x) < \exp\left(-c_2\sqrt{x}e^{\sqrt{2\pi x}}\right) \quad (13.48)$$

for $x \geq 2$ where c_1 and c_2 are positive absolute constants.

Concerning the mean square distribution of primes in short intervals, Selberg (1943) showed (assuming RH) that

$$\int_0^X (\psi((1+\delta)x) - \psi(x) - \delta x)^2 \frac{dx}{x^2} \ll \delta(\log X)^2$$

uniformly for $1/X \leq \delta \leq 1/\log X$. Theorem 13.7 and Corollary 13.8 are due to Titchmarsh (1930). Corollary 13.10 is due to Turán (1937). Theorem 13.11, in the case of the Legendre symbol, is due to Ankeny (1952), who used deeper estimates of Selberg (1946b) found in Exercise 13.1.11. Our simpler proof, and the extension to general non-principal characters, is from Montgomery (1971, p. 120). Theorem 13.12 is from Montgomery (1994, p. 164). See also Lagarias, Montgomery & Odlyzko (1979).

Section 13.2. All results here from Theorem 13.13 through Theorem 13.21 are due to Littlewood (1922, 1924b, 1926, 1928), although our proofs are much simpler than in the original ones. Indeed, referring to Theorem 13.21, Littlewood commented that, 'The proof of this theorem is long and difficult, and depends on a singularly varied set of ideas.' Precursors to Theorem 13.21 were established by Bohr, Landau & Littlewood (1913), Cramér (1918), and Landau (1920). See Titchmarsh (1927) for an alternative proof. Our simpler approach is that of Selberg (1944). Littlewood (1928) not only established Corollary 13.17, but also showed (assuming RH) that

$$|\zeta(1+it)| \geq \frac{\pi^2}{12e^{C_0}\log\log\tau} + O((\log\log\tau)^{-2}).$$

In the opposite direction, Titchmarsh (1928) showed (unconditionally) that

$$\limsup_{t\to+\infty} \frac{|\zeta(1+it)|}{\log\log t} \geq e^{C_0}.$$

Also, Titchmarsh (1933) showed (unconditionally) that

$$\liminf_{t\to+\infty} |\zeta(1+it)|\log\log t \geq \frac{\pi^2}{6e^{C_0}}.$$

Here we see a factor of 2 between the two sets of bounds. The same factor of 2 arises when we consider what is known concerning large values of the zeta function in the critical strip. Let $\alpha(\sigma)$ denote the least number such that

$$\zeta(\sigma + it) \ll \exp\left((\log \tau)^{\alpha(\sigma)+\varepsilon}\right)$$

as $t \to \infty$. From Corollary 13.16 we see that $\alpha(\sigma) \le 2 - 2\alpha$, assuming RH. In the opposite direction, Titchmarsh (1928) showed (unconditionally) that $\alpha(\sigma) \ge 1 - \alpha$. More precisely, it is known that if $1/2 \le \sigma < 1$, then there is a $c(\sigma) > 0$ such that

$$|\zeta(\sigma + it)| = \Omega\left(\exp\left(\frac{c(\sigma)(\log \tau)^{1-\sigma}}{(\log\log \tau)^{\sigma}}\right)\right).$$

For $1/2 < \sigma < 1$ this is due to Montgomery (1977); the case $\sigma = 1/2$ is due to Balasubramanian & Ramachandra (1977). Opinions as to where the truth lies between these bounds vary widely among experts. For more on the value distribution of the zeta function and L-functions, see Titchmarsh (1986), Joyner (1986), and Laurinčikas (1996).

That the estimate $M(x) \ll x^{1/2+\varepsilon}$ is equivalent to RH was proved by Littlewood (1912). Theorems 13.22 through 13.24 are due to Titchmarsh (1927). Theorem 13.24 has been improved upon by Maier & Montgomery (2006), who showed (assuming RH) that

$$M(x) \ll x^{1/2} \exp\left((\log x)^{39/61}\right).$$

13.4 References

Ankeny, N. C. (1952). The least quadratic non residue, *Ann. of Math.* **55**, 65–72.

Axer, A. (1911). Über einige Grenzwertsätze, *S.-B. Wiss. Wien* IIa **120**, 1253–1298.

Balasubramanian, R. & Ramachandra, K. (1977). On the frequency of Titchmarsh's phenomenon for $\zeta(s)$, III, *Proc. Indian Acad. Sci. Sect.* A **86**, 341–351.

Bohr, H., Landau, E., & Littlewood, J. E. (1913). Sur la fonction $\zeta(s)$ dans le voisinage de la droite $\sigma = 1/2$, *Acad. Roy. Belgique Bull. Cl. Sci.*, 1144–1175; *Bohr's Collected Works*, Vol. 1. København: Dansk Mat. Forening, 1952, B.2; *Landau's Collected Works*, Vol. 6. Essen: Thales Verlag, 1986, pp. 61–93; *Littlewood's Collected Papers*, Vol. 2. Oxford: Oxford University Press, 1982, pp. 797–828.

Cramér, H. (1918). Über die Nullstellen der Zetafunktion, *Math. Z.* **2**, 237–241; *Collected Works*, Vol. 1. Berlin: Springer-Verlag, 1994, 92–96.

—— (1921). Some theorems concerning prime numbers, *Arkiv för Mat. Astr. Fys.* **15**, no. 5, 33 pp.; *Collected Works*, Vol. 1. Berlin: Springer-Verlag, 1994, pp. 138–170.

—— (1922). Ein Mittelwertsatz der Primzahltheorie, *Math. Z.* **12**, 147–153; *Collected Works*, Vol. 1. Berlin: Springer-Verlag, 1994, pp. 229–235.

Goldston, D. A. (1982). On a result of Littlewood concerning prime numbers, *Acta Arith.* **40**, 263–271.

Joyner, D. (1986). *Distribution Theorems of L-functions*, Pitman Research Notes in Math. 142. Harlow: Longman.

von Koch, H. (1901). Sur la distribution des nombres premiers, *Acta Math.* **24**, 159–182.

Lagarias, J. C., Montgomery, H. L., & Odlyzko, A. M. (1979). A bound for the least prime ideal in the Chebotarev density theorem, *Invent. Math.* **54**, 271–296.

Landau, E. (1920). Über die Nullstellen der Zetafunktion, *Math. Z.* **6**, 151–154; *Collected Works*, Vol. 7. Essen: Thales Verlag, 1986, pp. 226–229.

Laurinčikas, A. (1996). *Limit Theorems for the Riemann Zeta-function*, Mathematics and its Applications 352. Dordrecht: Kluwer.

Littlewood, J. E. (1912). Quelques conséquences de l'hypothèse que la fonction $\zeta(s)$ de Riemann n'a pas de zéros dans le demi-plan $R(s) > \frac{1}{2}$, *Comptes Rendus Acad. Sci. Paris* **154**, 263–266; *Collected Papers*, Vol. 2. Oxford: Oxford University Press, 1882, pp. 793–796.

(1922). Researches in the theory of the Riemann ζ-function, *Proc. London Math. Soc.* (2) **20**, xxii–xxviii; *Collected Papers*, Vol. 2. Oxford: Oxford University Press, 1982, pp. 844–850.

(1924a). Two notes on the Riemann Zeta-function, *Proc. Cambridge Philos. Soc.* **22**, 234–242; *Collected Papers*, Vol. 2. Oxford: Oxford University Press, 1982, pp. 851–859.

(1924b). On the zeros of the Riemann zeta-function, *Proc. Cambridge Philos. Soc.* **22**, 295–318; *Collected Papers*, Vol. 2. Oxford: Oxford University Press, 1982, pp. 860–883.

(1926). On the Riemann zeta function, *Proc. London Math. Soc.* (2) **24**, 175–201; *Collected Papers*, Vol. 2. Oxford: Oxford University Press, 1982, pp. 844–910.

(1928). Mathematical Notes (5): On the function $1/\zeta(1 + ti)$, *Proc. London Math. Soc.* (2) **27**, 349–357; *Collected Papers*, Vol. 2, Oxford: Oxford University Press, 1982, pp. 911–919.

Maier, H. & Montgomery, H. L. (2006). *On the sum of the Möbius function*, to appear, 16 pp.

Montgomery, H. L. (1971). *Topics in Multiplicative Number Theory*, Lecture Notes in Math. 227. Berlin: Springer-Verlag.

(1977). Extreme values of the Riemann zeta-function, *Comment. Math. Helv.* **52**, 511–518.

(1994). *Ten lectures on the interface between analytic number theory and harmonic analysis*, CMBS 84. Providence: Amer. Math. Soc.

Montgomery, H. L. & Vaughan, R. C. (1981). The distribution of square-free numbers, *Recent Progress in Analytic Number Theory* (Durham, 1979), Vol. 1. London: Academic Press, pp. 247–256.

Selberg, A. (1943). On the normal density of primes in small intervals, *Arch. Math. Natur-vid.* **47**, 87–105; *Collected Papers*, Vol. 1, New York: Springer Verlag, 1989, pp. 160–178.

(1944). On the Remainder in the Formula for $N(T)$, the Number of Zeros of $\zeta(s)$ in the Strip $0 < t < T$. *Avhandl. Norske Vid.-Akad. Oslo I. Mat.-Naturv. Kl.*, no. 1; *Collected Papers*, Vol. 1, New York: Springer Verlag, 1989, pp. 179–203.

(1946a). Contributions to the Theory of the Riemann zeta-function, *Arch. Math. Naturvid.* **48**, 89–155; *Collected Papers*, Vol. 1, New York: Springer Verlag, 1989, pp. 214–280.

(1946b). Contributions to the Theory of Dirichlet's L-functions, *Skrifter Norske Vid.-Akad. Oslo I. Mat.-Naturvid. Kl.*, no. 3; *Collected Papers*, Vol. 1, New York: Springer Verlag, 1989, pp. 281–340.

Titchmarsh, E. C. (1927). A consequence of the Riemann hypothesis, *J. London Math. Soc.* **2**, 247–254.

(1928). On an inequality satisfied by the zeta-function of Riemann, *Proc. London Math. Soc.* (2) **28**, 70–80.

(1930). A divisor problem, *Rend. Circ. Mat. Palermo* **54**, 414–429.

(1933). On the function $1/\zeta(1 + it)$, *Quart. J. Math.* Oxford **4**, 64–70.

(1986). *The Theory of the Riemann Zeta-function*, Second edition. Oxford: Oxford University Press.

Turán, P., (1937). Über die Primzahlen der Arithmetischen Progression, I, *Acta Sci. Szeged* **8**, 226–235; *Collected Papers*, Vol. 1. Budapest: Akadémiai Kiadó, 1990, pp. 64–73.

Vorhauer, U. M. A. (2006). *The Hadamard product formula for Dirichlet L-functions*, to appear.

Wintner, A. (1941). On the distribution function of the remainder term of the Prime Number Theorem, *Amer. J. Math.* **63**, 233–248.

14

Zeros

14.1 General distribution of the zeros

If $T > 0$ is not the ordinate of a zero of the zeta function, then we let $N(T)$ denote the number of zeros $\rho = \beta + i\gamma$ of $\zeta(s)$ in the rectangle $0 < \beta < 1, 0 < \gamma < T$. If T is the ordinate of a zero, then we set $N(T) = (N(T^+) + N(T^-))/2$. By the argument principle we obtain

Theorem 14.1 *For any real t, put*

$$S(t) = \frac{1}{\pi} \arg \zeta(1/2 + it). \tag{14.1}$$

If $T > 0$, then

$$N(T) = \frac{1}{\pi} \arg \Gamma(1/4 + iT/2) - \frac{T}{2\pi} \log \pi + S(T) + 1. \tag{14.2}$$

Proof Since

$$N(T) = \frac{1}{2}(N(T^+) + N(T^-)), \qquad S(T) = \frac{1}{2}(S(T^+) + S(T^-)),$$

it suffices to prove (14.2) when T is not the ordinate of a zero. Let $\mathcal{C}$ denote the contour that proceeds by straight lines from 2 to $2 + iT$ to $-1 + iT$ to -1 to 2. Then by the argument principle,

$$N(T) = \frac{1}{2\pi i} \int_{\mathcal{C}} \frac{\xi'}{\xi}(s) \, ds.$$

Now let $\mathcal{C}_1$ denote the contour that proceeds by line segments from $1/2$ to 2 to $2 + iT$ to $1/2 + iT$, and let $\mathcal{C}_2$ be the contour that proceeds from $1/2 + iT$ to $-1 + iT$ to -1 to $1/2$. Thus $\int_{\mathcal{C}} = \int_{\mathcal{C}_1} + \int_{\mathcal{C}_2}$. For $s \in \mathcal{C}_2$ we use the identity

$$\frac{\xi'}{\xi}(s) = -\frac{\xi'}{\xi}(1 - s),$$

and thus we see that

$$\int_{C_2} \frac{\xi'}{\xi}(s)\,ds = -\int_{C_2} \frac{\xi'}{\xi}(1-s)\,ds = \int_{C_3} \frac{\xi'}{\xi}(s)\,ds$$

where C_3 proceeds from $1/2 - iT$ to $2 - iT$ to 2 to $1/2$. On adding this to the integral over C_1, we see that the contribution of the interval $[1/2, 2]$ cancels, and hence

$$N(T) = \frac{1}{2\pi i} \int_{C_4} \frac{\xi'}{\xi}(s)\,ds$$

where C_4 runs from $1/2 - iT$ to $2 - iT$ to $2 + iT$ to $1/2 + iT$. By (10.25) we see that the above is

$$= \frac{1}{2\pi i} \left[\log s + \log(s-1) + \log \zeta(s) + \log \Gamma(s/2) - \frac{s}{2} \log \pi \right] \Big|_{1/2-iT}^{1/2+iT}.$$

By the Schwarz reflection principle, the real parts cancel and the imaginary parts reinforce. Thus the above is

$$= \frac{1}{\pi} \Bigg(\arg(1/2 + iT) + \arg(-1/2 + iT) + \arg \zeta(1/2 + iT)$$

$$+ \arg \Gamma(1/4 + iT/2) - \frac{T}{2} \log \pi \Bigg).$$

Here $\arg(1/2 + iT) + \arg(-1/2 + iT) = \pi$, so we have the stated result. □

By Stirling's formula (Theorem C.1) we know that

$$\log \Gamma(s) = (s - 1/2) \log s - s + \frac{1}{2} \log 2\pi + O(1/|s|). \qquad (14.3)$$

By using this, we obtain

Corollary 14.2 *For $T \geq 2$,*

$$N(T) = \frac{T}{2\pi} \log \frac{T}{2\pi} - \frac{T}{2\pi} + \frac{7}{8} + S(T) + O(1/T).$$

Proof Clearly

$$\Im\big((-1/4 + iT/2) \log(1/4 + iT/2) - (1/4 + iT/2)\big)$$

$$= -\frac{1}{4} \arg\big(\tfrac{1}{4} + i\tfrac{T}{2}\big) + \frac{T}{4} \log\big(\tfrac{1}{16} + \tfrac{T^2}{4}\big) - \frac{T}{2}.$$

But $\arg(1/4 + iT/2) = \pi/2 + O(1/T)$, and $\log(1/16 + T^2/4) = 2\log T/2 + O(1/T^2)$, so we obtain the stated result. □

By combining the above with Lemma 12.3 or Theorem 13.20, we obtain

Corollary 14.3 *For $T \geq 4$,*

$$N(T) = \frac{T}{2\pi} \log \frac{T}{2\pi} - \frac{T}{2\pi} + O(\log T).$$

Corollary 14.4 *If the Riemann Hypothesis is true, then*

$$N(T) = \frac{T}{2\pi} \log \frac{T}{2\pi} - \frac{T}{2\pi} + O\left(\frac{\log T}{\log \log T}\right).$$

Note that these estimates imply the estimates of Theorem 10.13 and Lemma 13.18, respectively. In addition, from the first estimate above we see that there is an absolute constant $C > 0$ such that

$$N(T + h) - N(T) \asymp h \log T \tag{14.4}$$

uniformly for $C \leq h \leq T$. Similarly, there is an absolute constant $C > 0$ such that if RH is true, then (14.4) holds for $C/\log \log T \leq h \leq T, T \geq 4$. By modifying our method we obtain corresponding estimates for the number of zeros of a Dirichlet L-function.

Theorem 14.5 *Let χ be a primitive character modulo q, with $q > 1$. For $T > 0$, let $N(T, \chi)$ denote the number of zeros $\rho = \beta + i\gamma$ of $L(s, \chi)$ with $0 < \beta < 1$ and $0 \leq \gamma \leq T$. Any zeros with $\gamma = 0$ or $\gamma = T$ should be counted with weight $1/2$. Also, for any real number T, put*

$$S(T, \chi) = \frac{1}{\pi} \arg L(1/2 + iT, \chi). \tag{14.5}$$

Then

$$N(T, \chi) = \frac{1}{\pi} \arg \Gamma(1/4 + \kappa/2 + iT/2) + \frac{T}{2\pi} \log \frac{q}{\pi} + S(T, \chi) - S(0, \chi)$$

where $\kappa = 0$ or 1 according as $\chi(-1) = 1$ or -1.

There is no need to establish a separate result pertaining to zeros with $\gamma < 0$, since the number of zeros of $L(s, \chi)$ with $-T \leq \gamma \leq 0$ is $N(T, \overline{\chi})$.

Proof We may assume that T is not the ordinate of a zero, for if it were, then we have only to replace T by $T^{\pm}$, and average. However, we must take some precautions against the possibility that $L(s, \chi)$ has a zero on the real axis in the interval $(0, 1)$. Let $\mathcal{C}^{\pm}$ be the contour from $2 \pm i\varepsilon$ to $2 + iT$ to $-1 + iT$ to $-1 \pm i\varepsilon$ to $2 \pm i\varepsilon$, let $\mathcal{C}_1^{\pm}$ be the contour from $1/2 \pm i\varepsilon$ to $2 \pm i\varepsilon$ to $2 + iT$ to $1/2 + iT$, let $\mathcal{C}_2^{\pm}$ be the path from $1/2 + iT$ to $-1 + iT$ to $-1 \pm i\varepsilon$ to $1/2 \pm i\varepsilon$, and let $\mathcal{C}_3^{\pm}$ be the path from $1/2 - iT$ to $2 - iT$ to $2 \mp i\varepsilon$ to $1/2 \mp i\varepsilon$. By the argument principle, the number of zeros with $0 < \gamma \leq T$ is

$$\frac{1}{2\pi i} \int_{\mathcal{C}^+} \frac{\xi'}{\xi}(s, \chi) \, ds = \frac{1}{2\pi i} \int_{\mathcal{C}_1^+} \frac{\xi'}{\xi}(s, \chi) \, ds + \frac{1}{2\pi i} \int_{\mathcal{C}_2^+} \frac{\xi'}{\xi}(s, \chi) \, ds.$$

For $s \in \mathcal{C}_2^+$ we write

$$\frac{\xi'}{\xi}(s, \chi) = -\frac{\xi'}{\xi}(1 - s, \overline{\chi}),$$

and thus we find that

$$\int_{\mathcal{C}_2^+} \frac{\xi'}{\xi}(s, \chi)\, ds = -\int_{\mathcal{C}_2^+} \frac{\xi'}{\xi}(1 - s, \overline{\chi})\, ds = \int_{\mathcal{C}_3^+} \frac{\xi'}{\xi}(s, \overline{\chi})\, ds.$$

By (10.33), it follows that

$$\int_{\mathcal{C}_1^+} \frac{\xi'}{\xi}(s, \chi)\, ds = \left[\log L(s, \chi) + \log \Gamma((s + \kappa)/2) + \frac{s}{2} \log q/\pi \right]\Big|_{1/2+i\varepsilon}^{1/2+iT}$$

$$= \log L(1/2 + iT, \chi) - \log L(1/2 + i\varepsilon, \chi)$$
$$+ \log \Gamma(1/4 + \kappa/2 + iT/2) - \log \Gamma(1/4 + \kappa/2 + i\varepsilon/2)$$
$$+ i\frac{T - \varepsilon}{2} \log \frac{q}{\pi},$$

and that

$$\int_{\mathcal{C}_3^+} \frac{\xi'}{\xi}(s, \overline{\chi})\, ds = \left[\log L(s, \overline{\chi}) + \log \Gamma((s + \kappa)/2) + \frac{s}{2} \log q/\pi \right]\Big|_{1/2-iT}^{1/2-i\varepsilon}$$

$$= \log L(1/2 - i\varepsilon, \overline{\chi}) - \log L(1/2 - iT, \overline{\chi})$$
$$+ \log \Gamma(1/4 + \kappa/2 - i\varepsilon/2) - \log \Gamma(1/4 + \kappa/2 - iT/2)$$
$$+ i\frac{T - \varepsilon}{2} \log \frac{q}{\pi}.$$

When these quantities are added, the real parts cancel and the imaginary parts are doubled, so after dividing by $2\pi i$ we find that the number of zeros with $0 < \gamma \le T$ is

$$\frac{1}{\pi} \arg \Gamma(1/4 + \kappa/2 + iT/2) + S(T, \chi) - S(0^+, \chi) + \frac{T}{2\pi} \log \frac{q}{\pi}.$$

By proceeding similarly with the opposite sign, we find that the number of zeros with $0 \le \gamma \le T$ is

$$\frac{1}{\pi} \arg \Gamma(1/4 + \kappa/2 + iT/2) + S(T, \chi) - S(0^-, \chi) + \frac{T}{2\pi} \log \frac{q}{\pi}.$$

We form the average of these two identities to obtain the stated result. $\qquad\square$

Corollary 14.6 *Let χ be a primitive character modulo q, with $q > 1$. Then for $T > 0$,*

$$N(T, \chi) = \frac{T}{2\pi} \log \frac{qT}{2\pi} - \frac{T}{2\pi} + S(T, \chi) - S(0, \chi) - \chi(-1)/8 + O(1/(T + 1)).$$

Proof If $0 < T \le 2$, then $\arg \Gamma(1/4 + \kappa/2 + iT/2) \ll 1$ and $T \log T/2 - T \ll 1$, so the estimate is immediate in this case. Suppose that $T \ge 2$.

Clearly

$$\Im((-1/4+\kappa/2+iT/2)\log(1/4+\kappa/2+iT/2)-(1/4+\kappa/2+iT/2))$$
$$=(-1/4+\kappa/2)\arg(1/4+\kappa/2+iT/2)+\frac{T}{4}\log((1/4+\kappa/2)^2+T^2/4)-\frac{T}{2}.$$

Here $\arg(1/4+\kappa/2+iT/2)=\pi/2+O(1/T)$, $\log((1/4+\kappa/2)^2+T^2/4)=2\log T/2+O(1/T^2)$, and $2\kappa-1=-\chi(-1)$, so the result follows by Stirling's formula in the form (14.3). $\square$

By combining the above with Lemma 12.8 we obtain

Corollary 14.7   *Let* χ *be a primitive character modulo* q, $q > 1$. *Then for* $T \geq 4$,

$$N(T,\chi)=\frac{T}{2\pi}\log\frac{qT}{2\pi}-\frac{T}{2\pi}+O(\log qT).$$

14.1.1   Exercise

1. Let χ be a primitive character modulo q with $q > 1$. Show that if $L(s,\chi)\neq 0$ for $\sigma > 1/2$, then

$$N(T,\chi)=\frac{T}{2\pi}\log\frac{qT}{2\pi}-\frac{T}{2\pi}+O\left(\frac{\log qT}{\log\log qT}\right)$$

for $T \geq 2$.

14.2   Zeros on the critical line

At present we are unable to prove the Riemann Hypothesis, which asserts that all non-trivial zeros of the zeta function lie on the critical line $\sigma = 1/2$. However, we are able to show that infinitely many zeros lie on this line.

Theorem 14.8   (Hardy) *There exist infinitely many real numbers* γ *such that* $\zeta(1/2+i\gamma)=0$.

For real t, let

$$Z(t)=\zeta(1/2+it)\frac{\Gamma(1/4+it/2)\pi^{-1/4-it/2}}{|\Gamma(1/4+it/2)\pi^{-1/4-it/2}|}. \tag{14.6}$$

Thus, as depicted in Figure 14.1, $Z(t)$ is real-valued, $|Z(t)|=|\zeta(1/2+it)|$, and $Z(t)$ changes sign at γ if and only if $\zeta(s)$ has a zero at $1/2+i\gamma$ of odd

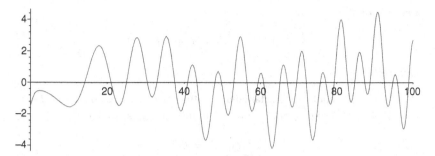

Figure 14.1 Graph of $Z(t)$ for $0 \leq t \leq 100$.

multiplicity. If $T > 0$ is a real number such that

$$\left| \int_T^{2T} Z(t)\,dt \right| < \int_T^{2T} |Z(t)|\,dt, \tag{14.7}$$

then $Z(t)$ is not of constant sign in the interval $(T, 2T)$, which is to say that $\zeta(s)$ has at least one zero $1/2 + i\gamma$ of odd multiplicity, with $T < \gamma < 2T$. Although it is possible to show that (14.7) holds for all large T, the requisite arguments involve technical tools that we have not yet developed. Fortunately, there is a family of weights $W(t)$ such that the integral $\int W(t)Z(t)\,dt$ can be evaluated by interpreting it as an inverse Mellin transform with a familiar kernel. Thus we are able to establish a weighted variant of (14.7), which suffices for our purpose. In preparation for the main argument, we establish two preliminary results.

Lemma 14.9 *If* $\Re z > 0$ *and* $\sigma_0 > 1$, *then*

$$\frac{1}{2\pi i} \int_{\sigma_0-i\infty}^{\sigma_0+i\infty} \zeta(s)\Gamma(s/2)(\pi z)^{-s/2}\,ds = 2\sum_{n=1}^{\infty} e^{-\pi n^2 z}.$$

This is the inverse of the Mellin transform relationship (10.7) that Riemann used to establish the functional equation.

Proof By Theorem C.4 we see that if $\Re w > 0$ and $\sigma_0 > 0$, then

$$\frac{1}{2\pi i} \int_{\sigma_0-i\infty}^{\sigma_0+i\infty} \Gamma(s/2)w^{-s/2}\,ds = 2e^{-w}.$$

We take $w = \pi n^2 z$, and sum over n, to obtain the desired identity. Here the exchange of summation and integration is permissible since the Dirichlet series for $\zeta(s)$ is uniformly convergent on the abscissa σ_0, and since

$$\int_{-\infty}^{\infty} \left| \Gamma((\sigma_0 + it)/2)(\pi z)^{-s/2} \right| dt < \infty.$$

$\square$

Lemma 14.10 *We have*

$$\int_1^T \zeta(1/2 + it)\, dt = T + O(T^{1/2})$$

uniformly for $T \geq 2$.

Proof Let $\mathcal{C}$ denote the rectangular contour with vertices $1/2 + i$, $2 + i$, $2 + iT$, $1/2 + iT$. Since $\zeta(s)$ is analytic in this rectangle, we have

$$\int_{\mathcal{C}} \zeta(s)\, ds = 0$$

by Cauchy's theorem. The integral from $1/2 + i$ to $2 + i$ is an absolute constant, and by Corollary 1.17 the integral from $1/2 + iT$ to $2 + iT$ is

$$\ll \int_{1/2}^2 \left(1 + T^{1-\sigma}\right)(\log T)\, d\sigma \ll T^{1/2}.$$

Thus

$$\int_1^T \zeta(1/2 + it)\, dt = \int_1^T \zeta(2 + it)\, dt + O(T^{1/2}).$$

This latter integral is

$$= \sum_{n=1}^\infty n^{-2} \int_1^T n^{-it}\, dt = T - 1 + \sum_{n=2}^\infty \frac{n^{-i} - n^{-iT}}{in^2 \log n} = T + O(1),$$

so we have the stated result. □

Proof of Theorem 14.8 The integrand in Lemma 14.9 has a pole at $s = 1$ with residue $z^{-1/2}$, but is otherwise analytic for $\sigma > 0$. We move the path of integration to the line $\sigma = 1/2$, and multiply both sides by $z^{1/4}$ to see that

$$\frac{1}{2\pi} \int_{-\infty}^\infty \zeta(1/2 + it)\Gamma(1/4 + it/2)\pi^{-1/4 - it/2} z^{-it/2}\, dt$$

$$\tag{14.8}$$

$$= -z^{-1/4} + 2z^{1/4} \sum_{n=1}^\infty e^{-\pi n^2 z}.$$

Here the left-hand side is of the form $\int_{-\infty}^\infty W(t)Z(t)\, dt$ with

$$W(t) = \frac{|\Gamma(1/4 + it/2)|}{2\pi^{5/4} z^{it/2}}.$$

Write z in polar coordinates, $z = re^{i\theta}$. Then $z^{-it/2} = r^{-it/2} e^{\theta t/2}$. For our approach to work, $W(t)$ must have constant argument. Accordingly, we take $r = 1$, and set $\theta = \pi/2 - \delta$ where δ is small and positive. By (C.19) we see that

$$|\Gamma(s/2)| \asymp \tau^{(\sigma-1)/2} e^{-\pi\tau/4}.$$

Hence

$$W(t) \asymp \tau^{-1/4} e^{\pi(t-\tau)/4} e^{-\delta t/2} \asymp \begin{cases} \tau^{-1/4} e^{-(\pi-\delta)\tau/2} & \text{if } t \geq 0, \\ \tau^{-1/4} e^{-(1-\delta)\pi\tau/2} & \text{if } t \leq 0. \end{cases}$$

Thus $W(t)$ tends to 0 very rapidly as $t \to -\infty$, but relatively slowly as $t \to +\infty$. In particular,

$$W(t) \asymp \tau^{-1/4}$$

uniformly for $0 \leq t \leq 1/\delta$.

By the above and Lemma 14.10 we see that

$$\int_{-\infty}^{\infty} W(t)|Z(t)| \, dt \gg \delta^{1/4} \int_{1/(2\delta)}^{1/\delta} |Z(t)| \, dt = \delta^{1/4} \int_{1/(2\delta)}^{1/\delta} |\zeta(1/2 + it)| \, dt$$
$$\gg \delta^{-3/4}.$$

In order to exhibit a disparity, we must show that the right-hand side of (14.8) is $o(\delta^{-3/4})$. To this end it suffices to argue fairly crudely. Since $z = ie^{-i\delta} = \sin\delta + i \cos\delta$, by the triangle inequality the right-hand side of (14.8) is

$$\ll \sum_{n=1}^{\infty} e^{-\pi n^2 \sin\delta}.$$

By the integral test this is

$$\leq \int_0^{\infty} e^{-\pi u^2 \sin\delta} \, du = (\sin\delta)^{-1/2} \int_0^{\infty} e^{-\pi v^2} \, dv \ll \delta^{-1/2}.$$

If $\zeta(s)$ had only finitely many zeros on the critical line, then we would have

$$\left| \int_{-\infty}^{\infty} W(t)Z(t) \, dt \right| = \int_{-\infty}^{\infty} W(t)|Z(t)| \, dt + O(1)$$

uniformly as $\delta \to 0^+$. On the contrary, we have shown that

$$\int_{-\infty}^{\infty} W(t)Z(t) \, dt \ll \delta^{-1/2}, \qquad \int_{-\infty}^{\infty} W(t)|Z(t)| \, dt \gg \delta^{-3/4},$$

so the theorem is proved. □

14.2.1 Exercise

1. (a) Show that the right-hand side of (14.8) is

$$= -z^{-1/4} - z^{1/4} + z^{1/4}\vartheta(z),$$

in the notation of (10.8).

(b) Show that if $z = ie^{-i\delta} = \sin \delta + i \cos \delta$, then

$$\vartheta(z) = \sum_{n=-\infty}^{\infty} (-1)^n (1 + O(n^2\delta^2))e^{-\pi n^2 \sin \delta}.$$

(c) Show that

$$\sum_{n=-\infty}^{\infty} n^2 e^{-\pi n^2 \sin \delta} \asymp \delta^{-3/2}$$

for $0 < \delta \leq 1$.

(d) By taking $\alpha = 1/2$ in Theorem 10.1, or otherwise, show that

$$\sum_{n=-\infty}^{\infty} (-1)^n e^{-\pi n^2 x} \asymp x^{-1/2} e^{-\pi/(4x)}$$

uniformly for $0 < x \leq 1$.

(e) Show that if z is taken as in (b), then $\vartheta(z) \ll \delta^{1/2}$.

(f) Conclude that the right-hand side of (14.8) is $= -2 \cos \pi/8 + O(\delta^{1/2})$.

14.3   Notes

Section 14.1. Theorem 14.1 and Corollary 14.2 are due to Backlund (1914, 1918), and this gave a shorter proof of Corollary 14.3 which had been obtained by von Mangoldt (1905). Earlier von Mangoldt (1895) had the error term $O((\log T)^2)$. Riemann (1859) proposed Corollary 14.3 but with no indication of a proof. It is remarkable that Corollary 14.3 is perhaps the only theorem on the Riemann zeta function that has not seen some significant improvement in the last 100 years.

Although the maximum order of $S(t)$ is unclear, even assuming the Riemann Hypothesis, we have considerable (unconditional) knowledge of its moments and distribution. Selberg (1944) showed that if k is a fixed non-negative even integer, then

$$\int_0^T S(t)^k \, dt = \frac{k!}{(k/2)!(2\pi)^k} T(\log\log T)^{k/2} + O(T(\log\log T)^{k/2-1}).$$

Although Selberg did not mention it, his techniques can also be used to show that

$$\int_0^T S(t)^k \, dt = o(T(\log\log T)^{k/2})$$

when k is odd. From these estimates it follows that the distribution of $S(t)$ is asymptotically normal, in the sense that

$$\lim_{T\to\infty}\frac{1}{T}\operatorname{meas}\{t\in[0,T] \,:\, 2\pi S(t)\le c\log\log T\}=\frac{1}{\sqrt{2\pi}}\int_\infty^c e^{-t^2/2}\,dt$$

for any given real number c. Similar results apply to the distribution of the real part of $\log\zeta(1/2+it)$, and indeed Selberg (unpublished) showed that the real and imaginary parts can be treated simultaneously. Specifically,

$$\int_0^T (\log\zeta(1/2+it))^h(\log\zeta(1/2-it))^k dt = \delta_{h,k}k!T(\log\log T)^k$$
$$+ O_{h,k}\left(T(\log\log T)^{(h+k-1)/2}\right)$$

where

$$\delta_{h,k}=\begin{cases}1 & \text{if } h=k,\\ 0 & \text{otherwise.}\end{cases}$$

From this it follows that $\log\zeta(1/2+it)$ is asymptotically normally distributed in the complex plane, in the sense that if Ω is a set in the complex plane with Jordan content, then

$$\lim_{T\to\infty}\frac{1}{T}\operatorname{meas}\left\{t\in[4,T] \,:\, \frac{\log\zeta(1/2+it)}{\sqrt{\log\log t}}\in\Omega\right\}=\frac{1}{\pi}\iint_\Omega e^{-|z|^2}\,dx\,dy.$$

Section 14.2. Theorem 14.8 was announced and a proof sketched in Hardy (1914). Further details are given in Hardy & Littlewood (1917). Let $N_0(T)$ denote the number of zeros of the form $1/2+i\gamma$ with $0<\gamma\le T$. Hardy & Littlewood (1921) showed that $N_0(T)\gg T$. Later Selberg improved this, first (1942a) to $N_0(T)\gg T\log\log T$ and then (1942b) to $N_0(T)\gg T\log T$, so that a positive proportion of the zeros are on the $\frac{1}{2}$-line. Levinson (1974) introduced an alternative method that enabled him to show that at least one-third of the non-trivial zeros are on the $\frac{1}{2}$-line. Selberg's method detects only zeros of odd multiplicity. This should not be a handicap, since presumably all zeros are simple. Heath-Brown (1979) has observed that Levinson's method detects only simple zeros. Conrey (1989) used Levinson's method to show that $N_0(T)\gtrsim\frac{2}{5}N(T)$.

The proof we have given of Hardy's Theorem 14.8 is but one of several described by Titchmarsh (1986, Chapter 10).

14.4 References

Backlund, R. J. (1914). Sur les zéros de la fonction $\zeta(s)$ de Riemann, *C. R. Acad. Sci. Paris* **158**, 1979–1981.

(1918). Über die Nullstellen der Riemannschen Zetafunktion, *Acta Math.* **41**, 345–
375.

Conrey, J. B. (1989). More than two fifths of the zeros of the Riemann zeta function are
on the critical line, *J. Reine Angew. Math.* **399**, 1–26.

Hardy, G. H. (1914). Sur les zéros de la fonction $\zeta(s)$ de Riemann, *C. R. Acad. Sci. Paris*
158, 1012–1014; *Collected Papers*, Vol. 2, Oxford: Oxford University Press, 1967,
pp. 6–8.

Hardy, G. H. & Littlewood, J. E. (1917). Contributions to the theory of the Riemann
Zeta-function and the theory of the distribution of primes, *Acta Math.* **41**, 119–196;
Collected Papers, Vol. 2, Oxford: Oxford University Press, 1967, pp. 20–97.

(1921). The zeros of Riemann's zeta-function on the critical line, *Math. Z.* **10**, 283–
317; *Collected Papers*, Vol. 2, Oxford: Oxford University Press, 1967, pp. 115–149.

Heath–Brown, D. R. (1979). Simple zeros of the Zeta-function on the critical line, *Bull.
London Math. Soc.* **11**, 17–18.

Levinson, N. (1974). More than one third of zeros of Riemann's zeta-function are on
$\sigma = 1/2$, *Adv. Math.* **13**, 383–436.

von Mangoldt, H. (1895). Zu Riemann's Abhandlung "Ueber die Anzahl der Primzahlen
unter einer gegebenen Grösse", *J. Reine Angew. Math.* **114**, 255–305.

(1905). Zur Verteilung der Nullstellen der Riemannschen Funktion $\xi(t)$, *Math. Ann.*
60, 1–19.

Riemann, B. (1859). Ueber die Anzahl der Primzahlen unter eine gegebenen Grösse,
Monatsber. Kgl. Preuss. Akad. Wiss. Berlin, 671–680; *Werke*, Leipzig: Teubner,
1876, pp. 3–47. Reprint: New York: Dover, 1953.

Selberg, A. (1942a). On the zeros of Riemann's zeta-function on the critical line, *Arch.
Math. Naturvid.* **45**, 101–114; *Collected Papers*, Vol. 1, New York: Springer Verlag,
1989, pp. 142–155.

(1942b). On the Zeros of Riemann's Zeta-function, *Skr. Norske Vid. Akad. Oslo* I.,
no. 10; *Collected Papers*, Vol. 1, New York: Springer Verlag, 1989, pp. 156–159.

(1944). On the Remainder in the Formula for $N(T)$, the Number of Zeros of $\zeta(s)$ in
the Strip $0 < t < T$, *Avh. Norske Vid. Akad. Oslo.* I, no. 1; *Collected Papers*, Vol.
1, New York: Springer Verlag, 1989, pp. 179–203.

Titchmarsh, E. C. (1986). *The Theory of the Riemann Zeta-function*, Second edition.
New York: Oxford University Press.

15

Oscillations of error terms

15.1   Applications of Landau's theorem

In this section we make repeated use of the following simple analogue of Landau's theorem (Theorem 1.7) concerning Dirichlet series with non-negative coefficients.

Lemma 15.1 *Suppose that $A(x)$ is a bounded Riemann-integrable function in any finite interval $1 \le x \le X$, and that $A(x) \ge 0$ for all $x > X_0$. Let σ_c denote the infimum of those σ for which $\int_{X_0}^{\infty} A(x)x^{-\sigma}\, dx < \infty$. Then the function*

$$F(s) = \int_1^{\infty} A(x)x^{-s}\, dx$$

is analytic in the half-plane $\sigma > \sigma_c$, but not at the point $s = \sigma_c$.

Proof   Write

$$F(s) = \int_1^{X_0} A(x)x^{-s}\, dx + \int_{X_0}^{\infty} A(x)x^{-s}\, dx = F_1(s) + F_2(s),$$

say. Then the function $F_1(s)$ is entire, and the proof of Theorem 1.7 can be adapted to $F_2(s)$ to give the stated result.   □

In Exercise 13.1.1 we saw that if Θ denotes the supremum of the real parts of the zeros of the zeta function, then $\psi(x) = x + O(x^{\Theta}(\log x)^2)$. Conversely, if $\psi(x) = x + O(x^{\alpha+\varepsilon})$, then by Theorem 1.3 the Dirichlet series $\sum_{n=1}^{\infty}(\Lambda(n) - 1)n^{-s}$ converges for $\sigma > \alpha$, and hence $\zeta(s) \ne 0$ in this half-plane. That is, $\psi(x) - x = \Omega(x^{\Theta-\varepsilon})$. We now sharpen this, by showing that $\psi(x) - x$ must be large in both signs.

463

Theorem 15.2 *Let Θ denote the supremum of the real parts of the zeros of the zeta function. Then for every $\varepsilon > 0$,*

$$\psi(x) - x = \Omega_\pm(x^{\Theta - \varepsilon}) \tag{15.1}$$

and

$$\pi(x) - \mathrm{li}(x) = \Omega_\pm(x^{\Theta - \varepsilon}) \tag{15.2}$$

as $x \to \infty$.

Proof By Theorem 1.3 we have

$$-\frac{\zeta'}{\zeta}(s) = s \int_1^\infty \psi(x) x^{-s-1}\, dx$$

for $\sigma > 1$. Hence

$$-\frac{\zeta'(s)}{s\zeta(s)} - \frac{1}{s-1} = \int_1^\infty (\psi(x) - x) x^{-s-1}\, dx$$

for $\sigma > 1$. Suppose that

$$\psi(x) - x < x^{\Theta - \varepsilon} \text{ for all } x > X_0(\varepsilon). \tag{15.3}$$

Then we apply Lemma 15.1 to the function

$$\frac{1}{s - \Theta + \varepsilon} + \frac{\zeta'(s)}{s\zeta(s)} + \frac{1}{s-1} = \int_1^\infty (x^{\Theta - \varepsilon} - \psi(x) + x) x^{-s-1}\, dx.$$

Here the left-hand side has a pole at $\Theta - \varepsilon$, but is analytic for real $s > \Theta - \varepsilon$, in view of Corollary 1.14. Hence the above identity holds for $\sigma > \Theta - \varepsilon$, and both sides are analytic in this half-plane. But by the definition of Θ, the function ζ'/ζ has poles with real part $> \Theta - \varepsilon$. From this contradiction we deduce that the assertion (15.3) is false. That is, $\psi(x) - x = \Omega_+(x^{\Theta - \varepsilon})$. To obtain the corresponding Ω_- estimate we argue similarly using the identity

$$\frac{1}{s - \Theta + \varepsilon} - \frac{\zeta'(s)}{s\zeta(s)} - \frac{1}{s-1} = \int_1^\infty (x^{\Theta - \varepsilon} + \psi(x) - x) x^{-s-1}\, dx.$$

In contrast to the situation of Corollary 2.5 or Theorem 13.2, it does not seem possible to derive (15.2) from (15.1) by integrating by parts. Instead, we pursue an argument modelled on the one just given. First we examine the Mellin transform of $\mathrm{li}(x)$. By integrating by parts we see that

$$s \int_2^\infty \mathrm{li}(x) x^{-s-1}\, dx = \int_2^\infty \frac{dx}{x^s \log x} = \int_{(s-1)\log 2}^\infty e^{-u}\, \frac{du}{u}.$$

Clearly this is

$$= \int_1^\infty e^{-u} \frac{du}{u} + \int_{(s-1)\log 2}^1 \frac{e^{-u} - 1}{u} \, du - \log(s-1) - \log\log 2.$$

By (7.31) we see that this is

$$= -\int_0^{(s-1)\log 2} \frac{e^{-u} - 1}{u} \, du - C_0 - \log(s-1) - \log\log 2.$$

Thus we find that

$$s \int_2^\infty \mathrm{li}(x) x^{-s-1} \, dx = -\log(s-1) + r(s)$$

where $r(s)$ is an entire function. Put

$$\Pi(x) = \sum_{n \le x} \frac{\Lambda(n)}{\log n}.$$

By Theorem 1.3 we know that

$$s \int_2^\infty \Pi(x) x^{-s-1} \, dx = \log \zeta(s)$$

for $\sigma > 1$. Hence

$$\frac{1}{s - \Theta + \varepsilon} - \frac{1}{s} \log(\zeta(s)(s-1)) + \frac{r(x)}{s}$$
$$= \int_2^\infty (x^{\Theta - \varepsilon} - \Pi(x) + \mathrm{li}(x)) x^{-s-1} \, dx$$

for $\sigma > 1$. We observe that this function is analytic on the real axis for $s > \Theta - \varepsilon$. Thus by Lemma 1, if $\Pi(x) - \mathrm{li}(x) < x^{\Theta - \varepsilon}$ for all sufficiently large x, then the identity above holds in the half-plane $\sigma > \Theta - \varepsilon$. However, we are assuming that the zeta function has a zero $\rho = \beta + i\gamma$ with $\beta > \Theta - \varepsilon$, and the left-hand side above has a logarithmic singularity at $s = \rho$. Thus we have a contradiction, and so $\Pi(x) - \mathrm{li}(x) = \Omega_+(x^{\Theta - \varepsilon})$. Since $\pi(x) = \Pi(x) + O(x^{1/2}/\log x)$, and since $\Theta \ge 1/2$, it follows that $\pi(x) - \mathrm{li}(x) = \Omega_+(x^{\Theta - \varepsilon})$. For the corresponding Ω_- estimate, we argue similarly from the identity

$$\frac{1}{s - \Theta + \varepsilon} + \frac{1}{s} \log(\zeta(s)(s-1)) - \frac{r(x)}{s}$$
$$= \int_2^\infty (x^{\Theta - \varepsilon} + \Pi(x) - \mathrm{li}(x)) x^{-s-1} \, dx.$$

$\square$

Next we show that if there is a zero of $\zeta(s)$ on the line $\sigma = \Theta$, then we may draw a stronger conclusion.

Theorem 15.3 *Suppose that Θ is the supremum of the real parts of the zeros of $\zeta(s)$, and that there is a zero ρ with $\Re\rho = \Theta$, say $\rho = \Theta + i\gamma$. Then*

$$\limsup_{x\to\infty} \frac{\psi(x) - x}{x^\Theta} \geq \frac{1}{|\rho|}, \tag{15.4}$$

and

$$\liminf_{x\to\infty} \frac{\psi(x) - x}{x^\Theta} \leq -\frac{1}{|\rho|}. \tag{15.5}$$

Proof Suppose that $\psi(x) \leq x + cx^\Theta$ for all $x \geq X_0$. Then by Lemma 15.1,

$$\frac{c}{s - \Theta} + \frac{\zeta'(s)}{s\zeta(s)} + \frac{1}{s - 1} = \int_1^\infty (cx^\Theta - \psi(x) + x)x^{-s-1}\,dx \tag{15.6}$$

for $\sigma > \Theta$. Call this function $F(s)$. Then

$$F(s) + \frac{1}{2}e^{i\phi}F(s + i\gamma) + \frac{1}{2}e^{-i\phi}F(s - i\gamma)$$
$$= \int_1^\infty (cx^\Theta - \psi(x) + x)(1 + \cos(\phi - \gamma\log x))x^{-s-1}\,dx$$

for $\sigma > \Theta$. We now consider the behaviour of these two expressions as s tends to Θ from above through real values. On the right-hand side, the integral from 1 to X_0 is uniformly bounded, while the integral from X_0 to ∞ is non-negative. Thus the lim inf of the right-hand side is $> -\infty$ as $s \to \Theta^+$. On the other hand, the left-hand side is a meromorphic function that has a pole at $s = \Theta$ with residue

$$c + \frac{me^{i\phi}}{2\rho} + \frac{me^{-i\phi}}{2\overline{\rho}}$$

where $m \geq 1$ denotes the multiplicity of the zero ρ. We choose ϕ so that $e^{i\phi}/\rho = -1/|\rho|$. Then the above is $c - m/|\rho|$. This quantity must be non-negative, for if it were negative, then the left-hand side would tend to $-\infty$ as $s \to \Theta^+$. Hence $c \geq 1/|\rho|$, and we have (15.4). The proof of (15.5) is similar. $\square$

Corollary 15.4 *As x tends to $+\infty$,*

$$\psi(x) - x = \Omega_\pm(x^{1/2}), \tag{15.7}$$
$$\vartheta(x) - x = \Omega_-(x^{1/2}), \tag{15.8}$$

and

$$\pi(x) - \mathrm{li}(x) = \Omega_-(x^{1/2}(\log x)^{-1}). \tag{15.9}$$

The problem of proving Ω_+ companions of (15.8) and (15.9) is more difficult, and is dealt with in the next section.

Proof We first prove (15.7). If RH is false, then $\Theta > 1/2$, and we have a stronger result by Theorem 15.2. If RH holds, then we have (15.7) by Theorem 15.3, and the remaining assertions follow by Theorem 13.2. ☐

Many similar results can be proved using the above ideas. For example, for $M(x) = \sum_{n \le x} \mu(n)$ we find, in the manner of Theorem 15.2, that

$$M(x) = \Omega_\pm(x^{\Theta - \varepsilon}). \qquad (15.10)$$

In analogy to (15.6) we put

$$G(s) = \frac{1}{s\zeta(s)} - \frac{c}{s - \Theta} = \int_1^\infty (M(x) - cx^\Theta)x^{-s-1}\,dx.$$

Then in the manner of the proof of Theorem 15.3, we find that if $\Theta + i\gamma$ is a zero of $\zeta(s)$, then

$$\limsup_{x \to \infty} \frac{M(x)}{x^\Theta} \ge \frac{1}{|\rho\zeta'(\rho)|}, \qquad (15.11)$$

and

$$\liminf_{x \to \infty} \frac{M(x)}{x^\Theta} \le -\frac{1}{|\rho\zeta'(\rho)|}. \qquad (15.12)$$

Here we are assuming that $\zeta'(\rho) \ne 0$. In the contrary case ρ would be a multiple zero of $\zeta(s)$, and our method would allow us to replace the right-hand side of (15.11) by $+\infty$ and that of (15.12) by $-\infty$. In fact we can prove still more, by considering the function

$$H(s) = \frac{1}{s\zeta(s)} - \frac{c(m-1)!}{(s-\Theta)^m} = \int_1^\infty (M(x) - cx^\Theta(\log x)^{m-1})x^{-s-1}\,dx.$$

Then our method allows us to deduce that if $\Theta + i\gamma$ is a zero of multiplicity $m \ge 1$, then

$$M(x) = \Omega_\pm(x^\Theta(\log x)^{m-1}).$$

Then in the manner of Corollary 15.4 we find that in any case

$$M(x) = \Omega_\pm\big(x^{1/2}\big), \qquad (15.13)$$

and that if $\zeta(s)$ has a multiple zero, then

$$M(x) = \Omega_\pm\big(x^{1/2}\log x\big). \qquad (15.14)$$

In the explicit formula for $\psi(x) - x$, or for $M(x)$, the arguments of the terms in the sum over the zeros are governed by the quantities $x^{i\gamma}$. If the ordinates $\gamma > 0$ are linearly independent over $\mathbb{Q}$, then these arguments will tend to be statistically independent as x runs over a long range. Numerical experiments have failed to disclose any linear dependences, and in the absence of any indication to the contrary, we presume that the ordinates $\gamma > 0$ are linearly independent. Under this assumption, we can improve on the estimate (15.13).

Theorem 15.5   *Let* $0 < \gamma_1 < \gamma_2 < \cdots < \gamma_K$ *and* γ *be ordinates of zeros of* $\zeta(s)$. *For* $1 \le k \le K$ *let* ε_k *take one of the values* $-1, 0, 1$. *Suppose that*

$$\sum_{k=1}^{K} \varepsilon_k \gamma_k = 0 \tag{15.15}$$

for such ε_k *only when* $\varepsilon_k = 0$ *for all* k. *Suppose also that the equation*

$$\sum_{k=1}^{K} \varepsilon_k \gamma_k = \gamma \tag{15.16}$$

has a solution only if γ *is one of the* γ_k, *say* $\gamma = \gamma_{k_0}$ *and that in this case the only solution is obtained by taking* $\varepsilon_{k_0} = 1$, $\varepsilon_k = 0$ *for* $k \ne k_0$. *Then*

$$\limsup_{x \to \infty} \frac{M(x)}{x^{1/2}} \ge \sum_{k=1}^{K} \frac{1}{|\rho_k \zeta'(\rho_k)|} \tag{15.17}$$

and

$$\liminf_{x \to \infty} \frac{M(x)}{x^{1/2}} \le -\sum_{k=1}^{K} \frac{1}{|\rho_k \zeta'(\rho_k)|}. \tag{15.18}$$

Proof   In view of (15.10) and (15.14), we may assume that RH holds and that all zeros of the zeta function are simple. We suppose that $M(x) \le cx^{1/2}$ for all large x and consider the integral

$$I(s) = \int_1^{\infty} \frac{M(x) - cx^{1/2}}{x^{s+1}} \prod_{k=1}^{K} (1 + \cos(\phi_k - \gamma_k \log x)) \, dx.$$

With $G(s)$ defined as above (with $\Theta = 1/2$), we multiply out the product to see that this integral is a linear combination of G at various arguments. More precisely, we see that

$$I(s) = G(s) + \frac{1}{2} \sum_{k=1}^{K} (e^{i\phi_k} G(s + i\gamma_k) + e^{-i\phi_k} G(s - i\gamma_k)) + J(s)$$

where $J(s)$ is a linear combination of G at arguments of the form

$$s + i \sum_{k=1}^{K} \varepsilon_k \gamma_k$$

with more than one of the ε_k non-zero. The function $G(s)$ is analytic in the half-plane $\sigma > 0$, except for poles at $s = 1/2$ and at the non-trivial zeros ρ.

Hence by Landau's theorem we see that $I(s)$ converges for $\sigma > 1/2$, and our hypotheses (15.15), (15.16) imply that $J(s)$ is analytic at the point $s = 1/2$. Thus the integral $I(s)$ has a pole at $s = 1/2$ with residue

$$-c + \Re \sum_{k=1}^{K} \frac{e^{i\phi_k}}{\rho_k \zeta'(\rho_k)}.$$

We choose the ϕ_k so that the summands here are positive real. Since $I(s)$ is bounded above uniformly for $s > 1/2$, by letting s tend to $1/2$ from above we deduce that

$$c \geq \sum_{k=1}^{K} \frac{1}{|\rho_k \zeta'(\rho_k)|}.$$

This gives (15.17), and the proof of (15.18) is similar. $\qquad\square$

It is not known whether it is possible to choose zeros ρ in such a way that the hypotheses (15.15), (15.16) hold, and for which the sum in (15.17) and (15.18) is large, but at least we are able to establish

Theorem 15.6 *Suppose that the Riemann Hypothesis is true and that the zeros of the zeta function are simple. Then*

$$\sum_{0 < \gamma \leq T} \frac{1}{|\zeta'(\rho)|} \gg T$$

as $T \to \infty$.

From this it follows by partial summation that

$$\sum_{0 < \gamma \leq T} \frac{1}{|\rho \zeta'(\rho)|} \gg \log T$$

as $T \to \infty$. Thus by combining Theorems 15.5 and 15.6 we have

Corollary 15.7 *If the ordinates $\gamma > 0$ of the Riemann zeta function are linearly independent over $\mathbb{Q}$, then*

$$\limsup_{x \to \infty} \frac{M(x)}{x^{1/2}} = +\infty$$

and

$$\liminf_{x \to \infty} \frac{M(x)}{x^{1/2}} = -\infty.$$

Proof of Theorem 15.6   It is enough to prove the inequality with T restricted to the special sequence of values T_ν of Theorem 13.21, for which $|\zeta(s)| \gg \tau^{-\varepsilon}$

uniformly for $-1 \le \sigma \le 2$. By the calculus of residues we see that

$$\sum_{0<\gamma\le T_v} \frac{1}{\zeta'(\rho)} = \frac{1}{2\pi i} \int_C \frac{1}{\zeta(s)} \, ds$$

where C is the rectangular contour with vertices $2+i$, $2+iT_v$, $-1+iT_v$, $-1+i$. The top of this rectangle contributes an amount $\ll T_v^\varepsilon$. For s on the left side of this contour, $|\zeta(s)| \asymp \tau^{3/2}$ by Corollary 10.5, so that the integral along the left-hand side is $\ll 1$. The integral along the bottom of the rectangle is clearly $\ll 1$ as well. To estimate the integral along the right-hand side, we expand $1/\zeta(s)$ in its Dirichlet series, and integrate term by term. The integral of 1 contributes $T_v - 1$, while for $n > 1$ the integral of n^{-2-it} is $\ll n^{-2}/\log n$. On summing over n we find that the integral of $1/\zeta(s)$ over the right-hand side of the rectangle is $T_v + O(1)$. On combining these estimates we see that the sum above is $T_v + O(T_v^\varepsilon)$, and this gives the stated result. □

15.1.1   Exercises

1. (a) Suppose that ε is small and positive, and let Li(x) be defined as in Exercise 6.2.22. Explain why

$$s\int_{1+\varepsilon}^{\infty} \mathrm{Li}(x)x^{-s-1}\,dx = \mathrm{Li}(1+\varepsilon)(1+\varepsilon)^{-s} + \int_{1+\varepsilon}^{\infty} \frac{dx}{x^s \log x} = T_1 + T_2.$$

(b) Show that Li$(1 - \varepsilon) = $ Li$(1 + \varepsilon) + O(\varepsilon)$.

(c) Show that

$$\mathrm{Li}(1-\varepsilon) = -\int_{\varepsilon}^{\infty} e^{-v}\frac{dv}{v}.$$

(d) Show that Li$(1+\varepsilon) \ll \log 1/\varepsilon$.

(e) Deduce that

$$T_1 = -\int_{\varepsilon}^{\infty} e^{-v}\frac{dv}{v} + O\left(\varepsilon \log \frac{1}{\varepsilon}\right).$$

(f) Show that

$$T_2 = \int_{(s-1)\log(1+\varepsilon)}^{\infty} e^{-v}\frac{dv}{v}.$$

(g) Show that

$$T_2 = \int_{(s-1)\varepsilon}^{\infty} e^{-v}\frac{dv}{v} + O(\varepsilon).$$

(h) Show that

$$T_1 + T_2 = -\log(s-1) - \int_\varepsilon^{(s-1)\varepsilon} (e^{-v} - 1)\frac{dv}{v} + O(\varepsilon \log 1/\varepsilon).$$

(i) Conclude that

$$s \int_1^\infty \mathrm{Li}(x)x^{-s-1}\, dx = -\log(s-1)$$

for $\sigma > 1$.

2. Let $\psi_1(x) = \sum_{n \le x} \Lambda(n)(x-n)$. Show that $\psi_1(x) - \frac{1}{2}x^2 = \Omega_\pm(x^{3/2})$.

3. Show that $\psi(2x) - 2\psi(x) = \Omega_\pm(x^{1/2})$.

4. (a) Show that as $x \to \infty$,

$$\sum_{n \le x}(1 - n/x)\mu(n) = \Omega_\pm\big(x^{1/2}\big).$$

(b) Show that as $x \to \infty$,

$$\sum_{n \le x}\mu(n)/n = \Omega_\pm\big(x^{-1/2}\big).$$

(c) Show that as $x \to \infty$,

$$\sum_{n=1}^\infty \mu(n)e^{-n/x} = \Omega_\pm\big(x^{1/2}\big).$$

5. Let $Q(x)$ denote the number of square-free numbers not exceeding x.

(a) Show that

$$Q(x) - \frac{6}{\pi^2}x = \Omega_\pm\big(x^{1/4}\big).$$

(b) Show that

$$Q(2x) - 2Q(x) = \Omega_\pm\big(x^{1/4}\big).$$

6. (a) Suppose that $\zeta(1/2 + i\gamma) = 0$ and that $\zeta(1/2 + 2i\gamma) \ne 0$. Show that

$$\limsup_{x \to \infty} \frac{\psi(x) - x}{x^{1/2}} \ge \frac{4}{3|\rho|}$$

and that

$$\liminf_{x \to \infty} \frac{\psi(x) - x}{x^{1/2}} \le -\frac{4}{3|\rho|}.$$

(b) Show that if $\zeta(1/2 + i\gamma_1) = \zeta(1/2 + i\gamma_2) = 0$ but $\zeta(1/2 + i(\gamma_1 + \gamma_2)) \ne 0$ and $\zeta(1/2 + i(\gamma_1 - \gamma_2)) \ne 0$, then

$$\limsup_{x \to \infty} \frac{\psi(x) - x}{x^{1/2}} \ge \frac{1}{|1/2 + i\gamma_1|} + \frac{1}{|1/2 + i\gamma_2|}$$

and that

$$\liminf_{x\to\infty} \frac{\psi(x)-x}{x^{1/2}} \le -\frac{1}{|1/2+i\gamma_1|} - \frac{1}{|1/2+i\gamma_2|}.$$

7. Show that $\sum_{n\le x}(-1)^{\omega(n)} \ll x^{1/2+\varepsilon}$ if and only if $(3^s-2)/\zeta(s)$ is analytic for $\sigma > 1/2$.

8. (Ingham 1942; cf. Haselgrove 1958) Let $L(x) = \sum_{n\le x} \lambda(n)$.

   (a) Show that if $\Theta > 1/2$, then for every $\varepsilon > 0$, $L(x) = \Omega_\pm(x^{\Theta-\varepsilon})$ as $x\to\infty$.

   (b) Show that $\liminf_{x\to\infty} L(x)/x^{1/2} \le 1/\zeta(1/2) \,(= -0.685\ldots)$.

   (c) Show that if $\zeta(s)$ has a multiple zero, then $L(x) = \Omega_\pm(x^{1/2}\log x)$.

   (d) Show that if RH holds and σ is fixed, $1/4 < \sigma < 1/2$, then $|\zeta(2s)/\zeta(s)| = \tau^{\sigma-1/2+o(1)}$.

   (e) Show that if RH holds, then there is a sequence of $T_\nu \to \infty$ in such a way that $T_{\nu+1} \le T_\nu + 2$, and
   $$\sum_{0<\gamma\le T_\nu} \frac{\zeta(2\rho)}{\zeta'(\rho)} = T_\nu + O\big(T_\nu^{3/4+\varepsilon}\big).$$

   (f) Show that if RH holds and the ordinates $\gamma > 0$ of the zeros of the zeta function are linearly independent over $\mathbb{Q}$, then
   $$\limsup_{x\to\infty} \frac{L(x)}{x^{1/2}} = +\infty$$
   and
   $$\liminf_{x\to\infty} \frac{L(x)}{x^{1/2}} = -\infty.$$

9. (Turán 1948; cf. Haselgrove 1958)

   (a) Show that if $\sum_{n\le x} \lambda(n)/n \ge 0$ for all $x \ge 1$, then the Riemann Hypothesis is true.

   (b) Show that
   $$\sum_{n\le x} \lambda(n)/n = \Omega_+\big(x^{-1/2}\big)$$
   as $x\to\infty$.

10. Let the positive integer q be fixed. Suppose that if χ is a character (mod q), then $L(\sigma,\chi) \ne 0$ for $0 < \sigma < 1$. Suppose also that a and b are integers such that $(ab,q)=1$ and $a \not\equiv b \pmod q$.

   (a) Let $\Theta = \Theta(q; a, b)$ denote the supremum of the real parts of the poles of the function
   $$\sum_\chi (\overline{\chi}(a) - \overline{\chi}(b)) \frac{L'}{L}(s,\chi).$$

Show that

$$\psi(x;q,a) - \psi(x;q,b) = \Omega_\pm(x^{\Theta-\varepsilon})$$

for any $\varepsilon > 0$.

(b) Let $r(a)$ denote the number of solutions of the congruence $x^2 \equiv a$ (mod q). Show that

$$\vartheta(x;q,a) = \psi(x;q,a) - \frac{r(a)}{\varphi(q)}x^{1/2} + o(x^{1/2}).$$

(c) Show that if $\Theta(q;a,b) > 1/2$, then

$$\vartheta(x;q,a) - \vartheta(x;q,b) = \Omega_\pm(x^{\Theta-\varepsilon}),$$
$$\pi(x;q,a) - \pi(x;q,b) = \Omega_\pm(x^{\Theta-\varepsilon})$$

for any $\varepsilon > 0$.

(d) Show that $\Theta(q;a,b) \geq 1/2$.

(e) Show that

$$\psi(x;q,a) - \psi(x;q,b) = \Omega_\pm(x^{1/2}).$$

(f) Show that if $r(a) \geq r(b)$, then

$$\vartheta(x;q,a) - \vartheta(x;q,b) = \Omega_-(x^{1/2}),$$
$$\pi(x;q,a) - \pi(x;q,b) = \Omega_-(x^{1/2}/\log x).$$

(g) Show that if $r(a) \leq r(b)$, then

$$\vartheta(x;q,a) - \vartheta(x;q,b) = \Omega_+(x^{1/2}),$$
$$\pi(x;q,a) - \pi(x;q,b) = \Omega_+(x^{1/2}/\log x).$$

(h) Show that

$$\pi(x;4,1) - \pi(x;4,3) = \Omega_-(x^{1/2}/\log x).$$

11. (Hardy & Littlewood 1918; Landau 1918a, b) Let $\chi_{-4}(n) = (\frac{-4}{n})$ denote the non-principal character modulo 4, and let

$$T_1(x) = \sum_{n \leq x} \Lambda(n)\chi_{-4}(n)(x - n).$$

(a) Show that

$$T_1(x) = -\sum_\rho \frac{x^{\rho+1}}{\rho(\rho + 1)} + O(x)$$

where ρ runs over the non-trivial zeros of $L(s, \chi_{-4})$. In parts (b)–(l) below, assume that all these zeros lie on the line $\sigma = 1/2$.

(b) Show that

$$\sum_{\rho} \frac{1}{|\rho|^2} = 2\log 2 - \log \pi - C_0 + 2\frac{L'}{L}(1, \chi_{-4}).$$

(c) Show that $L(1, \chi_{-4}) = \pi/4$.

(d) Show that

$$L'(1, \chi_{-4}) = \frac{\log 3}{6} + \sum_{k=2}^{\infty} \frac{(-1)^k}{2}\left(\frac{\log 2k - 1}{2k - 1} - \frac{\log 2k + 1}{2k + 1}\right),$$

and apply the alternating series test to show that $0.19 < L'(1, \chi_{-4}) < 0.196$.

(e) Deduce that

$$0.148 < \sum_{\rho} \frac{1}{|\rho|^2} < 0.164.$$

(f) Show that $|T_1(x)| < (0.165)x^{3/2}$ for all large x.

(g) Show that

$$\sum_{p \leq x^{1/2}} (\log p)(x - p^2) = \frac{2}{3}x^{3/2} + o\left(x^{3/2}\right).$$

(h) Let $T_2(x) = \sum_{2 < p \leq x}(\log p)(-1)^{(p-1)/2}(x - p)$. Show that

$$-\frac{5}{6}x^{3/2} < T_2(x) < -\frac{1}{2}x^{3/2}$$

for all large x.

(i) Let $T_3(x) = \sum_{2 < p \leq x}(-1)^{(p-1)/2}(x - p)$. Show that

$$T_3(x) = \frac{T_2(x)}{\log x} + \int_3^x \frac{T_2(u)}{u^2(\log u)^2}\left(x + \frac{2(x - u)}{\log u}\right) du$$

$$= \frac{T_2(x)}{\log x} + O\left(\frac{x^{3/2}}{(\log x)^2}\right).$$

(j) Let $P(x) = \sum_{p > 2}(-1)^{(p-1)/2}e^{-p/x}$. Show that

$$P(x) = \frac{1}{x^2}\int_0^{\infty} T_3(u)e^{-u/x}\,du.$$

(k) Show that

$$\int_2^{\infty} u^{3/2}(\log u)^{-1}e^{-u/x}\,du = \frac{3}{4}\sqrt{\pi}x^{5/2}(\log x)^{-1} + O\left(x^{5/2}(\log x)^{-2}\right).$$

(l) Deduce that

$$P(x) < -\frac{3}{5}\frac{x^{1/2}}{\log x}$$

for all large x.

(m) Chebyshev (1853) proposed that $P(x) < 0$ for all sufficiently large x. Conclude that Chebyshev's conjecture is equivalent to the assertion that $L(s, \chi_{-4}) \neq 0$ for $\sigma > 1/2$.

15.2 The error term in the Prime Number Theorem

We have seen that $\psi(x) - x$ changes sign infinitely often. We now show that these sign changes can be localized if there is a zero on the abscissa Θ.

Theorem 15.8 *Let Θ denote the supremum of the real parts of the zeros of $\zeta(s)$. If $\zeta(s)$ has a zero with real part Θ, then there exists a constant $C > 0$ such that $\psi(x) - x$ changes sign in every interval $[x, Cx]$ for which $x \geq 2$.*

Proof For each integer $k \geq 0$, put

$$R_k(y) = \frac{1}{k!}\sum_{n \leq e^y}(y - \log n)^k \Lambda(n) - e^y.$$

We see easily that $R_k(y)$ is differentiable for $k > 1$, and that $R_k'(y) = R_{k-1}(y)$. By the method used to prove explicit formulæ we see also that

$$R_k(y) = -\sum_\rho \frac{e^{\rho y}}{\rho^{k+1}} + O(y^{k+1}).$$

Suppose that the numbers γ_j are determined, $0 < \gamma_1 < \gamma_2 < \ldots$ so that the numbers $\Theta \pm i\gamma_j$ constitute all the zeros of $\zeta(s)$ on the line $\sigma = \Theta$, and let m_j denote the multiplicity of the zero $\rho_j = \Theta + i\gamma_j$. Since $\sum_\rho |\rho|^{-\alpha} < \infty$ for $\alpha > 1$, we see that if $k \geq 1$, then

$$R_k(y) = -2e^{\Theta y}\Re\sum_j \frac{m_j e^{i\gamma_j y}}{\rho_j^{k+1}} + o(e^{\Theta y}) \tag{15.19}$$

as $y \to \infty$. Let K be the least number for which

$$\frac{m_1}{|\rho_1|^K} > \sum_{j>1}\frac{m_j}{|\rho_j|^K}.$$

Choose ϕ so that $e^{i\gamma_1\phi}/\rho_1^K > 0$. By taking $k = K$ in (15.19) and using the above inequality, we see that for all large numbers n, $R_K(\phi + \pi n/\gamma_1)$ is positive or negative according as n is odd or even. Take $C = \exp(\pi(K+2)/\gamma_1)$. Then any interval $[y_0, y_0 + \log C]$ contains at least $K + 2$ points of the form $\phi + \pi n/\gamma_1$. Thus if y_0 is large, then such an interval contains $K + 2$ points at which $R_K(y)$ alternates in sign. By the mean value theorem for derivatives we know that if f is differentiable on an interval $[\alpha, \beta]$ and $f(\alpha) < 0$, $f(\beta) > 0$, then there must be a number $\xi, \alpha < \xi < \beta$, such that $f'(\xi) > 0$. Thus we can choose $K + 1$ points in the interval $[y_0, y_0 + \log C]$ at which $R_{K-1}(y)$ alternates in sign. Continuing in this manner, we conclude that we can find three points in this interval at which $R_1(y)$ alternates in sign. Now $R_1(y)$ is continuous, and $R_1'(y) = R_0(y)$ in intervals containing no prime power, so that $R_1(y)$ is an indefinite integral of $R_0(y)$. Thus, although $R_0(y)$ is not everywhere differentiable, it is nevertheless true that R_1 will be monotonic in any interval in which R_0 is of constant sign. Since R_1 is not monotonic in the interval in question, we deduce that R_0 changes sign. $\qquad\square$

The method used to prove Corollary 15.7 could be applied to $\psi(x) - x$, but for this function we have a different approach that succeeds without any unproved hypothesis. In view of Theorem 15.2 we may assume that the Riemann Hypothesis is true. By substituting e^y for x in the explicit formula for $\psi(x)$, we see that

$$\frac{\psi(e^y) - e^y}{e^{y/2}} = -\sum_{\rho} e^{i\gamma y}/\rho + O\left(e^{-y/2}\right)$$

uniformly for $y \geq 1$. Since $1/\rho = 1/(i\gamma) + O(1/\gamma^2)$ and $\sum 1/\gamma^2 < \infty$, the above is

$$-2 \sum_{\gamma > 0} \frac{\sin \gamma y}{\gamma} + O(1).$$

Here each term in the sum is periodic, and if γ is large, then both the period and the amplitude of the term are small. The sum is not absolutely convergent, but by suitably averaging this with respect to y we may arrange that the γ beyond a chosen point make a small contribution. Suppose, for simplicity, that by such an averaging we could truncate the sum, which would leave us to consider the partial sum

$$-2 \sum_{0 < \gamma \leq T} \frac{\sin \gamma y}{\gamma}. \tag{15.20}$$

Here the sum of the absolute values of the coefficients is $\asymp (\log T)^2$, and the sum will be of this order of magnitude if we can find a y for which the fractional parts $\{\gamma y/(2\pi)\}$ are approximately $1/4$ for all the above γ. This, however, is an inhomogeneous problem of Diophantine approximation, and in general such a problem has a solution only if the coefficients γ are linearly independent over $\mathbb{Q}$. Moreover, in order to obtain a quantitative result it would be necessary to have quantitative lower bounds for the absolute values of linear forms in the γ. Since we have no such information, we are confined to homogeneous approximation. Dirichlet's theorem assures us that there exist large y for which each of the numbers $\gamma y/(2\pi)$ is near an integer. That is, $\|\gamma y/(2\pi)\|$ is small for $0 < \gamma \leq T$, where $\|\theta\|$ denotes the distance from θ to the nearest integer, $\|\theta\| = \min_{n\in\mathbb{Z}} |\theta - n|$. However, the sum (15.20) vanishes when $y = 0$, and will therefore be small when the numbers $\|\gamma y/(2\pi)\|$ are small. On the other hand, if we take $y = \pi/T$ in (15.20), then $\sin\gamma y \asymp \gamma/T$, and the sum is $\asymp N(T)/T \asymp \log T$. While this is smaller than the $(\log T)^2$ that we might have hoped for, it is definitely large. This y is small, but by Dirichlet's theorem there exists a large number y_0 for which the numbers $\|\gamma y_0/(2\pi)\|$ are small, and then we may take $y = y_0 \pm \pi/T$ to make the sum (15.20) large in either sign.

The truth of the matter is that the sum (15.20) is not an average of the error term in the Prime Number Theorem, but we can form a weighted sum that resembles (15.20).

Lemma 15.9 *If the Riemann Hypothesis is true, then*

$$\frac{1}{(e^\delta - e^{-\delta})x} \int_{e^{-\delta}x}^{e^\delta x} (\psi(u) - u)\,du = -2x^{1/2}\sum_{\gamma>0} \frac{\sin\gamma\delta}{\gamma\delta} \cdot \frac{\sin(\gamma\log x)}{\gamma} + O\left(x^{1/2}\right)$$

uniformly for $x \geq 4$, $1/(2x) \leq \delta \leq 1/2$.

The first factor in the sum is near 1 if γ is small compared to $1/\delta$, and then becomes small for larger γ. Thus, despite its more complicated appearance, the above sum behaves like the partial sum (15.20) with $T \asymp 1/\delta$.

Proof We recall that

$$\int_0^x (\psi(u) - u)\,du = -\sum_\rho \frac{x^{\rho+1}}{\rho(\rho+1)} - \frac{\zeta'}{\zeta}(0)x + O(1)$$

for $x \geq 2$. We replace x by $e^{\pm\delta}x$ and difference to see that the left-hand side in the lemma is

$$-\frac{\delta}{\sinh\delta} \sum_\rho \frac{(e^{\delta(\rho+1)} - e^{-\delta(\rho+1)})x^\rho}{2\delta\rho(\rho+1)} + O(1). \qquad (15.21)$$

We appeal to RH, and observe that $e^{\pm\delta(\rho+1)} = e^{\pm i\gamma\delta}(1 + O(\delta)) = e^{\pm i\gamma\delta} + O(\delta)$. Since $N(T+1) - N(T) \ll \log T$, we see easily that $\sum_\gamma \gamma^{-2} \ll 1$. Thus when we replace $e^{\pm\delta(\rho+1)}$ by $e^{\pm i\gamma\delta}$ in (15.21), we introduce an error term that

478 Oscillations of error terms

is $\ll x^{1/2}$. Hence the expression (15.21) is

$$-ix^{1/2}\left(\frac{\delta}{\sinh\delta}\right)\sum_{\rho}\frac{\sin\gamma\delta}{\delta}\cdot\frac{x^{i\gamma}}{\rho(\rho+1)}+O\left(x^{1/2}\right).$$

The factor in parentheses is $1+O(\delta^2)$, and the sum over ρ is

$$\ll\sum_{0<\gamma\leq1/\delta}\frac{1}{\gamma}+\frac{1}{\delta}\sum_{\gamma>1/\delta}\frac{1}{\gamma^2}\ll(\log 1/\delta)^2,$$

so our expression is

$$-ix^{1/2}\sum_{\rho}\frac{\sin\gamma\delta}{\delta}\cdot\frac{x^{i\gamma}}{\rho(\rho+1)}+O\left(x^{1/2}\right).$$

Now $1/\rho = 1/(i\gamma)+O(1/\gamma^2)$, and the first factor in the above sum is $\ll |\gamma|$, so that if we replace $1/\rho$ by $1/(i\gamma)$, then we introduce an error term that is $\ll x^{1/2}\sum_\gamma 1/\gamma^2 \ll x^{1/2}$. Similarly we may replace $1/(\rho+1)$ by $1/(i\gamma)$. Thus we see that the above sum is

$$-x^{1/2}\sum_{\rho}\frac{\sin\gamma\delta}{\gamma\delta}\cdot\frac{x^{i\gamma}}{i\gamma}+O\left(x^{1/2}\right).$$

We now obtain the stated result by combining the contributions of γ and $-\gamma$. $\qquad\square$

We now formulate a simple form of Dirichlet's theorem that is suitable for our use.

Lemma 15.10 (Dirichlet) *If $\theta_1,\ldots,\theta_K$ are real numbers, and N is a positive integer, then there is a positive integer $n \leq N^K$ such that $\|\theta_k n\| < 1/N$ for $1 \leq k \leq K$.*

Proof The point $\boldsymbol{p}(n) = (\{\theta_1 n\},\ldots,\{\theta_K n\})$ lies in the hypercube $[0,1)^K$. We partition this hypercube into N^K hypercubes of side length $1/N$. We allow n to take the values $0, 1, \ldots, N^K$, which gives us $N^K + 1$ points. Hence by the pigeon-hole principle there are two values of n, say $0 \leq n_1 < n_2 \leq N^K$, for which the points $\boldsymbol{p}(n_1)$, $\boldsymbol{p}(n_2)$ lie in the same hypercube. Thus

$$\|\theta_k n_1 - \theta_k n_2\| \leq |\{\theta_k n_1\} - \{\theta_k n_2\}| < 1/N$$

for $1 \leq k \leq K$. We take $n = n_2 - n_1$ to obtain the desired result. $\qquad\square$

Theorem 15.11 (Littlewood) *As $x \to \infty$,*

$$\psi(x) - x = \Omega_\pm\left(x^{1/2}\log\log\log x\right), \tag{15.22}$$

and

$$\pi(x) - \mathrm{li}(x) = \Omega_\pm \left(x^{1/2} (\log x)^{-1} \log \log \log x \right). \qquad (15.23)$$

Proof We consider (15.22). If RH is false, then Theorem 15.2 is stronger. Thus it remains to prove (15.22) if RH holds. Let N be a large integer. We apply Lemma 15.10 to those numbers $\gamma (\log N)/(2\pi)$ for which $0 < \gamma \le T = N \log N$. Thus in Lemma 15.10 we have $K = N(T) \asymp T \log T$, and there exists an integer n, $1 \le n \le N^K$ such that

$$\left\| \frac{\gamma n}{2\pi} \log N \right\| < \frac{1}{N}$$

for $0 < \gamma \le T$. We take $x = N^n e^{\pm 1/N}$, $\delta = 1/N$ in Lemma 15.9. From the general inequality $|\sin 2\pi\alpha - \sin 2\pi\beta| \le 2\pi \|\alpha - \beta\|$ we see that

$$|\sin(\gamma \log x) \mp \sin \gamma/N| \le 2\pi/N.$$

Since

$$\sum_\gamma \left| \frac{\sin \gamma/N}{\gamma/N} \cdot \frac{1}{\gamma} \right| \ll (\log N)^2$$

and $\sum_{\gamma > T} 1/\gamma^2 \ll T^{-1} \log T \ll 1/N$, we deduce that the right-hand side in Lemma 15.9 is

$$\mp 2x^{1/2} N^{-1} \sum_{\gamma > 0} \left(\frac{\sin \gamma/N}{\gamma/N} \right)^2 + O\left(x^{1/2} \right).$$

The sum over γ is $\asymp N \log N$. But $x \le N^{N^K} e^{1/N}$ and $K = N(T) \asymp T \log T \asymp N(\log N)^2$, so that

$$\log \log x \ll N (\log N)^3,$$

and hence $\log N \ge (1 + o(1)) \log \log \log x$. The left-hand side in Lemma 15.9 is simply the average of $\psi(u) - u$ over a neighbourhood of x. Since $x \gg N$ and N is arbitrarily large, we have (15.22).

As for (15.23), we note that if RH holds, then (15.22) and (15.23) are equivalent, in view of Theorem 13.2. If RH is false, then Theorem 15.2 gives a stronger result. $\qquad \square$

15.2.1 Exercises

1. Show that

$$\pi(x; 4, 1) - \pi(x; 4, 3) = \Omega_\pm \left(x^{1/2} (\log x)^{-1} \log \log \log x \right)$$

as $x \to \infty$.

2. (a) Show that if $f^{(k-1)}(x)$ is continuous in $[a, a + kh]$ and if $f^{(k)}(x)$ exists throughout $(a, a + kh)$, then there exists a $\xi \in (a, a + kh)$ such that

$$h^k f^{(k)}(\xi) = \sum_{j=0}^{k} (-1)^k \binom{k}{j} f(a + jh).$$

(b) Show that there exist constants $C > 0$, $c > 0$ such that if RH holds, then for all $x \geq 2$,

$$\sup_{x \leq u \leq Cx} (\psi(u) - u) \geq cx^{1/2}$$

and

$$\inf_{x \leq u \leq Cx} (\psi(u) - u) \leq -cx^{1/2}.$$

3. Show that for every $C > 1$ there is a $\delta = \delta(C) > 0$ such that if RH holds, then

$$\sup_{x \leq u \leq Cx} |\psi(u) - u| \geq \delta x^{1/2}$$

for all $x \geq 2$.

4. (Ingham 1936)
   (a) Let N be a positive integer, Y a positive real number, and let $\theta_1, \ldots, \theta_K$ be arbitrary real numbers. By using Dirichlet's theorem, or otherwise, show that there is a real number y, $Y \leq y \leq YN^K$ such that $\|\theta_k y\| < 1/N$ for $1 \leq k \leq K$.
   (b) Let N be an integer > 1, Y a positive real number. Show that there exist real numbers $\theta_1, \ldots, \theta_K$ such that $\max_k \|\theta_k y\| \geq 1/N$ uniformly for all real y in the interval $Y \leq y \leq Y(N - 1)^K$.
   (c) Suppose that RH holds. Show that there exists an absolute constant $c > 0$ such that for any real numbers $X \geq 2$ and $Z \geq 16$ there exists an x, $X \leq x \leq XZ$, for which

$$\pi(x) - \mathrm{li}(x) > cx^{1/2}(\log x)^{-1} \log \log \log Z,$$

and an x' in the same interval for which

$$\pi(x) - \mathrm{li}(x) < -cx^{1/2}(\log x)^{-1} \log \log \log Z.$$

   (d) Deduce that there is an absolute constant $C > 0$ such that if RH holds, then $\pi(x) - \mathrm{li}(x)$ changes sign in every interval $[X, CX]$ for $X \geq 2$.

5. Show that the implicit constant in Littlewood's theorem can be taken to be $1/2$. That is,

$$\limsup_{x \to \infty} \frac{\psi(x) - x}{x^{1/2} \log \log \log x} \geq 1/2,$$

with similar inequalities for the lim inf and for $\pi(x) - \text{li}(x)$.

6. Suppose that q is an integer such that $\prod_\chi L(\sigma, \chi) \neq 0$ for $\sigma > 1/2$. Show that if $(b, q) = 1, b \not\equiv 1 \pmod{q}$, then

$$\pi(x; q, 1) - \pi(x; q, b) = \Omega_\pm \left(x^{1/2} (\log x)^{-1} \log \log \log x \right).$$

7. Suppose that $\sum_n |c_n| < \infty$, and put $g(y) = \sum_n c_n e^{i\lambda_n y}$ where the λ_n are real. Show that for any y_0 and any $\varepsilon > 0$, there exist arbitrarily large numbers y such that $|g(y) - g(y_0)| < \varepsilon$.

8. Suppose that $g(y) = \sum_n c_n e^{i\lambda_n y}$ is uniformly convergent for y in a neighbourhood of y_0, and put

$$M_\delta = \frac{1}{\delta} \int_{-\delta}^{\delta} \left(1 - \frac{|y|}{\delta} \right) g(y_0 + y) \, dy.$$

(a) Show that

$$M_\delta = \sum_n c_n \left(\frac{\sin \lambda_n \delta/2}{\lambda_n \delta/2} \right)^2 e^{i\lambda_n y_0}$$

for all small positive δ.

(b) Show that $M_\delta \to g(y_0)$ as $\delta \to 0^+$.

9. (Jurkat 1973, Anderson 1991) Suppose that there is a constant K such that $M(x) \leq K x^{1/2}$ for all $x \geq 1$, or that there is a constant K such that $-K x^{1/2} \leq M(x)$ for all $x \geq 1$.

(a) Show that the Riemann Hypothesis is true, that the zeros of $\zeta(s)$ are simple, and that $|\zeta'(\rho)| \gg 1/|\rho|$.

(b) Show that there is a sequence of T_ν tending to infinity such that

$$M(x) = \lim_{\nu \to \infty} \sum_{|\gamma| \leq T_\nu} \frac{x^\rho}{\rho \zeta'(\rho)} - 2 + \sum_{n=1}^{\infty} \frac{(-1)^{n-1}(2\pi/x)^{2n}}{(2n)! n \zeta(2n+1)}$$

for $x > 0$, and that the convergence is uniform in intervals that do not contain a square-free number.

(c) Let

$$g(y) = \lim_{\nu \to \infty} \sum_{|\gamma| \leq T_\nu} \frac{e^{i\gamma y}}{\rho \zeta'(\rho)}.$$

Show that if $g(y)$ is continuous at y_0, then for any $\varepsilon > 0$ there exist arbitrarily large y such that $|g(y) - g(y_0)| < \varepsilon$.

(d) Show that $g(0^+) - g(0^-) = 1$.

(e) Deduce that $\lim \sup_{x \to \infty} |M(x)|/x^{1/2} \geq 1/2$.

10. (a) Let $h(x) = (M(2x) - M(x))/x^{1/2}$. Show that $h(1^+) = -1$ and that $h(1^-) = 1$.

(b) Show that

$$\lim_{x \to \infty} \sup \left| \sum_{x < n \leq 2x} \mu(n) \right| x^{-1/2} \geq 1.$$

15.3 Notes

Theorems 15.2 and 15.3, and Corollary 15.4, are due in substance to E. Schmidt (1903). Mertens (1897) conjectured that $|M(x)| \leq x^{1/2}$ for all $x \geq 1$. This 'Mertens Hypothesis' was disproved by Odlyzko and te Riele (1984), who showed that

$$\lim_{x \to \infty} \sup \frac{M(x)}{x^{1/2}} \geq 1.06$$

and that

$$\lim_{x \to \infty} \inf \frac{M(x)}{x^{1/2}} \leq -1.009.$$

One would expect that here the lim sup is $+\infty$ and the lim inf is $-\infty$, but neither of these assertions has been proved. Ingham (1942) proved Theorem 15.5 under the stronger hypothesis that the ordinates $\gamma > 0$ are joined by at most a finite number of linear relations. That one may restrict the coefficients of the linear relations, and thus in principle verify the hypothesis for the first several zeros, was shown by Bateman *et al.* (1971). The product used in the proof of Theorem 15.5 is very similar to the Riesz products used in the study of lacunary Fourier series (see Zygmund 1959, pp. 208–212).

The method used to prove Theorem 15.8 was introduced by Littlewood (1927) for the purpose of providing a simple proof of Theorem 15.3.

Theorem 15.11 was announced by Littlewood (1914), who sketched the proof. Full details were given later by Hardy and Littlewood (1918). The initial proofs depended on an appeal to the Phragmén–Lindelöf principle. Ingham (1936) found that this could be dispensed with. Ingham considered a more complicated weighted average of $\psi(u) - u$ which led to the simpler weighted partial sum

$$\sum_{0 < \gamma \le T} (1 - \gamma/T) \frac{\sin \gamma y}{\gamma}$$

of the sum (15.20). The present exposition was inspired by Ingham's editorial remark in Hardy's Collected Works (1967, p. 99).

The proof given of Theorem 15.11 is non-effective in the sense that it does not permit one to determine an explicit constant c about which one can assert that $\pi(x) > \text{li}(x)$ for some $x < c$. Skewes (1933, 1955) formulated a slightly different division into cases (RH 'nearly true' vs. RH 'significantly false'), which permitted him to show that one can take

$$c = \exp(\exp(\exp(\exp(7.705)))).$$

One of the problems here is to construct a function $f(x)$ about which one can assert that in any interval $[x_0, f(x_0)]$ there exist x for which the sum over the non-trivial zeros is not highly cancelling. That is, the conclusion of Theorem 15.2 must be put in a more quantitative, localized form. In this connection, Littlewood (1937) was led to consider a question concerning a sum of cosines. Turán (1946) discovered that the theorem formulated by Littlewood is false – the argument provided establishes a weaker result than claimed. Turán undertook a detailed study of such power sums. His 'power sum method' has many important applications to the oscillatory error terms that arise in analytic number theory (see Turán 1984). In particular, Knapowski (1961) used Turán's method to show, without need of extensive numerical calculations, that an effective upper bound for the constant c can be determined. Subsequently, Lehman (1966) used extensive numerical information concerning the zeros ρ to show that one can take $c = 1.65 \times 10^{1165}$. Using the same method te Riele (1989) shows that $\pi(x) > \text{li}x$ for at least 10^{180} consecutive integers in the interval $[6.627\ldots \times 10^{370}, 6.687\ldots \times 10^{370}]$. More recently Bays & Hudson (2000) have given some new regions where $\pi(x) > \text{li}(x)$, the first of these being around 1.39×10^{316}. An extension of Littlewood's theorem to Beurling primes has been given by Kahane (1999).

Monach & Montgomery (cf. Monach 1980) have conjectured that for every $\varepsilon > 0$ and every $K > 0$ there is a $T_0(\varepsilon, K)$ such that

$$\left| \sum_{0 < \gamma \le T} k_\gamma \gamma \right| > \exp(-T^{1+\varepsilon}) \tag{15.24}$$

whenever $T \ge T_0$ and the k_γ are integers, not all 0, for which $|k_\gamma| \le K$. From this they have shown that

$$\limsup_{x \to \infty} \frac{\psi(x) - x}{x^{1/2}(\log\log\log x)^2} \geq \frac{1}{2\pi}, \tag{15.25}$$

and that

$$\liminf_{x \to \infty} \frac{\psi(x) - x}{x^{1/2}(\log\log\log x)^2} \leq \frac{-1}{2\pi}. \tag{15.26}$$

In view of (13.48), it is plausible that equality holds in (15.25) and (15.26).

Let $L(x) = \sum_{n \leq x} \lambda(n)$. It was conjectured by Pólya (1919) that $L(x) \leq 0$ for all $x \geq 2$, and it has been verified that this inequality holds for $2 \leq x \leq 10^6$. Pólya's conjecture was disproved by Haselgrove (1958), whose extensive computer calculations led to the conclusion that

$$\limsup_{x \to \infty} \frac{L(x)}{x^{1/2}} > 0.$$

Subsequently Lehman (1960) found that $L(906,180,359) = 1$.

15.4 References

Anderson, R. J. (1991). On the Möbius sum function, *Acta Arith.* **59**, 205–213.

Bateman, P. T., Brown, J. W., Hall, R. S., Kloss, K. E., Stemmler, R. M. (1971). Linear relations connecting the imaginary parts of the zeros of the zeta function, *Computers in Number Theory*. New York: Academic Press, pp. 11–19.

Bays, C. & Hudson, R. H. (2000). A new bound for the smallest x with $\pi(x) > \text{li}(x)$, *Math. Comp.* **69**, 1285–1296.

Chebyshev, P. L. (1853). On a new theorem concerning prime numbers of the forms $4n + 1$ and $4n + 3$, *Bull. Acad. Imp. Sci. St. Petersburg, Phys.-Mat. Kl.* **11**, 208; *Collected Works*, Vol. 1. Moscow-Leningrad: Akad. Nauk SSSR.

Hardy, G. H. (1967). *Collected Papers of G. H. Hardy*, Vol. 2, Oxford: Clarendon Press.

Hardy, G. H. & Littlewood, J. E. (1918). Contributions to the theory of the Riemann zeta-function and the theory of the distribution of primes, *Acta Math.* **41**, 119–196; *Collected Papers*, Vol. 2. Oxford: Clarendon Press, 1967, pp. 20–97.

Haselgrove, C. B. (1958). A disproof of a conjecture of Pólya, *Mathematika* **5**, 141–145.

Ingham, A. E. (1936). A note on the distribution of primes, *Acta Arith.* **1**, 201–211.

— (1942). On two conjectures in the theory of numbers, *Amer. J. Math.* **64**, 313–319.

Jurkat, W. B. (1973). On the Mertens Conjecture and Related General Ω-theorems, *Analytic Number Theory* (St. Louis, 1972), Proc. Sympos. Pure Math. 24. Providence: Amer. Math. Soc., pp. 147–158.

Kahane, J.-P. (1999). Un théorème de Littlewood pour les nombres premiers de Beurling, *Bull. London Math. Soc.* **31**, 424–430.

Knapowski, S. (1961). On sign-changes in the remainder-term in the prime-number formula, *J. London Math. Soc.* **36**, 451–460.

Landau, E. (1905). Über einen Satz von Tschebyscheff, *Math. Ann.* **61**, 527–550; *Collected Works*, Vol. 2. Essen: Thales Verlag, 1986, pp. 206–229; Commentary, *Collected Works*, Vol. 3. pp. 72–75.

(1918a). Über einige ältere Vermutungen und Behauptungen in der Primzahlentheorie, *Math. Z.* **1**, 1–24; *Collected Works*, Vol. 6. Essen: Thales Verlag, 1986, pp. 469–492.

(1918b). Über einige ältere Vermutungen und Behauptungen in der Primzahlentheorie, Zweite Abhandlung, *Math. Z.* **1**, 213–219; *Collected Works*, Vol. 6. Essen: Thales Verlag, 1986, pp. 506–512.

Lehman, R. S. (1960). On Liouville's function, *Math. Comp.* **14**, 311–320.

(1966). On the difference $\pi(x) - \mathrm{li}(x)$, *Acta Arith.* **11**, 397–410.

Littlewood, J. E. (1914). Sur la distribution des nombres premiers, *C. R. Acad. Sci. Paris* **158**, 1869–1872; *Collected Papers*, Vol. 2. Oxford: Oxford University Press, 1982, pp. 829–832.

(1927). Mathematical notes (3): On a theorem concerning the distribution of prime numbers, *J. London Math. Soc.* **2**, 41–45; *Collected Papers*, Vol. 2. Oxford: Oxford University Press, 1982, pp. 833–837.

(1937). Mathematical notes. XII.: An inequality for a sum of cosines, *J. London Math. Soc.* **12**, 217–221; *Collected Papers*, Vol. 2. Oxford: Oxford University Press, 1982, pp. 838–842.

Mertens, F. (1897). Über eine zahlentheoretische Funktion, *Sitz. Akad. Wiss. Wien* **106**, 761–830.

Monach, W. R. (1980). *Numerical Investigation of Several Problems in Number Theory*, Doctoral Thesis. Ann Arbor: University of Michigan.

Odlyzko, A. M. & te Riele, H. J. J. (1984). Disproof of the Mertens conjecture, *J. Reine Angew. Math.* **357**, 138–160.

Pólya, G. (1919). Verschiedene Bermerkungen zur Zahlentheorie, *Jahresbericht Deutsche Math.–Ver.* **28**, 31–40.

te Riele, H. J. J. (1989). On the sign of the difference $\pi(x) - \mathrm{li}x$, *Math. Comp.* **48**, 323–328.

Schmidt, E. (1903). Über die Anzahl der Primzahlen unter gegebener Grenze, *Math. Ann.* **57**, 195–204.

Skewes, S. (1933). On the difference $\pi(x) - \mathrm{li}x$, *J. London Math. Soc.* **8**, 277–283.

(1955). On the difference $\pi(x) - \mathrm{li}x$, II, *Proc. London Math. Soc.* (3) **5**, 48–69.

Turán, P. (1946). On a theorem of Littlewood, *J. London Math. Soc.* **21**, 268–275; *Collected Papers*, Vol. 1. Budapest: Akad Kiadó, 1990, pp. 284–293.

(1948). On some approximative Dirichlet polynomials in the theory of the zeta-function of Riemann, *Danske Vid. Selsk. Mat.-Fys. Medd.* **24**, no. 17, 36 pp.; *Collected Papers*, Vol. 1. Budapest: Akad Kiadó, 1990, pp. 369–402.

(1984). *On a New Method of Analysis and its Applications*, New York: Wiley-Interscience.

Zygmund, A. (1959). *Trigonometric Series*, Vol. 1. Cambridge: Cambridge University Press.

Appendix A

The Riemann–Stieltjes integral

We generalize the Riemann integral $\int_a^b f(x)\,dx$ by defining an integral $\int_a^b f(x)\,dg(x)$ as a limit of Riemann sums $\sum_n f(\xi_n)\Delta g(x_n)$. More precisely, for $a < b$ suppose that we have a partition

$$a = x_0 \leq x_1 \leq \cdots \leq x_N = b. \tag{A.1}$$

For ξ_n in the interval $x_{n-1} \leq \xi_n \leq x_n$ we form the sum

$$S(x_n, \xi_n) = \sum_{n=1}^{N} f(\xi_n)(g(x_n) - g(x_{n-1})).$$

We say that the Riemann–Stieltjes integral $\int_a^b f(x)\,dg(x)$ exists and has the value I if for every $\varepsilon > 0$ there is a $\delta > 0$ such that

$$|S(x_n, \xi_n) - I| < \varepsilon$$

whenever the x_n and the ξ_n are as above and

$$\text{mesh}\{x_n\} = \max_{1 \leq n \leq N} (x_n - x_{n-1}) \leq \delta.$$

The values taken on by f and g may be either real or complex. We do not determine precisely the pairs (f, g) for which the Riemann–Stieltjes integral exists. For our purposes it is enough to prove

Theorem A.1 *The Riemann–Stieltjes integral $\int_a^b f(x)\,dg(x)$ exists if f is continuous on $[a, b]$ and g is of bounded variation on $[a, b]$.*

Proof We recall that by definition

$$\text{Var}_{[a,b]}(g) = \sup \sum_{n=1}^{N} |g(x_n) - g(x_{n-1})|$$

486

where the supremum is taken over all $\{x_n\}$ satisfying (A.1). Since f is uniformly continuous on $[a, b]$, there is a $\delta > 0$ such that $|f(\xi) - f(\xi')| < \varepsilon$ whenever $|\xi - \xi'| \le \delta$. We show that

$$|S(x_n, \xi_n) - S(x_n', \xi_n')| \le 2\varepsilon \mathrm{Var}_{[a,b]}(g) \tag{A.2}$$

provided that mesh$\{x_n\} \le \delta$ and that mesh$\{x_n'\} \le \delta$. This clearly suffices.

Suppose first that the partition $\{x_n\}$ is a subsequence of a second partitioning $\{x_n''\}$. Let $\mathcal{M}(n) = \{m : x_{n-1} < x_m'' \le x_n\}$. The sets $\mathcal{M}(n)$ partition the set $\{1, 2, \ldots, M\}$, so we may write

$$S(x_n, \xi_n) - S(x_m'', \xi_m'')$$
$$= \sum_{n=1}^{N} \left(f(\xi_n)(g(x_n) - g(x_{n-1})) - \sum_{m \in \mathcal{M}(n)} f(\xi_m'')(g(x_m'') - g(x_{m-1}'')) \right).$$

Since the sequence $\{x_n\}$ is an increasing subsequence of the increasing sequence $\{x_m''\}$, it follows that

$$g(x_n) - g(x_{n-1}) = \sum_{m \in \mathcal{M}(n)} g(x_m'') - g(x_{m-1}'').$$

On inserting this in the former expression, we find that it is

$$\sum_{n=1}^{N} \sum_{m \in \mathcal{M}(n)} (f(\xi_n) - f(\xi_m''))(g(x_m'') - g(x_{m-1}'')).$$

Since $|\xi_n - \xi_m''| \le \delta$, it follows that

$$|S(x_n, \xi_n) - S(x_m'', \xi_m'')| \le \varepsilon \sum_{n} \sum_{m \in \mathcal{M}(n)} |g(x_m'') - g(x_{m-1}'')|$$
$$= \varepsilon \sum_{m=1}^{M} |g(x_m'') - g(x_{m-1}'')|$$
$$\le \varepsilon \mathrm{Var}_{[a,b]} g. \tag{A.3}$$

We now take $\{x_m''\}$ to be the union of $\{x_n\}$ and $\{x_n'\}$, so that both $\{x_n\}$ and $\{x_n'\}$ are subsequences of $\{x_m''\}$. Since

$$|S(x_n, \xi_n) - S(x_n', \xi_n')| = |S(x_n, \xi_n) - S(x_m'', \xi_m'') + S(x_m'', \xi_m'') - S(x_n', \xi_n')|$$
$$\le |S(x_n, \xi_n) - S(x_m'', \xi_m'')| + |S(x_m'', \xi_m'') - S(x_n', \xi_n')|$$

by the triangle inequality, the desired bound (A.2) follows by applying (A.3) twice. $\qquad\square$

The main negative feature of the Riemann–Stieltjes integral is that $\int_a^b f\,dg$ does not exist if f and g have a common discontinuity in (a, b). However, if f is continuous, the Riemann–Stieltjes integral enables us to express the sum $\sum_{n=1}^{N} a_n f(n)$ in terms of the unweighted partial sums $A(x) = \sum_{1 \le n \le x} a_n$. Indeed,

$$\sum_{n=1}^{N} a_n f(n) = \int_0^N f(x)\,dA(x). \tag{A.4}$$

There is some freedom in the interval of integration, since the left endpoint can be any number in $[0, 1)$, and the right endpoint can be any number in $[N, N+1)$ without affecting the value of the integral. Frequently it is useful to integrate from 1^- to N, i.e. to consider $\lim_{\varepsilon \to 0^+} \int_{1-\varepsilon}^N$. Some care must be exercised in choosing the endpoints of integration, since for example

$$\int_1^N f(x)\,dA(x) = \sum_{n=2}^{N} a_n f(n).$$

Theorem A.2 *If $\int_a^b f\,dg$ exists, then $\int_a^b g\,df$ also exists, and*

$$\int_a^b g\,df = f(b)g(b) - f(a)g(a) - \int_a^b f\,dg.$$

As we see in the above, we lose no information by writing $\int_a^b f\,dg$ instead of the longer $\int_a^b f(x)\,dg(x)$. On combining Theorems A.1 and A.2 we see that $\int_a^b f\,dg$ exists if f is of bounded variation on $[a, b]$ and g is continuous on $[a, b]$.

Proof Put $\xi_0 = a$ and $\xi_{N+1} = b$. Then

$$\sum_{n=1}^{N} g(\xi_n)(f(x_n) - f(x_{n-1}))$$

$$= f(b)g(b) - f(a)g(a) - \sum_{n=1}^{N+1} f(x_{n-1})(g(\xi_n) - g(\xi_{n-1})).$$

Here the sum on the right-hand side is a Riemann–Stieltjes sum $S(\xi_n, x_{n-1})$ approximating to $\int_a^b f\,dg$, since $x_{n-1} \in [\xi_{n-1}, \xi_n]$. Moreover, mesh$\{\xi_n\} \le$ 2mesh$\{x_n\}$, so that the sum on the right tends to $\int_a^b f\,dg$ as mesh$\{x_n\}$ tends to 0. $\square$

This proof displays the close relation between partial summation and integration by parts. Rather than sum the series $\sum a_n f(n)$ by parts, we can integrate by parts in (A.4) to see that

$$\sum_{n=1}^{N} a_n f(n) = A(N)f(N) - \int_0^N A(x)\,df(x). \tag{A.5}$$

It is to be expected that if g is differentiable, then $\int_a^b f \, dg$ should resemble $\int_a^b f g' \, dx$. In this direction we establish

Theorem A.3 *If g' is continuous on $[a, b]$, then*

$$Var_{[a,b]} g = \int_a^b |g'(x)| \, dx.$$

If in addition f is Riemann integrable, then

$$\int_a^b f(x) \, dg(x) = \int_a^b f(x) g'(x) \, dx.$$

Proof By the mean value theorem there is a $\zeta_n \in [x_{n-1}, x_n]$ such that

$$g(x_n) - g(x_{n-1}) = g'(\zeta_n)(x_n - x_{n-1}).$$

Hence

$$\sum_{n=1}^{N} |g(x_n) - g(x_{n-1})| = \sum_{n=1}^{N} |g'(\zeta_n)|(x_n - x_{n-1}),$$

which tends to $\int_a^b |g'| \, dx$ as mesh$\{x_n\}$ tends to 0. Since $g'(x)$ is uniformly continuous on $[a, b]$, there is a $\delta > 0$ such that $|g'(\xi) - g'(\zeta)| < \varepsilon$ whenever $|\xi - \zeta| < \delta$. Clearly

$$\sum_{n=1}^{N} f(\xi_n)(g(x_n) - g(x_{n-1})) = \sum_{n=1}^{N} f(\xi_n) g'(\zeta_n)(x_n - x_{n-1})$$

$$= \sum_{n=1}^{N} f(\xi_n) g'(\xi_n)(x_n - x_{n-1})$$

$$+ \sum_{n=1}^{N} f(\xi_n)(g'(\zeta_n) - g'(\xi_n))(x_n - x_{n-1})$$

$$= \Sigma_1 + \Sigma_2,$$

say. The function $f g'$ is Riemann integrable, and hence Σ_1 tends to $\int_a^b f g' \, dx$ as mesh$\{x_n\}$ tends to 0. Suppose that M is chosen so that $|f(x)| \leq M$ for all $x \in [a, b]$. If mesh$\{x_n\} < \delta$, then $|\Sigma_2| \leq M\varepsilon(b - a)$. Hence $\int_a^b f \, dg$ exists and has the value $\int_a^b f g' \, dx$. □

Continuing from (A.4), we see that if f' is continuous, then

$$\sum_{n=1}^{N} a_n f(n) = A(N) f(N) - \int_0^N A(x) f'(x) \, dx. \tag{A.6}$$

This useful identity can be verified without mention of Riemann–Stieltjes integration, but its formulation and derivation is most natural through (A.4) and (A.5).

Suppose that f is Riemann integrable. A version of the triangle inequality asserts that $|\int_a^b f| \le \int_a^b |f|$. We now derive an analogue of this for the Riemann–Stieltjes integral.

Theorem A.4 *Suppose that g has bounded variation, and put $g^*(x) = \mathrm{Var}_{[a,x]} g$. Then*

$$\left| \int_a^b f(x)\, dg(x) \right| \le \int_a^b |f(x)|\, dg^*(x).$$

provided that both integrals exist.

Proof   Clearly

$$|S(x_n, \xi_n)| \le \sum_{n=1}^{N} |f(\xi_n)||g(x_n) - g(x_{n-1})|$$

$$\le \sum_{n=1}^{N} |f(\xi_n)|(g^*(x_n) - g^*(x_{n-1})),$$

which gives the result.   □

The differential dg^* is sometimes abbreviated $|dg|$. From Theorem A.4 we see that if $|f(x)| \le M$ for $a \le x \le b$ and g is of bounded variation, then

$$\left| \int_a^b f(x)\, dg(x) \right| \le M \mathrm{Var}_{[a,b]} g \qquad (A.7)$$

provided that the integral exists. As with Riemann integrals, we set $\int_a^a f\, dg = 0$. If $a > b$ we set $\int_a^b f\, dg = -\int_b^a f\, dg$, so that $\int_a^c + \int_c^b = \int_a^b$ for any real numbers a, b, c. Finally, improper Riemann–Stieltjes integrals are defined as limits of proper integrals, e.g.

$$\int_a^\infty f(x)\, dg(x) = \lim_{b \to \infty} \int_a^b f(x)\, dg(x).$$

Exercises

1. Suppose that $\varphi(t)$ is continuous and strictly increasing for $\alpha \le t \le \beta$, and that $\varphi(\alpha) = a$, $\varphi(\beta) = b$. Put $F(t) = f(\varphi(t))$, $G(t) = g(\varphi(t))$. Show that

$$\int_a^b f(x)\, dg(x) = \int_\alpha^\beta F(t)\, dG(t)$$

provided that either integral exists.

2. Let f and g be continuous, and h have bounded variation. Put $I(x) = \int_a^x g\,dh$. Show that

$$\int_a^b f(x)g(x)\,dh(x) = \int_a^b f(x)\,dI(x).$$

3. The proof of Theorem A.2 depends on summation by parts. We now show that, conversely, summation by parts can be recovered from Theorem A.2. Suppose that the numbers $a_1, \ldots, a_N$ and $b_1, \ldots, b_N$ are given. Put $A_n = a_1 + \cdots + a_n$ for $1 \le n \le N$. For $1 \le x < N+1$ put $A(x) = A_{[x]}$; set $A(x) = 0$ for $x < 1$. For $1/2 \le x \le N + 1/2$ let $B(x) = b_{[x+1/2]}$. (The discontinuities of $B(x)$ are displaced in order to ensure that $A(x)$ and $B(x)$ do not have a common discontinuity.)

(a) Show that

$$\sum_{n=1}^{N} a_n b_n = \int_{1^-}^{N} B(x)\,dA(x).$$

(b) Show that

$$\sum_{n=1}^{N-1} A_n(b_n - b_{n+1}) = -\int_{1^-}^{N} A(x)\,dB(x).$$

(c) Use Theorem 2 to derive Abel's lemma:

$$\sum_{n=1}^{N} a_n b_n = A_N b_N + \sum_{n=1}^{N-1} A_n(b_n - b_{n+1}).$$

4. Show that

$$\left| \int_a^b fg\,dh \right|^2 \le \left(\int_a^b |f|^2\,|dh| \right) \left(\int_a^b |g|^2\,|dh| \right)$$

provided that these integrals exist.

5. Suppose that f is non-negative and decreasing, that $g(a) = h(a)$, and that $g(x) \le h(x)$ for $a \le x \le b$. Show that

$$\int_a^b f\,dg \le \int_a^b f\,dh$$

provided that these integrals exist.

6. (First mean value theorem) Suppose that f and g are real-valued functions with f continuous on $[a, b]$, and g weakly increasing on this interval. Put $m = \min_{x \in [a,b]} f(x)$, $M = \max_{x \in [a,b]} f(x)$.

(a) Show that

$$m(g(b) - g(a)) \le \int_a^b f\,dg \le M(g(b) - g(a)).$$

(b) Show that there is an $x_0 \in [a, b]$ such that

$$\int_a^b f\,dg = f(x_0)(g(b) - g(a)).$$

7. (Second mean value theorem) Suppose that f and g are real-valued functions with f weakly increasing on $[a, b]$, and g continuous on this interval. Show that there is an $x_0 \in [a, b]$ such that

$$\int_a^b f\,dg = f(a)(g(x_0) - g(a)) + f(b)(g(b) - g(x_0)).$$

8. (Darst & Pollard 1970) Suppose that f and g are real-valued functions with f of bounded variation on $[a, b]$, and g continuous on this interval. (a) Show that if $\xi \in [a, b]$ and $f(\xi) = 0$, then

$$\int_\xi^b f\,dg \le \mathrm{Var}_{[\xi,b]}(f)\max_{\xi \le x \le b}(g(b) - g(x)),$$

$$\int_a^\xi f\,dg \le \mathrm{Var}_{[a,\xi]}(f)\max_{\xi \le x \le b}(g(x) - g(a)).$$

(b) Show that if $\inf_{a \le x \le b} f(x) = 0$, then

$$\int_a^b f\,dg \le \mathrm{Var}_{[a,b]}(f)\max_{a \le \alpha \le \beta \le b}(g(\beta) - g(\alpha)).$$

(c) Show that in general,

$$\int_a^b f\,dg \le (g(b) - g(a))\inf_{a \le x \le b} f(x) + \mathrm{Var}_{[a,b]}(f)\max_{a \le \alpha \le \beta \le b}(g(\beta) - g(\alpha)).$$

9. Suppose that

$$f(x) = \begin{cases} 1 & \text{if } 0 < x \le 1, \\ 0 & \text{otherwise}; \end{cases} \qquad g(x) = \begin{cases} 1 & \text{if } 0 \le x \le 1 \\ 0 & \text{otherwise}. \end{cases}$$

Show that $\int_{-1}^0 f\,dg$ and $\int_0^1 f\,dg$ both exist, but that $\int_{-1}^1 f\,dg$ does not exist.

A.1 Notes

Our treatment follows that of Ingham in his lectures at Cambridge University. Several variants of the Riemann–Stieltjes (R-S) integral have been proposed. The integral as we have defined it is known as the *uniform* Riemann–Stieltjes integral. A slightly more powerful variant is the *refinement* Riemann–Stieltjes integral, in which $\int_a^b f\,dg$ is said to have the value I if for every $\varepsilon > 0$ there is a partition $\{x_n\}$ such that if $\{x'_m\}$ is a refinement of $\{x_n\}$, then $|S(x'_m, \xi'_m) - I| < \varepsilon$

for all choices of $\xi_m' \in [x_{m-1}', x_m']$. The refinement Riemann–Stieltjes integral is developed in considerable detail by Apostol (1974, Chapter 9) and Bartle (1964, Section 22), and is used by Bateman & Diamond (2004). If $\int_a^b f\,dg$ exists in the sense of uniform R–S integration, then it also exists in the refinement R–S sense, and has the same value. The refinement integral has the attractive property that if $a < b < c$, and if $\int_a^b f\,dg$, $\int_b^c f\,dg$ both exist, then $\int_a^c f\,dg$ exists and

$$\int_a^c f\,dg = \int_a^b f\,dg + \int_b^c f\,dg.$$

This is not true for the uniform R–S integral, as we see by the example in Exercise A.9.

We mention without proof two more advanced properties of the Riemann–Stieltjes integral: If f is continuous on $[a, b]$, and if g is absolutely continuous on the same interval, then

$$\int_a^b f\,dg = \int_a^b fg'$$

where the integral on the right is a Lebesgue integral. Secondly, the *Riesz representation theorem*, which is fundamental to functional analysis, asserts that if G is a positive bounded linear functional on the space $C[a, b]$ of continuous functions on $[a, b]$, then there exists a weakly increasing function g on $[a, b]$ such that

$$G(f) = \int_a^b f\,dg$$

for all $f \in C[a, b]$. An account of this is given in Kestelman (1960, pp. 265–269).

For more extensive accounts of Riemann–Stieltjes integration, see Apostol (1974, Chapter 9), Hildebrandt (1938), Kestelman (1960, Chapter 11), Rankin (1963, Section 29), or Widder (1946, Chapter 1).

A.2   References

Apostol, T. M. (1974). *Mathematical Analysis*, Second edition. Menlo Park: Addison–Wesley.

Bartle, R. G. (1964). *The Elements of Real Analysis*. New York: Wiley.

Bateman, P. T. & Diamond, H. G. (2004). *Analytic number theory. An introductory course*, Hackensack: World Scientific.

Darst, R. & Pollard, H. (1970). An inequality for the Riemann–Stieltjes integral, *Proc. Amer. Math. Soc.* **25**, 912–913.

Hildebrandt, T. H. (1938). Stieltjes integrals of the Riemann type, *Amer. Math. Monthly* **45**, 265–277.

Kestelman, H. (1960). *Modern Theories of Integration*. New York: Dover.

Rankin, R. A. (1963). *An Introduction to Mathematical Analysis*. Oxford: Pergamon.

Widder, D. V. (1946). *The Laplace Transform*, Princeton: Princeton University Press.

Appendix B

Bernoulli numbers and the Euler–Maclaurin summation formula

Suppose that f is a continuous function on an interval $[a, b]$. Then by Theorem A.1,

$$\sum_{a < n \leq b} f(n) = \int_a^b f(x) \, d[x] = \int_a^b f(x) \, dx - \int_a^b f(x) d\{x\},$$

since $[x] = x - \{x\}$. On integrating the last integral by parts (recall Theorem A.2), we find that the right-hand side above is

$$\int_a^b f(x) \, dx - f(b)\{b\} + f(a)\{a\} + \int_a^b \{x\} \, df(x).$$

The familiar 'integral test' is an immediate corollary of this identity, and indeed the last term on the right gives an explicit representation of the difference between $\sum f(n)$ and $\int f(x)$. If f has a continuous first derivative then (by Theorem A.3) we may replace $df(x)$ by $f'(x) \, dx$ in the last integral, so that

$$\sum_{a < n \leq b} f(n) = \int_a^b f(x) \, dx - f(b)\{b\} + f(a)\{a\} + \int_a^b \{x\} f'(x) \, dx. \quad \text{(B.1)}$$

Of course this elementary identity can be verified easily without reference to Riemann–Stieltjes integration. If f has derivatives of higher order, then the last integral may be repeatedly integrated by parts. In order to systematize this we introduce the Bernoulli polynomials.

We define the Bernoulli polynomials $B_k(x)$ inductively. We begin by setting

$$B_0(x) = 1. \quad \text{(B.2)}$$

If $B_{k-1}(x)$ is given, then $B_k(x)$ is determined, apart from its constant term, by the differential equation

$$\frac{d}{dx} B_k(x) = k B_{k-1}(x) \qquad (k \geq 1). \quad \text{(B.3)}$$

The Bernoulli number B_k is the constant term of $B_k(x)$. Its value is determined by the condition

$$\int_0^1 B_k(x)\,dx = 0 \qquad (k \geq 1). \tag{B.4}$$

From (B.2) and (B.3) we see that $B_1(x) = x + B_1$, and from (B.4) we deduce that $B_1 = -1/2$. Hence $B_2(x) = x^2 - x + B_2$, and then we find that $B_2 = 1/6$. These polynomials and numbers have many significant properties, a few of which we now investigate.

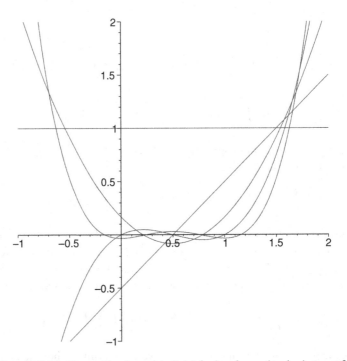

Figure B.1 The Bernoulli polynomials $B_k(x)$ for $k = 0, \ldots, 4$ and $-1 \leq x \leq 2$.

By using (B.3) inductively it is evident that

$$B_k(x) = \sum_{j=0}^{k} \binom{k}{j} x^j B_{k-j} \qquad (k \geq 0). \tag{B.5}$$

In view of (B.3), the integral (B.4) is $(B_{k+1}(1) - B_{k+1}(0))/(k + 1)$. Thus (B.4) is equivalent to the assertion that

$$B_k(0) = B_k(1) \qquad (k \geq 2). \tag{B.6}$$

By taking $x = 1$ in (B.5) it then follows that

$$B_k = \sum_{j=0}^{k} \binom{k}{j} B_{k-j} \qquad (k \geq 2). \qquad (B.7)$$

After subtracting B_k from both sides, this identity provides a formula for B_{k-1} in terms of $B_0, B_1, \ldots, B_{k-2}$.

Next we determine a power series generating function for the B_k. The function $z/(e^z - 1)$ is analytic except at the points $z = 2\pi k i$, $k \neq 0$. In particular, this function is analytic in the disc $|z| < 2\pi$, and we may write its power series in the form

$$\frac{z}{e^z - 1} = \sum_{k=0}^{\infty} \frac{c_k}{k!} z^k.$$

After multiplying both sides by $e^z - 1$ and equating power series coefficients, we see not only that $c_0 = 1$ but also that the c_k satisfy the recurrence (B.7). Consequently $c_k = B_k$ for all k. That is,

$$\frac{z}{e^z - 1} = \sum_{k=0}^{\infty} \frac{B_k}{k!} z^k \qquad (|z| < 2\pi). \qquad (B.8)$$

Theorem B.1 *If k is odd, then*

$$B_k = 0 \qquad\qquad (k \geq 3), \qquad\qquad (B.9)$$
$$B_k(x) = -B_k(1 - x) \qquad (k \geq 1), \qquad\qquad (B.10)$$
$$\operatorname{sgn} B_k(x) = (-1)^{(k+1)/2} \qquad (k \geq 1,\ 0 < x < 1/2). \qquad (B.11)$$

If k is even, then

$$(-1)^{k/2} B_k(x) \quad \uparrow \qquad (k \geq 2,\ 0 < x < 1/2), \qquad (B.12)$$
$$B_k(x) = B_k(1 - x) \qquad (k \geq 0), \qquad\qquad (B.13)$$
$$\operatorname{sgn} B_k = (-1)^{(k/2)+1} \qquad (k \geq 2). \qquad\qquad (B.14)$$

From (B.10) and (B.13) we see that $B_k(x + 1/2)$ is an odd function for odd k, and an even function for even k. From (B.10) it follows that the sign is reversed in (B.11) if the interval $0 < x < 1/2$ is replaced by $1/2 < x < 1$, and similarly from (B.12) and (B.13) we see that $(-1)^{k/2} B_k(x)$ is strictly decreasing for $1/2 \leq x \leq 1$ when k is even, $k \geq 2$. Such properties are evident in the graphs of Figure B.1.

Proof These assertions are evident for $k = 0,\ 1,\ 2$. We proceed by induction.

CASE I. *k odd.* We integrate by parts in (B.4) and use (B.3) to see that

$$0 = B_k - k \int_0^1 x B_{k-1}(x) \, dx.$$

Table B.1

k		B_k
0	$1/1 =$	1.00000 00000
1	$-1/2 =$	$-0.50000\ 00000$
2	$1/6 =$	0.16666 66667
4	$-1/30 =$	$-0.03333\ 33333$
6	$1/42 =$	0.02380 95238
8	$-1/30 =$	$-0.03333\ 33333$
10	$5/66 =$	0.07575 75758
12	$-691/2730 =$	$-0.25311\ 35531$
14	$7/6 =$	1.16666 66667
16	$-3617/510 =$	$-7.09215\ 68627$
18	$43867/798 =$	54.97117 79449
20	$-174611/330 =$	$-529.12424\ 24242$

From (B.13)$_{k-1}$ we see that this integral is $\frac{1}{2}\int_0^1 B_{k-1}$. By (B.4) this integral vanishes, so we have (B.9). To prove (B.10), let

$$f_k(x) = B_k(x) + B_k(1-x).$$

Then (B.3) gives $f_k'(x) = k(B_{k-1}(x) - B_{k-1}(1-x))$, which vanishes by (B.13)$_{k-1}$. Thus $f_k(x)$ is a constant. To determine its value we note that by (B.6) and (B.9), $f_k(0) = 2B_k = 0$. Thus we have (B.10). To prove (B.11) we first note that $B_k(0) = B_k(1/2) = 0$ by (B.9) and (B.10). Suppose that $k \equiv 1 \pmod 4$. It now suffices to show that $B_k(x)$ is convex for $0 < x < 1/2$. But this follows from (B.3) and (B.12)$_{k-1}$. If $k \equiv 3 \pmod 4$, then $B_k(x)$ is concave for $0 < x < 1/2$, and (B.11) again follows.

CASE 2. k even. The assertion (B.12) is immediate from (B.3) and (B.11)$_{k-1}$. To prove (B.13), take

$$g_k(x) = B_k(x) - B_k(1-x).$$

Then by (B.3) we have $g_k'(x) = kf_{k-1}(x) = 0$ by (B.10)$_{k-1}$. Thus $g_k(x)$ is a constant. But $g_k(0) = 0$ by (B.6). To prove (B.14) we note by (B.4) and (B.13) that

$$\int_0^{1/2} B_k(x)\,dx = 0.$$

From this and (B.12) it follows that $(-1)^{k/2}B_k(0) < 0$, $(-1)^{k/2}B_k(1/2) > 0$. Thus we have (B.14), and the proof is complete. $\square$

The first Bernoulli numbers are easily calculated; in Table B.1 we display only the non-zero values.

For even k, the identity (B.13) contains (B.6) as a special case. For odd k, (B.6) is similarly contained in (B.10), in view of (B.9). The identity (B.6) can be generalized in other ways. For example,

$$\frac{B_{k+1}(x+1) - B_{k+1}(x)}{k+1} = x^k \quad (k \geq 0). \tag{B.15}$$

This is obvious for $k = 0$; to prove this for larger k we argue by induction. By the inductive hypothesis we see that the derivatives of the two sides are equal. Thus the two sides differ by at most a constant. We set $x = 0$ and use (B.6) to see that this constant is 0.

Suppose that a and b are integers with $a < b$. In (B.15) we let x take on the values $a, a+1, \ldots, b$, and sum, to obtain the important corollary

$$\sum_{n=a}^{b} n^k = \frac{B_{k+1}(b+1) - B_{k+1}(a)}{k+1} \quad (k \geq 0). \tag{B.16}$$

Apart from the value of the constant term, there can be at most one polynomial with this property. Hence this identity provides a further characterization of the polynomials $B_k(x)$.

When (B.1) is integrated by parts repeatedly the functions $B_k(\{x\})$ arise. Since these latter functions have period 1, it is natural to consider their expansions in Fourier series. In general, if f has period 1 we define the Fourier coefficient $f(m)$ by the formula

$$\widehat{f}(m) = \int_0^1 f(x)e(-mx)\,dx$$

where $e(\theta) = e^{2\pi i\theta}$. From (B.4) we see that $\widehat{B}_k(0) = 0$ for all $k \geq 1$. By integrating by parts we find that if $m \neq 0$, then $\widehat{B}_1(m) = -1/(2\pi im)$. If F has period 1 and $F' = f \in L^1(\mathbb{T})$, then $\widehat{F}(m) = \widehat{f}(m)/(2\pi im)$ for $m \neq 0$. Hence by (B.3) we see that $\widehat{B}_k(m) = k\widehat{B}_{k-1}(m)/(2\pi im)$ and hence that $\widehat{B}_k(m) = -k!/(2\pi im)^k$ for $m \neq 0$. Now $B_1(\{x\})$ has a jump discontinuity at the integers, but since it has bounded variation on $[0, 1]$ the symmetric partial sums of its Fourier series will converge to $B_1(\{x\})$ when x is not an integer. For $k > 1$ the function $B_k(\{x\})$ is continuous and its Fourier series is absolutely convergent, so the series converges uniformly to $B_k(\{x\})$. Thus we have proved

Theorem B.2 *If $x \notin \mathbb{Z}$, then*

$$B_1(\{x\}) = -\frac{1}{\pi} \sum_{m=1}^{\infty} \frac{1}{m} \sin 2\pi mx. \tag{B.17}$$

If k > 1, then

$$B_k(\{x\}) = -k! \sum_{m \neq 0} (2\pi i m)^{-k} e(mx) \tag{B.18}$$

uniformly in x.

A self-contained proof of (B.17), with particular attention to the rate of convergence, is given in Appendix D.1. Since only the defining properties (B.3) and (B.4) were used in deriving the above, these formulæ provide a second means of proving the earlier assertions (B.6), (B.9), (B.10), (B.13), (B.14). These formulæ have many applications. For example, we may take $x = 0$ in (B.18) to obtain

Corollary B.3 *For any integer $k \geq 1$,*

$$\zeta(2k) = (-1)^{k-1} 2^{2k-1} \pi^{2k} B_{2k}/(2k)!. \tag{B.19}$$

Hence $\zeta(2) = \pi/6$, $\zeta(4) = \pi^4/90$, $\zeta(6) = \pi^6/945$, and in general $\zeta(2k)$ is a rational multiple of π^{2k}.

Since $1 < \zeta(2k) < 1 + 2^{2-2k}$ for $k \geq 1$, this gives not only the sign of B_k but also a very precise estimate of its size, namely

$$2(2k)!(2\pi)^{-2k} < |B_{2k}| < 2(2k)!(2\pi)^{-2k}(1 + 2^{2-2k}) \quad (k \geq 1). \tag{B.20}$$

We may similarly derive from Theorem B.2 an estimate for the Bernoulli polynomials in the interval $0 \leq x \leq 1$.

Corollary B.4 *Suppose that $0 \leq x \leq 1$. Then $|B_1(x)| \leq 1/2$, and*

$$|B_k(x)| \leq k! 2^{1-k} \pi^{-k} \zeta(k) \quad (k \geq 2). \tag{B.21}$$

If k is even, then this takes the simpler form $|B_k(x)| \leq |B_k|$, and equality is achieved when $x = 0$ or 1. For odd $k \geq 3$ the inequality can be improved slightly (see Exercise B.5(e)).

We are now in a position to formulate the Euler–Maclaurin summation formula.

Theorem B.5 (Euler–Maclaurin) *Suppose that K is a positive integer and that f has continuous derivatives through the Kth order on the interval $[a, b]$ where a and b are real numbers with $a < b$. Then*

$$\sum_{a < n \leq b} f(n) = \int_a^b f(x)\,dx$$

$$+ \sum_{k=1}^{K} \frac{(-1)^k}{k!} \left(B_k(\{b\}) f^{(k-1)}(b) - B_k(\{a\}) f^{(k-1)}(a) \right)$$

$$- \frac{(-1)^K}{K!} \int_a^b B_K(\{x\}) f^{(K)}(x)\,dx.$$

In most applications the last term is treated as an error term that is only crudely bounded. For example, by Corollary B.4 above we see that the modulus of this term does not exceed

$$\frac{2\zeta(K)}{(2\pi)^K} \int_a^b |f^{(K)}(x)| \, dx. \tag{B.22}$$

Further observations concerning this term are derived in Exercise B.16.

Proof We induct on K. The identity (1) gives the case $K = 1$. From (B.4), and then (B.3), we see that

$$\int_0^x B_K(\{u\}) \, du = \int_0^{\{x\}} B_K(u) \, du = \frac{B_{K+1}(\{x\}) - B_{K+1}}{K+1}.$$

Hence by integrating by parts we find that the last integral in Theorem B.5 is

$$\frac{1}{K+1} \left(B_{K+1}(\{b\}) f^{(K)}(b) - B_{K+1}(\{a\}) f^{(K)}(a) \right)$$

$$- \frac{1}{K+1} \int_a^b B_{K+1}(\{x\}) f^{(K+1)}(x) \, dx,$$

which gives the inductive step. $\qquad\square$

The Euler–Maclaurin formula provides a means of deriving useful identities and asymptotic estimates, and it is also important in numerical calculations.

We now use Theorem B.5 to derive some interesting formulæ for $\zeta(s)$. We assume initially that $\sigma > 1$, and take $f(x) = x^{-s}$. Then $f^{(k)}(x) = k! \binom{-s}{k} x^{-s-k}$, and on taking $a = 1$ and letting b tend to infinity we find that

$$\zeta(s) = \frac{1}{1^s} + \frac{1}{s-1} - \sum_{k=1}^K (-1)^k \binom{-s}{k-1} \frac{B_k}{k}$$

$$- (-1)^K \binom{-s}{K} \int_1^\infty B_K(\{x\}) x^{-s-K} \, dx. \tag{B.23}$$

Here the second term has a pole at $s = 1$, but the integral converges for $\sigma > 1 - K$, and hence this formula provides an analytic continuation of $\zeta(s)$ into this larger half-plane. Since K can be taken arbitrarily large, it follows that $\zeta(s)$ is analytic in the entire plane, apart from the pole at $s = 1$. Moreover, the factor $\binom{-s}{K}$ has zeros at $s = 0$, $s = -1, \ldots, s = 1 - K$, and so the last term vanishes when s is a non-positive integer and K is sufficiently large. Let n denote a non-negative integer, and set $s = -n$. If $K \geq n + 2$, then we find that

$$\zeta(-n) = 1 - \frac{1}{n+1} - \sum_{k=1}^K (-1)^k \binom{n}{k-1} \frac{B_k}{k}.$$

Here the sum may be restricted to $1 \leq k \leq n + 1$, since the binomial coefficient vanishes when $k > n$. Thus we obtain an expression for $\zeta(-n)$ that is independent of K. Since there are only finitely many terms on the right-hand side above, and since each term is rational, it is at once clear that $\zeta(-n)$ is a rational number. However, by making use of the properties of Bernoulli polynomials we can make this more precise. First we use the identity $(n+1)\binom{n}{k-1} = k\binom{n+1}{k}$, and then we observe that the second term on the right supplies an amount that would arise if we allowed $k=0$ in the sum. Thus we see that

$$\zeta(-n) = 1 - \frac{1}{n+1}\sum_{k=0}^{n+1}(-1)^k\binom{n+1}{k}B_k.$$

By taking $x=-1$ in (B.5), we see that the above is

$$= 1 + \frac{(-1)^n}{n+1}B_{n+1}(-1).$$

By taking $x=-1$ in (B.15) we see that $B_{n+1}(-1) = B_{n+1} - (-1)^n(n+1)$. Hence we conclude that

$$\zeta(-n) = (-1)^n\frac{B_{n+1}}{n+1}.$$

In conjunction with the values provided by Theorem B.1, this may be formulated as follows.

Theorem B.6 *Apart from a simple pole at $s=1$, the zeta function is analytic in the complex plane. Moreover, $\zeta(0) = -1/2$, $\zeta(-2n) = 0$ for $n = 1, 2, \ldots$, and $\zeta(1-2n) = -B_{2n}/(2n)$ for $n = 1, 2, \ldots$.*

The functional equation of the zeta function (Corollary 10.3) relates $\zeta(s)$ to $\zeta(1-s)$, so that for many purposes it suffices to consider $\zeta(s)$ for $\sigma \geq 1/2$. In this half-plane, the formula (B.23) is not very useful, since the terms in the sum are far larger than $\zeta(s)$ when $|s|$ is large. This is due to the fact that in our application of the Euler–Maclaurin summation formula, the numbers $f^{(k)}(1)$ increase rapidly in size with k. It is in situations in which the values $f^{(k)}(x)$ decreases rapidly in size as k increases that the Euler–Maclaurin formula provides accurate estimates. With this in mind we break the defining series $\sum n^{-s}$ into two ranges, $n \leq N$ and $n > N$, and apply the sum formula only in the second range. Taking $a=N$ and letting b tend to infinity, we find that

$$\zeta(s) = \sum_{n=1}^{N}n^{-s} + \frac{N^{1-s}}{s-1} + N^{-s}\sum_{k=1}^{K}\binom{s+k-2}{k-1}B_k N^{-k+1}/k$$

$$-\binom{s+K-1}{K}\int_N^\infty B_K(\{x\})x^{-s-K}\,dx.$$
(B.24)

The initial derivation of this is carried out under the assumption that $\sigma > 1$, but then one sees that the above provides a valid formula for $\zeta(s)$ throughout the half-plane $\sigma > 1 - K$. The earlier formula (B.23) is recovered by taking $N = 1$. The above formula is useful even in the half-plane $\sigma > 1$, in which the defining series of $\zeta(s)$ is absolutely convergent. Suppose, for example, that we wish to estimate $\zeta(3/2)$ to within 10^{-10}. If we were to use only the defining series, it would be necessary to sum the first $4 \cdot 10^{20}$ terms. In contrast to this, if we take $s = 3/2$, $N = 5$, $K = 15$ in (B.24), then by (B.22) we find that the last term has modulus $< 0.5 \cdot 10^{-10}$. Since the term $n = N$ in the first sum can be combined with the term $k = 1$ in the second sum, this leaves us only 13 non-zero quantities to evaluate, and we find that $\zeta(3/2) = 2.6123753487$ to 10 decimal places.

By applying the Euler–Maclaurin formula to $f(x) = \log x$ we obtain an approximation to $n!$. For example, with $a = 1$, $b = n$, $K = 2$, we find that

$$\log(n!) = n \log n - n + \frac{1}{2} \log n + c + \frac{1}{12n} - \frac{1}{2} \int_n^\infty B_2(\{x\}) x^{-2} \, dx \quad \text{(B.25)}$$

where

$$c = \frac{11}{12} + \frac{1}{2} \int_1^\infty B_2(\{x\}) x^{-2} \, dx.$$

From (B.22) we see that the last term in (B.25) has modulus less than $1/(12n)$. In addition we describe below how it may be shown that $c = \frac{1}{2} \log 2\pi$, so that on exponentiating we obtain Stirling's formula

$$n! = \left(\frac{n}{e}\right)^n \sqrt{2\pi n} (1 + O(1/n)). \quad \text{(B.26)}$$

More accurate approximations can be derived by using larger values of K. The value of c can be determined by appealing to Wallis's formula, which asserts that

$$\frac{2}{\pi} = \prod_{n=1}^\infty \left(1 - \frac{1}{4n^2}\right). \quad \text{(B.27)}$$

Here the product of the first N terms is

$$\frac{(2N + 1)(2N)!^2}{2^{4N} N!^4},$$

and on invoking (B.26) we see that this tends to $4e^{-2c}$, so that $e^c = \sqrt{2\pi}$. A simple proof of (B.27) is outlined in Exercise B.17 below. A determination of c by use of an inverse Mellin transform and properties of the zeta function is outlined in Exercise B.23 below. In the next appendix we extend our application of the Euler–Maclaurin summation formula to give an asymptotic estimate of the gamma function in the complex plane.

Exercises

1. Show that $(-1)^k B_k(-x) = B_k(x) + kx^{k-1}$ for all $k \geq 0$.
2. Prove the following generalization of (B.5):

$$B_k(x + h) = \sum_{j=0}^{k} \binom{k}{j} B_{k-j}(x) h^j \quad (k \geq 0).$$

3. Show that if $|z| < 2\pi$, then

$$\frac{ze^{xz}}{e^z - 1} = \sum_{k=0}^{\infty} B_k(x) z^k / k!.$$

4. Show that if $k \geq 3$ is odd, then $B_k(x)$ has simple zeros at 0, 1/2, and 1, and no other zeros in $[0, 1]$. Show that if $k \geq 2$ is even, then $B_k(x)$ has one simple zero in $(0, 1/2)$ and another in $(1/2, 1)$, and no other zeros in $[0, 1]$.

5. (Lehmer 1940)
   (a) Show that $\max_{0 \leq x \leq 1} |B_3(x)| = \sqrt{3}/36 < 3/(2\pi^3)$.
   (b) Deduce that

   $$\max_x \sum_{m=1}^{\infty} m^{-3} \sin 2\pi mx = \sqrt{3}\pi^3/54 = 0.994527\ldots.$$

   (c) Show that

   $$\max_{0 \leq x \leq 1} |B_5(x)| = \sqrt{1 - \frac{2}{15}\sqrt{30}} \left(2 + \frac{2}{3}\sqrt{30}\right) / 120 < 15/(2\pi)^5.$$

   (d) Using Theorem B.2, or otherwise, show that if k is odd, $k \geq 3$, then

   $$\max_{0 \leq x \leq 1} |B_k(x)| = k! 2^{1-k} \pi^{-k} (1 - 3^{-k} + O(4^{-k})).$$

   (e) Show that if k is odd, $k \geq 3$, then

   $$\max_{0 \leq x \leq 1} |B_k(x)| < k! 2^{1-k} \pi^{-k}.$$

6. Show that if $j \geq 1$ and $k \geq 1$, then

$$\int_0^1 B_j(x) B_k(x) \, dx = (-1)^{k-1} \frac{j! k!}{(j + k)!} B_{j+k}.$$

7. Show that

$$B_k(1/2) = -(1 - 2^{1-k}) B_k \quad (k \geq 0).$$

8. Show that

$$\sum_{m=0}^{\infty} \frac{(-1)^m}{(2m + 1)^3} = \frac{\pi^3}{32}.$$

9. Show that if $k \geq 0$ and $q \geq 1$, then

$$B_k(qx) = q^{k-1} \sum_{a=0}^{q-1} B_k(x + a/q).$$

(Suggestion: Suppose first that $0 < x < 1/q$, and use Theorem B.2.)

10. Show that if a and b are positive integers, then

$$\int_0^1 B_1(\{ax\}) B_1(\{bx\}) \, dx = \frac{(a,b)^2}{12ab}.$$

11. Using (8), or otherwise, show that

$$z \cot z = \sum_{k=0}^{\infty} (-1)^k \frac{B_{2k}}{(2k)!} (2z)^{2k}$$

for $|z| < \pi$, and that

$$\tan z = \sum_{k=1}^{\infty} (-1)^{k-1} \frac{B_{2k}}{(2k)!} (2^{4k} - 2^{2k}) z^{2k-1}$$

for $|z| < \pi/2$. Show that all coefficients in the latter series are positive.

12. (a) Suppose that $A(z) = \sum_{n=0}^{\infty} a_n z^n / n!$ and $B(z) = \sum_{n=0}^{\infty} b_n z^n / n!$ are power series with positive radii of convergence, and put $C(z) = A(z)B(z)$. Show that $C(z) = \sum_{n=0}^{\infty} c_n z^n / n!$ has positive radius of convergence, and that

$$c_n = \sum_{k=0}^{\infty} \binom{n}{k} a_k b_{n-k}. \qquad (B.28)$$

(b) Suppose that $B(z) = \sum_{n=0}^{\infty} b_n z^n / n!$ and $C(z) = \sum_{n=0}^{\infty} c_n z^n / n!$ are power series with positive radii of convergence, and that $b_0 \neq 0$. Deduce that $A(z) = C(z)/B(z) = \sum_{n=0}^{\infty} a_n z^n / n!$ has positive radius of convergence, and that (B.28) holds.

(c) In the above situation, suppose that the b_n and c_n are all integers, and that $b_0 = \pm 1$. Deduce that the a_n are all integers.

13. Put

$$T_k = (-1)^{k-1} \frac{B_{2k}}{2k} (2^{4k} - 2^{2k}).$$

These are called the 'tangent coefficients' because

$$\tan z = \sum_{k=1}^{\infty} T_k \frac{z^{2k-1}}{(2k-1)!}$$

for $|z| < \pi/2$ (cf. Exercise 11). By taking $C(z) = \sin z$, $B(z) = \cos z$ in the preceding exercise, or otherwise, show that the T_k are all positive integers.

14. (a) By suitable applications of the identity of Exercise 3, or otherwise, show that

$$\frac{e^{3z/4} - e^{z/4}}{e^z - 1} = -2 \sum_{k=0}^{\infty} B_{2k+1}(1/4) \frac{z^{2k}}{(2k+1)!}$$

for $|z| < 2\pi$.

(b) By the substitution $z = 4iw$, show that

$$\sec w = \sum_{k=0}^{\infty} (-1)^{k+1} 4^{2k+1} \frac{B_{2k+1}(1/4)}{(2k+1)!} w^{2k}$$

for $|w| < \pi/2$.

(c) Put

$$E_k = (-1)^{k+1} 4^{2k+1} \frac{B_{2k+1}(1/4)}{2k+1}.$$

These are called the 'Euler numbers' or 'secant coefficients', since

$$\sec z = \sum_{k=0}^{\infty} E_k \frac{z^{2k}}{(2k)!}$$

for $|z| < \pi/2$. Show that $E_k > 0$ for all $k \geq 0$.

(d) By taking $C(z) = 1$, $B(z) = \cos z$ in Exercise 12, or otherwise, show that the E_k are all integers.

15. With the Euler numbers defined as above, show that

$$L(2k+1, \chi_{-4}) = \frac{E_k}{(2k)! 2^{2k+2}} \pi^{2k+1}$$

for all non-negative integers k.

16. Suppose that a and b are integers and that K is even.

(a) Show that if $f^{(K)}(x)$ is of constant sign in (a, b), then the modulus of the last term in the Euler–Maclaurin formula does not exceed that of the term $k = K$ in the sum.

(b) Show that

$$\int_a^b B_{K+1}(\{x\}) f^{(K+1)}(x) \, dx = \int_0^{1/2} B_{K+1}(x) g(x) \, dx$$

where

$$g(x) = \sum_{r=1}^{b-a} \left(f^{(K+1)}(a+r-1+x) - f^{(K+1)}(a+r-x) \right).$$

(c) Show that if $f^{(K+1)}(x)$ exists and is monotonically decreasing in $[a, b]$, then

$$\operatorname{sgn} \int_a^b B_K(\{x\}) f^{(K)}(x) \, dx = -\operatorname{sgn} B_K.$$

(d) Show that if $f^{(K)} < 0$, $f^{(K+1)} > 0$, $f^{(K+2)} < 0$ throughout $[a, b]$, then the last term in the Euler–Maclaurin formula has smaller modulus than, and opposite sign to, the term $k = K$ in the sum.

(e) Show that

$$1 < \frac{n!}{(n/e)^n \sqrt{2\pi n}} < e^{1/(12n)}.$$

17. For $n \geq 0$, let $I_n = \int_0^\pi (\sin x)^n \, dx$.

(a) Show that $I_0 = \pi$, $I_1 = 2$.

(b) Show that $I_{n+2} = \frac{n+1}{n+2} I_n$.

(c) Show that $I_n/I_{n+1} \to 1$ as $n \to \infty$.

(d) Deduce the formula (B.27) of Wallis (1656).

18. Show that if $0 < x < 1$, then

$$\sum_{n=-\infty}^{\infty} \frac{e(n\alpha)}{x^2 - n^2} = \frac{\pi}{n} \cdot \frac{\sin 2\pi\alpha x - \sin 2\pi(\alpha - 1)x}{1 - \cos 2\pi x}.$$

19. Let C_0 denote Euler's constant. Show that if N and K are positive integers, then

$$\sum_{n=1}^{N} \frac{1}{n} = \log N + C_0 + \frac{1}{2N} - \sum_{k=1}^{K-1} \frac{B_{2k}}{2k N^{2k}} - \theta \frac{B_{2K}}{2K N^{2K}}$$

for some $\theta \in (0, 1)$.

20. Let t be real, fixed. Show that $\sum_{n \leq x} (-1)^{n-1} n^{-it}$ is boundedly oscillating.

21. (Carlitz 1964)

(a) Choose $\sigma_0 > 1$ so that $\log \zeta(\sigma_0) = 2\pi$. By substituting $z = \log \zeta(s)$ in (B.8), show that

$$\frac{\log \zeta(s)}{\zeta(s) - 1} = \sum_{k=0}^{\infty} \frac{B_k}{k!} (\log \zeta(s))^k$$

for $\sigma > \sigma_0$.

(b) Choose $\sigma_1 > 1$ so that $\zeta(\sigma_1) = 2$. By writing $\log \zeta(s) = \log(1 + (\zeta(s) - 1))$, show that

$$\frac{\log \zeta(s)}{\zeta(s) - 1} = \sum_{k=0}^{\infty} (-1)^k \frac{(\zeta(s) - 1)^k}{k + 1}$$

for $\sigma > \sigma_1$.

(c) Show that there exist rational numbers $b(n)$ such that

$$\frac{\log \zeta(s)}{\zeta(s) - 1} = \sum_{n=1}^{\infty} b(n) n^{-s}$$

is absolutely convergent for $\sigma > \sigma_1$.

(d) Show that $b(1) = 1$.

(e) Show that $b(p^k) = -1/(k(k+1))$ for $k \geq 1$.

(f) Show that if n is square-free, then $b(n) = B_{\omega(n)}$.

22. Show that $\zeta'(0) = -\frac{1}{2}\log 2\pi$. (Suggestion: Differentiate both sides of (B.24), set $s = 0$, and then compare with (B.26).)

23. (a) Let $F_0(x) = \sum_{n \leq x} \log n$. Show that

$$F_0(x) = x \log x - x + c - B_1(x)\log x + O(1/x)$$

for $x \geq 1$ where c is the constant in (B.25).

(b) Let $F_1(x) = \sum_{n \leq x}(x-n)\log n = \int_1^x F_0(u)\,du$. Show that

$$F_1(x) = \frac{1}{2}x^2 \log x - \frac{3}{4}x^2 + cx + O(\log x)$$

for $x \geq 1$.

(c) By (5.19), show that

$$F_1(x) = \frac{-1}{2\pi i}\int_{\sigma_0 - i\infty}^{\sigma_0 + i\infty} \zeta'(s)\frac{x^{s+1}}{s(s+1)}\,ds .$$

(d) Show that the residue of the above at $s = 1$ is $\frac{1}{2}x^2 \log x - \frac{3}{4}x^2$, and at $s = 0$ is $-\zeta'(0)x$.

(e) Use Corollary 10.5, and Cauchy's formula with a circular contour of radius $1/\log \tau$ to show that $\zeta'(s) \ll \tau^{1/2-\sigma}\log \tau$ uniformly for $-A \leq \sigma \leq -\varepsilon$.

(f) Take the contour to the abscissa $-1/2 + \varepsilon$ to show that

$$F_1(x) = \frac{1}{2}x^2 \log x - \frac{3}{4}x^2 - \zeta'(0)x + O\left(x^{1/2+\varepsilon}\right).$$

(g) By combining the above with the preceding exercise, show that $c = \frac{1}{2}\log 2\pi$.

24. Show that $1^1 2^{1/2} \cdots n^{1/n} \sim c n^{(\log n)/2}$ as $n \to \infty$, where $c > 0$ is an absolute constant.

25. (Kinkelin 1860) Show that

$$1^1 2^2 \cdots n^n = C n^{n^2/2 + n/2 + 1/12} e^{-n^2/4}(1 + O(1/n^2))$$

as $n \to \infty$, where c is a positive constant.

26. (Glaisher 1895)

(a) Let $A_0(x) = \sum_{n \leq x} n \log n$. Show that

$$A_0(x) = \frac{1}{2}x^2 \log x - \frac{1}{4}x^2 - B_1(x)x \log x$$
$$+ \frac{1}{2}B_2(x)(\log x + 1) + \log C - \frac{1}{12} + O(1/x)$$

for $x \geq 1$ where C is the constant in the preceding exercise.

(b) Put $A_1(x) = \sum_{n \leq x}(x - n)n \log n = \int_1^x A_0(u) \, du$. Show that

$$A_1(x) = \frac{1}{6}x^3 \log x - \frac{5}{36}x^3 - \frac{1}{2}B_2(x)x \log x$$
$$+ (\log C - 1/12)x + O(\log x)$$

for $x \geq 1$.

(c) Put $A_2(x) = \frac{1}{2}\sum_{n \leq x}(x - n)^2 n \log n = \int_1^x A_1(u) \, du$. Show that

$$A_2(x) = \frac{1}{24}x^4 \log x - \frac{13}{288}x^4 + \frac{1}{2}(\log C - 1/12)x^2 + O(x \log x)$$

for $x \geq 1$.

(d) By using (5.19), show that

$$A_2(x) = \frac{-1}{2\pi i} \int_{\sigma_0-i\infty}^{\sigma_0+i\infty} \zeta'(s - 1)\frac{x^{s+2}}{s(s + 1)(s + 2)} \, ds .$$

(e) Show that the residue at $s = 2$ in the above integral is $\frac{1}{24}x^4 \log x - \frac{13}{288}x^4$, and that the residue at $s = 0$ is $\frac{-1}{2}\zeta'(-1)x^2$.

(f) By taking the contour to the abscissa $\sigma = -1/2 + \varepsilon$, and using the result of Exercise 23(e), show that

$$A_2(x) = \frac{1}{24}x^4 \log x - \frac{13}{288}x^4 - \frac{1}{2}\zeta'(-1)x^2 + O\left(x^{3/2+\varepsilon}\right)$$

for $x \geq 1$.

(g) Show that $\Gamma'(2) = 1 - C_0$.

(h) By differentiating both sides of (10.9), show that

$$\zeta'(-1) = \frac{\zeta'(2)}{2\pi^2} + \frac{1}{12}(1 - C_0 - \log 2\pi).$$

(i) Conclude that

$$\log C = \frac{1}{12}\log 2\pi + \frac{1}{12}C_0 - \frac{\zeta'(2)}{2\pi^2}$$

where C is the constant in Exercise 25.

27. (a) Integrate by parts to show that

$$\int_0^1 x B_k(x) \, dx = \frac{B_{k+1}(1)}{k + 1}.$$

(b) Use (B.5) to show that

$$\int_0^1 x B_k(x) \, dx = \sum_{j=0}^k \binom{k}{j}\frac{B_{k-j}}{j + 2}.$$

(c) Conclude that if $k > 0$, then

$$\sum_{j=0}^{k} \binom{k}{j} \frac{B_j}{k-j+2} = \frac{B_{k+1}(1)}{k+1}.$$

In the next exercise we develop some of the 'calculus of finite differences', which we then use to derive an explicit formula for $B_{k+1}(x)$, and hence for B_k.

28. For a given function f we let Δf denote the function $f(x+1) - f(x)$, and we put $\Delta^{(n)} f = \Delta(\Delta^{(n-1)} f)$.

   (a) Show that

   $$\Delta^{(n)} f(x) = \sum_{i=0}^{n} (-1)^i \binom{n}{i} f(x+n-i).$$

   (b) Suppose that $f(x)$ is a polynomial expressed in the form

   $$f(x) = \sum_{r=0}^{k} c_r \binom{x}{r} \tag{B.29}$$

   where $\binom{x}{r} = x(x-1)\cdots(x-r+1)/r!$ for $r > 0$, and $\binom{x}{0} = 1$.
   Show that

   $$\Delta f(x) = \sum_{r=1}^{k} c_r \binom{x}{r-1}.$$

   (c) In the above notation, show that

   $$\Delta^{(n)} f(x) = \sum_{r=n}^{k} c_r \binom{x}{r-n}.$$

   (d) Deduce that

   $$c_r = \Delta^{(r)} f(x)\Big|_{x=0} = \sum_{i=0}^{r} (-1)^i \binom{r}{i} f(r-i).$$

   (e) Suppose that f is defined as in (B.29), and put

   $$F(x) = \sum_{r=0}^{k} c_r \binom{x}{r+1}.$$

   Show that $\Delta F = f$.

   (f) Let f and F be as above, and suppose that G is a further function such that $\Delta G = f$. Show that $F - G$ is periodic with period 1, and hence that if G is a polynomial then $G = F + C$ for some constant C.

(g) Let f and F be as above, and suppose that a and b are integers such that $a \le b$. Show that

$$\sum_{j=a}^{b} f(x+j) = F(x+b+1) - F(x+a).$$

29. Suppose that numbers a_{rk} are chosen so that

$$x^k = \sum_{r=0}^{k} a_{rk}x(x-1)\cdots(x-r+1).$$

(a) Explain why the a_{rk} are integers.
(b) Show that

$$a_{rk}r! = \sum_{i=0}^{r}(-1)^i \binom{r}{i}(r-i)^k.$$

(c) Put

$$F(x) = \sum_{r=0}^{k} a_{rk}r!\binom{x}{r+1}.$$

Show that $F(x+1) - F(x) = x^k$.
(d) Show that $F(0) = 0$.
(e) Deduce that

$$F(x) = \frac{B_{k+1}(x) - B_{k+1}}{k+1}.$$

(f) Note that the coefficient of x on the right-hand side above is B_k.
(g) Show that

$$\frac{d}{dx}\binom{x}{r+1}\bigg|_{x=0} = \frac{(-1)^r}{r+1}.$$

(h) Conclude that

$$B_k = \sum_{r=0}^{k} \frac{(-1)^r a_{rk}r!}{r+1} = \sum_{r=0}^{k} \frac{1}{r+1}\sum_{i=0}^{r}(-1)^i\binom{r}{i}i^k. \qquad \text{(B.30)}$$

30. (a) Show that if $r+1$ is composite and $r+1 > 4$, then $(r+1)|r!$.
(b) Show that if $k > 0$, then $a_{3k}3! = 3^k - 3\cdot2^k + 3$, and that this is a multiple of 4 if k is even.
(c) Deduce that if k is positive and even, then

$$B_k \equiv \sum_{p\le k+1}\frac{1}{p}\sum_{i=0}^{p-1}(-1)^i\binom{p-1}{i}i^k \pmod{1}.$$

31. Put $S_k(p) = \sum_{a=1}^{p} a^k$.
    (a) Show that $S_0(p) \equiv 0 \pmod{p}$.
    (b) Show that if $(p-1)|k$ and $k > 0$, then $S_k(p) \equiv -1 \pmod{p}$.
    (c) Show that if $(c, p) = 1$, then $c^k S_k(p) \equiv S_k(p) \pmod{p}$.
    (d) Show that if $(p-1) \nmid k$, then there is a c, $(c, p) = 1$, such that $c^k \not\equiv 1 \pmod{p}$.
    (e) Deduce that if $(p-1) \nmid k$, then $S_k(p) \equiv 0 \pmod{p}$.
    (f) Summarize:
$$S_k(p) \equiv \begin{cases} -1 \pmod{p} & \text{if } (p-1)|k, k > 0; \\ 0 \pmod{p} & \text{otherwise.} \end{cases}$$

32. (von Staudt 1840, Clausen 1840, cf. Lucas 1891, Carlitz 1960/61) By combining the preceding two exercises, deduce the von Staudt–Clausen theorem: If k is positive and even, then
$$B_k + \sum_{(p-1)|k} \frac{1}{p}$$
is an integer.

33. (a) Let $S_k(p)$ be defined as in Exercise 29. Use the binomial theorem to show that
$$\sum_{k=0}^{n-1} \binom{n}{k} S_k(p) \equiv 0 \pmod{p}.$$

    (b) Deduce that
$$\sum_{\substack{0<k<n \\ (p-1)|k}} \binom{n}{k} \equiv 0 \pmod{p}.$$

34. (Bartz & Rutkowski 1993)
    (a) Suppose that q is a positive integer, and that a is a non-negative integer. Explain why
$$q^k B_k((a+1)/q) = \sum_{j=0}^{k} \binom{k}{j} B_j(a/q) q^j .$$

    (b) Suppose that $k = 1$ or that k is a positive even integer, and let q be a positive integer. By using the von Staudt–Clausen theorem, or otherwise, show that
$$q^k B_k + \sum_{\substack{(p-1)|k \\ p \nmid q}} \frac{1}{p}$$
is an integer.

(c) Suppose that $k = 1$ or that k is a positive even integer, and let q be a positive integer. By inducting on a, show that

$$q^k B_k(a/q) + \sum_{\substack{(p-1)|k \\ p \nmid q}} \frac{1}{p}$$

is an integer.

(d) Suppose that k is odd, $k \geq 3$, and that q is a positive integer. By inducting on a, show that $q^k B_k(a/q)$ is an integer, for all non-negative integers a.

35. (Almkvist & Meurman 1991) Suppose that q and k are positive integers. Show that $q^k(B_k(a/q) - B_k)$ is an integer for all integers a.

36. Suppose that $0 < \alpha \leq 1$, and recall that the Hurwitz zeta function is defined to be $\zeta(s, \alpha) = \sum_{n=0}^{\infty}(n + \alpha)^{-s}$ for $\sigma > 1$.

(a) Show that

$$\zeta(s, \alpha) = \frac{1}{\alpha^s} + \frac{1}{s - 1} - \sum_{k=1}^{K}(-1)^k \binom{-s}{k} \frac{B_k(1 - \alpha)}{k}$$

$$- (-1)^K \binom{-s}{K} \int_1^{\infty} B_K(\{x - \alpha\})x^{-s-K}\,dx$$

for $\sigma > 1 - K$.

(b) Deduce that $\zeta(s, \alpha)$ is an analytic function of s throughout the complex plane, except for a simple pole with residue 1 at $s = 1$.

(c) Let n denote a non-negative integer. Show that

$$\zeta(-n, \alpha) = \alpha^n - \frac{1}{n + 1}\sum_{k=0}^{n+1}(-1)^k \binom{n + 1}{k} B_k(1 - \alpha).$$

(d) By (B.10), (B.13), (B.15), and Exercise 2, deduce that

$$\zeta(-n, \alpha) = -\frac{B_{n+1}(\alpha)}{n + 1}.$$

B.1 Notes

Although the notation we have adopted here is quite common, other (conflicting) notations for the Bernoulli numbers are to be found in the literature. Thus it is important to recognize the notational conventions when comparing texts.

The basic facts concerning the Bernoulli numbers and polynomials can be derived in many ways, so the approach depends on one's motivation. Other expositions of note are found in Borevich & Shafarevich (1966, Section 5.8), Rademacher (1973, Chapters 1, 2), and Boas (1977). The proof of the von

Staudt–Clausen theorem sketched in Exercises B.28–B.32 is due to Lucas (1891). The critical identity (B.30) can also be derived by using the generating function (B.8) (cf. Carlitz 1960/61). Borevich & Shafarevich (1966, pp. 384–385) and Cassels (1986, pp. 7–10) give p-adic proofs, the latter of which is due to Witt. The Bernoulli numbers possess a number of further arithmetic properties, such as the Kummer congruences, which are best viewed from a p-adic perspective (cf. Koblitz 1977, p. 44).

The fact that $\zeta(2k)$ is a rational multiple of π^{2k} was discovered by Euler. As reported by Whittaker & Watson (1927, p. 127) and Barnes (1905, p. 253), the Euler–Maclaurin sum formula was discovered by Euler in 1732, but not published by him until 1738. Euler (9 June, 1736) wrote to Stirling of his formula. Stirling (16 April, 1738) responded that Euler's formula included his own as a special case, but that the more general formula had been discovered by Maclaurin. Euler then wrote to Stirling, waiving any claim of priority. Maclaurin published the formula in 1742. Proofs of the formula have been given by Jacobi (1834), Kronecker (1889, 1901, pp. 317–319), Wirtinger (1902), Barnes (1903), Jordan (1922), and Hardy (1949, Chapter 13).

Euler invented a number of methods for accelerating the convergence of series. Such methods (described in Hardy 1949, pp. 7–8, 23–29, 70–73) can be applied to the zeta function. For example, the formula of Apéry (1979),

$$\zeta(3) = \frac{5}{2} \sum_{n=1}^{\infty} \frac{(-1)^{n-1}}{n^3 \binom{2n}{n}},$$

can be derived in this way. Apéry (cf. van der Poorten (1978/79), (1980), Beukers (1979), Ball & Rivoal (2001)) used this formula to prove that $\zeta(3)$ is irrational. It still is not known whether $\zeta(2k+1)$ is irrational when $k \geq 2$, nor is it known whether $\zeta(2k+1)/\pi^{2k+1}$ is irrational. (In this latter connection see Grosswald (1970) and Terras (1976).) Presumably Euler's constant $C_0 = 0.577215664901532\ldots$ and Catalan's constant

$$L(2, \chi_{-4}) = \sum_{m=0}^{\infty} (-1)^m / (2m+1)^2 = 0.915965594\ldots$$

are irrational as well, but this has not been proved.

The value of $\zeta(-n)$ can be determined in a variety of ways. For example, the values given in Theorem B.4 can be arrived at by combining the functional equation of the zeta function (Theorem 10.4) with Corollary B.1 above. Alternatively, by taking $a_n = 1$ in (5.23) we find that

$$\zeta(s) = \frac{1}{\Gamma(s)} \int_0^{\infty} \frac{x^{s-1}}{e^x - 1} \, dx$$

for $\sigma > 1$. Now suppose that the complex plane is slit along the positive real axis, and that C is the 'Hankel path' that starts at $+\infty$ on the positive side of the slit, and follows the slit to the origin, circles the origin in the positive sense, and then returns to $+\infty$ along the negative side of the slit. Set

$$I(s) = \int_C \frac{z^{s-1}}{e^z - 1} \, dz.$$

This integral is uniformly convergent in any compact portion of the plane, and therefore defines an entire function. Suppose that $\sigma > 1$. We shrink the path C until it coincides with the slit. The integral along the first leg of the path is then

$$-\int_0^\infty \frac{x^{s-1}}{e^x - 1} \, dx.$$

The portion of the path that circles the origin becomes negligible, and the integral along the second leg is

$$\int_0^\infty \frac{(xe^{2\pi i})^{s-1}}{e^x - 1} \, dx.$$

On combining these results and using the fact that $\Gamma(s)\Gamma(1-s) = \pi/\sin \pi s$ (see Appendix C), we find that

$$\zeta(s) = e^{-\pi i s} \Gamma(1-s) I(s)/(2\pi i).$$

Although we have derived this under the assumption that $\sigma > 1$, by the uniqueness of analytic continuation it remains valid throughout the complex plane. In general the integrand in $I(s)$ has a branch point at the origin, but if s is a negative integer then the singularity is merely a pole, the residue can then be calculated using the power series (B.8), and we obtain Theorem B.4 once more. See Apostol (1951) for a discussion of the values of the Lerch zeta functions.

By means of the Euler–Maclaurin formula one can calculate $\zeta(s)$ and its derivatives, when $|s|$ is not too large. Let $S(t)$ and $Z(t)$ be defined as in Chapter 14. As long as $\zeta(1/2 + it)$ is calculated sufficiently accurately to allow the sign of $Z(t)$ to be determined, one can prove the existence of zeros on the critical line by detected changes of sign of $Z(t)$. Let $H(n)$ denote the assertion that the first n zeros lie on the critical line and are simple. Gram (1903) established $H(10)$, Backlund (1914) $H(79)$, and Hutchinson (1925) $H(138)$, all using the Euler–Maclaurin formula. Since the amount of computation to evaluate $Z(t)$ for a single value of t is comparable to t by this method, it would be slow work to continue this for larger t. However, in unpublished notes of Riemann, Siegel (1932) discovered indications of a more rapidly convergent formula, known today as the Riemann–Siegel formula: Let $\theta = \theta(t) = -\frac{1}{2}t \log \pi +$

$\arg\Gamma(1/4 + it/2)$, $m = [\sqrt{t/(2\pi)}]$. Then

$$Z(t) = 2\sum_{n=1}^{m} n^{-1/2}\cos(\theta - t\log n) + R(t)$$

where the remainder $R(t)$ has an asymptotic expansion that is rapidly convergent when t is large. The most trivial estimate is that $R(t) \ll t^{-1/4}$, but if this is not sufficient one can write

$$R(t) = \frac{(-1)^{m-1}h\left(\sqrt{t/(2\pi)} - m\right)}{(t/(2\pi))^{1/4}} + O\left(t^{-3/4}\right)$$

where $h(u) = (\cos 2\pi(u^2 - u - 1/16)/\cos 2\pi u$ for $0 \le u < 1$. Titchmarsh (1935, 1936) used the above to establish $H(1041)$. All such calculations fall into two parts. First one calculates $Z(t)$; by detecting sign changes one obtains a lower bound for $N(t)$. Secondly, one computes $S(t)$, so that $N(t)$ is known via Theorem 14.1. Titchmarsh argued that if $\Re\zeta(\sigma + it) > 0$ for $\sigma \ge 1/2$, then $N(t)$ is the integer nearest to

$$\frac{1}{\pi}\arg\Gamma(1/4 + it/2) - \frac{t}{2\pi}\log\pi + 1.$$

Values of t for which this works are rare when t is large, but Turing (1953) devised an alternative procedure that depends on the estimate

$$\int_0^T S(t)\,dt \ll \log T, \tag{B.31}$$

which is due to Littlewood (1924). Turing (1953) was the first to employ a digital computer as an aid to the computation; he achieved $H(1104)$. To be useful in numerical calculations, estimates need to be constructed for the various implicit constants. For the Riemann–Siegel formula this was done by Titchmarsh. For (B.31) this was done by Turing. Titchmarsh's analysis contained errors that were later corrected by Rosser, Yohe & Schoenfeld (1969). Turing's argument also contained errors, which were repaired by Lehman (1970). Subsequently, Lehmer (1956a,b) achieved $H(25,000)$, Meller (1958) $H(35,337)$, Lehman (1966) $H(250,000)$, Rosser, Yohe & Schoenfeld (1969) $H(3,500,000)$, Brent (1979) $H(81,000,001)$, Brent, van de Lune, te Riele & Winter (1982a,b) $H(200,000,001)$, van de Lune & te Riele (1983) $H(300,000,001)$, van de Lune, te Riele & Winter (1986) $H(1,500,000,001)$ and Wedeniwski $H(9 \cdot 10^{11})$ (cf http://www.zetagrid.net). The evaluation of $\zeta(1/2 + it)$ by means of the Riemann–Siegel formula involves $\asymp t^{1/2}$ arithmetic operations, which is a big improvement over the Euler–Maclaurin method. Odlyzko & Schönhage (1988) have shown that if multiple evaluations are to be made, the amount of calculation per evaluation can be reduced to t^{ε}. This new algorithm was implemented by Gourdon & Demichel (2004), who used it to establish $H\left(10^{13}\right)$.

B.2 References

Almkvist, G. & Meurman, A. (1991). Values of Bernoulli polynomials and Hurwitz's zeta function at rational points, *C. R. Math. Rep. Acad. Sci. Canada* **13**, 104–108.

Apéry, R. (1979). Irrationalité de $\zeta(2)$ et $\zeta(3)$, *Astérisque* **61**, 11–13.

Apostol, T. M. (1951). On the Lerch zeta functions, *Pacific J. Math.* **1**, 161–167.

Backlund, R. (1914). Sur les zéros de la fonction $\zeta(s)$ de Riemann, *C. R. Acad. Sci. Paris*, **158**, 1979–1982.

Ball, K. & Rivoal, T. (2001). Irrationalité d'une infinité de valeurs de la fonction zeta aux entiers impairs, *Invent. Math.* **146**, 193–207.

Barnes, E. W. (1903). The generalisation of the Maclaurin sum formula, and the range of its applicability, *Quart. J.* **35**, 175–188.

(1905). The Maclaurin sum-formula *Proc. London Math. Soc.* (2) **3**, 253–272.

Bartz, K. & Rutkowski, J. (1993). On the von Staudt–Clausen theorem, *C. R. Math. Rep. Acad. Sci. Canada* **15**, 46–48.

Beukers, F. (1979). A note on the irrationality of $\zeta(2)$ and $\zeta(3)$, *Bull. London Math. Soc.* **11**, 268–272.

Boas, R. P. (1977). Partial sums of infinite series, and how they grow, *Amer. Math. Monthly* **84**, 237–258.

Borevich, Z. I. & Shafarevich, I. R. (1966). *Number Theory*. New York: Academic Press.

Brent, R. (1979). On the zeros of the Riemann zeta function in the critical strip, *Math. Comp.* **33**, 1361–1372.

Brent, R. P., van de Lune, J., te Riele, H. J. J., Winter, D. T. (1982a). The first 200,000,001 zeros of Riemann's zeta function, *Computational Methods in Number Theory*, Part II, Math. Centre Tracts 155. Amsterdam: Math. Centrum, 389–403.

(1982b). On the zeros of the Riemann zeta function in the critical strip. II, *Math. Comp.* **39**, 681–688; Corrigenda, **46** (1986), 771.

Carlitz, L. (1960/1961). The Staudt–Clausen theorem, *Math. Mag.* **34**, 131–146.

(1964). Extended Bernoulli and Eulerian numbers, *Duke Math. J.* **31**, 667–689.

Cassels, J. W. S. (1986). *Local Fields*, London Math Soc. Student Texts 3, Cambridge: Cambridge University Press.

Clausen, Th. (1840). Theorem, *Astronomische Nachrichten* **17**, 351.

Euler, L. (1732/33). *Comm. Petropol.* **6**, 68–97; *Opera*, Vol. 1, 15, pp. 42–72.

Glaisher, J. W. L. (1895). On the constant which occurs in the formula for $1^1 2^2 \cdots n^n$, *Messenger of Math.* **24**, 1–16.

Gourdon, X. & Demichel, P. (2004). The 10^{13} first zeros of the Riemann zeta function, and zeros computation at very large height, http://numbers.computation.free.fr/Constants/Miscellaneous/zetazeros1e13–1e24.pdf.

Gram, J. (1903). Sur les zéros de la fonction $\zeta(s)$ de Riemann, *Acta Math.* **27**, 289–304.

Grosswald, E., (1970). Die Werte der Riemannschen Zetafunktion an ungeraden Argumentstellen, *Nachr. Akad. Wiss. Göttingen Math.–Phys.* Kl. II, 9–13.

Hardy, G. H., (1949). *Divergent Series*. London: Oxford University Press.

Hutchinson, J. I. (1925). On the roots of the Riemann zeta function, *Trans. Amer. Math. Soc.* **27**, 49–60.

Jacobi, C. G. J. (1834). De usu legitimo formulae summatoriae Maclaurinianae, *J. Reine Angew. Math.* **12**, 263–272; *Gesammelte Werke*, Vol. 6. Berlin: Reimer, 1891, pp. 64–75.

Jordan, C. (1922). On a new demonstration of Maclaurin's or Euler's summation formula, *Tôhoku Math. J.* **21**, 244–246.

Kinkelin, H. (1860). Ueber eine mit der Gammafunction verwandte Transcendente und deren Anwendung auf die Integralrechnung, *J. Reine Angew. Math.* **57**, 122–138.

Koblitz, N. (1977). *p-adic Numbers, p-adic Analysis, and Zeta Functions*, Graduate Texts Math. 58. New York: Springer-Verlag.

Kronecker, L. (1889). Bemerkungen über die Darstellung von Reihen durch Integrale, *J. Reine Angew. Math.* **105**, 157–159, 345–354; *Werke*, Vol. 5. Leipzig: Teubner, 1939, pp. 327–342.

—— (1901). *Vorlesungen über Zahlentheorie*, Vol. 1. Leipzig: Teubner.

Lehman, R. S. (1966). Separation of the zeros of the Riemann zeta function, *Math. Comp.* **20**, 523–541.

—— (1970). On the distribution of zeros of the Riemann zeta-function, *Proc. London Math. Soc.* (3) **20**, 303–320.

Lehmer, D. H. (1940). On the maxima and minima of Bernoulli polynomials, *Amer. Math. Monthly* **47**, 533–538.

—— (1956a). Extended computation of the Riemann zeta-function, *Mathematika* **3**, 102–108; MTAC **11** (1957), 273.

—— (1956b). On the roots of the Riemann zeta-function, *Acta Math.* **95**, 291–298; MTAC **11** (1957), 107–108.

Littlewood, J. E. (1924). On the zeros of the Riemann zeta-function, *Proc. Cambridge Philos. Soc.* **22**, 295–318.

Lucas, É. (1891). *Théorie des Nombres*. Paris: Gauthier–Villars.

van de Lune, J. & te Riele, H. J. J. (1983). On the zeros of the Riemann zeta function in the critical strip. III, *Math. Comp.* **41**, 759–767; Corrigenda, **46** (1986), 771.

van de Lune, J., te Riele, H. J. J., & Winter, D. T. (1981). Rigorous high speed separation of zeros of Riemann's zeta function, *Afdeling Numerieke Wiskunde* **113**, Amsterdam: Mathematisch Centrum.

—— (1986). On the zeros of the Riemann zeta function in the critical strip. IV, *Math. Comp.* **46**, 667–681.

Maclaurin, C. (1742). *Treatise of Fluxions*. Edinburgh, p. 672.

Meller, N. A. (1958). Computation connected with the check of Riemann's hypothesis, *Dokl. Akad. Nauk SSSR* **123**, 246–248.

Nielsen, N. (1923). *Traité élémentaire des nombres de Bernoulli*, Paris: Gauthier–Villars.

Odlyzko, A. M. & Schönhage, A. (1988). Fast algorithms for multiple evaluations of the Riemann zeta function, *Trans. Amer. Math. Soc.* **309**, 797–809.

van der Poorten, A. (1978/79). A proof that Euler missed . . . Apéry's proof of the irrationality of $\zeta(3)$, An informal report, *Math. Intelligencer* **1**, 195–203.

—— (1980). Some wonderful formulae . . . footnotes to Apéry's proof of the irrationality of $\zeta(3)$, *Séminaire Delange–Pisot–Poitou, Théorie des nombres*, Fasc. 2, Exp. No. 29, Paris: Secrétariat Math. 7 pp.

Rademacher, H. (1973). *Topics in Analytic Number Theory*. New York: Springer-Verlag.

Rosser, J. B., Yohe, J. M. & Schoenfeld, L. (1969). Rigorous computation and the zeros of the Riemann zeta-function, *Information Processing 68* (Proc. IFIP Congress,

Edinburgh, 1968), Vol. 1: *Mathematics, Software*, Amsterdam: North-Holland, pp. 70–76; Errata, *Math. Comp.* **29** (1975), 243.

Siegel, C. L. (1932). Über Riemanns Nachlaß zur analytischen Zahlentheorie, *Quellen Studien Gesch. Math. Astro. Phys.* **2**, 45–80; *Gesammelte Abhandlungen*, Vol. 1. Berlin: Springer-Verlag, 1966, pp. 275–310.

von Staudt, K. G. C. (1840). Beweis eines Lehresatzes, die Bernoullischen Zahlen betreffend, *J. Reine Angew. Math.* **21**, 372–374.

Terras, A. (1976). Some formulas for the Riemann zeta function at odd integer argument resulting from Fourier expansions of the Epstein zeta function, *Acta Arith.* **29**, 181–189.

Titchmarsh, E. C. (1935). The zeros of the Riemann zeta function, *Proc. Royal Soc. London Ser.* A **151**, 234–255.

(1936). The zeros of the Riemann zeta function, *Proc. Roy. Soc. London Ser.* A **157**, 261–263.

Turing, A. (1953). Some calculations of the Riemann zeta-function, *Proc. London Math. Soc.* (3) **3**, 99–117.

Wallis, J. (1656). *Arithmetica Infinitorum*, Oxford.

Whittaker, E. T. & Watson, G. N. (1927). *A Course of Modern Analysis*, Fourth edition. Cambridge: Cambridge University Press.

Wirtinger, W. (1902). Einige Anwendungen der Euler–Maclaurin'schen Summenformel, insbesondere auf eine Aufgabe von Abel, *Acta Math.* **26**, 255–271.

Appendix C
The gamma function

For any complex number s not equal to a non-positive integer we define the gamma function by its Weierstrass product,

$$\Gamma(s) = \frac{e^{-C_0 s}}{s} \prod_{n=1}^{\infty} \frac{e^{s/n}}{1 + s/n}. \tag{C.1}$$

Here C_0 is Euler's constant, and we recall from Corollary 1.14 or Exercise B.15 that this constatnt is determined by the relation

$$\sum_{n=1}^{N} \frac{1}{n} = \log N + C_0 + O(1/N). \tag{C.2}$$

From (C.1) it is evident that $1/\Gamma(s)$ is an entire function with simple zeros at the non-positive integers, which is to say that $\Gamma(s)$ is a non-vanishing meromorphic function with simple poles at the non-positive integers as depicted in Figure C.1. On considering the Nth partial product in (C.1) and appealing to (C.2), we obtain *Gauss's formula*,

$$\Gamma(s) = \lim_{N \to \infty} \frac{N^s N!}{s(s+1) \cdots (s+N)}. \tag{C.3}$$

By taking $s = 1$ we see that $\Gamma(1) = 1$. Moreover, from (C.3) it is also immediate that

$$s\Gamma(s) = \Gamma(s+1). \tag{C.4}$$

Hence by induction we find that

$$\Gamma(n+1) = n! \tag{C.5}$$

for non-negative integers n. As will become apparent, the gamma function not only interpolates the values of the factorial, but does so quite smoothly.

The function $\Gamma(s)\Gamma(1-s)$ has a simple pole at every integer. Since the same can be said for $1/\sin \pi s$, it is reasonable to investigate the relation between these

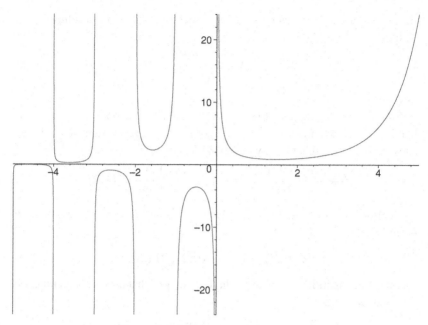

Figure C.1 Graph of $\Gamma(s)$ for $-5 < s \le 5$.

two functions. To this end we let $p_N(s)$ denote the expression on the right in (C.3), and note that

$$p_N(s)p_N(1-s) = \frac{N}{s(N+1-s)} \prod_{n=1}^{N}(1-(s/n)^2)^{-1}.$$

On the other hand, we recall that the Weierstrass product for the sine function may be written

$$\sin s = s \prod_{n=1}^{\infty}\left(1-\frac{s^2}{(\pi n)^2}\right).$$

On comparing these formulæ we conclude that

$$\Gamma(s)\Gamma(1-s) = \frac{\pi}{\sin \pi s}. \tag{C.6}$$

We take $s = 1/2$ to see that $\Gamma(1/2)^2 = \pi$. But from (C.1) it is clear that $\Gamma(1/2) > 0$, so we have

$$\Gamma(1/2) = \sqrt{\pi}. \tag{C.7}$$

From (C.1) we see that $\Gamma(s)$ never takes the value 0, and that it has simple poles at the non-positive integers. Let k be a non-negative integer. Since

$\sin \pi s \sim (-1)^k \pi (s+k)$ as $s \to -k$, and since $\Gamma(k+1) = k!$, it follows from (C.6) that

$$\Gamma(s) \sim \frac{(-1)^k}{k!(s+k)} \tag{C.8}$$

as $s \to -k$.

Similarly we observe that $\Gamma(s)\Gamma(s+1/2)$ has a simple pole at $0, -1/2, -1, -3/2, -2, \ldots$, and that the same is true of $\Gamma(2s)$. We now establish a relation between these two functions by observing that

$$\frac{p_N(s)p_N(s+1/2)}{p_{2N}(2s)} = 2^{1-2s} \frac{N+1/2}{N+s+1/2} p_N(1/2).$$

On letting $N \to \infty$ and using (C.7) we obtain *Legendre's duplication formula*,

$$\Gamma(s)\Gamma(s+1/2) = \sqrt{\pi} 2^{1-2s} \Gamma(2s). \tag{C.9}$$

On taking logarithmic derivatives in (C.1) we find that the *digamma function* $\frac{\Gamma'}{\Gamma}(s)$ can be written

$$\frac{\Gamma'}{\Gamma}(s) = -\frac{1}{s} - C_0 - \sum_{n=1}^{\infty} \left(\frac{1}{s+n} - \frac{1}{n}\right). \tag{C.10}$$

Setting $s = 1$, we see in particular that

$$\frac{\Gamma'}{\Gamma}(1) = -C_0. \tag{C.11}$$

Since $\Gamma(1) = 1$, this is equivalent to

$$\Gamma'(1) = -C_0. \tag{C.12}$$

We write $z = re(\theta)$ in the power series expansion $\log(1-z)^{-1} = \sum_{n=1}^{\infty} z^n/n$, let $r \to 1^-$, and apply Abel's theorem to see that

$$\sum_{n=1}^{\infty} \frac{e(n\theta)}{n} = -\log(1 - e(\theta)) \tag{C.13}$$

provided that $\theta \notin \mathbb{Z}$. By applying this formula for various rational values of θ we can express the series in (C.10) in closed form, for any rational value of s. For example, by taking $\theta = 1/2$ we find that

$$1 - \frac{1}{2} + \frac{1}{3} - \frac{1}{4} + \cdots = \log 2,$$

which with (C.10) gives

$$\frac{\Gamma'}{\Gamma}(1/2) = -C_0 - 2\log 2. \tag{C.14}$$

Also, since

$$\frac{-1-i}{4}e(n/4) - \frac{1}{2}e(n/2) + \frac{-1+i}{4}e(3n/4) = \begin{cases} 1 & \text{if } n \equiv 1 \ (\text{mod } 4), \\ -1 & \text{if } n \equiv 0 \ (\text{mod } 4), \\ 0 & \text{otherwise}, \end{cases}$$

by taking $\theta = 1/4, 1/2, 3/4$ in (C.13) we deduce via (C.10) that

$$\frac{\Gamma'}{\Gamma}(1/4) = -C_0 - 3\log 2 - \pi/2. \tag{C.15}$$

Similarly,

$$\frac{\Gamma'}{\Gamma}(3/4) = -C_0 - 3\log 2 + \pi/2. \tag{C.16}$$

We now consider the asymptotic behaviour of the gamma function.

Theorem C.1 *Let $\delta > 0$ be given, and let $\mathcal{R} = \mathcal{R}(\delta)$ be the set of those complex numbers s for which $|s| \geq \delta$ and $|\arg s| < \pi - \delta$. Then*

$$\frac{\Gamma'}{\Gamma}(s) = \log s + O(1/|s|) \tag{C.17}$$

and

$$\Gamma(s) = \sqrt{2\pi} s^{s-1/2} e^{-s} (1 + O(1/|s|)) \tag{C.18}$$

uniformly for $s \in \mathcal{R}$.

The second estimate here is *Stirling's formula* for the gamma function, which generalizes his estimate (B.26) for $n!$. From this we see that

$$|\Gamma(s)| \asymp \tau^{\sigma-1/2} e^{-\pi\tau/2} \tag{C.19}$$

as $|t| \to \infty$ with σ uniformly bounded.

Proof From (C.2) and (C.10) we see that if $N > |s|$, then

$$\frac{\Gamma'}{\Gamma}(s) = \log N - \sum_{n=0}^{N} \frac{1}{n+s} + O(|s|/N).$$

By the Euler–MacLaurin summation formula (Theorem B.5) with $f(x) = 1/(x+s)$, $a = 0^-$, $b = N$, $K = 2$ we find that

$$\sum_{n=0}^{N} \frac{1}{n+s} = \log(N+s) - \log s + \frac{1}{2s} + \frac{1}{2(s+N)} + O(|s|^{-2}).$$

On combining these estimates and letting N tend to infinity we find that

$$\frac{\Gamma'}{\Gamma}(s) = \log s - \frac{1}{2s} + O(|s|^{-2}). \tag{C.20}$$

This estimate is more precise than (C.17), and still greater accuracy can be obtained by choosing a larger value of K.

To derive (C.18) we begin by taking logarithms in (C.3) and applying the Euler–MacLaurin summation formula, or we integrate (C.20) from s to $s + \infty$ along a ray parallel to the real axis. In either case we find that

$$\log \Gamma(s) = s \log s - s - \frac{1}{2} \log s + c + O(1/|s|),$$

and it remains to determine the value of the constant c. This may be done in a number of ways. For example, we could appeal to (C.5) and (B.26). Alternatively, we can take logarithms in (C.9) and apply the above to see that $c = (\log 2\pi)/2$. Then (C.18) follows by exponentiating. □

The gamma function can be expressed as a definite integral in various ways. We now establish two important integral representations for the gamma function.

Theorem C.2 (Euler's integral) *If* $\Re s > 0$, *then*

$$\int_0^\infty e^{-x} x^{s-1}\, dx = \Gamma(s). \tag{C.21}$$

Proof By integrating by parts repeatedly it is easy to verify that

$$\frac{N!}{s(s+1)\cdots(s+N)} = \int_0^1 (1-y)^N y^{s-1}\, dy.$$

We make the change of variable $x = Ny$ and recall Gauss's formula (C.3) to find that

$$\Gamma(s) = \lim_{N\to\infty} \int_0^\infty f_N(x)\, dx$$

where

$$f_N(x) = \begin{cases} (1 - x/N)^N x^{s-1} & \text{for } 0 \le x \le N, \\ 0 & \text{for } x > N. \end{cases}$$

To complete the proof we employ the dominated convergence theorem. Put $f(x) = e^{-x} x^{\sigma-1}$. Then $\int_0^\infty f(x)\, dx < \infty$ when $\sigma > 0$, and $|f_N(x)| \le f(x)$ uniformly in N and x. Since

$$\lim_{N\to\infty} f_N(x) = e^{-x} x^{s-1}$$

for each fixed x, the formula (C.21) now follows. □

Let $C(\rho)$ denote the circular arc $\{z = \rho e(\theta) : 0 \le \theta \le 1/4\}$. It is easy to verify that

$$\int_{C(\rho)} |e^{-z} z^{s-1}| \, |dz| \to 0$$

as $\rho \to \infty$. Thus by Cauchy's theorem the formula (C.21) still holds if x is replaced by a complex variable z that goes to infinity along a ray from the origin, $z = \rho e(\theta)$, $0 \le \rho < \infty$, provided that $-1/4 \le \theta \le 1/4$.

For $r > 0$ we let $\mathcal{H} = \mathcal{H}(r)$ denote the Hankel contour, which consists of a path that passes from $-ir - \infty$ to $-ir$ along the ray $x - ir$, $-\infty < x \le 0$, and then from $-ir$ to ir along the semicircle $re(\theta)$, $-1/4 \le \theta \le 1/4$, and then from ir to $ir - \infty$ along the ray $x + ir$, $-\infty < x \le 0$.

Theorem C.3 (Hankel) *For any complex number s,*

$$\frac{1}{2\pi i} \int_{\mathcal{H}} e^z z^{-s} \, dz = \frac{1}{\Gamma(s)}. \tag{C.22}$$

Here z^{-s} is assumed to have its principal value.

As in the preceding theorem, the contour of integration may be altered substantially without changing the value of the integral. For example, the ray from ir to $-\infty + ir$ may be replaced by a ray in the direction $e(\theta)$, provided that $1/4 < \theta < 1/2$.

Proof It is clear that the left-hand side is an entire function of s. Thus it suffices to prove the identity when $\sigma < 1$. For such s we let $r \to 0^+$, and note that the integral along the semicircle tends to 0. The remaining integrals tend to

$$e^{i\pi s} \int_0^\infty e^{-x} x^{-s} \, dx - e^{-i\pi s} \int_0^\infty e^{-x} x^{-s} \, dx = 2i(\sin \pi s)\Gamma(1 - s)$$

by (C.21). To complete the proof it suffices to appeal to (C.6). $\qquad \square$

Euler's formula asserts that the gamma function is the Mellin transform of the function e^{-x}. We now establish the inverse.

Theorem C.4 (Mellin) *If $\Re z > 0$ and $c > 0$, then*

$$\frac{1}{2\pi i} \int_{c-i\infty}^{c+i\infty} \Gamma(s) z^{-s} \, ds = e^{-z}.$$

Proof From Stirling's formula we see that

$$\int_{-K+iK}^{c+iK} |\Gamma(s) z^{-s}| \, |ds| \longrightarrow 0$$

as $K \to \infty$, and similarly for the integral from $-K - iK$ to $c - iK$. Moreover, if we first apply (C.6) and then Stirling's formula, we find that

$$\int_{-K-iK}^{-K+iK} |\Gamma(s)z^{-s}| \, |ds| \longrightarrow 0$$

as $K \to \infty$ through values of the form $K = n + 1/2, n \in \mathbb{Z}$. (We are assuming here that the path of integration is a line segment joining the two endpoints.) Thus by the calculus of residues

$$\frac{1}{2\pi i} \int_{c-i\infty}^{c+i\infty} \Gamma(s)z^{-s} \, ds = \sum_{k=0}^{\infty} Res\left(\Gamma(s)z^{-s}\right)\Big|_{s=-k}.$$

From (C.8) we see that the above is

$$\sum_{k=0}^{\infty} \frac{(-1)^k}{k!} z^k = e^{-z}.$$

$\square$

The digamma function can be examined in a similar way. In view of (C.17), this function is not absolutely integrable on the line $\sigma = c$, and thus we cannot define its Fourier transform in the classical manner. We now formulate a useful substitute.

Theorem C.5 *Let $a > 0$ and $b > 0$ be fixed. If $x < 0$ and $T \geq 1$, then*

$$\int_{-T}^{T} \frac{\Gamma'}{\Gamma}(a + ibt)e(-xt)\, dt = -\frac{\Gamma'}{\Gamma}(a + ibT)\frac{e(-xT)}{2\pi ix} + \frac{\Gamma'}{\Gamma}(a - ibT)\frac{e(xT)}{2\pi ix}$$
$$- 2\pi b^{-1} e^{2\pi ax/b}(1 - e^{2\pi x/b})^{-1} + O(x^{-2}T^{-1}),$$

while if $x > 0$ and $T \geq 1$, then

$$\int_{-T}^{T} \frac{\Gamma'}{\Gamma}(a + ibt)e(-xt)\, dt$$
$$= -\frac{\Gamma'}{\Gamma}(a + ibT)\frac{e(-xT)}{2\pi ix} + \frac{\Gamma'}{\Gamma}(a - ibT)\frac{e(xT)}{2\pi ix} + O(x^{-2}T^{-1}).$$

Proof We write the integral as

$$\frac{1}{i} \int_{-iT}^{iT} \frac{\Gamma'}{\Gamma}(a + bs)e^{-2\pi xs} \, ds.$$

Suppose that $x < 0$. Let $\mathcal{C}$ be the contour passing by line segment from $-\infty - iT$ to $-iT$ to iT to $-\infty + iT$. By the calculus of residues and (C.10) we find that

$$\int_{\mathcal{C}} \frac{\Gamma'}{\Gamma}(a + bs)e^{-2\pi xs} \, ds = -\frac{2\pi i}{b} \sum_{n=0}^{\infty} e^{2\pi x(n+a)/b}$$
$$= -\frac{2\pi i}{b} e^{2\pi ax/b}(1 - e^{2\pi x/b})^{-1}.$$

We parametrize the integral $\int_{-\infty-iT}^{-iT}$, and integrate by parts, to see that it is

$$\int_{-\infty}^{0} \frac{\Gamma'}{\Gamma}(a + b\sigma - ibT)e(xT)e^{-2\pi x\sigma}\, d\sigma$$

$$= -\frac{\Gamma'}{\Gamma}(a - ibT)\frac{e(xT)}{2\pi x} + \frac{be(xT)}{2\pi x}\int_{-\infty}^{0} \left(\frac{\Gamma'}{\Gamma}\right)'(a + b\sigma - ibT)e^{-2\pi x\sigma}\, d\sigma.$$

But

$$\left(\frac{\Gamma'}{\Gamma}\right)'(s) = \sum_{n=0}^{\infty}(n + s)^{-2} \ll 1/|t|$$

for $|t| \geq 1$, and hence the last integral above is $\ll x^{-2}T^{-1}$. Similarly,

$$\int_{iT}^{-\infty+iT} \frac{\Gamma'}{\Gamma}(a + bs)e^{-2\pi xs}\, ds = \frac{\Gamma'}{\Gamma}(a + ibT)\frac{e(-xT)}{2\pi x} + O(x^{-2}T^{-1}).$$

We obtain the stated result on combining these estimates. The case $x > 0$ is treated similarly, but with a contour from $+\infty - iT$ to $-iT$ to iT to $+\infty + iT$. $\qquad\square$

Exercises

1. Show:
   (a)
   $$|\Gamma(it)|^2 = \frac{\pi}{t \sinh \pi t};$$
   (b)
   $$|\Gamma(1/2 + it)|^2 = \frac{\pi}{\cosh \pi t};$$
   (c)
   $$\Im\frac{\Gamma'}{\Gamma}(s) > 0 \text{ if } t > 0;$$
   (d)
   $$\frac{\partial}{\partial t} \log |\Gamma(s)| < 0 \text{ when } t > 0;$$
   (e) For any given σ, $|\Gamma(s)|$ is a strictly decreasing function of t on the interval $0 < t < \infty$.

2. (Gauss 1812) Prove *Gauss's multiplication formula*:
   $$\prod_{a=0}^{q-1} \Gamma(s + a/q) = (2\pi)^{(q-1)/2}q^{1/2-qs}\Gamma(qs).$$

3. Show:
   (a)
   $$\frac{\Gamma'}{\Gamma}(1 - s) - \frac{\Gamma'}{\Gamma}(s) = \pi \cot \pi s;$$
   (b)
   $$\frac{\Gamma'}{\Gamma}(s + 1) = \frac{1}{s} + \frac{\Gamma'}{\Gamma}(s);$$
   (c) If n is an integer, $n > 1$, then
   $$\frac{\Gamma'}{\Gamma}(n) = -C_0 + \sum_{k=1}^{n-1}\frac{1}{k}.$$

4. (Gauss 1812) Using additive characters (as discussed in Chapter 4), or otherwise, show that if $0 < a \le q$, then

$$\frac{\Gamma'}{\Gamma}(a/q) = -C_0 - \log q + \sum_{h=1}^{q-1} e(-ah/q)\log(1 - e(h/q)).$$

5. Show that $\frac{\Gamma'}{\Gamma}(1/3) = -C_0 - \frac{3}{2}\log 3 - \pi\sqrt{3}/6$.
6. Show that

$$\frac{\Gamma'}{\Gamma}(s) = -C_0 + \sum_{n=1}^{\infty}(-1)^{n+1}\zeta(n+1)(s-1)^n$$

for $|s - 1| < 1$.
7. Show:
   (a)
$$\left(\frac{\Gamma'}{\Gamma}\right)'(s) = \sum_{n=0}^{\infty}(s+n)^{-2};$$
   (b)
$$\frac{\Gamma''(s)}{\Gamma(s)} = \frac{\Gamma'}{\Gamma}(s)^2 + \sum_{n=0}^{\infty}(s+n)^{-2};$$
   (c) The functions $\Gamma(\sigma)$, $\Gamma''(\sigma)$ have the same sign for all real σ.
8. Show that if $x > 0$ and $y \ge 1$, then

$$\frac{\Gamma(x+y)}{\Gamma(x)} \ge x^y.$$

9. (Hermite 1881) Let x_n denote the unique critical point of $\Gamma(\sigma)$ in the interval $(-n, -n+1)$. Show that $x_n = -n + (\log n)^{-1} + O((\log n)^{-2})$ for $n \ge 2$.
10. Show that $\left(\frac{\Gamma'}{\Gamma}\right)'(s) = s^{-1} + \frac{1}{2}s^{-2} + O(|s|^{-3})$ uniformly in the region $\mathcal{R}$ of Theorem C.1.
11. (a) Show that $\int_1^\infty e^{-x}x^{s-1}\,dx$ is an entire function.

   (b) Show that if $\sigma > 0$, then

$$\int_0^1 e^{-x}x^{s-1}\,dx = \sum_{n=0}^{\infty}\frac{(-1)^n}{n!(s+n)}.$$

   (c) Show that if s is not a non-positive integer, then

$$\Gamma(s) = \int_1^\infty e^{-x}x^{s-1}\,dx + \sum_{n=0}^{\infty}\frac{(-1)^n}{n!(s+n)}.$$

12. (a) Show that if $\sigma > 0$, then

$$\Gamma^{(k)}(s) = \int_0^\infty e^{-x}x^{s-1}(\log x)^k\,dx.$$

(b) Show that

$$\int_0^\infty e^{-x} \log x \, dx = -C_0.$$

13. (Cauchy 1827; Saalschütz 1887, 1888) Show that if $-1 < \sigma < 0$, then

$$\Gamma(s) = \int_0^\infty (e^{-x} - 1)x^{s-1} \, dx.$$

14. Let s be fixed with $\sigma > 0$, and let $f_N(x)$ be the function defined in the proof of Theorem C.2. Show that

$$\int_0^\infty f_N(x) \, dx = \Gamma(s) - \Gamma(s+2)/(2N) + O(N^{-2}).$$

15. (Mellin 1883a, b) Let $P(z)$ and $Q(z)$ be relatively prime polynomials over $\mathbb{C}$, with roots $\alpha_1, \ldots, \alpha_m$ and $\beta_1, \ldots, \beta_n$, respectively, and suppose that none of these roots is a positive integer.
    (a) Suppose that $\prod_{k=1}^\infty \frac{P(k)}{Q(k)}$ converges. Show:
        (i) $m = n$;
        (ii) P and Q have the same leading coefficient;
        (iii) $\sum \alpha_i = \sum \beta_i$.
    (b) Show conversely that if conditions (i)–(iii) hold, then the product converges, and has the value

$$\prod_{i=1}^m \frac{\Gamma(1 - \beta_i)}{\Gamma(1 - \alpha_i)}.$$

    (c) Show that if a and b are complex numbers such that none of $a, b, a+b$ is a negative integer, then

$$\prod_{n=1}^\infty \frac{n(n+a+b)}{(n+a)(n+b)} = \frac{\Gamma(a+1)\Gamma(b+1)}{\Gamma(a+b+1)}.$$

16. (Liouville 1852) Show that if q is an integer, $q > 1$, then

$$\prod_{n=1}^\infty (1 - (z/n)^q)^{-1} = -z^q \prod_{a=1}^q \Gamma(-ze(a/q)).$$

17. (Mellin 1891, p. 324)
    (a) Show that

$$\frac{\Gamma(\sigma)^2}{|\Gamma(s)|^2} = \prod_{n=0}^\infty \left(1 + \frac{t^2}{(n+\sigma)^2}\right).$$

    (b) Give a second derivation of the assertion of Exercise 1(e).
18. (Gram 1899) Show that

$$\prod_{n=2}^\infty \frac{(n^3 - 1)}{(n^3 + 1)} = \frac{2}{3}.$$

19. Show that if $\sigma > 0$, then

$$\Gamma(s) = \int_0^1 (\log 1/x)^{s-1}\, dx,$$

and

$$\Gamma(s) = \int_{-\infty}^\infty e^{-e^x} e^{sx}\, dx.$$

20. (Euler 1794)
    (a) Show that if $-1 < \sigma < 1$, then

$$\int_0^\infty (\sin x) x^{s-1}\, dx = \Gamma(s) \sin \frac{1}{2}\pi s.$$

    (b) Show that if $0 < \sigma < 1$, then

$$\int_0^\infty (\cos x) x^{s-1}\, dx = \Gamma(s) \cos \frac{1}{2}\pi s.$$

21. For $\Re a > 0$, $\Re b > 0$ let the *beta function* $B(a, b)$ be defined to be

$$B(a, b) = \int_0^1 x^{a-1}(1-x)^{b-1}\, dx.$$

    (a) Write

$$\Gamma(a)\Gamma(b) = \int_0^\infty \int_0^\infty e^{-u-v} u^{a-1} v^{b-1}\, du\, dv$$

    and make the change of variables $u = rx$, $v = r(1-x)$ to show that

$$B(a, b) = \frac{\Gamma(a)\Gamma(b)}{\Gamma(a+b)}.$$

    (b) Show that if $\Re a > 0$ and $\Re b > 0$, then

$$\int_0^\infty x^{2a-1}(1-x^2)^{b-1}\, dx = \frac{1}{2} B(a, b).$$

    (c) Show that if $\Re a > 0$ and $\Re b > 0$, then

$$\int_0^{\pi/2} (\sin \theta)^{2a-1}(\cos \theta)^{2b-1}\, d\theta = \frac{1}{2} B(a, b).$$

    (d) By writing $t = \tan^2 \theta$, or otherwise, show that if $\Re a > 0$ and $\Re b > 0$, then

$$\int_0^\infty \frac{t^{a-1}}{(1+t)^{a+b}}\, dt = B(a, b).$$

22. (Dirichlet 1839; Liouville 1839) Let $f(x)$ be a continuous function defined on $[0, 1]$. Let $\mathcal{R}$ denote that portion of $\mathbb{R}^n$ for which $x_i \geq 0$ and $\sum x_i \leq 1$.

Show that

$$\int_{\mathcal{R}} f(x_1 + \cdots + x_n)x_1^{a_1-1}\cdots x_n^{a_n-1}\,dx_1\cdots dx_n$$
$$= \frac{\Gamma(a_1)\cdots\Gamma(a_n)}{\Gamma(a_1+\cdots+a_n)}\int_0^1 f(x)x^{a-1}\,dx$$

where $a = \sum a_i$ and $\Re a_i > 0$ for all i.

23. (Mellin 1902) Suppose that z lies in the slit plane formed by deleting the negative real axis. Show that if $0 < c < \Re a$, then

$$\frac{\Gamma(a)}{(1+z)^a} = \frac{1}{2\pi i}\int_{c-i\infty}^{c+i\infty} \Gamma(s)\Gamma(a-s)z^{-s}\,ds.$$

(This is the inverse of the Mellin transform in Exercise 21(d).)

24. (Raabe 1844) Show that if s is not a negative real number or 0, then

$$\int_s^{s+1} \log\Gamma(z)\,dz = s\log s - s + \frac{1}{2}\log 2\pi.$$

25. (Barnes 1900) Let

$$G(s+1) = (2\pi)^{s/2}\exp\left(-\frac{1}{2}(C_0+1)s^2 - \frac{1}{2}s\right)\prod_{n=1}^{\infty}\left(\left(1+\frac{s}{n}\right)^n e^{-s-s^2/(2n)}\right).$$

Show:
(a) $G(s)$ is an entire function.
(b) $G(1) = 1$.
(c) $G(s+1) = \Gamma(s)G(s)$.
(d)
$$G(n+1) = \frac{(n!)^n}{1^1 2^2 3^3 \cdots n^n}.$$

26. Show that

$$\sum_{n=1}^{\infty}\frac{(-1)^n n^2}{n^3+1} = \frac{1}{3}\ln 2 - \frac{1}{3} - \frac{\pi}{3\cosh(\pi\sqrt{3}/2)}.$$

C.1 Notes

Euler, in a letter of 1729 to Goldbach (cf. Fuss 1843, p. 3) gave the formula

$$\Gamma(s) = \frac{1}{s}\prod_{n=1}^{\infty}\left(\left(1+\frac{1}{n}\right)^s\left(1+\frac{s}{n}\right)^{-1}\right).$$

This is substantially the same as the formula (C.3) that Gauss (1812) took to be fundamental. Based on the above definition of the gamma function, the formula

(C.1) was proved by Schlömilch (1844) and Newman (1848). Weierstrass (1856) took (C.1) to be the definition of the gamma function. Euler had given the special value (C.7) already in his letter to Goldbach. Euler (1771) also discovered the reflection formula (C.6). The duplication formula (C.9) of Legendre (1809) is a special case of the multiplication formula of Gauss (1812), given in Exercise C.3. Stirling (1730, p. 135) gave the series expansion

$$\log \Gamma(s) = \left(s - \frac{1}{2}\right) \log s - s + \frac{1}{2} \log 2\pi + \sum_{n=2}^{\infty} \frac{B_n}{n(n-1)s^{n-1}}.$$

This series diverges, but a partial sum provides an asymptotic expansion. The approximation (C.17) is a weak form of this. To calculate $\Gamma(s)$ numerically, it suffices to consider $\sigma \geq 1/2$, in view of (C.6). If $|s|$ is small then (C.4) should be used repeatedly. Thus it remains to evaluate $\Gamma(s)$ when $\sigma \geq 1/2$ and $|s|$ is large, and this is quickly achieved by using the expansion above. By these means it may be found that the sole minimum of $\Gamma(\sigma)$ for $\sigma > 0$ is at $\sigma_0 = 1.4616321\ldots$, and that $\Gamma(\sigma_0) = 0.88560319\ldots$. The convenient estimate (C.19) was noted by Pincherle (1888). Theorems C.1 and C.2 may be established in several ways. An instructive collection of such proofs is found in Sections 8.4, 8.5, 11.1, 11.11, and 12.12 of Henrici (1977). Euler (1730) gave the formula of Theorem C.2, expressed in the form $n! = \int_0^1 (\log 1/y)^n \, dy$, and subsequently found many other integral formulæ involving the gamma function. Thus Euler was led in quite a different direction than Gauss (1812), whose independent investigations were more directly related to Gauss's formula (C.3). Legendre (1809) called the formula (C.21) the 'Euler integral of the second kind', and introduced the notation $\Gamma(z)$. The 'Euler integral of the first kind' is known today as the beta function (see Exercise C.21). Theorem C.3 is due to Hankel (1864), and Theorem C.4 to Mellin (1896, p. 76, 1899, p. 39).

Simple proofs of Stirling's formula for $n!$, using a minimum of tools, have been given by Robbins (1955) and Feller (1965).

For more extensive expositions of the subject the reader is referred to Artin (1964), Henrici (1977), Jensen (1916), Nielsen (1906), and to Whittaker & Watson (1950, Chapter 12). The related Mellin–Barnes integrals are discussed in Section 8.8 of Henrici (1977).

Gauss and Binet established several useful formulæ for $\log \Gamma(s)$ and for $\frac{\Gamma'}{\Gamma}(s)$. Kummer (1847) proved that if $0 < \sigma < 1$, then

$$\log \Gamma(\sigma) = (C_0 + \log 2) \left(\frac{1}{2} - \sigma\right) + (1 - \sigma) \log \pi - \frac{1}{2} \log \sin \pi \sigma$$

$$+ \sum_{n=1}^{\infty} \frac{\log n}{\pi n} \sin 2\pi n\sigma.$$

In conjunction with the analysis of Chapter 9, this gives

$$\sum_{a=1}^{q} \chi(a) \log \Gamma(a/q) = -(C_0 + \log 2\pi) \sum_{a=1}^{q} a\chi(a) - \frac{\sqrt{q}}{\pi} L'(1, \chi)$$

where χ is a primitive character (mod q) for which $\chi(-1) = -1$.

Artin (1931, 1964; p. 14) showed that if $f(x)$ is positive and $\log f(x)$ is convex for $x > 0$, if $xf(x) = f(x + 1)$ for all $x > 0$, and $f(1) = 1$, then $f(x) = \Gamma(x)$.

Hölder (1886) showed that $\Gamma(s)$ does not satisfy an algebraic differential equation. Additional proofs of this have been given by Moore (1897), Jensen (1916, pp. 103–112) and Ostrowski (1919).

C.2 References

Artin, E. (1931). *Einführung in die Theorie der Gamma-Funktion*. Hamburger math. Einzelschriften 11. Leipzig: Teubner.

(1964). *The Gamma Function*. New York: Holt, Reinhart and Winston.

Barnes, E. W. (1900). The theory of the G-function, *Quart. J. Math.* **31**, 264–314.

Cauchy, A. L. (1827). *Exercices de Math.* Vol. 2. Paris: de Buse Frèses, pp. 91–92.

Lejeune–Dirichlet, P. G. (1839). Sur une nouvelle methode pour la détermination des intégrales multiples, *J. Math. pures appl.* **4**, 164–168; *Werke* I, pp. 375–380.

Euler, L. (1730). De Progressionibus transcendemibus seu quarum termini generales algebraice dari nequennt, *Comment. Acad. Sci. Petropolitanae* **5**, 36–57; *Opera Omnia*, Ser 1, Vol. 14, Teubner, 1924, pp. 1–14.

(1771). Evolutio formulae integralis $\int x^{f-1}(\log x)^{m/n} dx$ integratione a valore $x = 0$ ad $x = 1$ extensa, *Novi Comment. Acad. Petropol.* **16**, 91–139.

(1794). *Institutiones calculi integralis*, Vol. 4, p. 342.

Feller, W. (1965). A direct proof of Stirling's formula, *Amer. Math. Monthly* **74**, 1223–1225.

Fuss, P.-H. (1843). Correspondence Mathématique et Physique de quelques célèbres géomètres du XVIIème siècle, Vol. 1. St. Petersburg: Acad. Impér. Sci.

Gauss, C. F. (1812). Disquisitiones generales circa seriem infinitam etc., *Comment. Gott.* **2**, 1–46; *Werke*, Vol. 3. Berlin: Deutsch von H. Simon, 1888, pp. 123–162.

Gram, J. P. (1899). *Nyt Tidsskrift Mat.* **10B**, 96.

Hankel, H. (1864). Die Eulerschen Integrale bei unbeschränkter Variabilität des Arguments, *Zeit. Math. Phys.* **9**, 1–21.

Henrici, P. (1977). Applied and Computational Complex Analysis, Vol. 2. New York: Wiley.

Hermite, Ch. (1881). Sur l'intégrale Eulérienne de seconde espèce, *J. Reine Angew. Math.* **90**, 332–338.

Hölder, O. (1886). Über die Eigenschaft der Gammafunktion keiner algebraischen Differentialgleichung zu genügen, *Math. Ann.* **28**, 1–13.

Jensen, J. L. W. V. (1916). An elementary exposition of the theory of the Gamma function, *Annals of Math.* (2) **17**, 124–166.

Kummer, E. E. (1847). Beitrage zur Theorie der Funktion $\Gamma(x)$, *J. Reine Angew. Math.* **35**, 1–4.

Legendre, A. M. (1809). Recherches sur diverses sortes d'intégrales définies, *Mémoires de l'Institut de France* **10**, 416–509.

Liouville, J. (1839). Note sur quelques intégrales définies, *J. Math. Pures Appl.* **4**, 225–235.

— (1852). Note sur la fonction gamma de Legendre, *J. Math. Pures Appl.* **17**, 448–453.

Mellin, H. (1883a). Eine Verallgemeinerung der Gleichung $\Gamma(x)\Gamma(1-x) = \pi : \sin \pi x$, *Acta Math.* **3**, 102–104.

— (1883b). Über gewisse durch die Gammafunktion ausdrückbare Produkte, *Acta Math.* **3**, 322–324.

— (1891). Zur Theorie der linearen Differenzengleichungen erster Ordnung, *Acta Math.* **15**, 317–384.

— (1896). Über die fundamentale Wichtigkeit des Satzes von Cauchy für die Theorien der Gamma- und hypergeometrischen Funktionen, *Acta Soc. Fennicae* **21**, no. 1, p. 76.

— (1899). Über eine Verallgemeinerung der Riemannschen Funktion $\zeta(s)$, *Acta Soc. Fennicae* **24**, 50 pp.

— (1902). Über den Zusammenhang zwischen den linearen Differential- und Differenzengleichungen, *Acta Math.* **25**, 139–164.

Moore, E. H. (1897). Concerning transcendentally transcendental functions, *Math. Ann.* **48**, 49–74.

Newman, F. W. (1848). On Γa, especially when a is negative, *Cambridge and Dublin Math. J.* **3**, 57–60.

Nielsen, N. (1906). *Handbuch der Theorie der Gammafunktion*. Leipzig: Teubner.

Ostrowski, A. (1919). Neuer Beweis des Hölderschen Satzes daß die Gammafunktion keiner algebraischen Differentialgleichung genügt, *Math. Ann.* **79**, 286–288.

Pincherle, S. (1888). Sulle funzioni ipergeometriche generalizzate, *Rend. Reale Accad. Lincei* (4) **4**, 694–700; 792–799.

Raabe, J. (1844). Angenäherte Bestimmung der Faktorenfolge $n!$, wenn n eine sehr große ganze Zahl ist, *J. Reine Angew. Math.* **28**, 12–14.

Robbins, H. (1955). A remark on Stirling's formula, *Amer. Math. Monthly* **62**, 26–29.

Saalschütz, L. (1887). Bemerkungen über die Gammafunktionen mit negativem Argument, *Zeit. Math. Phys.* **32**, 246–250.

— (1888). Bemerkungen über die Gammafunktionen mit negativem Argument, *Zeit. Math. Phys.* **33**, 362–371.

Schlömilch, O. (1844). Über einige merkwürdige bestimmte Integrale, *Grunert Archiv* **5**, 204–212.

Stirling, J. (1730). *Methodus differentialis: sive, Tractatus de sommationes et interpolationes serium infinitorum*. London: G. Strahan.

Weierstrass, K. (1856). Über die Theorie der analytischen Fakultäten, *J. Reine Angew. Math.* **51**, 1–60; *Werke*, Vol. 1. pp. 153–211.

Whittaker, E. T. & Watson, G. N. (1950). *A Course of Modern Analysis*, Fourth edition. Cambridge: Cambridge University Press.

Appendix D

Topics in harmonic analysis

D.1 Pointwise convergence of Fourier series

Let $f \in L^1(\mathbb{T})$, and suppose that

$$\widehat{f}(k) = \int_{\mathbb{T}} f(x)e(-kx)\,dx \tag{D.1}$$

are the Fourier coefficients of f. Here $e(\theta) = e^{2\pi i\theta}$ is the complex exponential with period 1. It is a familiar fact in the theory of Fourier series that if f has bounded variation on $\mathbb{T}$, then

$$\lim_{K \to \infty} \sum_{k=-K}^{K} \widehat{f}(k)e(k\alpha) = \frac{f(\alpha^+) + f(\alpha^-)}{2}. \tag{D.2}$$

Less familiar is the strong quantitative version of this that we now derive.

Let $D_K(x) = \sum_{k=-K}^{K} e(kx)$. This is the *Dirichlet kernel*. We multiply both sides of (D.1) by $e(k\alpha)$ and sum, to see that

$$\sum_{k=-K}^{K} \widehat{f}(k)e(k\alpha) = \int_{\mathbb{T}} f(x)D_K(\alpha - x)\,dx = \int_{\mathbb{T}} D_K(x)f(\alpha - x)\,dx.$$

Since D_K is an even function, the above is

$$= \int_{\mathbb{T}} D_K(x)f(\alpha + x)\,dx. \tag{D.3}$$

Clearly $D_K(0) = 2K + 1$. If $x \notin \mathbb{Z}$, then $D_K(x)$ is the sum of a segment of a geometric progression, which permits us to write D_K in closed form,

$$D_K(x) = \frac{e((K+1)x) - e(-Kx)}{e(x) - 1} = \frac{e\left(\left(K + \frac{1}{2}\right)x\right) - e\left(-\left(K + \frac{1}{2}\right)x\right)}{e(x/2) - e(-x/2)}$$

$$= \frac{\sin(2K+1)\pi x}{\sin \pi x}. \tag{D.4}$$

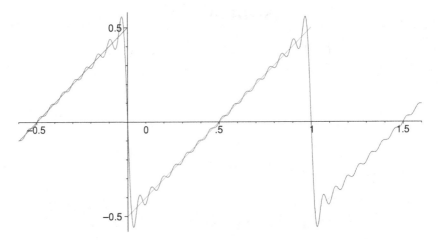

Figure D.1 Graph of $s(x)$ and its Fourier approximation $-\sum_{k=1}^{15}\sin 2\pi kx/(\pi k)$.

Our analysis of the pointwise convergence of Fourier series is based on the behaviour of the the Fourier series of one particular function, namely the 'saw-tooth function' $s(x)$ given by

$$s(x) = \begin{cases} \{x\} - \frac{1}{2} & (x \notin \mathbb{Z}), \\ 0 & (x \in \mathbb{Z}) \end{cases}.$$

Lemma D.1 *Let*

$$E_K(x) = s(x) + \sum_{k=1}^{K} \frac{\sin 2\pi kx}{\pi k}.$$

Then $|E_K(x)| \le \min\left(1/2,\, 1/((2K+1)\pi|\sin \pi x|)\right).$

It is easy to compute the Fourier coefficients of $s(x)$; we find that $\widehat{s}(0) = 0$, and that $\widehat{s}(k) = -1/(2\pi ik)$ for $k \ne 0$. Thus the above lemma constitutes a quantitative form of (D.2), for the function $s(x)$. A numerical example of Lemma D.1 is graphed in Figure D.1.

Proof All terms comprising $E_K(x)$ are odd, and hence E_K is odd. Thus we may suppose that $0 \le x \le 1/2$. The case $x = 0$ is clear. We observe that if $x \notin \mathbb{Z}$, then

$$E_K'(x) = 1 + 2\sum_{k=1}^{K} \cos 2\pi kx = D_K(x).$$

Hence if $0 < x \le 1/2$, then by (D.4) we see that

$$E_K(x) = -\frac{1}{2} \int_x^{1-x} D_K(z)\,dz$$

$$= \frac{-1}{2} \int_x^{1-x} \frac{\sin(2K+1)\pi z}{\sin \pi z}\,dz$$

$$= \frac{i}{2} \int_x^{1-x} \frac{e\left(\left(K+\frac{1}{2}\right)z\right)}{\sin \pi z}\,dz.$$

The integrand is analytic in the rectangle $x \le \Re z \le 1 - x$, $0 \le \Im z \le Y$, so by letting $Y \to \infty$ and applying Cauchy's theorem we see that the above is

$$= \frac{i}{2} \int_x^{x+i\infty} \frac{e\left(\left(K+\frac{1}{2}\right)z\right)}{\sin \pi z}\,dz - \frac{i}{2} \int_{1-x}^{1-x+i\infty} \frac{e\left(\left(K+\frac{1}{2}\right)z\right)}{\sin \pi z}\,dz.$$

On writing $z = x + iy$ in the first integral, and $z = 1 - x + iy$ in the second, we see that the above is

$$= \frac{-1}{2} \int_0^\infty \left(\frac{e\left(\left(K+\frac{1}{2}\right)x\right)}{\sin \pi (x+iy)} - \frac{e\left(-\left(K+\frac{1}{2}\right)x\right)}{\sin \pi (1-x+iy)} \right) e^{-(2K+1)\pi y}\,dy. \quad \text{(D.5)}$$

But $\sin \pi (x+iy) = (\sin \pi x)\cosh \pi y - i(\cos \pi x)\sinh \pi y$, so that $|\sin \pi (x+iy)| \ge \sin \pi x$ for all real y. Hence the expression above has absolute value not exceeding

$$\frac{1}{\sin \pi x} \int_0^\infty e^{-(2K+1)\pi y}\,dy = \frac{1}{(2K+1)\pi \sin \pi x}.$$

This gives the second part of the bound. The first bound, $|E_K(x)| \le 1/2$, is weaker if $1/(2K+1) \le x \le 1/2$, since $\sin \pi x \ge 2x$ in this range. Thus it suffices to show that $|E_K(x)| \le 1/2$ when $0 < x < 1/(2K+1)$. Since $0 < \sin u < u$ for $0 \le u \le \pi$, it follows from the definition of $E_K(x)$ that

$$x - \frac{1}{2} \le E_K(x) \le (2K+1)x - \frac{1}{2}$$

for $0 \le x \le 1/(2K+1)$. This gives the desired bound. $\qquad\square$

We now establish an analogue of Lemma D.1 for arbitrary functions of bounded variation.

Theorem D.2 *If f has bounded variation on $\mathbb{T}$, with $\widehat{f}(k)$ given by (D.1), then for any α,*

$$\left| \frac{f(\alpha^+) + f(\alpha^-)}{2} - \sum_{k=-K}^{K} \widehat{f}(k)e(k\alpha) \right|$$
$$\leq \int_{0^+}^{1^-} \min\left(\frac{1}{2}, \frac{1}{(2K+1)\pi \sin \pi x} \right) |df(\alpha + x)|.$$

Since the right-hand side here tends to 0 as $K \to \infty$, this inequality implies the qualitative relation (D.2).

Proof As $E_K'(x) = D_K(x)$ when $x \notin \mathbb{Z}$, the integral (D.3) is

$$\int_{0^+}^{1^-} E_K'(x)f(\alpha + x)\,dx = \int_{0^+}^{1^-} f(\alpha + x)\,dE_K(x),$$

by Theorem A.3. But $E_K(0^+) = -1/2$, $E_K(1^-) = 1/2$. Hence by integrating by parts (as in Theorem A.2) we see that the above is

$$\frac{1}{2}f(\alpha^+) + \frac{1}{2}f(\alpha^-) - \int_{0^+}^{1^-} E_K(x)\,df(\alpha + x).$$

To complete the proof it suffices to apply the triangle inequality (as in Theorem A.4) and the bound of Lemma D.1. $\square$

D.2   The Poisson summation formula

The formula in question asserts that under suitable conditions,

$$\sum_{n=-\infty}^{\infty} f(n) = \sum_{k=-\infty}^{\infty} \widehat{f}(k) \tag{D.6}$$

where f is a function of a real variable, and $\widehat{f}$ is its Fourier transform,

$$\widehat{f}(t) = \int_{\mathbb{R}} f(x)e(-tx)\,dx. \tag{D.7}$$

To ensure that $\widehat{f}$ is well-defined, we impose the condition $f \in L^1(\mathbb{R})$, i.e., that the integral $\int_{\mathbb{R}} |f(x)|\,dx$ is finite. Put

$$F(\alpha) = \sum_{n \in \mathbb{Z}} f(n + \alpha). \tag{D.8}$$

This sum is absolutely convergent for almost all α, since

$$\int_0^1 \sum_{n \in \mathbb{Z}} |f(n + \alpha)|\,d\alpha = \sum_{n \in \mathbb{Z}} \int_n^{n+1} |f(\alpha)|\,d\alpha = \int_{\mathbb{R}} |f(\alpha)|\,d\alpha < \infty.$$

Moreover, $F(\alpha)$ has period 1, $\int_{\mathbb{T}} |F(\alpha)|\, d\alpha < \infty$, and F has Fourier coefficients

$$
\begin{aligned}
\widehat{f}(k) = \int_0^1 F(\alpha)e(-k\alpha)\, d\alpha &= \sum_{n\in\mathbb{Z}} \int_0^1 f(n+\alpha)e(-k\alpha)\, d\alpha \\
&= \int_{\mathbb{R}} f(x)e(-kx)\, dx \qquad (D.9) \\
&= \widehat{f}(k).
\end{aligned}
$$

Here the interchange of the integral and the sum is justified by absolute convergence. Thus the Fourier expansion of F is

$$
\sum_{k\in\mathbb{Z}} \widehat{f}(k)e(k\alpha).
$$

The Poisson summation formula (D.6) is simply the assertion that this Fourier expansion converges to $F(\alpha)$ when $\alpha = 0$. Our hypotheses thus far do not ensure this, but in this direction we establish the following two precise results.

Theorem D.3 *Suppose that $f \in L^1(\mathbb{R})$, and that f is of bounded variation on $\mathbb{R}$. Then*

$$
\sum_{n\in\mathbb{Z}} \frac{f(n^+) + f(n^-)}{2} = \lim_{K\to\infty} \sum_{k=-K}^{K} \widehat{f}(k).
$$

If in addition f is continuous, then we have a result which is close to (D.6), although it is still necessary to restrict ourselves to symmetric partial sums on the right-hand side.

Proof We first note that if $n \le \alpha \le n+1$, then

$$
f(\alpha) = \int_n^{n+1} f(x)\, dx + \int_n^{\alpha} (x-n)\, df(x) + \int_{\alpha}^{n+1} (x-n-1)\, df(x),
$$

as can readily be seen by integration by parts. Hence

$$
|f(\alpha)| \le \int_n^{n+1} |f(x)|\, dx + \mathrm{var}_{[n,n+1]} f, \qquad (D.10)
$$

and it follows from our hypotheses that the sum

$$
\sum_{n\in\mathbb{Z}} f(n+\alpha)
$$

is absolutely convergent for all α, and uniformly convergent in compact regions. Hence $F(\alpha)$ can be taken to be the value of this sum for all α, not merely for almost all α. By the triangle inequality, $\mathrm{var}_{\mathbb{T}} F \le \mathrm{var}_{\mathbb{R}} f$, so that F is of bounded variation on $\mathbb{T}$, and hence the relation (D.2) applies to F. Thus we see that the Fourier series of F converges to $(F(\alpha^+) + F(\alpha^-))/2$ for all α. Using the fact that

f is of bounded variation once more, we see that $F(\alpha^+) = \sum_{n \in \mathbb{Z}} f((n + \alpha)^+)$, and similarly for $F(\alpha^-)$. Hence we have the stated result. □

Theorem D.4 *Suppose that f is continuous, and that the series $\sum_{n \in \mathbb{Z}} f(n + \alpha)$ is uniformly convergent for $0 \le \alpha \le 1$. Then*

$$\sum_{n \in \mathbb{Z}} f(n) = \lim_{K \to \infty} \sum_{k=-K}^{K} \left(1 - \frac{|k|}{K} \right) \widehat{f}(k).$$

Proof Clearly $F(\alpha)$ given in (D.8) is continuous. Since we have not assumed that $f \in L^1(\mathbb{R})$, the Fourier transform $\widehat{f}(t)$ may not exist. However, if k is an integer, then $\widehat{f}(k)$ exists as a convergent improper integral. To see this we first note that $\sum_{n=M}^{N} f(n + \alpha)$ is small if M and N are large integers and $0 \le \alpha \le 1$. Then

$$\int_0^1 \sum_M^N f(n + \alpha)e(-k\alpha)\,d\alpha = \int_M^{N+1} f(x)e(-kx)\,dx$$

is small. The hypothesis that $\sum_n f(n + \alpha)$ converges uniformly implies that $f(x) \to 0$ as $|x| \to \infty$. Hence $\int_u^v f(x)e(-kx)\,dx \to 0$ as u, v tend to infinity through real values. The calculation of $\widehat{f}(k)$ in (D.9) is still valid, but is now justified by uniform convergence. Next we appeal to a theorem of Fejér, which asserts that the Fourier series of a continuous function $F(\alpha)$ with period 1 is uniformly $(C, 1)$-summable to F (see Katznelson (2004), p.19). That is,

$$\sum_{k=-K}^{K} \left(1 - \frac{|k|}{K} \right) \widehat{f}(k)e(k\alpha) \longrightarrow F(\alpha)$$

uniformly as $K \to \infty$. The stated identity follows on taking $\alpha = 0$. □

Exercises

1. Show that if f satisfies the hypotheses of Theorem D.2, and α and β are real numbers, then the function $f(x + \alpha)e(\beta x)$ does also. Specify conditions under which

$$\sum_n f(n + \alpha)e(\beta n) = \sum_k \widehat{f}(k - \beta)e((k - \beta)\alpha).$$

2. Suppose that f has bounded variation on $[-A, A]$, for every $A > 0$. Show that

$$\lim_{N \to \infty} \sum_{n=-N}^{N} f(n) = \lim_{T \to \infty} \sum_{k=-\infty}^{\infty} \int_{-T}^{T} f(x)e(-kx)\,dx$$

provided that either limit exists.

3. Suppose that $f \in L^1(\mathbb{R}^n)$, and for $x \in \mathbb{T}^n$ put

$$F(x) = \sum_{\lambda \in \mathbb{Z}^n} f(\lambda + x).$$

(a) Show that the sum $F(x)$ is absolutely convergent for almost all x.

(b) Show that $F \in L^1(\mathbb{T}^n)$ and that $\|F\|_{L^1(\mathbb{T}^n)} \le \|f\|_{L^1(\mathbb{R}^n)}$.

(c) Define the Fourier transform of f, and the Fourier coefficient of F, respectively, to be $\widehat{f}(t) = \int_{\mathbb{R}^n} f(x)e(-t \cdot x)\,dx$, $\widehat{F}(k) = \int_{\mathbb{T}^n} F(x)e(-k \cdot x)\,dx$. Show that $\widehat{F}(k) = \widehat{f}(k)$.

4. (a) Suppose that there is a $\delta > 0$ such that $c(k) \ll (1 + |k|)^{-n-\delta}$. Show that

$$\sum_{k \in \mathbb{Z}^n} c(k)e(k \cdot x)$$

is a continuous function of $x \in \mathbb{T}^n$.

(b) Suppose that there is a $\delta > 0$ such that $f(x) \ll (1 + |x|)^{-n-\delta}$ for $x \in \mathbb{R}^n$. Suppose also that $f(x)$ is continuous. Show that

$$F(x) = \sum_{\lambda \in \mathbb{Z}^n} f(\lambda + x)$$

is a continuous function for $x \in \mathbb{T}^n$.

(c) Suppose that in addition to the hypotheses in (b), the function f also has the property that $\widehat{f}(t) \ll (1 + |t|)^{-n-\delta}$. Show that

$$\sum_{\lambda \in \mathbb{Z}^n} f(\lambda + x) = \sum_{k \in \mathbb{Z}^n} \widehat{f}(k)e(k \cdot x)$$

for all $x \in \mathbb{T}^n$.

5. A *lattice* in $\mathbb{R}^n$ is a set of points of the form $A\mathbb{Z}^n$ where A is a non-singular $n \times n$ matrix. Thus $\mathbb{Z}^n$ is an example of a lattice, called the *lattice of integral points*.

(a) Suppose that $\Lambda_1 = A\mathbb{Z}^n$ and $\Lambda_2 = B\mathbb{Z}^n$ are two lattices. Show that $\Lambda_2 \subseteq \Lambda_1$ if and only if there is an $n \times n$ matrix K with integral entries such that $B = AK$.

(b) An $n \times n$ matrix U is said to be *unimodular* if (i) its entries are integers, and (ii) $\det U = \pm 1$. Show that if $\Lambda_1 = A\mathbb{Z}^n$ and $\Lambda_2 = B\mathbb{Z}^n$ are two lattices, then $\Lambda_1 = \Lambda_2$ if and only if there is a unimodular matrix U such that $B = AU$.

(c) Let $a_1, \ldots, a_n$ denote the columns of A. These vectors are said to form a basis for Λ_1, because every member of Λ_1 has a unique representation in the form $c_1 a_1 + \cdots c_n a_n$ where the c_i are integers. If $\Lambda = A\mathbb{Z}^n$, we say that the *determinant* of Λ is $d(\Lambda) = |\det A|$. Show that the determinant of a lattice is independent of the basis by which it is presented.

(d) Suppose that $\Lambda = A\mathbb{Z}^n$ is a lattice in $\mathbb{R}^n$. Let Λ^* be the set of all those points $\mu \in \mathbb{R}^n$ such that $\mu \cdot \lambda \in \mathbb{Z}$ for all $\lambda \in \Lambda$. Show that Λ^* is a lattice, and indeed that $\Lambda^* = \left(A^{-1}\right)^{\mathrm{T}}\mathbb{Z}^n$.

(e) Suppose that f is a continuous function on $\mathbb{R}^n$ such that

$$f(x) \ll (1 + |x|)^{-n-\delta},$$
$$\widehat{f}(t) \ll (1 + |t|)^{-n-\delta}$$

for some $\delta > 0$. Let $\Lambda = A\mathbb{Z}^n$ be a lattice. Show that

$$\sum_{\lambda \in \Lambda} f(\lambda + x) = \frac{1}{d(\Lambda)} \sum_{\mu \in \Lambda^*} \widehat{f}(\mu)e(\mu \cdot x)$$

for all x.

D.3 Notes

Section D.1. The relation (D.2) is the famous Dirichlet–Jordan test, which is usually derived with much less effort. Theorem D.2 generalizes and refines an argument of Pólya (1918), who estimated the rate of convergence of the Fourier series (9.18). For more on the convergence of Fourier series, see Katznelson (2004, Chapter 2), Körner (1988, Part I), or Zygmund (2002, Chapter II).

Section D.2. For more on the Poisson summation formula, see Katznelson (2004, VI.1.15), Körner (1988, Section 27), or Zygmund (2002, Chapter 2, Section 13). For a discussion of the Poisson summation formula in higher dimensions, see Stein & Weiss (1971, Chapter VII Section 2). Siegel (1935) showed that Minkowski's convex body theorem could be derived by applying the Poisson summation formula. Cohn & Elkies (2003), Cohn (2002) and Cohn & Kumar (2004) have applied the Poisson summation formula in $\mathbb{R}^n$ to limit the density of sphere packings.

D.4 References

Cohn, H. (2002). New upper bounds on sphere packings, II, *Geom. Topol.* **6**, 329–353.

Cohn, H. & Elkies, N. (2003). New upper bounds on sphere packings, I, *Ann. of Math.* (2) **157**, 689–714.

Cohn, H. & Kumar, A. (2004). The densest lattice in twenty-four dimensions, *Electron. Res. Announc. Amer. Math. Soc.* **10**, 58–67.

Katznelson, Y. (2004). *An Introduction to Harmonic Analysis*, Third edition. Cambridge: Cambridge University Press.

Körner, T. W. (1988). *Fourier Analysis*, Second edition. Cambridge: Cambridge University Press.

Pólya, G. (1918). Über die Verteilung der quadratischen Reste und Nichtreste, *Nachr. Akad. Wiss. Göttingen*, 21–29.

Siegel, C. L. (1935). Über Gitterpunkte in convexen Körpern und ein damit zusammen-hängendes Extremalproblem, *Acta Math.* **65**, 307–323; *Gesammelte Abhandlungen*, Vol. I. Berlin: Springer-Verlag, 1966, 311–325.

Stein, E. & Weiss, G. (1971). *Introduction to Fourier analysis on Euclidean spaces*, Princeton Math. Series 32. Princeton: Princeton University Press.

Zygmund, A. (2002). *Trigonometric Series*, Third edition, Vol. I. Cambridge: Cambridge University Press.

Name index

Subject index

Printed in the United States
By Bookmasters